Integrated Local Energy Communities

Integrated Local Energy Communities

From Concepts and Enabling Conditions to Optimal Planning and Operation

Edited by Marialaura Di Somma, Christina Papadimitriou, Giorgio Graditi, and Koen Kok

WILEY-VCH

Editors

Marialaura Di Somma
Department of Industrial Engineering
University of Naples Federico II
Piazzale Tecchio, 80
Naples, 80125
Italy

Christina Papadimitriou
Eindhoven University of Technology-TU/e
Groene Loper 19
Eindhoven
5612 AP
The Netherlands

Giorgio Graditi
ENEA
Lungotevere Thaon di Revel 76
Rome, 00196
Italy

Koen Kok
Eindhoven University of Technology-TU/e
Groene Loper 19
Eindhoven
5612 AP
The Netherlands

Cover Image: © Adyna/Getty Images

Library of Congress Card No.: applied for

British Library Cataloguing-in-Publication Data
A catalogue record for this book is available from the British Library.

Bibliographic information published by the Deutsche Nationalbibliothek The Deutsche Nationalbibliothek lists this publication in the Deutsche Nationalbibliografie; detailed bibliographic data are available on the Internet at <http://dnb.d-nb.de>.

Print ISBN: 978-3-527-35235-7
ePDF ISBN: 978-3-527-84330-5
ePub ISBN: 978-3-527-84329-9
oBook ISBN: 978-3-527-84328-2

Typesetting Straive, Chennai, India

Contents

Preface

Decarbonization of the energy system requires an integrated approach across all different sectors, such as electricity, heating, cooling, and transportation. Sector coupling, through the integrated use of different energy carriers, both at the supply side (e.g. by converting (surplus) electricity into other forms, such as hydrogen) and at the demand side (e.g. by using residual heat from power generation processes for district heating), is represented as a key for reaching the ambitious climate neutrality by 2050. The main benefit associated to sector coupling is related to the increase of flexibility of the energy system through the coordinated management and operation of different sectors and the exploitation of interplay of multiple energy carriers and related technologies. On the other hand, the ongoing transition from traditional centralized energy systems to decentralized schemes brings new opportunities for distributed energy resources integration and for the evolution of the role of final users from passive consumers to active users through energy communities. Combining sector coupling at the local level with energy communities leads to the innovative concept of integrated local energy communities (ILECs), which represent an effective way of managing available energy resources at local level by also fostering consumer engagement and empowerment. The ILEC concept may refer to a set of energy users deciding to make common choices in terms of satisfying their energy needs, in order to maximize the benefits deriving from this collegial approach, thanks to the implementation of a variety of electricity and thermal technologies and energy storages and the optimized management of energy flows. The aim of this book is to present in a thorough and comprehensive way all the critical aspects that are needed when designing, planning, and operating an ILEC from end to end. This book's objective and ambition are timely, as the integrated energy system is an important means to achieve the energy transition and minimize dependence on fossil fuels.

To this end, the following key topics are comprehensively discussed throughout the book:

- Conceptualization of ILECs with analysis of key features, enabling technologies including ICT, actors, business models and key issues for their implementation, and validation of ILEC solutions through simulation and testing in a lab environment and real-world applications.

- Presentation of innovative approaches for the coordinated planning and operation of ILECs, with integrated flexibility identification and employment, and for peer-to-peer (P2P) energy trading.
- Analysis of key social aspects related to the reorganization of the energy system according to the energy community paradigm.
- Definition of guidelines and recommendations for optimal implementation of ILECs.

The book supports readers in finding innovative solutions and detailed insights for the planning and operation of ILECs while fostering research advances to the state of the art on this topic. The book does this by presenting approaches, methodologies, critical assessments, real-time applications, as well as efficient optimization models and algorithms for MCES and emerging technologies/carriers including hydrogen and electric vehicles. The proposed optimization frameworks are scalable and flexible for adaptation to several real contexts thus representing valid tools to provide support to decision-makers for ILECs planning and operational aspects.

August 2024

Marialaura Di Somma
University of Naples Federico II

Christina Papadimitriou
Eindhoven University of Technology-TU/e

Giorgio Graditi
ENEA

Koen Kok
Eindhoven University of Technology-TU/e

1

Introduction: The Need for Sector Coupling and the Energy Transition Goals

Marialaura Di Somma[1], Christina Papadimitriou[2], Giorgio Graditi[1], and Koen Kok[2]

[1] *Italian National Agency for New Technologies, Energy and Sustainable Economic Development, Department of Energy Technologies and Renewable Sources, Lungotevere Thaon di Revel, 76, 00196, Rome, Italy*
[2] *Eindhoven University of Technology, Electrical Engineering Department, Groene Loper 3, 5612 AE, Eindhoven, the Netherlands*

1.1 Introduction

1.1.1 The Needs and Challenges of the Current Energy System

The present energy system faces several pressing needs. With a growing demand for energy driven by population growth, industrialization, and improved living standards, there is an urgency to meet it in a sustainable manner. Simultaneously, addressing climate change requires a transition to low-carbon or carbon-neutral energy sources. However, integrating renewable energy sources (RESs) poses challenges due to their converter-based, intermittent, and variable nature. To ensure a reliable and resilient power system, aging infrastructure needs to be upgraded and modernized. Grid resilience must also be enhanced to withstand extreme weather events, cyberattacks, and other disruptions. Furthermore, supporting the electrification of different sectors, e.g. heating, cooling, and transportation, necessitates infrastructure development and adequate grid capacity that is now lacking. To this end, advancements in energy storage technologies are also crucial. Lastly, improving energy efficiency across sectors is vital to reduce overall energy demand and greenhouse gas (GHG) emissions, which requires a combination of technological advancements and policy measures.

Addressing the aforementioned needs is not trivial and numerous challenges are present. Balancing energy supply with the ever-increasing demand is a hard task as it does not only presuppose upgrades in related infrastructure and equipment but also changes on how the operators schedule and manage the grid. Transitioning to low-carbon energy sources while ensuring a reliable and uninterrupted power supply poses a complex task that requires careful planning and investment. Integrating RES into the grid requires addressing the intermittency and variability associated with them, by necessitating innovative solutions for effective integration.

Integrated Local Energy Communities: From Concepts and Enabling Conditions to Optimal Planning and Operation, First Edition. Edited by Marialaura Di Somma, Christina Papadimitriou, Giorgio Graditi, and Koen Kok.

Moreover, upgrading and maintaining aging infrastructure presents financial and logistical challenges. Safeguarding the power grid against risks such as extreme weather events and cyberattacks requires robust strategies and investments in cybersecurity measures but also affects the "business as usual" in the operational planning of the operators. Advancing energy storage technologies is crucial for both dealing with the excess renewable energy and providing flexibility for balancing supply and demand. The electrification of various sectors presents several challenges that impact the power system overall. First, to satisfy all types of energy demands resulting from electrification, a significant amount of additional power network capacity is required. This includes accommodating the increased electricity consumption from transportation, buildings, and industry. Second, contingencies in the power system can have far-reaching consequences. Any disruption or failure can lead to power outages, affecting not only the resilience of the power system and of the electrified sectors but also the overall functioning of society. Grid stability and management become crucial factors in ensuring a reliable and resilient power supply in this case. As electrification changes the energy landscape, operators face new challenges in planning and scheduling. They must consider the increased complexity of managing diverse energy sources, grid capacity, and demand patterns to optimize system performance and ensure uninterrupted power supply.

As such, the distribution operators' planning and scheduling processes need to adapt to the evolving requirements of an electrified system to maintain efficient operations. Promoting energy-efficient practices and technologies across sectors requires changes in consumer behavior as well as supportive policies and incentives.

Finally, creating and implementing effective policies and regulations to facilitate the transition to a sustainable and resilient power system necessitates balancing the interests of different stakeholders and ensuring fair market competition.

Addressing these needs and challenges requires collaborative efforts from governments, energy providers, technology developers, researchers, and consumers to create a sustainable, secure, and affordable energy power system for the future.

1.1.2 What Is Sector Coupling?

Sector coupling originally referred to the electrification of end-use sectors such as heating, cooling, and transport, aiming at increasing the RES share in these sectors, based on the assumption that the electricity supply can be mostly renewable. More recently, the concept has been widened by also including supply-side sector coupling, integrating, for instance, power and gas sectors through power-to-gas (P2G) technologies. It must be highlighted that sector coupling is very similar to that of integrated energy systems, introduced by ETIP SNET Vision 2050 [1–3], defined as a *system of systems*. Namely, an integrated energy system is an integrated infrastructure for all energy carriers with the electrical system as a backbone, characterized by a high level of integration between all networks of energy carriers, coupling electrical networks with gas networks, heating, and cooling, supported by energy storage and conversion processes. The creation of these systems is based on the coordination of the planning and operation key processes. Within these processes, different types of energy systems across multiple geographical scales are considered

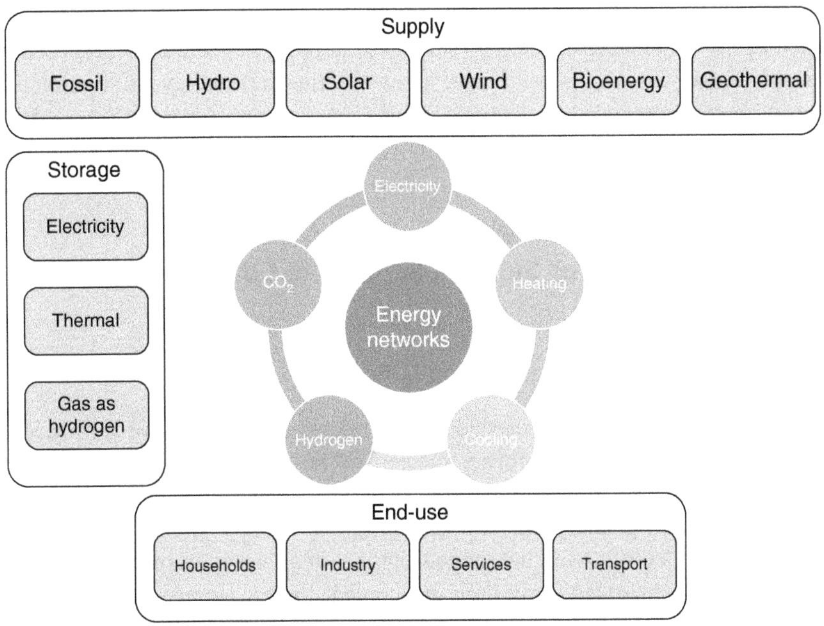

Figure 1.1 Representation of sector coupling concept. Source: Adapted from Van Nuffel et al. [4].

to foster reliability and efficiency in energy services while also minimizing negative environmental impacts [4]. The different sectors that can be involved under the concept of sector coupling are represented in Figure 1.1.

Two different strategies are considered under the concept of sector coupling, namely [5]:

- **"End-user" sector coupling** aiming at the electrification of end-use sectors and consisting of energy conversion technologies for electrification of final users' energy demand, thus enabling flexibility at the final users/prosumers level. An example of these technologies is well represented by electric vehicles (EVs) allowing for the electrification of the transport sector.
- **"Cross-vector" sector coupling** aiming at integrating multiple energy carriers mainly linking electricity and gas sectors through P2G technologies that can be used to produce hydrogen or synthetic methane when excess renewable electricity is available. The produced gas can be then stored for later re-conversion into electricity when renewable electricity supply is insufficient (and hence high electricity prices), by using the so-called power-to-gas-to-power process. On the other hand, electricity can be produced by hydrogen through fuel cells. Another alternative is that the hydrogen produced can be processed into methane or liquid fuel like methanol by making it reacts with CO or CO_2, the so-called power-to-liquid route. These fuels can be used in transport sectors such as shipping.

The combination of these two strategies allows increasing the flexibility of the energy system, while also supporting RES integration through optimal use strategies.

A good example was already provided above, but there is another key example represented by Power-to-Heat technology such as heat pumps. These latter, especially when combined with thermal storage, allow for thermal energy production in periods with excess renewable electricity which can be then stored and re-used in periods with insufficient renewable electricity, thereby representing a cost-effective and efficient solution.

1.2 Opportunities for Sector Coupling to Contribute to Decarbonization

1.2.1 Electrification and Sector Coupling

Electrification in power systems refers to the process of transitioning from traditional, fossil fuel-based energy sources to electrical power for various applications. It involves replacing the direct use of fossil fuels, such as gasoline and natural gas, with electricity as the primary source of energy.

The concept of electrification has gained significant attention in recent years due to its potential to reduce GHG emissions and combat climate change. So, electrification is one of the main drivers of energy transition as it is perceived nowadays and a reliable solution for effective decarbonization at the end user's side.

Therefore, the electrification scenario can be applied in different sectors. Some examples that can foster electrification are given below:

Transportation: Electrification of transportation involves transitioning from conventional internal combustion engines (ICEs) to EVs. This shift reduces reliance on fossil fuels, decreases air pollution, and offers opportunities for smart charging and integration with the grid, e.g. with vehicle-to-grid (V2G) services. EVs can expand services to other vectors/domains as well through the so-called V2X services. For example, in a V2Home scenario, EVs can supply power to homes during power outages or peak demand periods.

Residential and commercial buildings: Electrifying buildings involves replacing fossil fuel-based heating systems, such as oil or natural gas furnaces, with electric heat pumps. An instance is provided in Figure 1.2. This approach reduces carbon emissions, improves energy efficiency, and enables demand response programs.

Industrial processes: Electrification can also be applied to various industrial processes. For example, using electric furnaces instead of traditional fuel-based furnaces in manufacturing reduces emissions and provides more precise temperature control. Electrification can also power other industrial equipment, such as pumps and motors.

Although electrification presents direct benefits such as reduced GHG emissions and air quality improvement, it also presents challenges that are difficult to overcome. The most persistent challenges of electrification are discussed below.

One significant challenge is the need for additional power network capacity to accommodate the increased demand from electrified sectors related to transportation and buildings. This requires substantial investments in grid infrastructure,

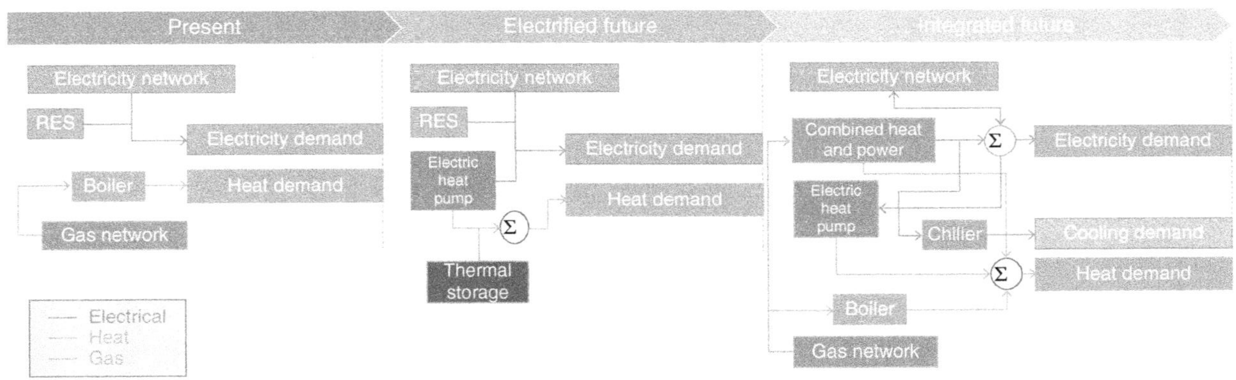

Figure 1.2 The transition from the present to the Integrated Systems approach.

such as new transmission lines, substations, and distribution networks, to ensure a reliable and resilient electricity supply. Another challenge lies in maintaining stability in the power system, as the integration of RES and the growth of decentralized generation introduce variability and uncertainty. Robust contingency plans and advanced grid management techniques are necessary to handle potential disruptions and ensure system reliability. Grid stability and energy balancing also become crucial considerations, as the intermittent nature of RES and the varying electricity consumption patterns of electrified sectors can impact the balance between electricity supply and demand. Effective energy storage solutions, demand response programs, and grid control mechanisms are required to stabilize the grid and optimize energy utilization. Finally, operators' planning and scheduling become more complex due to the increased number of distributed energy resources, EVs, and flexible loads. Advanced modeling, forecasting, and optimization tools are essential for operators to efficiently plan, schedule, and manage the operation of the power system while considering factors such as demand fluctuations, charging infrastructure availability, and grid constraints.

Nevertheless, if electrification is enhanced by the sector coupling approach that fosters coordinated management and operation of different sectors, then the challenges identified above can be lifted to a certain extent. In Figure 1.2, the current energy system is compared with the electrified and the integrated energy system paradigm. In specific, an example of electrifying the heat demand is shared. In the present, heat demand is covered by gas-fired boilers, whereas in an electrified future the heat demand is covered by electric heat pumps powered by the electricity network. Through the sector coupling approach, the electrified system is further enhanced. In fact, in such a system, multiple hybrid energy technologies are managed with high synergy to satisfy the multi-energy demand, and services can be provided with the most convenient energy carrier and sector. Moreover, sector coupling allows increasing efficiency in the energy resources use through exploiting synergies coming from interplay of different energy carriers and reduction of RES curtailment. In practice, for instance, in case of excess electricity from RES, it can be converted into gas as hydrogen or synthetic methane through P2G technologies, stored and/or transported by existing gas infrastructures for immediate or later usage, or re-converted again into electricity when renewable electricity supply is insufficient to satisfy the loads.

1.2.2 Enhancing System Stability and Reliability at the Grid Level Through Sector Coupling

As mentioned, sector coupling refers to the seamless integration, coordination, and operation of different energy sectors, leveraging the synergies between them. By combining the power grid with heating/cooling systems, transportation infrastructure, and other sectors, sector coupling offers numerous benefits that contribute to enhanced power system stability and reliability. An integrated grid can be seen in Figure 1.3. Five different carriers (water, electricity, heating, cooling, and gas) are

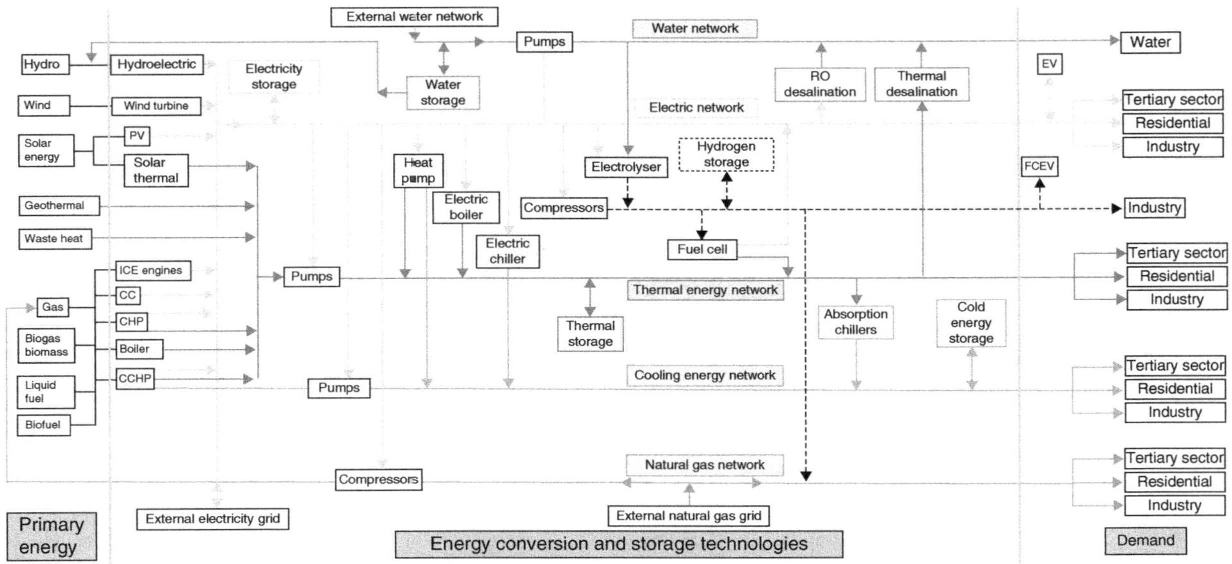

Figure 1.3 Overview of an integrated grid through sector coupling. Source: Adapted from Papadimitrou et al. [6].

seen integrated through the existence of conversion and hybrid technologies along with storage (thermal, electrical, hydrogen, EVs), allowing the interaction and collaboration of the different carriers when needed.

One of the key advantages of sector coupling is improved grid stability through flexibility and redundancy. The integration of diverse energy sources, such as wind and solar, with complementary technologies like heat pumps or combined heat and power (CHP), enables the balancing of fluctuating supply and demand. This flexibility allows for more efficient management of energy flows and reduces the risk of grid instability, ensuring a reliable power supply even in the presence of intermittent RES.

Additionally, sector coupling enhances power quality and resilience. By integrating energy storage systems into the grid, excess renewable energy can be stored and released when needed to the electricity grid or other carrier as already explained, smoothing out variations and mitigating voltage fluctuations. Furthermore, the coupling of heating, cooling, and power systems enables the utilization of waste heat from power generation, improving overall energy efficiency and reducing reliance on fossil fuels.

Another benefit of sector coupling is its ability to facilitate demand response schemes and load balancing. By integrating intelligent demand response mechanisms, consumers can adjust their energy consumption from technologies residing in different carriers based on grid conditions, helping stabilize the system during peak demand periods. This dynamic interaction between the power system and end-users – that can expand to all carriers – contributes to grid reliability and reduces the need for costly infrastructure upgrades.

Furthermore, the integration of EVs plays a crucial role in sector coupling. EVs can act as mobile energy storage units, offering grid support through V2G technology. During times of high electricity demand, EVs can supply power back to the grid, supporting grid stability and reducing stress on the power system. This bidirectional flow of energy optimizes resource utilization and enhances the reliability of the grid.

However, the successful implementation of sector coupling is not without its challenges. Interoperability and system integration pose significant technical hurdles. Different sectors often use diverse technologies, protocols, and communication systems, requiring seamless coordination and interoperability to ensure efficient energy sharing and control. Standardization efforts and collaboration among stakeholders are crucial to overcome these barriers. Furthermore, the maturity of technologies and the availability of sustainable business models are critical considerations. While some sector coupling technologies, such as heat pumps or CHPs, have matured, others, like P2G, may still be in the early stages of development. Policy and regulatory frameworks also need to adapt to support sector coupling. Clear guidelines and incentives are required to encourage cross-sector integration and investment. Additionally, social acceptance and public engagement are vital to address concerns, educate the public, and promote behavioral changes necessary for successful sector coupling implementation.

In conclusion, by integrating different sectors and leveraging the synergies between them, sector coupling enables flexibility, resilience, and optimized resource utilization. Overcoming technical, regulatory, and social challenges will be crucial to realizing the full potential of sector coupling and building a more stable, reliable, and sustainable power system for the future.

1.2.3 Decarbonization of Heating and Cooling in Building Environment (End-User Level) Through Sector Coupling

Today, energy consumption for heating and cooling in the building environment at the end-user level is responsible for a notable share of GHG in industrialized countries such as Europe. Sector coupling offers a great opportunity for decarbonization of final energy use for heating and cooling purposes through several technologies that can be classified into three categories: (i) technologies making direct use of renewable energy such as solar thermal, geothermal energy, or biomass heating; (ii) technologies making indirect use of renewable energy through electrification, such as electric heat pumps; and (iii) cross-vector integration technologies such as CHP or combined cooling heat and power (CCHP). According to Ref. [4], electric reversible heat pumps represent the best option for decarbonizing this specific sector in European countries, and this is mainly due to the good technical performance of this technology represented by a high coefficient of performance, being then followed by CHP for large applications and district heating. The latter also represents one of the main solutions for the decarbonization of heating demand in the building environment especially when based on RES as biomass heating or solar thermal, while also offering a great potential for the sector coupling strategies presented in Section 1.2. In fact, heat pumps and CHPs can be used with optimized strategies in a complementary manner, with the former operating in periods with low electricity prices and the latter operating in periods with higher electricity prices. However, the lack and inadequacy of existing infrastructures is the main barrier to deployment of this technology as it may require significant investments in new assets.

Another benefit represented by heat pumps and CHPs when coupled with thermal storage is represented by the provision of ancillary services to the electrical network [7]. A practical example is given by heat pumps decreasing the produced thermal energy without compromising the user's comfort due to building thermal inertia, thereby providing ancillary services. On the other hand, CHPs also can provide flexibility services to the power system, by decoupling the production of electricity and heat through thermal storage depending on the demand [8].

Another promising option to electric heating for decarbonization of this sector is represented by small-scale micro-CHP that allows for the reduction of distribution network costs, by replacing gas-fired plants. This technology can provide firm capacity (assuming it is able to be managed to provide capacity during non-curtailable or non-shiftable peak demand occurrences) while improving conversion efficiency, since the thermal energy produced recovered from electricity generation is not wasted but used to meet local heat demand.

1.3 European Energy Legislation and Initiatives Supporting Sector Coupling

The European energy legislation on energy transition sets forth a visionary path, aiming to foster the transition from traditional fossil fuel-based energy sources to cleaner, renewable alternatives. Toward this, directives at the European Union (EU) level on energy transition were issued some years ago and have been regularly updated ever since. Directives are assumed to be transpositioned at the EU member state level and put in context. Policy measures and initiatives that are supportive are also propelling Europe toward a sustainable and low-carbon future. In the next subsections, the most recent developments that are sector coupling supportive are briefly discussed, whereas Chapter 2 dives deeper into the EU policy.

1.3.1 Directives

Clean Energy for All Europeans Package [9]: Since 2018, the EU adopted the Clean Energy Package, which consists of several directives and regulations aimed at accelerating the clean energy transition. The package includes measures promoting sector coupling by encouraging the integration of renewable energy in various sectors and promoting energy storage and demand response technologies. In specific, the package includes the Renewable Energy Directive (RED II) [10]. The RED II sets binding targets for renewable energy consumption in the EU. It promotes sector coupling by establishing a framework for supporting the use of renewable energy in heating and cooling as well as in the transport sector. It encourages the production and use of advanced biofuels and renewable gases, such as biomethane and hydrogen. The Energy Efficiency Directive (EED) [11] sets out binding energy efficiency targets and measures to promote energy efficiency across different sectors. It encourages the use of energy-efficient technologies and promotes the integration of energy systems through the utilization of waste heat, CHP systems, and district heating and cooling networks. Energy Performance of Buildings Directive (EPBD) [12]: The EPBD focuses on enhancing the energy performance of buildings by promoting energy-efficient renovations and setting minimum energy performance standards for new constructions. The new smart readiness indicator (SRI) for buildings addresses sector coupling through the promotion of flexibility and mobility integration in buildings. In addition, the Energy Market Directive – EMD II [13] aims to create a competitive and integrated electricity market within the EU. It emphasizes the integration of RESs and the improvement of cross-border electricity trading. Energy Infrastructure Regulation (TEN-E) aims to facilitate the development of cross-border energy infrastructure, including electricity and gas projects, to strengthen energy security and support the integration of RESs. The Alternative Fuels Infrastructure Directive (AFID) aims to facilitate the deployment of alternative fuels infrastructure, such as EV charging stations and refueling stations for hydrogen and natural gas.

1.3.2 Policy

EU has set some strategic plans for energy transition that are updated regularly and follow the latest sociopolitical developments. For example, RepowerEU [14] aims to reduce energy dependence from third countries and accelerate energy transition. The measures in the REPowerEU Plan can respond to this ambition, through energy savings, diversification of energy supplies, and accelerated roll-out of renewable energy to replace fossil fuels in homes, industry, and power generation. The diversification of energy supplies points out the need for the sector coupling. This comes on top of the comprehensive strategy of the European Green Deal [15], by outlining the EU's ambition to achieve climate neutrality by 2050. It encompasses various policy initiatives and legislative proposals to drive the transition to a sustainable, low-carbon economy without leaving anyone behind. Toward this direction, the emissions trading system (EU ETS) is a crucial policy instrument in the EU's energy transition efforts. It establishes a cap-and-trade system for GHG emissions from industries, incentivizing emission reductions and encouraging the transition to cleaner technologies.

There are also some funding schemes that promote the aforementioned frameworks through developments, innovation, and research. Connecting Europe Facility (CEF) is a funding program that supports the development of trans-European infrastructure, including energy infrastructure. It provides financial assistance for projects that promote sector coupling, such as interconnections between electricity and gas networks, smart grids, and cross-border renewable energy projects. Also, Horizon Europe is the EU's research and innovation funding program. It supports various projects related to sector coupling, including those focused on energy system integration, smart grids, energy storage, and advanced renewable energy technologies.

1.3.3 Initiatives

There are important initiatives that epitomize the EU's commitment to mitigating carbon emissions, promoting RES, and fostering energy efficiency across various sectors. In the next paragraph, active initiatives that foster the sector coupling approach are briefly discussed.

European Technology and Innovation Platforms (ETIPs) have been created by the European Commission in the framework of the new Integrated Roadmap Strategic Energy Technology Plan (SET Plan) by bringing together a multitude of stakeholders and experts from the energy sector. The ETIP Smart Networks for Energy Transition (SNET) role is to guide Research, Development & Innovation (RD&I) to support Europe's energy transition. The visionary idea of *system of systems* presented in ETIP SNET's "Vision 2050," [1] is elaborated on the concept of sector coupling. This approach places the electricity grid as the central backbone of the future energy system. It enables the switch of energy carriers through different conversion technologies, facilitating the transportation of vast amounts of energy across Europe.

The ETIP-SNET's vision has garnered support from multiple industrial associations representing key stakeholders in the European energy sector. Sector coupling is underscored as a crucial aspect in the RD&I Roadmap 2020–2030 by ENTSO-E [16], emphasizing cross-sector integration. Similarly, WindEurope [17] calls for a regulatory framework that fosters sector coupling, enabling market players to drive the development and operation of assets.

European Technology and Innovation Platform on Renewable Heating and Cooling (RHC-Platform) is also another ETIP that is sector coupling related. The RHC-Platform [18] is an industry-led initiative that brings together stakeholders to promote the use of renewable energy in heating and cooling systems. It supports research, innovation, and policy development in the field of renewable heating and cooling, which is an essential aspect of sector coupling.

The BRIDGE (Building the Research and Innovation for the Development of Green Energy) EU initiative, stemming from the Horizon projects mentioned above, stands as a pioneering force in advancing the concept of sector coupling within the EU. BRIDGE's mission is to facilitate collaboration among stakeholders, including researchers, industry players, policymakers, and communities, to unlock the full potential of an integrated approach for the energy transition. By fostering cross-sectoral cooperation, BRIDGE empowers the efficient utilization of RES, optimizing their deployment based on real-time demand and supply patterns. As BRIDGE continues to bridge the gaps between sectors and facilitate knowledge exchange, it is propelling the EU toward meeting its energy transition objectives.

Mission Innovation (MI) [19] is a global initiative of 23 countries and the European Commission (on behalf of the EU) on energy transition. Among other things, the activities of this initiative are broken down into missions. Missions bring together dynamic and delivery-focused high-ambition alliances between countries, corporations, investors, and research institutes that set global stretch goals and provide momentum to make sure more innovation happens, more rapidly. Missions focusing on urban transition, clean hydrogen production, and green-powered future are in high complementarity with the sector coupling approach.

In conclusion, European energy legislation, policy, and initiatives have made remarkable strides in promoting sector coupling and driving the energy transition toward a sustainable and low-carbon future. Through a series of directives, the EU has established binding targets and measures to encourage the integration of renewable energy in various sectors, including heating, cooling, and transportation.

1.4 Main Barriers to Implementation

Although the tangible benefits for the energy system are achievable with sector coupling, the concept is still addressed at the theoretical level, being far away from a large-scale implementation and deployment, due to several barriers. In detail, the bottlenecks to sector coupling can be classified into regulatory, economic, technical, technological, social, and market categories as discussed below.

Omissions in legislation/regulation pose a significant hurdle to implementing sector coupling. Existing regulations may not adequately address the integration of different sectors, such as power, heating, cooling, and transportation. The lack of clear guidelines and frameworks can discourage investments and hinder the development of cross-sector solutions and collaboration between the relevant stakeholders. Addressing this barrier requires policymakers to update and align regulations to accommodate the complex interactions and interdependencies between various sectors.

Economic barriers are another challenge. Implementing sector coupling often requires substantial upfront investments in infrastructure and technology across sectors. The costs associated with integrating different sectors, optimizing energy flows, and ensuring reliable and efficient management and operation can be high. In the absence of appropriate economic incentives and financial mechanisms, the business case for sector coupling may be weakened, hampering its widespread adoption.

Interoperability barriers arise from the diverse technologies, systems, and protocols used in different sectors. Achieving seamless integration and coordination between the power grid, heating systems, and transportation infrastructure can be technically challenging. Compatibility issues and data exchange protocols must be addressed to enable efficient energy sharing, control, and management across sectors. Collaboration among stakeholders, standardization efforts, and the development of interoperable solutions are necessary to overcome these barriers.

Technological barriers mainly arise from the need for further innovation in the various supply, demand, transmission, distribution, and storage technologies to improve the technical feasibility of the most relevant technologies. Moreover, performance aspects such as the efficiency, durability, and degradation of sector coupling technologies still represent a major barrier to their deployment. Most technologies are not fully mature as is the case for hydrogen technologies, and their performance must further improve to be worth deploying and integrating on a large scale. Then, it is worth highlighting that the large-scale deployment of these technologies only makes sense if large amounts of low-cost, carbon-free, surplus electricity are available, which usually implies that technologies harvesting RES are already deployed on a massive scale.

Social barriers are mainly related to the lack of citizens' engagement for the deployment of sector coupling technologies at the end-user level and increasing consumers' awareness about the benefits deriving from such types of implementation is thus necessary. Another non-negligible issue, which is strongly related to the public acceptance of sector coupling solutions, is the lack of economic incentives to invest in local energy-efficient and renewable projects, especially considering that the initial costs of these solutions can be very high. In such a context, priority should be always given to investments that facilitate the uptake of low-carbon technologies and flexibility solutions.

Beyond the bottlenecks to sector coupling identified, another criticality is related to the lack of methodological progress related to the various stages of energy system's

Table 1.1 Methodological progress needed to implement sector coupling at various stages of energy system management.

Stage of energy system management	Needed methodological changes to implement sector coupling
System planning	Power system planning needs to be addressed considering the electrification of different energy sectors like heating and transport. The main challenge is related to correctly account for the impacts that a strong electrified scenario may have on peak demands from households to national and continental systems and on energy supply from locally distributed energy resources to large generators, while also considering all direct and indirect effects.
System operation and optimization of energy carriers	The energy system operation needs to properly account for the electrification of different energy sectors, by optimizing and exploiting the potential flexibility coming from new emerging technologies such as EVs, electric pumps, storage, and demand-side management. The aim of this stage is to maximize RES penetration, by reducing curtailment and maximizing the usage of low-carbon energy technologies. Moreover, synergies coming from the interplay of multiple energy carriers should be managed by optimizing the cooperation of multiple energy carriers and the related conversion processes. It must be mentioned that to optimize conversion and storage processes across sectors, a coordinated infrastructure planning of all involved energy systems is required.
Market design	Configuration and mechanisms related to market should be updated to take into account the inclusion of emerging technologies related to sector coupling.
Ownership and business models	One of the key issues related to implementation of sector coupling at a large scale is related to the needed coordination among various actors with often conflicting interests. This type of collaboration needs to be regulated through specific governance rules that need to establish also ownership and operation roles. This requires also sustainable business models that demonstrate the economic viability of the different actors involved.

Source: Adapted from Munster et al. [7].

management. In this regard, the foreseen changes are related to the following stages defined in Table 1.1:

1.5 The Integrated Local Energy Community Concept to Foster Sector Coupling at the Local Level Through End-Users Engagement

As already mentioned, achieving an integrated grid based on cooperation of multiple energy carriers is a challenging task, and this is the main reason for which a bottom-up approach is preferable to demonstrate the economic and environmental sustainability of sector coupling solutions. This bottom-up approach is based on the assessment of local projects involving RES and multiple energy carriers. These local multi-carrier energy systems can potentially contribute to the EU energy and climate objectives, by helping reverse energy consumption and emission trends. As compared to traditional centralized energy systems, decentralized local energy systems promote an enhanced focus on local security of energy supply.

Combining the sector coupling concept at the local level with one of the energy communities leads to the innovative concept of integrated local energy communities (ILECs), that represent an efficient and sustainable way of managing energy at a local level by fostering consumer engagement and empowerment. The ILEC concept may refer to a set of energy users deciding to make common choices in terms of satisfying their energy needs, in order to maximize the benefits deriving from this collegial approach, thanks to the implementation of various electricity and heat technologies and energy storages and the optimized management of energy flows [20]. Indeed, ILECs are well placed to meet local energy needs, reduce the need for transmission infrastructure, and bring people together to achieve common goals for well-being, thereby fostering consumer engagement. By representing locally and collectively organized energy systems, ILECs are able to effectively integrate different sectors through various local generation of electricity, heating and cooling, flexible demand, storage, and electric mobility. As a result, ILECs are able to fulfill the multi-energy demand of the communities' users through the optimized operation of local distributed energy resources, by exploiting synergies and cooperation among the various energy carriers. In addition, ILECs may be able not only to efficiently self-provide for the communities' users but also to provide system services to neighboring systems such as balancing and ancillary services. In addition, exploiting the interplay among different energy sources at the local level may reduce RES curtailment, thereby supporting decarbonization targets. A graphical representation of the ILEC is presented in Figure 1.4. In the ILEC, prosumers can cooperate by sharing all energy carriers, with the aim to satisfy the energy needs of the entire local community. Community generation and community storage systems may be also involved through dedicated community energy management systems. The ILEC promotes local balancing as well as strategic exchanges with the external grids through coordination of exchange. In this way, the community can always have interactions with larger systems, sustaining access to the largest pool of external resources possible, while leaving open the possibility of local resource optimization and provision of services to the external system. The ILEC allows exploiting synergies between different sectors such as electricity, heating, cooling, and transport as well as between different technologies.

The aim of this book is to present in a thorough and comprehensive way all the critical aspects that are needed when designing, planning, and operating an ILEC from end to end and thus contribute to the needed methodological progress mentioned above. The book's objective and ambition is timely as the integrated energy system, i.e. the cross-sectoral and energy carriers' integration and interaction, is the means to achieve energy transition and accelerate the interdependence on fossil fuels as per the EU vision.

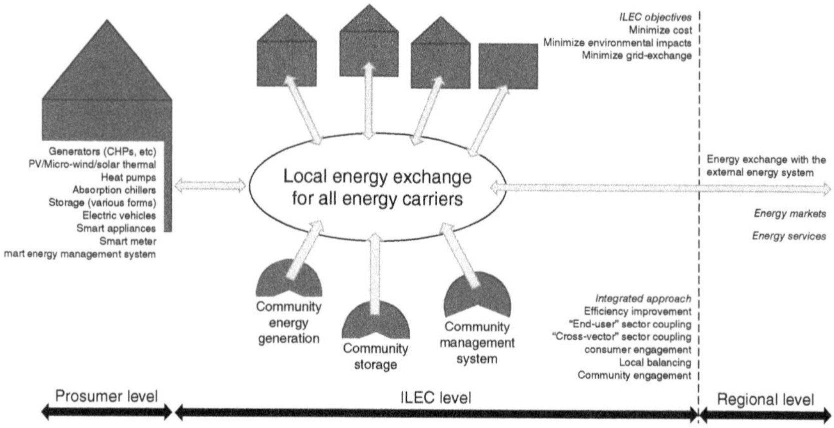

Figure 1.4 Graphical representation of an integrated local energy community.

To this end, this book addresses the regulatory and policy frameworks related to the deployment of local multi-carrier energy systems, the concept of ILEC with mapping of enabling technologies and key actors for the implementation of this energy paradigm at the local level, innovative approaches on investment decisions, and optimal planning in the long term but also operational analysis in both medium term and short term. Real-time operation and control is also addressed through consumer-centered approaches. On top of that, sustainable business models and the integrated market interaction with such systems under a realistic view are included.

List of Abbreviations

AFID	Alternative Fuels Infrastructure Directive
CCHP	combined cooling heat and power
CEF	Connecting Europe Facility
CHP	combined heat and power
EED	Energy Efficiency Directive
EPBD	Energy Performance of Buildings Directive
ETIP	European Technology and Innovation Platform
EU	European Union
EV	electric vehicle
GHG	greenhouse gas
ICE	internal combustion engine
ILEC	integrated local energy community
MI	Mission Innovation
P2G	power-to-gas
RED	Renewable Energy Directive
RESs	renewable energy sources

SNET Smart Networks for Energy Transition
SRI smart readiness indicator
V2G vehicle-to-grid

References

1 ETIP SNET "VISION 2050 (2018). Integrating smart networks for the energy transition: serving society and protecting the environment.

2 Di Somma, M. and Graditi, G. (2022). Integrated energy systems: the engine for energy transition. In: *Technologies for Integrated Energy Systems and Networks* (ed. G. Graditi and M. Di Somma), 15–40. Wiley.

3 Di Somma, M. and Graditi, G. (2022). Challenges and opportunities of the energy transition and the added value of energy systems integration. In: *Technologies for Integrated Energy Systems and Networks* (ed. G. Graditi and M. Di Somma), 1–14. Wiley.

4 Van Nuffel, L., Dedecca, J.G., Smit, T., and Rademaekers, K. (2018). Sector coupling: how can it be enhanced in the EU to foster grid stability and decarbonise?

5 Comodi, G., Spinaci, G., Di Somma, M., and Graditi, G. (2022). Transition potential of local energy communities. In: *Technologies for Integrated Energy Systems and Networks* (ed. G. Graditi and M. Di Somma), 275–304. Wiley.

6 Papadimitrou, C., Di Somma, M., Charalambous, C. et al. (2023). A comprehensive review of the design and operation optimization of energy hubs and their interaction with the markets and external networks. *Energies* 16 (10): 4018.

7 Munster, M., Moller Sneum, D., Bramstoft, R., et al. (2020). Sector coupling: concepts, state-of-the-art and perspectives.

8 Gao, S., Jurasz, J., Li, H. et al. (2022). Potential benefits from participating in day-ahead and regulation markets for CHPs. *Applied Energy* 306: 117974.

9 Clean energy for all Europeans package (2019). https://ec.europa.eu/energy/topics/energy-strategy/clean-energy-all-europeans_en (accessed November 2023).

10 Directive 2018/2001/EU on the promotion of the use of energy from renewable sources (2018). https://eur-lex.europa.eu/legal-content/EN/TXT/PDF/?uri=CELEX:32018L2001&from=EN (accessed November 2023).

11 Directive 2018/2002 on energy efficiency (2018). https://eur-lex.europa.eu/legal-content/EN/TXT/PDF/?uri=CELEX:32018L2002&from=EN (accessed November 2023).

12 Directive 2018/844 on the energy performance of buildings (2018). https://eur-lex.europa.eu/legal-content/EN/TXT/PDF/?uri=CELEX:32018L0844&from=IT (accessed November 2023).

13 Directive 2019/944/EU on the internal electricity market (2019). https://eur-lex.europa.eu/legal-content/EN/TXT/PDF/?uri=CELEX:32019L0944&from=EN (accessed November 2023).

14 REPowerEU: a plan to rapidly reduce dependence on Russian fossil fuels and fast forward the green transition (2022). https://ec.europa.eu/commission/presscorner/detail/en/IP_22_3131 (accessed November 2023).

15 A European Green Deal Striving to be the first climate-neutral continent (2019). https://ec.europa.eu/info/strategy/priorities-2019-2024/european-green-deal_en (accessed November 2023).

16 ENTSO-E Research, Development & Innovation Roadmap 2020 – 2030, October 2020 (2020). https://eepublicdownloads.entsoe.eu/clean-documents/Publications/RDC%20publications/entso-e-rdi_roadmap-2020-2030.pdf (accessed November 2023).

17 WindEurope. https://windeurope.org/

18 European Technology and Innovation Platform. https://www.rhc-platform.org/

19 Mission Innovation. http://mission-innovation.net/about-mi/overview/

20 Caliano, M., Buonanno, A., Di Somma, M. et al. (2022). Enabling conditions for the deployment of integrated local energy communities in Europe. *Workshop on Blockchain for Renewables Integration (BLORIN)*, Palermo, Italy. IEEE. pp. 144–148.

2

Current Status of Multi-carrier Energy Systems in Europe with Main Limitations and Shortcomings to the Optimal Use of Local Energy Resources

Andrei Morch[1], Hanne Sæle[1], Jesús Fraile Ardanuy[2], Giuseppe Conti[2], Gabriele Comodi[3], and Mosè Rossi[3]

[1] *SINTEF Energy Research, Department of Energy Systems, Sem Saelands vei 11, NO-7034, Trondheim, Norway*
[2] *Universidad Politécnica de Madrid, ETSI Telecomunicación, Avenida Complutense 30, 28040, Madrid, Spain*
[3] *Università Politecnica delle Marche, Faculty of Engineering, Department of Industrial Engineering and Mathematical Sciences (DIISM), Via Brecce Bianche 12, 60131, Ancona, Italy*

2.1 Introduction

Integrated local multi-energy systems are recognized as a promising bottom-up alternative to achieve the ambitious energy and climate goals set by the European Commission (EC) for 2030 [1]. The nature of integrated multi-energy systems presumes a very complex interaction of different aspects from various energy sectors, which historically worked almost independently. Therefore, configuration and development of an integrated local energy system under the energy community concept is an extremely demanding task, requiring a sound interdisciplinary knowledge base. In this sense, it becomes critically important to have a thorough understanding of the technologies that will make use of locally available resources and thus enable evolution and deployment of energy communities across Europe.

2.2 Methodology

The chapter includes three main parts related to the following studies:

- The initiating scoping study, which was based on the Pan-European decarbonization targets and did a screening of relevant selection of the documents. Several of the identified issues and shortcomings were further elaborated in the succeeding two parts.
- The second study focused on the enabling technologies. It further identifies and benchmarks the indicative status for the deployment of various technologies for integrated local multi-vector energy systems (including batteries and electric vehicles – EVs) and corresponding supporting mechanisms and tools in the member states (MSs).

Integrated Local Energy Communities: From Concepts and Enabling Conditions to Optimal Planning and Operation, First Edition. Edited by Marialaura Di Somma, Christina Papadimitriou, Giorgio Graditi, and Koen Kok.

- The third and final step identified the present limitations, shortcomings, and obstacles to innovation, which may prevent the intended transformation of the European energy landscape toward local multi-vector energy systems with a high level of decarbonization.

Considering that none of the existing formal definitions of energy communities [2] covers the multi-energy aspects, the H2020 project eNeuron introduced its own definition of *Integrated Local Energy Community (ILEC)* as a set of energy users deciding to make common choices in terms of satisfying their energy needs to maximize the benefits deriving from this collegial approach thanks to the implementation of various electricity and heat technologies, energy storages (ESs), and the optimized management of energy flows.

2.3 The Scoping Study: Road Maps and the Overall Pan-European Priorities

The selection of the documents is determined by the previously defined purpose of the study, and the documents considered in this study have been issued by several types of stakeholders, including:

- Governmental Organizations, including the EC, issuing Directives and Regulations.
- National Governments and National Regulatory Authorities (NRAs).
- Interest organizations and Industrial Associations, for example, WindEurope and Solar Power Europe.
- Worldwide international organizations such as the International Energy Agency (IEA) and similar.

To create a more systematic approach, a set of the so-called *topics of interest* has been identified. These topics represent either some key assumptions made within the project or some attributes essential for the project's outcomes. The identified topics of interest are presented in the following text as subsection titles.

Considering the focus on integrated local energy communities (LECs), the present study concentrated on the distribution level, including distribution system operators (DSOs), active customers, citizen energy communities (CECs), distributed generators, aggregators, and other relevant actors and roles. However, the study also considers important issues at the transmission level, if they may have any significant impacts on the primary scope of the analysis.

2.3.1 Energy Policies, Which Are Preferred in Europe (2020–2050)

Several of the screened documents contained information about *energy carriers* preferred in Europe for the time horizon 2020–2050.

Following the definition from Ref. [3] by energy carriers, the study means substance or phenomenon that can be used to produce mechanical work or heat or to operate chemical or physical processes. An energy carrier allows to transfer a quantity of energy in space and time.

The most recent road maps redefine the energy carriers preferred for decarbonization in Europe according to different sectors, such as electrification for personal transport, while hydrogen is a preferred technology for cargo transport.

2.3.1.1 The Pan-European Dimension

The latest recast of EU Internal Electricity Market (IEM) Directive 2019/944 [4] in many ways was a pivotal document for development of the European power industry, firmly connecting the political goals to specific actors, their roles, and responsibilities. Apart from redefining the well-established actors, it also introduced entirely different players as CECs, active customers, and closed distribution systems, thereby giving the MSs an opportunity to shape and develop these actors according to their local conditions, national needs, and requirements.

More specifically, the IEM Directive highlights moving toward decentralized generation and decarbonized markets as the main reason for changing rules for electricity trading [4]. In addition, the Directive also provides a guideline for the MSs, regulatory authorities, and transmission system operators (TSOs) to develop a common and interconnected network as well as an internal energy market, to deploy the renewable energy sources (RESs), free competition, and security of supply. The Directive defines tasks for the NRAs and mentions specifically facilitating access to the network for new generation capacity and ES facilities.

Furthermore, the Renewable Energy Directive (RED II) 2018/2001 on the promotion of the use of energy from RES [5] opens for the introduction of support schemes and modalities for integration of electricity from renewable sources but does not indicate any strong priorities among these resources. "Renewable energy" is used as a common term including wind, solar (thermal and photovoltaic) and geothermal energy, ambient energy, tide, wave and other ocean energy, hydropower, biomass, landfill gas, sewage treatment plant gas, and biogas.

Security of supply is important for different infrastructures present in diverse energy carriers and, according to the 2020 report on the "State of the Energy Union pursuant to Regulation (EU) 2018/1999 on Governance of the Energy Union and Climate Action" [6], the Risk Preparedness Regulation implemented in the electricity sector ensures cooperation between the MSs in relation to energy security. This Regulation ensures that MSs have sufficient tools for cooperation to prevent, prepare for, and mitigate electricity crisis. For the security of gas supply, preventive action and emergency plans have been developed, containing measures for mitigating the impact of gas supply disruption and risks identified at the national and regional level.

Important initiatives for integrating electricity and gas markets have been already performed, and more work is needed according to Ref. [6]. The document refers repeatedly to the recent EU's Communication "Annual Sustainable Growth Strategy 2021" [7], which defines several European flagships (coordinated investments and reforms):

- "Power-up flagship" frontloading future-proof clean technologies and, in particular, renewables and hydrogen.

- "Renovate flagship" to improve the energy and resource efficiency of buildings.
- "Recharge and refuel flagship" to accelerate the use of sustainable, accessible and smart transport, charging and refueling stations, and the extension of public transport.

More recently, Communication (2022/230) REPowerEU Plan [8] addressed an unexpected political and economic turmoil caused by the outbreak of the war in Ukraine, which resulted in unprecedented energy crises with extremely high prices for electricity in Europe. It proposed to increase the headline 2030 target for renewables from 40% to 45% under the Fit for 55 package [1].

2.3.1.2 The Reference Scenario Toward 2050

The EU Reference Scenario [9] is one of the EC's key analysis tools in energy, transport, and climate action, focusing on trend projections and providing a benchmark based on today's policy. The Reference Scenario reflects current trends and developments in the EU energy system and greenhouse gas (GHG) emissions. The horizon of the projection is 2050, and results are available in five-year time steps for each MS and the European Union of 27 states (after 1 February 2020).

According to the European Network of Transmission System Operators for Electricity (ENTSO-E's) "Energy Facts Europe," energy efficiency plays a crucial role in delivering cost-effective decarbonization, and energy consumption has fallen by 10% in primary energy and 7% in final energy terms since 2005 [10]. Accordingly, in Reference Scenario [9], the energy consumption is expected to reduce due to improved energy efficiency, also resulting in declining energy intensity to gross domestic product (GDP), even if the energy intensity per GDP varies per country. The gross inland consumption (GIC) and GDP continue to decouple. The main drivers of the decreasing trend of total primary energy requirements are the developments in final energy demand.

According to Ref. [9], steam and heat demand in EU28 will rise slightly toward 2025 and then remain approximately stable toward 2050. The main sources of demand are industry and households. It is projected that district heating will maintain its share in demand for heat. In the short and medium term, there is a gradual shift from solids and gas district heating boilers to biomass/waste boilers, and in the long term, electric boilers, heat pumps (HPs), geothermal, and thermal solar penetrate the district heating market and gain in market share. During the whole period, it is projected that electricity generation from combined heat and power (CHP) plants will increase, as well as the steam output, up to 2020 and will remain almost constant toward 2050.

The following changes for different types of energy carriers are estimated [9]:

- **Natural gas and nuclear:** Almost stable share in total primary energy requirements throughout the whole period.
- **Oil:** Represents the largest share in total primary energy requirements. This continues as the largest consumer is transport where substitution possibilities are limited.

- **Gas:** Maintains its share in total primary energy requirements due to convenience and low emissions relative to other fossil fuels. Some emergencies in transport and wide use in power generation.
- **Solids:** Decrease in share due to the decline of solid use in all sectors of demand and energy supply sectors.
- **Biomass and waste:** Increase in volume and share mainly due to increases in power generation and industrial uses.
- **Other renewables:** Increase steadily throughout the projection period toward a share as high as that of gas, driven mainly by the impressive developments in the power generation sector.

2.3.1.3 Vision 2050

The concept of sector coupling was elaborated by the European Technology and Innovation Platform Smart Networks for Energy Transition (ETIP-SNET) into the visionary concept of "System of Systems," which was presented in its "Vision 2050" [11]. According to the document, electrification of Europe's energy systems will be the backbone of its societies and markets. This will require an incremental coupling of electricity and gas networks via the production of carbon-neutral synthetic gases (methane) to ensure the long-term security of supply (seasonal storage) for an electricity system powered by RESs. Additionally, as a major carbon emission polluter, coal is completely phased out from European energy systems by 2050.

While solar and wind have the greatest potential at the EU level, all renewables such as hydropower, geothermal energy, ocean energy, or biomass/gas will be developed considering their potential, cost, and added value to the regional systems and grid stability [11]. Biomass and synthetic gas together with hydro, ocean energy, and nuclear (possibly fusion beyond 2050) will be used when the value of their use is high for embedding biomass or gas-fired flexible thermal generation and gas for certain industrial processes or for aviation, shipping, and long-distance trucking.

Daily or seasonal differences between energy demand will require a flexible system with injections in daily or seasonal ES such as pumped hydro, batteries, hot water reservoir thermal storage, or power-to-gas (P2G) conversion technologies. More, in general, locally available energy resources are used to their full economic potential, partly deferring needs for upgrades of the electricity transmission and distribution (T&D) networks and maximizing the resilience of supply channels for heating and cooling needs.

The 2050 Vision foresees that future integrated energy systems will mainly rely on

- Renewable electricity (hydro, solar, wind, geothermal, marine).
- Renewable electricity/heat (biomass, biogas, solar).
- Renewable heat and cooling (solar, geothermal, marine).
- Renewable gas (biomass, biogas).
- Renewable fuels as biofuels.
- Nuclear.

In "Power Facts Europe 2019" [10], the focus is on the future power system and the distributed generation that can be connected (for example, electricity produced by solar PV and wind). In this report, investment in power system infrastructure is mentioned as key to promote P2G and other Power-to-X (P2X) technologies.

Massive introduction of new technologies uncovers the necessity for several regulatory actions, and WindEurope [12] calls for a regulatory framework for sector coupling in which energy generation, conversion, storage, transmission, and distribution are unbundled and market players, not regulated entities like TSOs and DSOs, drive the development and operation of assets. Prioritizing the different technologies becomes important. In the same document, WindEurope calls for prioritizing direct electrification, and this also includes incentives for the use of heat pumps including direct heating and electrification of transport by building charging infrastructure.

2.3.2 Expected Roll-Out of the Different Technologies in the 2030 Horizon

This part of the study concentrates primarily on forthcoming technologies with an expected bigger impact on energy production; it can be new, or a combination of different technologies related to the *energy distribution* part. The electricity transmission part is also mentioned when it directly influences the distribution part.

According to RED II Directive [5], the lack of transparent rules and coordination between the different authorization bodies is reported as a critical issue that hampers energy deployment from renewable sources.

2.3.2.1 Deployment of Hydrogen Infrastructure

As it was mentioned in Section 2.3.1.1, hydrogen is a consistent part of the future decarbonization strategy. The importance of hydrogen production and deployment of P2X is provided in [13], referring to power-to-H2 (P2H2) as the most promising since the electrolysis of water can be performed through RES or other conventional systems, which are rearranged according to the used energy carrier. The same document estimates

- **The potential demand for P2H2 in 2050:** more than 500 TWh in industry and transportation.
- **Potential demand for power-to-methane (P2CH4) in 2050:** 900 TWh in industry, 70 TWh in transport, and 950 TWh in other uses, excluding power generation.
- **Potential demand for power-to-liquid (P2L) in 2050:** 60 TWh (industry), 2430 TWh (transport), and 100 TWh (other applications).

To meet these ambitious targets, it is required to establish the necessary infrastructure, which is almost nonexistent at the moment.

The EC's "Hydrogen Strategy" [14] defines several critical action points related to the deployment of hydrogen infrastructure:

- To produce renewable hydrogen, a huge investment in building electrolyzers is needed in Europe for the next ten years (up to 6 GW of renewable hydrogen electrolyzers by 2024) to reach the strategic objective of 40 GW electrolyzer capacity by 2030.
- The electrolyzers are planned to be built next to existing demand centers so they will be a distributed generation model (hydrogen valleys).
- To expand the coverage, infrastructure for the transport should be deployed (medium and backbone transmission).
- For a complete integration of hydrogen as an energetic solution in Europe, a Pan-European grid needs to be built, and infrastructures for transporting (via pipelines) and storage hydrogen accomplished, connecting supply and demand.
- For long-range transportation, the Trans-European Networks for Energy (TEN-E) should be revised. The Pan-European gas infrastructure could be reshaped to be used for large-scale cross-border transport of hydrogen.
- The transitional phase may accept the blending of hydrogen in the natural gas network at a limited percentage aiming at decentralized renewable hydrogen production in local networks, but this leads to practical difficulties and safety-related risks.

The "Hydrogen Roadmap Europe" expects that the roll-out of electrochemical hydrogen generation (electrolyzers) and utilization (fuel cells) technologies is expected to increase dramatically.

- In particular, electrolysis capacity is meant to increase by 15–40 GW deployed in 2030; scale-up to multi-MW systems is also foreseen.
- The deployment of more than 2.5 million fuel cell CHP units (considering each CHP unit of kW-scale results on the order of 1–10 GW) by 2040.

The document [15] argues the necessity for the development of charging infrastructure for fuel cell electric vehicles (FCEVs). By 2050, hydrogen could power a European fleet of approximately 45 million passenger cars, 6.5 million light commercial vehicles (LCVs), 250 000 buses, and 1.7 million trucks. With about 3700 large refueling stations in 2030, FCEVs could provide "full mobility" across Europe.

2.3.2.2 Energy Storage

The European Energy Storage Technology Development Roadmap [16] envisages that short-term balancing will become a major problem for the European power system, and the short-term electricity balancing market is where ES will be first applied based on commercial business cases. The need for additional balancing power will be substantiated already within the next five years. It is expected that from a long-term perspective (15–20 years) the ES will become an even more significant part of the electricity system. By this time, from the ES Technology Development Roadmap 2030 perspective, not only ancillary services but also energy arbitrage based on stored energy will be finding bridgeheads on the shore of the energy market. This will be a consequence of the increasing penetration of renewable supply sources alongside the corresponding withdrawal of fossil, dispatchable generation capacity. The same document underlines the importance of heat storage, which in terms of energy is by far the largest single ES application field in Europe. The document states that heat storage technologies will soon become even more important for balancing the electricity grids and for better and more economy-efficient heat management by the consumers.

"Study on Energy Storage" [17] estimates the optimal roll-out of diverse technology solutions capable of providing flexibility to the power system. According to the results shown in the document and the assumptions made by the authors of the study, a moderate/relevant increase in the installed capacity of pumped-hydro ES and Li-ion batteries is required to optimally deal with the system's flexibility needs in 2030. A substantial investment in combined and open-cycle gas turbines is required as well for such a purpose. According to the sensitivity analysis performed in the study, the investment in Li-ion batteries can be partly replaced by a higher penetration of flexible demands.

Related to battery storage, according to The ENTSO-E "Power Facts Europe" [10] in recent years, the cost of electrochemical battery storage has fallen significantly as interest among consumers, network operators, and policymakers has risen. The battery storage costs have fallen, and the uptake of storage has increased – also "behind-the-meter battery storage." The same document concludes that grid-connected storage (in front of consumers' meters) and behind-the-meter consumer storage devices are important flexibility resources, as storage technologies support the flexible operation of the power system, helping to balance out the peaks and troughs in supply and demand across different timescales. Energy storage systems (ESSs) have the potential to provide valuable services throughout the energy chain: from power generation, transportation, and distribution to the final consumer [16]. Examples in which electrical ES will add value to the electric chain are shown in Table 2.1.

Main characteristics and cost evolution will determine the future deployment of different storage technologies (it can be any storage technology, e.g. thermal storage and electricity).

Storage technology in its various forms is foreseen as a key enabler of the energy transition and a binding element of different energy carriers. However, it has been concluded that today only few ES applications can justify market-based business cases, and this is why many ES technologies have not spread into the market yet. There is a strong expectation that the short-term electricity balancing market is where ES will be first applied, based on commercial business cases, and we believe the need for additional balancing power will be substantiated already within the next five years. From a longer-term perspective (15–20 years), ES will become an even more significant part of the electricity system, providing more services.

Apart from electricity storage, heat storage will become increasingly important. Today more than 50% of the final energy demand in the EU is used for generating heat and already now heat storage is utilized in water-based systems for domestic and district heating. In terms of thermal energy, storage is by far the largest single ES application field in Europe.

Table 2.1 Energy storage segmentation.

Conventional generation	Transmission	Distribution	Customers services
Black start	Participation in the primary frequency control	Capacity support	End-user peak shaving
Arbitrage	Participation in the secondary frequency control	Dynamic, local voltage control	Time-of-use energy cost management
Support to conventional generation	Participation in the tertiary frequency control	Contingency grid support	Particular requirements in power quality
Renewable generation	Improvement of the frequency stability of weak grids	Intentional islanding	Continuity of energy supply
Distributed generation flexibility	Investment deferral	Reactive power compensation	Limitation of upstream disturbances
Capacity firming	Participation in angular stability	Distribution power quality	Compensation of the reactive power
Limitation of upstream perturbations	—	Limitation of upstream perturbations	—
Curtailment minimization	—	—	—

Source: Adapted from European Association for Storage of Energy [16].

2.3.2.3 Hydrogen Storage

Hydrogen storage is mentioned in several documents because of its specific properties, and in general is considered as a natural component of P2G scheme. "The Hydrogen Roadmap Europe" [15] highlights the properties of hydrogen storage. Hydrogen can be stored for long periods of time and at a large scale at competitive costs, compared to conventional large-scale ES, such as pumped hydro. Hydrogen storage form can be as gaseous/liquid hydrogen or bound to carrier molecules (ammonia, methanol, etc.). The storage capacity of hydrogen can be exploited not only for long-term unbalances but also to provide grid services or flexibility, despite the lower round-trip conversion efficiency. Hydrogen can become a favorable storage option for high variable renewable energy source (VRES) penetration regions, characterized by grid congestion, allowing to defer the upgrade of costly grid infrastructure. By 2030, power companies could store approximately 25 TWh of surplus renewable electricity in the form of hydrogen for use in other end-use segments. By 2050, this number could more than double to about 58 TWh.

2.3.2.4 Hybrid Solutions

Several RES technologies have inheritant technical limitations, especially when it comes to reliability and intermittance. Deployment of RESs, for example PV, can be significantly improved by coupling these with ES, as it is concluded by the Interreg Europe Final Report [18].

The EU Market Outlook for the Solar Power [19] connects PVs to production of renewable hydrogen, so it can help decarbonize the so-called hard-to-abate sectors and will play a key role in reaching carbon neutrality. The same document also expects that on-site PV Generation with storage, demand response, and smart applications can contribute to stable renewable energy supply and to provide flexibility in the electric system. The combination of PV and green roofs increases the building efficiency and provides valuable services to the electricity system.

Spain can be mentioned as an example of considering hybrid solutions in MSs. In their national "Integrated Plan for Energy and Climate" (2021–2030) [20], it is described actions to avoid the renewable energy spills by investing that energy in producing hydrogen. It also mentions the hybridization of renewables + storage to avoid unnecessary investments in the grid or thermal storage linked to thermo-solar power plants.

An increase in self-consumption is also mentioned as an additional benefit of combining different technologies (battery storage + PV generation) [18]. Depending upon different locations, it can be solar thermal with PV generation (examples from Portugal). Integration of PV generation with ES and demand-side response to increase the self-consumption degree.

2.3.2.5 Smart Metering System as a Key Enabling Technology

Implementation of Smart Metering solutions can be considered a first step and mandatory prerequisite for implementation of Smart Grid solutions, empowering end-users to follow dynamic prices and improving observability of distribution networks. The IEM Directive [21] specifically refers to Smart Metering, stating that the MSs should ensure its deployment.

2.3.3 Overview of Policies Supporting the Empowerment of the Final Consumer

This section refers to the regulatory aspects and road maps related to consumers' empowerment toward collective self-consumption and energy communities.

To achieve the planned energy transition, it is important to establish efficient energy markets [11] that provide a playing field for all involved stakeholders. This also includes energy communities, which need the right to produce, consume, store, and sell renewable energy. These communities also should be able to exchange, within the same community, the renewable energy produced by them and access all the appropriate electricity markets, directly or through aggregation, in a nondiscriminatory way. ENTSO-E RD&I Roadmap [22] states that during the overall process of system modernization, it is important to foster the engagement of stakeholders and end-users (prosumers) to raise awareness of the evolution of the energy system and the benefits associated with improved grid solutions and infrastructures. The IEM Directive mentions real-time or near real-time information about energy consumption as an important part of empowering consumers and engaging them in active participation in energy markets, and the deployment of smart meters is fundamental to achieve this goal [4].

Furthermore, to pave the way for this transition, the IEM Directive requires that the MSs shall ensure participation of consumers, including through demand response, through investments into variable and flexible energy generation, ES, or the deployment of electromobility. Recent European legislation requires that the ENTSO-E and a new EU DSO entity must involve more active citizens and energy communities in the generation, consumption, storage, and sell-off of electricity without facing disproportionate burdens.

2.3.4 Energy Communities – Different Actors and Operations

Following the Pan-European legislation for CECs [4], an array of possibilities appears to be open, e.g. it can undertake roles of final customers, producers, suppliers, or DSOs, engaging in energy generation, distribution, supply, ownership and management of batteries, EV charging points, etc. The IEM Directive also empowers the MSs to allow CECs to become DSOs either under the general regime or as "closed distribution system operators." Once a CEC is granted the status of a DSO, it should be treated as and be subject to the obligations of a DSO, with certain exemptions. At the very same time while undertaking these roles, CECs may have several exemptions from the responsibilities required from the full-scale operators.

For the time being, it is up to MSs to define national regulatory regimes, and there is an indication of considerable diversity among the forthcoming national models. However, simple inheritance of different roles, for example, the role of DSO, does not necessarily open all necessary functions for CECs, simply because these are non-existent at the DSO level.

2.3.5 Summarizing the Scoping Study

The EU has set the ambition to reduce GHG emissions to the point of becoming climate neutral by 2050 and prevent the negative and irreversible effects of climate change. Reaching this goal will require shifting the energy system to a renewable-based system and radical technological, behavioral, and organizational changes in the economy and society.

The present screening draws a rather complicated picture, based on a selection of legal documents, road maps, position papers, etc. from several types of stakeholders. It is once again necessary to remind the necessity to develop innovative tools for the optimal design and operation of LECs, integrating distributed energy resources, and multiple energy carriers at different scales. In view of this, several important things were learned.

The ETIP-SNETS's vision of the 2050 power system located the electricity grid in the center of the integrated power system to transport huge amounts of energy across Europe and enable switching between the energy carriers. The present study shows that now, several years after its publication, the vision becomes more strengthened, elaborated in more detail and substantiated in several road maps.

Achieving the Pan-European goals for decarbonization of the energy sector requires several radical steps following the energy transition path. Storage technology in its various forms is foreseen as a key enabler of the energy transition and a binding element of different energy carriers. Different types of ES will meet different needs and applications in decarbonized power sector. Introduction of RES brings a need for more efforts in balancing of the power system, which may become a major driving force for deployment of grid-connected ES, while other technologies such as pumped hydro and heat storage will become more important. Hydrogen storage has unique technical properties and may become a favorable storage option for high VRES penetration regions. However, very high costs, lack of infrastructure, and even more importantly lack of regulation are likely to postpone full-scale implementation of hydrogen as a storage and as an energy carrier in general. Uncertainty and especially the absence of clear regulatory provisions are possibly two of the most significant barriers to establishing new services, since this uncertainty could strongly discourage potential investors from developing the necessary infrastructure assets. Furthermore, to establish an operational environment, it can be equally important to indicate roles and responsibilities as well as any possible limitations of these to draw unambiguous legal borders.

Apart from electricity storage, heat storage will become increasingly important. Today more than 50% of the final energy demand in the EU is used for generating heat, and already now heat storage is utilized in water-based systems for domestic and district heating. In terms of energy, heat storage is by far the largest single ES application field in Europe.

Transport remains the sector with the lowest penetration of renewable energy in the EU energy system. Electric mobility is considered a key point for the decarbonization of the transport sector, especially in urban areas, and there is the need for policies to promote EVs that must also support charging points deployment and include new codes to install these charging points in new (or refurbishment) buildings.

2.4 Review of Sector Coupling Technologies for Integrated Local Energy Communities

To date, there are several energy conversion technologies that might be considered game changers for enabling local multi-carrier energy systems; thus, a proper review is crucial for their understanding and final use in particular energy contexts. In the view of having interconnections between different energy carriers, it is important to assess their correct operations and optimize them to fulfill end-users' energy demands in a sustainable (e.g. increase the share of renewables) and economic (e.g. increase the energy conversion systems efficiency) way. This is the approach of building up the so-called sector coupling concept.

The aim of this section is to provide a targeted review covering different factors that are essential for the optimal future development of ILECs in Europe. The ILEC concept is the bottom-up approach for contributing to the energy transition as per the EU vision. The section presents a broad overview of the technologies enabling the sector coupling in ILECs. Technologies are differentiated according to their end-use and implementation (e.g. connecting technologies at the prosumer and community level). Details about their description, technical information, level of maturity, and costs for enabling the sector coupling in ILECs at different scales are here provided.

This section is structured as follows: Section 2.4.1 provides the definition of connecting technologies, while Sections 2.4.2 and 2.4.3 present the technologies available at the prosumer and community level, respectively.

2.4.1 Definition of Connecting Technologies

In multi-carrier energy systems, connecting technologies connect at least two energy networks so that one energy carrier in input is converted into at least two different energy carriers operating as many energy networks as possible. There are different energy demands that must be fulfilled by different end-users (residential, industrial, etc.). Energy demands are obtained using energy carriers (electricity, hot/chilled water, natural gas, etc.). According to the technology, the primary energy demand will be consumed to obtain a determined energy carrier that will be used to satisfy a specific energy demand. For instance, hot water is the most common energy carrier to meet thermal energy demand for heating purposes. However, end-users can decide to fulfill this heating demand through a natural gas boiler, an engine operating in cogeneration mode, a heat pump (HP), or a solar thermal panel. In such a context, connecting technologies have a pivotal role in the energy transition since they enable potential synergies among different energy networks in the sector coupling approach.

In this regard, Table 2.2 lists the main connecting technologies at both the prosumer and community levels, which can provide more than one energy carrier as output while needing only one energy vector as input.

2.4.2 Connecting Technologies (Prosumer Level)

According to Table 2.2, the following equipment can be used as connecting technologies at the prosumer level: those that use electricity and natural gas to obtain more than one energy output carrier are discussed here in detail, thus focusing mainly on the end-user sector coupling.

Table 2.2 Overview of sector coupling technology and infrastructure lists for ILECs.

Connecting technologies (prosumer level)	Connecting technologies (community level)
• Heat pumps • Cogeneration and trigeneration • EV charging stations (wall-box)	• Electrolyzers • Electric planes, electric ferries, hydrogen transport such as trains, vessels, and lorries • Desalination

2.4.2.1 Heat Pump

A heat pump is an electricity-based space heating and cooling solution to produce hot water. Its function is based on the reverse vapor compression cycle and is classified by heat source and sink used like (i) air–water, (ii) air–air, (iii) water–water, (iv) water–air, and (v) ground source. Energy carriers involved are electricity (input) and hot/chilled water (output). According to the output energy carriers previously mentioned, a heat pump fulfills either heating or cooling energy demands depending on the season when it operates. This technology is already mature since it has a technology readiness level (TRL) of nine. Besides being used for pushing the electrification of both heating and cooling energy demands, heat pumps can be adopted in demand response applications.

2.4.2.2 Combined Cooling Heat and Power (CCHP)

Combined cooling heat and power (CCHP) stands for the simultaneous production of electricity and heat, where the latter is considered a useful effect. It is also possible to have cold energy through particular equipment (e.g. absorbers) that uses the heat from the congenator, thus constituting the so-called CCHP units. Main thermal machines used in cogeneration mode are (i) internal combustion engine (ICE), (ii) micro gas turbine (MGT), and (iii) fuel cell (FC). These technologies involve electricity, fuel, and hydrogen as input energy carriers to satisfy the electricity demand and, at the same time, heating/cooling ones. These technologies are currently mature since they have a TRL of nine.

2.4.2.3 EV Charging Station

The EV charging station is the main set of elements required to connect EVs to an electrical installation for charging purposes. Physically, EV charging stations can be found in two main configurations: wall box and charging post. The interface between them and the electric grid is directly hard-wired to a control device or a plug and protection box. The level of power that an EV charging station can deliver to an EV varies in each country depending on the distribution network and supply characteristics like voltage, current, frequency, and number of phases in the case of alternating current (AC) supply. This is defined mostly by national regulation and utilities procedures. The IEC61851 is an international standard dealing with EV conductive charging system, which has become a reference for the EV infrastructure. As future perspectives, EV charging stations allow to use locally available renewable sources as well as properly manage the electricity within the electric grid. Furthermore, EVs can be used as ESSs through the Vehicle-to-Grid (V2G) scheme, thus providing flexible services to the electric grid.

Finally, Table 2.3 sums up the main characteristics of the connecting technologies at the prosumer level analyzed so far.

2.4.3 Connecting Technologies (Community Level)

According to Table 2.2, the following equipment can be used as connecting technologies at the community level: those that use electricity, fuel, and water to obtain

Table 2.3 Main characteristics of the connecting technologies at the prosumer level.

Technology	Energy carriers involved	Energy demand satisfied	Energy networks involved	TRL	Costs	References
Heat pump	Electricity, hot water, chilled water	Heating, cooling, domestic hot water	Power grid	9	CAPEX: 450–3000 € kW^{-1} OPEX: 3–39 c€ kWh^{-1}	[23–25]
CCHP	Electricity, fuels, hot water/steam	Heating, cooling, electricity	Natural gas pipeline, fuel networks, power grid	9	CAPEX: 800–7500 € kW_e^{-1} OPEX: 0.8–9 € kWh_e^{-1}	[26–28]
EV charging station	Electricity	Mobility	Mobility, power grid	9	CAPEX: 600–6000 € OPEX: 90–900 € yr^{-1}	[29–31]

more than one energy output carrier are discussed here in detail like those employed at the prosumer level and discussed in Section 2.4.2.

2.4.3.1 Electrolyzers

Electrolysis is an electrochemical conversion process that allows to separate water into hydrogen and oxygen. A direct current (DC) is fed to the electrolyzer's core stack that is composed of typical electrodes and an electrolyte that catalytically activate the decomposition of the water molecule. Electrolyzers need electricity and water as input energy carriers to produce hydrogen and oxygen, where the former is considered the main output energy carrier.

The most widespread electrolyzers are low-temperature types (between 60 and 100 °C) and use an alkaline or polymeric electrolyte with noble catalyst at the electrodes. High-temperature electrolysis, namely steam electrolysis, can provide part of the required energy in the form of heat, strongly reducing the specific energy consumption (e.g. below 40 kWh kg^{-1} of hydrogen with respect to over 50–60 kWh kg^{-1} of hydrogen for low-temperature electrolysis). Focusing only on low-temperature electrolyzers, which have a TRL of nine and are more spread in the global market compared to high-temperature ones that are mostly at the lab scale, they can be differentiated as follows:

- **Alkaline electrolyzers:** it is the most mature and simplest technology. Most of the electrolyzers available in the market are alkaline-based and are commercially spread [32]. The cell has two electrodes, which are separated by a diaphragm filled with an alkaline solution (25 and 30 wt% of potassium hydroxide in water). Water is split at the cathode to produce hydrogen and releases hydroxide anions that pass through the diaphragm to form oxygen at the anode.
- **Proton exchange membrane (PEM) electrolyzers:** these kinds of electrolyzers do not have a liquid electrolyte, but a solid membrane. The use of a solid

membrane increases the safety operation and safeguards the reliability of the electrolyzer (e.g. potassium hydroxide is corrosive and potentially harmful for potential burns when it is in contact with human skin). PEM electrolyzers have further advantages such as lower gas permeability, high proton conductivity ($0.1 \pm 0.02\,S\,cm^{-1}$), lower thickness (20–300 μm), and high-pressure operations. Regarding the performance, PEM electrolyzers operate at high current density (above $2\,A\,cm^{-2}$), high efficiency at low current density (up to 80%), fast response, and have a small footprint for the environment [33, 34].

- **Anion exchange membrane (AEM) electrolyzers:** this is an emerging technology that uses an AEM as a separator and combines the benefits of both alkaline and PEM electrolyzers [35]. AEM electrolyzers overcome the low efficiency of alkaline water electrolysis (mainly at part-load operating conditions, reaching values up to 80%) and corrosivity problems; indeed, they operate at lower current density ($0.5–2\,A\,cm^{-2}$) than PEM electrolyzers [36].

2.4.3.2 Electric Planes, Electric Ferries, and Hydrogen Transport such as Trains, Vessels, and Lorries

Different types of fossil fuel-electric hybridization in the transportation sector involving trains, vessels, and lorries have been developed so far to reduce fuel consumption and carbon emissions, but fully electrified long means of transport previously mentioned are still under development [37].

Most of the European railways are already electrified, but there are new developments using fuel cell technology that will allow them to operate in both electrified and non-electrified rails [38]. Also, the maritime transport needs an upgrade in terms of sustainability: up to now, it generates around 940 million tonnes of CO_2 yearly, and it is responsible for about 2.5% of global GHGs emissions [39].

The TRL of the technologies previously mentioned that use either hydrogen or electricity as energy carriers is nine, but it differs when moving to specific applications. Even though electric ferries are already in operation (mainly in Norway), there are still undergoing developments for the use of hydrogen in trucks and maritime application, as well as electric batteries in planes [38].

2.4.3.3 Desalination

The desalination process removes the salt from water to make it ultrapure or drinkable. Different water qualities are obtained through membrane or thermal technologies, and the plants are designed according to the sector in which the desalination process is involved like industrial or municipal wastewater, seawater, brackish groundwater, surface water, or potable water [40]. Main solutions for water desalination are as follows:

- **Membrane processes:**
  - **Reverse osmosis (RO):** It consists of reversing the flow of osmosis using a hydraulic pressure higher than the osmotic one (related to a particular solution). The water moves from a high-solute region to a low-solute one through a semi-permeable non-porous membrane (no phase changes occur).

- **Thermal processes:**
  - **Multi-effect distillation (MED):** It consists of cells at decreasing pressure and temperature (from 90 to 65 °C). The lower pressure in each stage allows the water to evaporate at low temperatures as the boiling point of water decreases with pressure.
  - **Multi-stage flash (MSF):** Flash evaporation of part of the feed occurs as it moves through multiple stages that have sequentially higher pressure and temperature lower than the vapor one (in each stage).

Finally, Table 2.4 sums up the main characteristics of the connecting technologies at the community level analyzed so far.

2.5 Review of Limitations and Barriers for the Optimal Use of the Local Energy Resources

2.5.1 Introduction

The previous sections have presented a scoping study about the road maps and the overall Pan-European priorities related to ILECs (Section 2.3), and a description of the different technologies that enable the development of such communities (Section 2.4).

This section will present the main barriers, limitations, shortcomings, and obstacles that can limit the optimal use of the local energy resources. These constraints can be of various types, and in this work these limitations have been grouped into two main groups: technical and regulatory limitations.

2.5.2 Technical Limitations

ILECs have different multi-energy requirements to fulfill, from generation (such as heat-cold generation, electricity generation, and hydrogen generation) to storage (electricity storage, heat-cold storage, and hydrogen storage) to other complementary technologies involved in these communities such as mobility or control management systems, which have different technical barriers that can limit their future deployment.

2.5.2.1 Low Efficiency

The main technical limitations identified are related to the low efficiency of some of the technologies described in Section 2.4.

For example, in *thermal generation*, air-to-air heat pumps start losing efficiency when the outdoor temperature drops below −10 °C, reducing their coefficient of performance (CoP) from around 4.5 (at 7 °C) to 2.3 (at −7 °C) [46]. Electric boilers have a low heat capacity and their electricity consumption increases sharply, when it is required to maintain the thermal comfort throughout the coldest winter periods. Adsorption chillers, used to cool different rooms, have also low efficiencies, with an average CoP between 0.3 and 0.6, limiting the population of this technology [47].

Table 2.4 Main characteristics of the connecting technologies at the community level.

Technology	Energy carriers involved	Energy demand satisfied	Energy networks involved	TRL	Costs	References
Electrolyzer	Electricity, hydrogen	Electricity, hydrogen, water (non-energy network), oxygen (non-energy network)	Power grid, natural gas network, or dedicated hydrogen infrastructures	4 (high temperature)–9 (low temperature)	CAPEX: 500–3000 € kW^{-1} OPEX: 5–90 € kWh^{-1}	[32–36, 41]
Electric- and hydrogen-based transportation	Electricity, hydrogen	Mobility	Distribution network and local batteries as buffers	8–9	CAPEX: 100–3000 k€ OPEX: 6–270 k€ yr^{-1}	[37–39]
Desalination	Electricity, fuels, hot water, natural gas, renewables	Water	Water, district heating networks, power grids	9	CAPEX: 800–3100 € m^{-3} OPEX: 0.1–0.9 € m^{-3}	[29–31, 42–45]

In *electric generation*, efficiency is also a key issue. Photovoltaics (PVs) systems, which are one of the most used distributed generation technologies in ILECs, have low efficiency, with values around 20–23%, in current commercial PV panels [48]. There are other electric power generation technologies that can be used in certain types of energy communities (e.g. islands), which are still at an early stage of development, such as wave and tidal energy generation [49], and for that reason, their efficiency is still very low due to a lack of maturity. Fuel cells can be used as stationary electricity generators accommodating the local consumption of different loads in ILECs [50], but their efficiency is very low compared to other electricity systems storage such as Li-ion batteries [51].

Regarding *hydrogen production*, the efficiency depends on the different production technologies, varying from 35% when using gasification to 70–80%, when hydrogen is obtained through electrolysis or steam reforming [52].

When hydrogen is used to store excess renewable energy over long periods of time, the conversion from producing hydrogen, storing it, and converting back to power has a very low round-trip efficiency (around 18–46%) [51], three times less efficient than other mature long-duration technologies such as pumped-storage hydropower (with round-trip efficiencies over 70%). Other electricity storage technologies, such as compressed air energy storage (CAES) or liquid air energy storage (LAES), also present low round-trip efficiencies (around 50%) [53, 54].

2.5.2.2 Low Flexibility

Another technical barrier is the low flexibility of some technologies. These technologies are designed to operate at constant regimes, and their behavior worsens when operating at partial loads or having a low dynamic response, not being able to adapt to rapid load variations, reducing their controllability, and limiting the number of applications in which they can be used autonomously.

For example, biomass boilers are less responsible for changes in heat demand with respect to conventional gas or oil boilers in which ignition is instantaneous, and it can also be shut down instantaneously with little residual heat remaining. Steam boilers have also low controllability, having limited possibilities to regulate heat transfer. Heat pumps are normally operated in pre-defined cycles, and it is not possible to instantly switch off or on the device, making it difficult to use heat pumps as flexible loads, since changing their operational regime takes several minutes. Gas turbine combined heat and power (GT-CHP) systems are characterized by a significant reduction in efficiency at partial load. Fuel cells have also slow dynamics, and when these devices are used in mobility applications, they cannot handle power peaks during the acceleration phase or cannot recover energy produced during regenerative braking [55].

The easiest way to overcome this low flexibility limitation is by hybridizing these technologies that operate at constant load, with others that can handle the fast variation in thermal or electrical generation/consumption. For example, FCEVs hybridize fuel cells with an auxiliary Li-ion battery pack that compensates the lack of fast response of the fuel cells, absorbing the energy regenerated during braking and enhancing the overall performance of the system.

2.5.2.3 Security Concerns

There are some other technical limitations related to security concerns. Some technologies have explosion hazards due to the use of fuels (e.g. natural gas boilers and oil boilers) or flammable materials, such as H_2 handling or Li-ion batteries, which can explode due to thermal runaway [56].

2.5.2.4 Environmental Issues

Many of the analyzed technologies have several environmental issues, from high noise production, e.g. steam boilers, ICE used in combined heat and power (ICE-CHP) or in backup generators, etc., to emission of GHGs and other pollutants, due to the combustion of fossil fuels. For example, natural gas boilers are largely responsible for the carbon emissions used in domestic heating and are being replaced by low-carbon alternatives such as heat pumps or hydrogen boilers in several European countries. ICE diesel CHPs are characterized by high NO_x and particulate emissions. CAES systems use natural gas to preheat the compressed air before expanding it in the turbine, emitting CO_2. Other technologies use hazardous materials that if not handled and disposed of properly could pose serious environmental and public health threats. For example, PV cell manufacturing process uses toxic materials to clean and purify the semiconductor surface during its production. Thin-film PV cells contain a higher number of toxic materials compared to those used in traditional silicon PV, including gallium arsenide or cadmium-telluride. Rare-earth materials, such as neodymium, praseodymium, terbium, and dysprosium, are used in different types of electrical machines such as wind generators [57] and electric motors for EVs. These materials are mined by digging vast open pits in the ground, affecting the ecosystems and environment. Li-ion batteries currently mainly used in EVs are toxic, carcinogenic, or can undergo chemical reactions that produce hazardous heat or gases [58]. Finally, many of the materials required to manufacture thermochemical heat storage systems are also environmentally harmful.

Additionally, some of these technologies have an impact on wildlife. For example, wind turbines constitute a wildlife hazard, endangering birds flying too close to the rotating blades. Some types of solar thermal plants use hundreds of mirrors, leaving a negative impact on the animal wildlife. Large solar power plants have a similar impact, with a vast extension of surface covered with PV solar panels. Most of the hydropower plants require to build dams to create reservoirs that change the natural water temperatures, and the river flow characteristics, blocking or diverting the natural course of the river, impacting downstream wildlife populations, and affecting fish migration [59].

2.5.2.5 Spatial Issues

Other technologies require a large space to produce or store energy. For example, the land use per unit of generated electricity for a concentrated solar power (CSP) plant is around $24\ m^2\ MWh^{-1}$. The value for PV power plants installed on the ground is $22\ m^2\ MWh^{-1}$, while onshore wind reaches $0.8\ m^2\ MWh^{-1}$, when it only includes the area directly impacted by the excavation and insertion of the wind turbines [60].

It is common that land use of RESs can be used with other complementary activities, like agrovoltaic and farming [61]. Some ES technologies have also low energy densities (e.g. batteries, flywheels, and super-capacitors), requiring much space than others to store the same amount of energy. For example, current battery technologies still have energy densities between one or two orders of magnitude lower than hydrocarbons.

2.5.2.6 Cost and Cost-effectiveness

Cost and cost-effectiveness are two of the most limiting factors that affect several technologies. Some of them are still very expensive because of their current low level of maturity. It is important to highlight that as a technology develops, its cost is significantly reduced. For example, the price of the batteries has declined by 97% in the past three decades, going from \$7500 kWh^{-1} in 1991 to \$181 kWh^{-1} in 2018 [62]. Unfortunately, this fall has slowed down in recent years, and, in fact, the minimum was reached in 2021 (around \$130 kWh^{-1}) and, since then, the prices have gone up 7% in 2022–2023 due to a global battery supply chain constraints caused by COVID-19 and the current inflation [63]. The same behavior has been observed in the price of PV panels. The price of PV systems has exhibited a notable decrease over time, dropping from \$3 Wp^{-1} in 2009 to \$0.3 Wp^{-1} in 2021 [64]. However, over the recent it two year period, there has been a stabilization of the costs. New technologies such as CSP, wave and tidal energy, fuel cells H_2 production using electrolyzers, and thermochemical heat storage are still not cost-competitive.

The high cost of other technologies is due to high start-up and/or high operational costs, particularly those that require civil works (e.g. dams and other auxiliary civil infrastructures for hydropower plants, and installing CAES in caverns) or are installed in very extreme environments, such as offshore wind turbines and wave-tidal generation. All the components of these systems must be designed to be stress-resistant because they are not only faced to extreme atmospheric conditions but corrosion for long periods, with high humidity and salt concentration in the atmosphere and high winds, currents, or waves on a regular basis, making the installation of these generation systems a challenge due to the underwater location.

2.5.3 Regulatory Limitations

Most of the regulatory barriers related to *thermal generation* are related to environmental issues such as GHG emissions (in biomass boilers, natural gas boilers, hybrid HP, etc.), the toxicity of fluids (in HP, hybrid HP, geothermal HP, etc.), and security and installation procedures such as maximum noise level, vibrations, and resilience.

There are other regulatory limitations related to market design. For example, the current EU legislation separately covers the heat and electricity market, thereby reducing the development of co- and tri-generation since these devices suffer barriers in both markets [65].

The main regulatory principle that affects the electric generation is RED II [5], but there are several regulatory barriers at MS level policy and incentives which have an impact on the locally distributed electricity generation. For example, in recent years, several European countries have modified these government's contracts unilaterally, affecting investor confidence. For this reason, a stable framework is required to develop and deploy these new technologies in the market, avoiding these legal uncertainties.

For all-generation technologies, the administrative process for obtaining authorization/licenses is complex, which discourages investors. ENTSO-E has published a regulation for connection of generators [66] and demand connection code [67], but the administrative processes set by different DSOs for the connection of local generation technologies are still a barrier for the smaller domestic users.

2.5.3.1 Market-Related Issues

Other barriers are related to the electricity market, with some difficulties in the participation of distributed and intermittent renewable sources in the balancing markets. The provision of balancing energy is still mostly provided by conventional generators, as current balancing market arrangements are still in transition in some countries to accommodate other types of intermittent RESs such as PV systems and wind generators. Some countries already have the conditions to allow these generators to participate in the provision of energy on the balancing market, while other countries' balancing markets are still not developed to a stage where they can let these power generators participate. Even when these intermittent generators are allowed to provide balancing services, it is mostly for Replacement Reserve (RR) [68].

2.5.3.2 Need for Revising Grid Codes at the Distribution Level

In recent years, the deployment of DERs at the distribution level, which is medium voltage (MV) and low voltage (LV), is quickly increasing. The increasing penetration of renewables at lower levels of national grids has shed light on the need for revising grid codes at the distribution level as system stability is at risk.

The revision of grid codes regarding DER penetration at the distribution level is still in process in most European Countries. Eventually, the traditional view on a passive interconnection of renewable generation plants at MV and LV is now shifting to a new paradigm, where active interconnection is required for system stability.

Concerning the distribution network standards in the EC, the Directive 2014/35 about LV support is the only one existing [69]. It has been applicable since April 2016, and the target audience is European citizens, who are the end-users of electricity. The objective of such Directive is the harmonization of laws with regard to electrical equipment designed within certain voltage limits. The manufacturer must take all measures necessary to ensure compliance. This way protection against hazards of LV equipment is guaranteed. The electrical devices on which this Directive applies are those with a voltage range between 50 and 1000 V for AC or 75 and 1500 V for DC. As always, European Directives only specify principles and, therefore, the implementation and guidance are the responsibility of national authorities.

2.5.3.3 Definition of Roles and Responsibilities of Energy Communities

Related to CECs, IEM Directive 2019/944 [4] sets the rights and responsibilities of CECs in accordance with the roles they assume (final customers, producers, suppliers, or DSOs), but the national regulations transposing this have not yet been defined. Furthermore, at this moment, there are no common Pan-European rules that define all aspects of ownership and operation of these communities, and each MS has certain variation in the regulatory market framework.

Renewable energy communities (RECs) should be allowed to sell renewable energy production to their neighbors. For this purpose, it is necessary to define some crucial issues, e.g. the perimeter within which transactions are permitted and the interaction between individuals within the REC and between them and other neighboring RECs. These key components are now defined in national implementation in MSs [70].

Another important regulatory constraint is the diversity of regulations to be considered when installing local electric generation in LECs. These regulations are affected at different scales: European, national, regional, and local levels. Additionally, these regulations may be not directly related to electricity production. For example, the installation of solar PV rooftop or micro-wind turbines on buildings must also be compliant with several national and local building regulations. Marine renewable energy devices (specifically wave and tidal) have specific environmental requirements and must comply with the coastal legislation which is, in many cases, country specific [71]. Also, geothermal electricity generation must comply with several environmental regulations, focusing mainly on groundwater resources safeguarding, in the framework of the legislation regulating water resources utilization (as it happens in Germany, Netherlands, Austria, and Switzerland). Small-scale hydropower plants in, e.g. rural LECs are also affected by wildlife and river legislations that vary from country to country.

2.5.3.4 Efficiency Targets for Cogeneration

In the heat and power sector, the Energy Efficiency Directive (2018/2002) [72] and its recent recast [73] and the increase in energy efficiency targets represent strong development opportunities for hydrogen as renewable gas and for fuel cell micro-CHP. High-efficiency cogeneration is seen as an important technology to increase efficiency in the heating and cooling sector, and its deployment is fostered at various levels. The Directive promotes the use of technologies such as hydrogen FC micro-CHP.

This obligation for the EU MSs to lead a comprehensive assessment of the potential for high-efficiency cogeneration represents an opportunity for the fuel cell CHP and hydrogen industries to increase their visibility and demonstrate their technologies' potential for more energy-efficient and sustainable heating and cooling systems.

2.5.3.5 Regulatory Barriers Related to Hydrogen

There are several regulatory barriers related to hydrogen. The first regulatory limitation is the lack of a common definition and classification/taxonomy for hydrogen at the EU level (green, blue, pink, and brown hydrogen) and the difficulty of determining the renewable origin of the electricity used for H_2 production.

Another barrier is to develop a fair policy for H_2 production. For example, if green hydrogen production is supported by subsiding electrolyzers or electricity, it could give an unfair advantage compared to other types of H_2 production (e.g. generating H_2 from biomethane) [74].

Although some documentation is available at the international level (standards by ISO/TC 197 – Hydrogen technologies [75]), the documentation is still at an international standardization level which only provides technical and practical guidelines – which are implemented only on a voluntary basis [76].

There is no common EU framework (in terms of compulsory regulations/prescriptions) for installation of hydrogen generation – electrolyzer and their interconnection to the electricity grids or to the transport sector; therefore, each country relies on specific national regulations that set the prescriptions for the installation and national authorities having jurisdiction on the approval of the permitting procedures.

The roles and responsibilities of the market and regulated players in the production of hydrogen are not currently defined. Therefore, it is necessary to define regulations regarding ownership, production, and storage of H_2 installations. There are two alternatives related to the ownership and operation of electrolyzers: one is the naturally regulated monopoly and the other is a commercial activity.

In addition, challenges can be faced in the coupling of electrolyzer systems with RES assets, sometimes requiring double permitting from a grid connection point of view since the DSO/TSO consider the plants separately, although there is only one common Point of Connection.

In some European countries, green hydrogen production is considered an industrial activity and only can be done overland, which has been classified as "industrial." This fact limits the installations of electrolyzers at the domestic or district scale in LECs.

The installation of fuel cell-micro-CHP units in buildings requires compliance with the regulations for use, storage, and distribution of gaseous fuels and the Directive 2014/94 on the deployment of alternative fuels infrastructure [21]. The current regulations for the distribution and use of gaseous fuels and the regulation on the storage of chemical products do not meet with the hydrogen needs for residential or tertiary building installations that may arise if FC-micro-CHPs are used to cover part of the demand for heat and electricity in buildings. An update of these regulations on the production, storage, and distribution of gaseous fuels such as hydrogen should be required at the national level.

2.5.3.6 Integration of Energy Storage

ES can be integrated at different levels of the electricity system: generation, transmission, distribution, and user level. These different locations in the power system will involve different stakeholders and will have an impact on the types of services to be provided and different business models will be developed.

Regulatory limitations are linked to the consideration of the storage within national regulations more than to the specific technology. Regarding the IEM Directive [4], DSOs and TSOs shall not be allowed to own, develop, manage, or operate ES facilities unless:

- Other parties have not expressed their interest to own, develop, manage, or operate storage facilities.
- Such facilities are necessary for the DSOs and TSOs to fulfill their obligations, and they are not used to buy or sell electricity in the electricity markets.

The above limitation is for offering opportunities to other stakeholders to participate in the market and avoid market distortion and conflicts by having DSO and TSO sell or buy electricity coming from ES. This should not be a big limitation, but it is to be seen how it is materialized.

On the other hand, the rules for the connection of the generators to the grid [77] need to be fully transposed to national legislation of all MSs. When and how this is carried out in each country may have an impact on the deployment of storage in each region. One of the aspects to be considered is which network services are requested to be mandatory, without any monetary compensation, and which of them fall into the ancillary services market.

Linked to this, the development of markets and market products that permit the participation of storage, considering the intrinsic characteristics of the different technologies, will also play a very important role in monetizing the benefits that storage can bring to the owners and, in this way, to make storage investments cost-effective and push their deployment at the regional level. Capacity markets, local flexibility markets, and market products that take into consideration the state of charge of the batteries, etc. are examples of this. The main limitation for ES is the need to create appropriate market signals to incentivize the construction of storage capacity and provision of storage services, creating a European-level market and common balancing markets.

ESS policies, in general, play a major role in the development of green technologies that are good for low-carbon emissions. Most of the ESS policies revolve around battery storage as they can easily be integrated into the grid, renewable energy, used in EVs, and used as backup power. However, there is no clear legal framework for the interconnection of the different ES technologies with the electricity network.

There is an additional uncertainty related to the difficulties of classifying ES assets into a legal regime premised on three traditional categories of assets: (i) generation, (ii) transmission, and (iii) distribution. The most notable source of contention is whether ES constitutes a generation or a transmission asset [78].

At the European level, a clearer definition of role with regard to the ownership of hydrogen storage facilities is required. The RED includes a taxonomy of hydrogen, but it mainly focuses on the transport sector and not on the use of hydrogen in other important sectors, e.g. industry. More clarity is needed on the legal status of ES, with regard to P2G, which is likely restricted to the function of storing energy. Nevertheless, in practice, P2G could become an important part of a hybrid energy system. Therefore, its definition should not be limited to its role as ES from an electricity market perspective. This also has implications on the applicability of the unbundling. In response to this, a test framework could be applied to allow TSOs to invest in such facilities until the market is mature enough to attract more players. The term immature market refers to the case that no interest is expressed by market parties/investors and as a result TSOs/DSOs could be allowed to proceed in such investments [79].

2.5.3.7 Standardization of the Infrastructure for EVs

Regarding regulatory barriers for mobility, there is no clear standardization of the EV charging connectors, voltage levels, etc. in the EV charging stations, making interoperability difficult [80–82]. Even the payment process is different for different charging infrastructures, requiring specific cards, credit cards, specific mobile apps, etc.

A European standardization that allows a driver traveling with an EV throughout Europe to charge in any of the fast-charging points without having to register in any of the companies that offer this service will simplify the process significantly, promoting the electrification of the transportation sector.

The cost of an EV is largely determined by the high cost of its battery packs. To lower the entry barrier in transport electrification and reduce the cost of purchasing an EV, some manufacturers are promoting the leasing of batteries. The user buys the vehicle except for the battery, which still belongs to the manufacturer [83]. If EVs are used in the V2G applications, users can obtain economic rewards by participating in ancillary service markets, but depending on the battery ownership, the increment of the battery degradation by participating in V2G will impact this benefit. If the battery is leased and does not belong to the final users, manufacturers may not allow participation in V2G schemes because they would be penalized due to a reduction of the battery lifetime, but they would not receive any economic benefit, while if the battery is also bought by the final users, they have the rights to participate in this electricity market if the received benefit overcome the cost of the battery degradation. Additionally, batteries' guarantees, which are mainly based on the available range after a certain mileage, could be affected by atypical use of the batteries in V2G schemes.

2.5.3.8 Integration of EVs into the Electricity Market

EVs can contribute to active demand response through aggregators, avoiding charging during the peak demand or adjusting their consumption (V1G scheme) dynamically. Utilities may reward customers for the participation, offering financial incentives, cheaper energy prices, etc., in a similar way to current loyalty programs used by some homes where consumers receive an SMS message to reduce their consumption or other buildings with smart thermostats that adjust their set point remotely. This smart unidirectional charging process is currently available by many utilities.

EVs not only can be used to reduce the stress from the grid by V1G. Using bidirectional chargers in V2G applications can optimize the consumption on building or neighborhood level [84], provide power backup, contribute to local congestion management of the electric network, or maximize the use of renewable energies. However, implementing V2G technologies is more complex: requires bidirectional charging–discharging infrastructure, which is more expensive, and bidirectional and standardized communications between EVs, aggregators, and utilities. The current protocols used in the charging points – open charge point protocol (OCPP) [85] do not allow knowing the desired state of charge of the battery, the location of the EV, or the time that the vehicle will be connected. Additionally, grid information is not uniformly available throughout Europe, but this information should be shared in the coming years between DSOs and TSOs following the Clean Energy Package [86].

Energy regulation is complex and provides an obstacle for emerging technologies like V1G and V2G schemes which are challenged to compete in traditional energy markets that are not fully aligned with their capabilities. The current national and European regulatory framework for V2G is not mature yet [87] and should design and implement new flexibility markets, aggregation services, capacity mechanisms, and dynamic tariffing to promote V2G technology. For these V2X services, it can be understood that the charger point owner will be responsible but to provide support to the grid and participate in markets must be part of an aggregator; however, for distribution grids, it is not that clear who must be responsible for this system and become the owner.

2.6 Conclusions and Lessons Learned

This chapter aims to provide the readers with a contextualized overview and vision of the energy sector in Europe. The future energy production technologies will be mainly driven by RES, which will be spread out all over the continent and substitute those fueled by fossil fuels. The key enabler of the renewable deployment will be ES solutions and, in particular, hydrogen. Although technologies for producing/using hydrogen are still being further investigated (e.g. AEM and solid oxide), this element has been considered one of the main solutions for medium- and long-term ES and, at the same time, a clean energy carrier if produced by renewables.

The spread of renewables will lead to the electrification process of different sectors, constituting the so-called sector coupling approach. Besides electricity, multiple energy carriers will interact with each other in a wider context, thus allowing end-users to further enhance not only the efficiency of each technology but also the overall system in which they operate. Nevertheless, this ambitious goal has barriers and limitations that slow down the transition process of the European energy system. In particular, successful implementation of the new technologies and the energy transition process depends upon the support and engagement of the end-users. Several important steps in this direction have been recently made, more specifically the introduction of CECs and RECs, active customers, and other innovative concepts in the "Clean Energy for all Europeans" package. From the screened documents, it appears that the establishment of well-functioning local markets is a compulsory attribute of the successful energy transition.

The study creates an impression that despite a multitude of technologies that can contribute to decarbonization, the present situation of consumers' adoption of technologies is shaped by technology-specific interest organizations. The situation is complicated by the complexity of the problem and variety of the available solutions and combination of these both on a single consumer level as well as on an energy community level. This highlights the necessity of holistic approach with dedicated and tailorable optimization tools, which allow evaluating several decarbonization technologies. Furthermore, this requires versatile optimization tools, which will be essential for properly tailoring of functionality to any case study. Indeed, each case study has its own peculiarity that so far made it difficult to correctly resemble its current behavior and, moreover, forecast its operation with the implementation of new kinds of technologies/energy carriers. This aspect is mandatory to assess the viability of the ILECs in Europe and in other parts of the world as well.

Acknowledgment

The project eNeuron has received funding from the European Union's Horizon 2020 Research and Innovation Programme under Grant Agreement No. 957779.

List of Abbreviations

AC	alternating current
AEM	anion exchange membrane
CAES	compressed air energy storage
CCHP	combined heat cooling and power
CEC	citizen energy community
CoP	coefficient of performance
CSP	concentrated solar power
DC	direct current
DSO	distribution system operator
EC	European Commission
EHS	electrochemical hydrogen separation
ENTSO-E	European Network of Transmission System Operators for Electricity
ES	energy storage
ESSs	energy storage systems
ETIP-SNET	European Technology and Innovation Platform – Smart Networks for Energy Transition
EV	electric vehicle
FC	fuel cell
FCEVs	fuel cell electric vehicles
GDP	gross domestic product
GHG	greenhouse gas
GT-CHP	gas turbine combined heat and power

HP	heat pump
ICE	internal combustion engine
ICE-CHP	internal combustion engine combined heat and power
IEA	International Energy Agency
IEM	Internal Electricity Market
ILEC	integrated local energy community
ISO	International Organization for Standardization
LAES	liquid air energy storage
LCV	light commercial vehicle
LV	low voltage
MED	multi-effect distillation
MGT	micro gas turbine
MS	member state
MSF	multi-stage flash
MV	medium voltage
NRAs	National Regulatory Authorities
OCPP	open charge point protocol
P2CH4	power-to-methane
P2G	power-to-gas
P2H2	power-to-H2
P2L	power-to-liquid
P2X	power-to-X
PEM	proton exchange membrane
PV	photovoltaic
RD&I	Research Development and Innovation
REC	renewable energy community
RED	renewable energy directive
RESs	renewable energy sources
RO	reverse osmosis
RR	Replacement Reserve
SOs	system operators
T&D	transmission and distribution
TEN-E	Trans-European Networks for Energy
TRL	technology readiness level
V1G	unidirectional smart charging
V2G	vehicle-to-grid
V2X	vehicle-to-everything
VRES	variable renewable energy source

References

1 The European Commission (2021). Fit for 55. https://www.consilium.europa.eu/en/policies/green-deal/fit-for-55-the-eu-plan-for-a-green-transition/ (accessed 28 February 2021).

2 Roberts, J., Frieden, D., and d'Herbemont, S. (2019). Energy community definitions. https://main.compile-project.eu/wp-content/uploads/Explanatory-note-on-energy-community-definitions.pdf (accessed 25 March 2024).

3 International Organization for Standardization (ISO). ISO 13600:1997(en) Technical energy systems – basic concepts. https://www.iso.org/obp/ui/#iso:std:iso:13600:ed-1:v1:en (accessed 25 March 2024).

4 EU Commission (2019). Directive (EU) 2019/944 of the European Parliament and of the Council of 5 June 2019 on common rules for the internal market for electricity and amending directive 2012/27/EU. https://eur-lex.europa.eu/legal-content/EN/TXT/?uri=CELEX%3A32019L0944 (accessed 9 December 2021).

5 The European Commission (2018). Directive (EU) 2018/2001 of the European Parliament and of the Council of 11 December 2018 on the promotion of the use of energy from renewable sources. https://eur-lex.europa.eu/legal-content/EN/TXT/?uri=uriserv:OJ.L_.2018.328.01.0082.01.ENG&toc=OJ:L:2018:328:TOC https://eur-lex.europa.eu/eli/dir/2018/2001/oj (accessed 28 February 2021).

6 The European Commission (2020). 2020 report on the state of the energy union pursuant to regulation (EU) 2018/1999 on governance of the energy union and climate action. https://ec.europa.eu/energy/sites/ener/files/report_on_the_state_of_the_energy_union_com2020950.pdf https://energy.ec.europa.eu/topics/energy-strategy/energy-union/fifth-report-state-energy-union_en (accessed 07 May 2024).

7 The European Commission (2020). Communication from the commission: "annual sustainable growth strategy 2021" COM(2020) 575. https://eur-lex.europa.eu/legal-content/en/TXT/?uri=CELEX:52020DC0575 (accessed 25 March 2024).

8 The European Commission (2022). Communication REPowerEU Plan COM(2022)230. https://ec.europa.eu/info/publications/key-documents-repowereu_en (accessed 25 March 2024).

9 The European Commission (2016). EU reference scenario 2016: energy, transport and GHG emissions trends to 2050. https://climate.ec.europa.eu/system/files/2016-11/full_referencescenario2016report_en.pdf (accessed 25 March 2024).

10 ENTSO-E (2019). PowerFacts Europe 2019. https://eepublicdownloads.entsoe.eu/clean-documents/Publications/ENTSO-E%20general%20publications/ENTSO-E_PowerFacts_2019.pdf (accessed 25 March 2024).

11 Bacher, R., de Nigris, M., and Peirano, E. (2018). ETIP-SNET vision 2050. https://www.etip-snet.eu/wp-content/uploads/2018/06/VISION2050-DIGITALupdated.pdf (accessed 30 April 2020).

12 WindEurope (2019). Wind-to-X: a position paper on how to achieve net-zero emissions through a renewables-based electrification. https://windeurope.org/policy/position-papers/wind-to-x/ (accessed 25 March 2024).

13 Directorate-General for Energy (European Commission) (2019). The role and potential of power-to-X in 2050. https://op.europa.eu/en/publication-detail/-/publication/1e6b9012-6bbc-11e9-9f05-01aa75ed71a1/language-en/format-PDF/source-96288622 (accessed 28 February 2021).

14 The European Commission (2020). EC COM (2020) 301 final. A hydrogen strategy for a climate-neutral Europe. https://ec.europa.eu/energy/sites/ener/

files/hydrogen_strategy.pdf https://energy.ec.europa.eu/system/files/2020-07/hydrogen_strategy_0.pdf (accessed 07 May 2024).

15 Fuel Cells and Hydrogen 2 Joint Undertaking (2019). Hydrogen roadmap Europe. https://op.europa.eu/en/publication-detail/-/publication/0817d60d-332f-11e9-8d04-01aa75ed71a1/language-en (accessed 25 March 2024).

16 European Association for Storage of Energy (2015). EASE/EERA energy storage technology development roadmap towards 2030. https://ease-storage.eu/publication/easeeera-energy-storage-technology-development-roadmap-towards-2030 (accessed 25 March 2024).

17 Artelys, Directorate-General for Energy (European Commission), Enerdata, Trinomics (2020). Study on energy storage: contribution to the security of the electricity supply in Europe. https://op.europa.eu/en/publication-detail/-/publication/a6eba083-932e-11ea-aac4-01aa75ed71a1/language-en?WT.mc_id=Searchresult&WT.ria_c=37085&WT.ria_f=3608&WT.ria_ev=search (accessed 25 March 2024).

18 Interreg Europe (2020). Renewable energy self-consumption: a policy brief from the policy learning platform on low-carbon economy. https://www.interregeurope.eu/sites/default/files/inline/Energy_self-consumption__Policy_brief_final.pdf (accessed 25 March 2024).

19 Schmela, M., Rossi, R., Beauvais, A., et al. (2019). EU market outlook for solar power 2019–2023. https://www.solarpowereurope.org/insights/webinars/eu-market-outlook-solar-power-2019-2023 (accessed 25 March 2024).

20 Ministerio para la Transición Ecológica y el Reto Demográfico (2020). Plan Nacional Integrado de Energía y Clima (PNIEC) 2021–2030. https://www.idae.es/informacion-y-publicaciones/plan-nacional-integrado-de-energia-y-clima-pniec-2021-2030 / (accessed 25 March 2024).

21 EUR Lex (2014). Directive 2014/94/EU of the European parliament and of the council of 22 October 2014 on the deployment of alternative fuels infrastructure. https://eur-lex.europa.eu/legal-content/EN/TXT/PDF/?uri=CELEX:32014L0094 (accessed 16 February 2022).

22 ENTSO-E (2020). Research, development & innovation roadmap 2020–2030. https://www.entsoe.eu/2020/10/14/entso-e-research-development-innovation-roadmap-2020-2030/ (accessed 25 March 2024).

23 Fischer, D. and Madani, H. (2017). On heat pumps in smart grids: a review. *Renewable and Sustainable Energy Reviews* 70: 342–357.

24 Terreros, O., Spreitzhofer, J., Basciotti, D. et al. (2020). Electricity market options for heat pumps in rural district heating networks in Austria. *Energy* 196: 116875.

25 Rodríguez, L.R., Ramos, J.S., Domínguez, S.Á., and Eicker, U. (2018). Contributions of heat pumps to demand response: a case study of a plus-energy dwelling. *Applied Energy* 214: 191–204.

26 Ebrahimi, M. (2015). *Combined Cooling, Heating and Power: Decision-making, Design and Optimization*. Elsevier Ltd.

27 Pilavachi, P.A. (2002). Mini- and micro-gas turbines for combined heat and power. *Applied Thermal Engineering* 22 (18): 2003–2014.

28 Pachauri, R.K. and Chauhan, Y.K. (2015). A study, analysis and power management schemes for fuel cells. *Renewable and Sustainable Energy Reviews* 43: 1201–1319.

29 Martínez-Lao, J., Montoya, F.G., Montoya, M.G., and Manzano-Agugliaro, F. (2017). Electric vehicles in Spain: an overview of charging systems. *Renewable and Sustainable Energy Reviews* 77: 970–982.

30 Das, H.S., Rahman, M.M., Li, S., and Tan, C.W. (2020). Electric vehicles standards, charging infrastructure, and impact on grid integration: a technological review. *Renewable and Sustainable Energy Reviews* 120: 109618.

31 Shareef, H., Islam, M.M., and Mohamed, A. (2016). A review of the stage-of-the-art charging technologies, placement methodologies, and impacts of electric vehicles. *Renewable and Sustainable Energy Reviews* 64: 403–420.

32 Proost, J. (2019). State-of-the art CAPEX data for water electrolysers, and their impact on renewable hydrogen price settings. *International Journal of Hydrogen Energy* 44 (9): 4406–4413.

33 Kumar, S.S. and Himabindu, V. (2019). Hydrogen production by PEM water electrolysis – a review. *Materials Science for Energy Technologies* 2 (3): 442–454.

34 Cau, G., Cocco, D., Petrollese, M., and Kær, S.K. (2014). Energy management strategy based on short-term generation scheduling for a renewable microgrid using a hydrogen storage system. *Energy Conversion and Management* 87: 820–831.

35 Risbud, M., Kaur, G., Dhawale, D.S. et al. (2023). Chapter 14: Electrolyzer technologies for hydrogen economy. In: *Hydrogen Economy*, 2e (ed. A. Scipioni, A. Manzardo, and J. Ren), 459–485. Academic Press.

36 Scott, K. (2022). 4.01 – Introduction to hydrogen, electrolyzers and fuel cells science and technology. In: *Comprehensive Renewable Energy*, 2e (ed. T.M. Letcher), 1–28. Elsevier.

37 Flelez, J., Garcia-Sanchez, C., and Lozano, J.A. (2018). Control design for an articulated truck with autonomous driving in an electrified highway. *IEEE Access* 6: 60171–60186.

38 Lukic, M., Hebala, A., Klumpner, C., et al. (2018). State of the art of electric taxiing systems. *2018 IEEE International Conference on Electrical Systems for Aircraft, Railway, Ship Propulsion and Road Vehicles & International Transportation Electrification Conference (ESARS-ITEC)*, Nottingham, UK (07–09 November 2018). https://doi.org/10.1109/ESARS-ITEC.2018.8607786, Publisher IEEE.

39 Xing, H., Stuart, C., Spence, S., and Chen, H. (2021). Fuel cell power systems for maritime applications: progress and perspectives. *Sustainability* 13 (3): 1213.

40 Nassrullah, H., Shaheen, F.A., Hashaikeh, R., and Hilal, N. (2020). Energy for desalination: a state-of-the-art review. *Desalination* 491: 114569.

41 Haeolus, Hydrogen-Aeolic Energy with Optimised eLectrolysers Upstream of Substation. https://www.haeolus.eu. 2018–2023, Grant agreement ID: 779469, DOI: 10.3030/779469

42 Fehre, K., Oehmichen, J., Steinberg, P.J., and Widmann, B. (2023). The time for the future is now: CEO temporal focus and firms' identification and interpretation of grand challenges – the example of water scarcity. *Journal of Cleaner Production* 406: 137041.

43 Elsaid, K., Kamil, M., Sayed, E.T. et al. (2020). Environmental impact of desalination technologies: a review. *Science of the Total Environment* 748: 141528.

44 Sayed, E.T., Olabi, A.G., Elsaid, K. et al. (2023). Recent progress in renewable energy based-desalination in the Middle East and North Africa MENA region. *Journal of Advanced Research* 48: 125–156. https://doi.org/10.1016/j.jare.2022.08.016.

45 Feria-Díaz, J.J., Correa-Mahecha, F., López-Méndez, M.C. et al. (2021). Recent desalination technologies by hybridization and integration with reverse osmosis: a review. *Water* 13 (10): 1369.

46 Keogh, D., Saffari, M., de Rosa, M., and Finn, D. (2019). Energy assessment of hybrid heat pump systems as a retrofit measure in residential housing stock. CLIMA 2019 Congress, Bucharest, 26 May - 29 May 2019 *E3S Web of Conferences* 111 01064 (2019), https://doi.org/10.1051/e3sconf/201911101064.

47 Wang, R., Wang, L., and Wu, J. (2014). *Adsorption Refrigeration Technology: Theory and Application*. Singapore: Wiley.

48 Savrc, J. (2024). Most efficient solar panels 2024. https://www.cleanenergyreviews.info/blog/most-efficient-solar-panels (accessed 03 May 2024).

49 IRENA (2021). Offshore renewables: an action agenda for deployment. https://www.irena.org/publications/2021/jul/offshore-renewables-an-action-agenda-for-deployment (accessed 10 May 2023).

50 Sazali, N., Wan Salleh, W., Jamaludin, A., and Mhd Razali, M. (2020). New perspectives on fuel cell technology: a brief review. *Membranes* 5 (99): 10.

51 Plötz, P. (2022). Hydrogen technology is unlikely to play a major role in sustainable road transport. *Nature Electronics* 5: 8–10.

52 Younas, M., Shafique, S., Hafeez, A. et al. (2022). An overview of hydrogen production: current status, potential, and challenges. *Fuel* 316: 123317.

53 GridPIQ Energy storage system efficiency. https://gridpiq.pnnl.gov/v2-beta/doc/technologies/es/es-efficiency/ (accessed 10 May 2023).

54 She, X., Peng, X., Nie, B. et al. (2017). Enhancement of round trip efficiency of liquid air energy storage through effective utilization of heat of compression. *Applied Energy* 206: 1632–1642.

55 Ma, S., Lin, M., Lin, T.-E. et al. (2021). Fuel cell-battery hybrid systems for mobility and off-grid applications: a review. *Renewable and Sustainable Energy Reviews* 135: 110119.

56 Feng, X., Ren, D., He, X., and Ouyang, M. (2020). Mitigating thermal runaway of lithium-ion batteries. *Joule* 4 (4): 743–770.

57 Alves Dias, P., Bobba, S., Carrara, S., and Plazzotta, B. (2020). *The Role of Rare Earth Elements in Wind Energy and Electric Mobility: An Analysis of Future Supply/Demand Balances*. Luxembourg: Publications Office of the European Union, 2020, https://data.europa.eu/doi/10.2760/303258.

58 Vimmersted, L.J., Ring, S., and Hammel, C. (1995). Current status of environmental, health, and safety issues of lithium ion electric vehicle batteries. https://www.osti.gov/servlets/purl/135056 (accessed 10 May 2023).

59 Energy Information Administration (EIA) (2023). Hydropower explained. https://www.eia.gov/energyexplained/hydropower/#:~:text=At hydropower plants water flows,applies pressure on a turbine https://www.eia.gov/energyexplained/hydropower/ (accessed 10 May 2023).

60 World Economic Forum (2022). Energy: which electricity source uses the most land? *World Economic Forum*. https://www.weforum.org/agenda/2022/06/energy-electricity-sources-land (accessed 1 May 2023).

61 Iberdrola (2023). Agrovoltaic energy, when agriculture and renewables go hand in hand. https://www.iberdrola.com/innovation/agrovoltaics (accessed 10 May 2023).

62 Ritchie, H. (2021). The price of batteries has declined by 97% in the last three decades. *Our World Data*. https://ourworldindata.org/battery-price-decline (accessed 10 May 2023).

63 Colthorpe, A. (2022). Lithium battery pack prices go up for first time since Bloomberg NEF began annual survey. *Energy Storage News*. https://www.energy-storage.news/lithium-battery-pack-prices-go-up-for-first-time-since-bloombergnef-began-annual-survey/ (accessed 10 May 2023).

64 IRENA (2022). Renewable power generation costs in 2021. ISBN: 978-92-9260-452-3 https://www.irena.org/publications/2022/jul/renewable-power-generation-costs-in-2021 (accessed 10 May 2023).

65 Sokolowski, M. (2020). *European Law on Combined Heat and Power*. London: Routledge.

66 EUR-Lex (2016). Commission regulation (EU) 2016/631 of 14 April 2016 establishing a network code on requirements for grid connection of generators. https://eur-lex.europa.eu/legal-content/EN/TXT/?uri=OJ:JOL_2016_112_R_0001#d1e1043-1-1 (accessed 9 January 2022).

67 EUR-Lex (2016). Commission regulation (EU) 2016/1388 of 17 August 2016 establishing a network code on demand connection. https://eur-lex.europa.eu/legal-content/EN/TXT/?uri=uriserv:OJ.L_.2016.223.01.0010.01.ENG&toc=OJ:L:2016:223:TOC https://eur-lex.europa.eu/legal-content/EN/TXT/?uri=uriserv%3AOJ.L_.2016.223.01.0010.01.ENG(accessed 9 January 2022).

68 Holttinen, H., Ivanova, A., and Dominguez, J.L. (2016). Wind power in markets for frequency support services. *2016 13th International Conference on the European Energy Market (EEM)*, Porto, Portugal. 2016, pp. 1-5, doi: 10.1109/EEM.2016.7521310.

69 EUR-Lex (2014). Directive 2014/35/EU of the European parliament and of the council of 26 February 2014 on the harmonisation of the laws of the member states relating to the making available on the market of electrical equipment designed for use within certain voltage limits (recast) text with EEA relevance. https://eur-lex.europa.eu/legal-content/EN/TXT/?uri=celex%3A32014L0035 (accessed 8 December 2021).

70 IEA, International Energy Agency (2021). Snapshot of global PV markets. https://iea-pvps.org/snapshot-reports/snapshot-2021/ (accessed 8 December 2021).

71 Husseini, T. (2018). Riding the renewable wave: tidal energy advantages and disadvantages. *Power Technology*. https://www.power-technology.com/features/tidal-energy-advantages-and-disadvantages (accessed 5 December 2021).

72 The European Commission (2018). Directive (EU) 2018/2002 of the European Parliament and of the Council of 11 December 2018 amending Directive 2012/27/EU on energy efficiency. https://eur-lex.europa.eu/legal-content/EN/TXT/?uri=uriserv%3AOJ.L_.2018.328.01.0210.01.ENG (accessed 25 March 2024).

73 The European Commission (2021). Directive of the European Parliament and of the Council on energy efficiency (recast) (2021/0203). https://eur-lex.europa.eu/legal-content/EN/TXT/?uri=CELEX%3A52021PC0558 (accessed 25 March 2024).

74 Barnes, A. (2020). Can the current EU regulatory framework deliver decarbonisation of gas? https://www.oxfordenergy.org/wpcms/wp-content/uploads/2020/06/can-the-current-eu-regulatory-framework-deliver-decrbonisation-of-gas-insight-71.pdf (accessed 12 December 2021).

75 ISO (1990). ISO/TC 197 hydrogen technologies. https://www.iso.org/committee/54560.html (accessed 9 December 2021).

76 HyLAW Online DataBase (2018). Connection of the E-grid to the electrolyser. HyLAW Online DataBase. https://www.hylaw.eu/database/electricity-grid-issues-for-electrolysers/connection-of-the-e-grid-to-the-electrolyser https://www.hylaw.eu/database/?hyapplication=17#category6 (accessed 9 December 2021).

77 EU Commission (2016). Commission regulation (EU) 2016/631 of 14 April 2016 establishing a network code on requirements for grid connection of generators. https://op.europa.eu/es/publication-detail/-/publication/1267e3d1-0c3f-11e6-ba9a-01aa75ed71a1/language-en (accessed 9 December 2021).

78 Stein, A.L. (2013). Reconsidering regulatory uncertainty: making a case for energy storage. https://bear.warrington.ufl.edu/centers/purc/docs/papers/1414_Stein_Reconsidering%20Regulatory%20Uncertainty.pdf (accessed 9 December 2021).

79 ACER, European Union Agency for the Cooperation of Energy Regulators (2021). Possible regulation of hydrogen networks. http://extranet.acer.europa.eu/en/Gas/Documents/ACER%20H2%20Paper_%20vFinal_clean.pdf (accessed 9 December 2021).

80 EVExpert (2021). Connector types for EV charging around the world. EVExpert. https://www.evexpert.eu/eshop1/knowledge-center/connector-types-for-ev-charging-around-the-world (accessed 9 December 2021).

81 Rajendran, G., Vaithilingam, C.A., Misron, N. et al. (2021). A comprehensive review on system architecture and international standards for electric vehicle charging stations. *Journal of Energy Storage* 42: 103099.

82 EPRI, Electric Power Research Institute (2019). Interoperability of public electric vehicle charging infrastructure. https://www.epri.com/research/products/000000003002017164 (accessed 09 December 2021).

83 Huang, Y., Qian, L., Soopramanien, D., and Tyfield, D. (2021). Buy, lease, or share? Consumer preferences for innovative business models in the market for electric vehicles. *Technological Forecasting and Social Change* 166: 120639.
84 Nissan Motor Corporation. Official Global Newsroom (2018). Europe's largest energy storage system now live at the Johan Cruijff Arena. Nissan Motor Corporation. Official Global Newsroom. https://global.nissannews.com/en/releases/europes-largest-energy-storage-system-now-live-at-the-johan-cruijff-arena (accessed 9 December 2021).
85 Open Charge Alliance OCPP 2.0.1. open charge point protocol. Open Charge Alliance2021. https://www.openchargealliance.org/ (accessed 9 December 2021).
86 EU Commission (2021). The EU clean energy package. https://op.europa.eu/es/publication-detail/-/publication/7fa59d21-c7ff-11eb-a925-01aa75ed71a1 (accessed 9 December 2021).
87 EU Smart Cities Information Systems (2020). Electric vehicles and the grid. Solution booklet. https://smart-cities-marketplace.ec.europa.eu/sites/default/files/2021-02/D32.1D3_Solution%20Booklet_EVs%20and%20the%20Grid.pdf (accessed 9 December 2021).

3

The Concept of Integrated Local Energy Communities: Key Features and Enabling Technologies

Amedeo Buonanno[1], Martina Caliano[1], Gianfranco Chicco[2], Marialaura Di Somma[1], Giorgio Graditi[1], Valeria Palladino[1], Christina Papadimitriou[3], and Hanne Sæle[4]

[1] *Italian National Agency for New Technologies, Energy and Sustainable Economic Development, Lungotevere Thaon di Revel 76, 00196, Rome, Italy*
[2] *Politecnico di Torino, Dipartimento Energia "Galileo Ferraris", Corso Duca degli Abruzzi 24, 10129, Torino, Italy*
[3] *Eindhoven Technical University, Department of Electrical Engineering, Groene Loper 3, 5612 AE, Eindhoven, the Netherlands*
[4] *SINTEF Energy Research, Department of Energy Systems, Sem Saelands vei 11, NO-7034, Trondheim, Norway*

3.1 Introduction

Sector coupling solutions aim to foster carbon neutrality through the exploitation of synergies coming from the interplay of multiple energy carriers such as electricity, heating, cooling, mobility, and hydrogen based on an integrated infrastructure with the electrical system as backbone [1]. Among their main benefits, there is a possibility to increase the efficiency in energy resources use as well as an opportunity to reduce the renewable energy sources (RESs) curtailment, thanks to energy conversion processes sustained by *power-to-X* (P2X) solutions and energy storage in all forms including mobility acting as distributed storage. The large-scale deployment of sector coupling strategies at the regional or national level may represent a demanding task, due to the huge numbers of actors involved with different and often conflicting interests. This is the main reason for which interest of scientific community is growing toward local multienergy systems (MESs), thereby applying a bottom-up approach that can be easily replicated on large scales [2]. This is also in line with the recent EU policies fostering a bottom-up approach for tackling the energy transition as seen in Covenant of Mayors (https://eu-mayors.ec.europa.eu/en/home, 2008) or the most recent RePower EU Plan [3].

In such systems, the local renewable sources can be exploited at their full potential in combination with distributed fossil-based sources for the satisfaction of different types of end-users' demand, such as electricity, heat, cooling, and mobility, with the possible energy support from the upper distribution network. This energy carriers'

Integrated Local Energy Communities: From Concepts and Enabling Conditions to Optimal Planning and Operation, First Edition. Edited by Marialaura Di Somma, Christina Papadimitriou, Giorgio Graditi, and Koen Kok.

integration is made possible through the implementation of electrical, thermal, and other energy carrier-related technologies. These are further supported by energy storage and information and communication technologies (ICTs) that in a coordinated and optimized approach allow users' demand satisfaction in a cost- and sustainable-effective manner.

Integrated local energy communities (ILECs), combining the concept of sector coupling at the local level with one of the energy communities (ECs), represent an efficient and sustainable way of managing energy at a local level by also fostering consumer engagement and empowerment. The ILEC concept may refer to a set of energy users deciding to make common choices in terms of satisfying their energy needs, in order to maximize the benefits deriving from this collegial approach, thanks to the implementation of various electricity and heat technologies and energy storages and the optimized management of energy flows [4]. ILECs support not only integration among multiple energy carriers through P2X solutions but also electrification of the final uses in line with European policies, thus shifting the end user's demand toward the electricity sector fed by RES [5]. On the other hand, ILECs are also able to provide benefits to the larger system, for instance, by supporting the external grid through ancillary services and flexibility provision with different products such as demand response, congestion relief, and participation in local energy markets [6]. Therefore, ILECs represent a huge pool of opportunities for the different actors in the energy supply chain. For instance, local prosumers are benefiting from green and affordable energy from local resources and RES at lower costs, developers and solution providers can find new opportunities for technology within replicable business models, whereas grid operators can avoid grid congestion and defer network investments. Last but not least, policymakers can benefit from increasingly sustainable secure and reliable energy supply systems, increasingly independent of external sources and thus less vulnerable to global energy crisis and other geopolitical developments.

The aim of this chapter is to contextualize the ILEC concept first at the European level, through the detailed and critical analysis of the EU policy and regulatory frameworks in Section 3.2 and then from the technical point of view, by mapping the enabling technologies that can be employed in ILECs in Section 3.3. The latter are classified into three categories: (i) *demand-side flexibility technologies*, (ii) *distributed renewable energy technologies*, and (iii) *ICTs*. Also, for all of these technologies, the barriers to their implementation in ILEC context will be discussed, together with the potential recommendations to overcome them. Finally, the summary of the main barriers is provided in Section 3.4.

3.2 Key Features of ILECs

Based on a review performed in 2021 [7], different key features of energy communities (ECs) related to location, organization, and motivation for establishment and technologies are described in this section. In total, 76 ECs located in 11 different

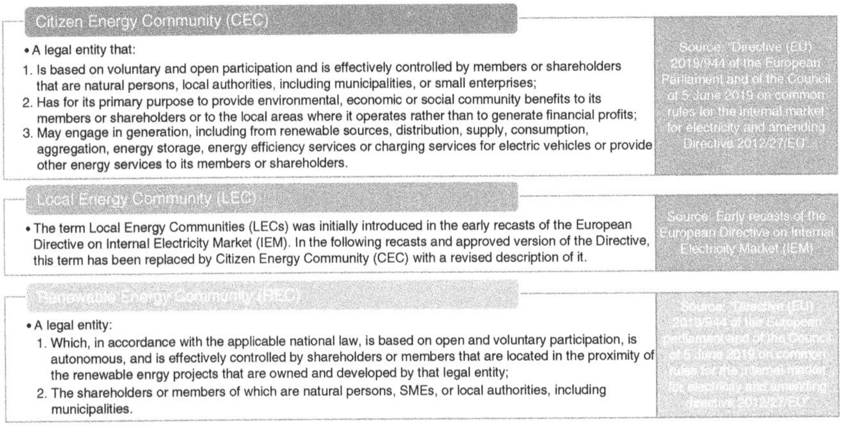

Figure 3.1 Overview of definitions for energy communities.

European countries are included in the review.[1] Additionally, the reviewed ECs are evaluated on a regional level: North,[2] Central,[3] and South[4] of Europe.

Note: The analysis described in this chapter is based on available open information about the 76 reviewed ECs. There is no intention to give a complete description of ECs in Europe.

3.2.1 Policy and Regulation Analysis for REC and CEC

"Energy community" (EC) is a term that conceptually has been introduced in recent European regulatory acts, with at least three different definitions. These are local energy communities (LECs), renewable energy communities (RECs), and citizen energy communities (CECs). These definitions are not synonyms in the European legislation; therefore, a proper understanding of the terms is important.

A formal definition of the term "Energy Community" was introduced in the following Directives in the "Clean Energy Package" [8]:

- Renewable Energy Directive (RED II) 2018/2001 [9]: Introduction and definition of main framework for RECs.
- Internal Electricity Market Directive (IEM) 2019/944 [10]: Introduction of the concept of CECs.

Initially, the term LECs was introduced in the early recasts of the European Directive on IEM. This term has been replaced by CEC in the following recasts and approved version of the Directive. Figure 3.1 gives an overview of the alternative definitions of ECs, with reference to relevant EU Directives.

---

1 The analysis described in this paragraph is based on available open information about the 76 reviewed ECs and it has not been intended as a complete description of status of ECs in Europe.
2 North = Denmark, Norway, and Sweden.
3 Central = Belgium, Germany, Ireland, Poland, and United Kingdom.
4 South = Greece, Italy, and Spain.

3.2.2 Differences Between CEC, REC, and ILEC

There are several similarities that can be observed between CECs and RECs. The first is that financial profit is not the primary purpose, but rather to provide environmental, economic, or social community benefits for its members or the local areas where they operate. Further, the definitions of CEC and REC emphasize participation and effective control by citizens, local authorities, and smaller businesses, whose primary economic activity is not in the energy sector. The third is that the participation in CECs and RECs must be open and voluntary.

Based on the definition of CEC and REC, the ILEC can be defined as a set of energy users who collectively make decisions regarding their energy needs. Its goal is to maximize the advantages that arise from this collaborative approach by utilizing diverse electricity and heat technologies, energy storage systems, and efficient energy flow management [4].

3.2.3 Organizational Structure

The organizational structure of the different ECs is identified in the review. The results are presented in Figure 3.2, both for the total review (on the left in the figure) and in a regional perspective (north, central, south – on the right in the figure). The figure shows that "Cooperative" is the largest part, representing 43% of the reviewed ECs. In a regional perspective, "Cooperative" is the most common model for all regions (north: 36%; central: 42%; south: 46%). The second most common organizational model is "Partnership," representing 25% of the total reviewed ECs (north: 9%; central: 42%; south: 22%).

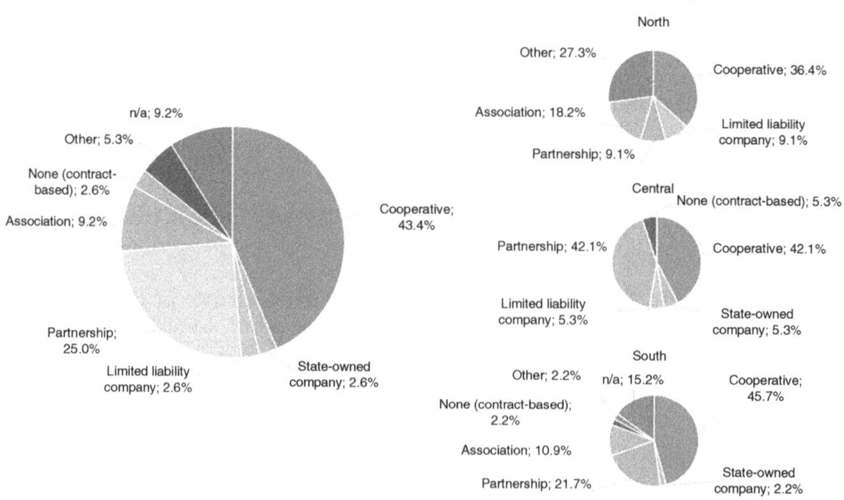

Figure 3.2 Organizational structure of reviewed ECs – in total and in a regional perspective (total on the left in the figure, divided by regions on the right in the figure).

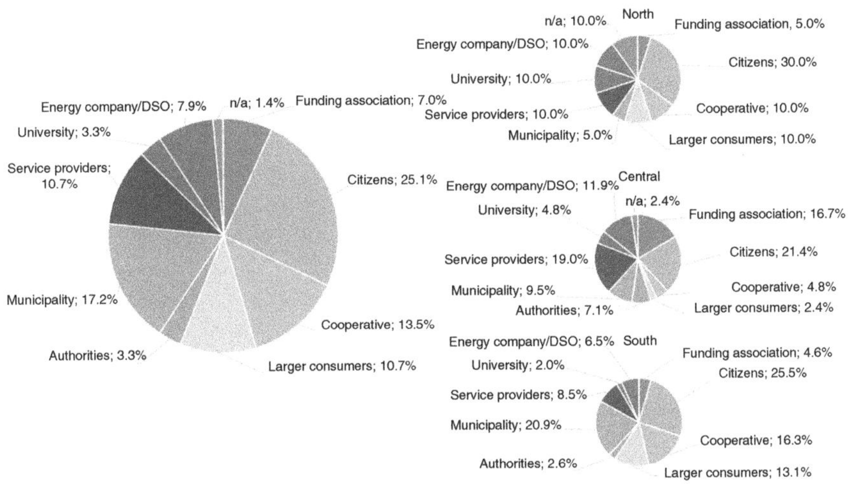

Figure 3.3 Involved stakeholders (total on the left in the figure, divided by regions on the right in the figure).

3.2.4 Involved Stakeholders

The review also evaluated involved stakeholders, and the results presented in Figure 3.3 show that "Citizens" is the largest part (25%), followed by "Municipality" (17%). In a regional perspective, "Citizens" is the main part for all regions (north: 30%; central: 21%; south: 25%). For the central region, "Service providers" (19%) and "Funding associations" (17%) are also important stakeholders. In the south region, "Municipalities" has a large share (21%).

3.2.5 Motivation for Establishment

The ECs included in the review are evaluated according to nine alternative motivations, and several motivations can be indicated for each community. The results are presented in Figure 3.4. The main motivations for the total group are "renewable energy" (23%) and "economical" (22%), followed by "environmental" (16%).

In the North region, the main motivation is "renewable energy" (47%), followed by "environmental" (20%), and both "energy efficiency" and "sustainability" have a share of 13% (only five different motivations are indicated for this region). Nine different motivations are presented in the central region and "renewable energy" is the largest one (19%), followed by "economical" (17%). Eight different motivations are specified in the south region, where "Economical" (26%) is the largest one, followed by "renewable energy" (22%).

The motivation "self-sufficient/energy security" (13–16%) and "energy poverty" (3–4%) are only specified for the central and south regions.

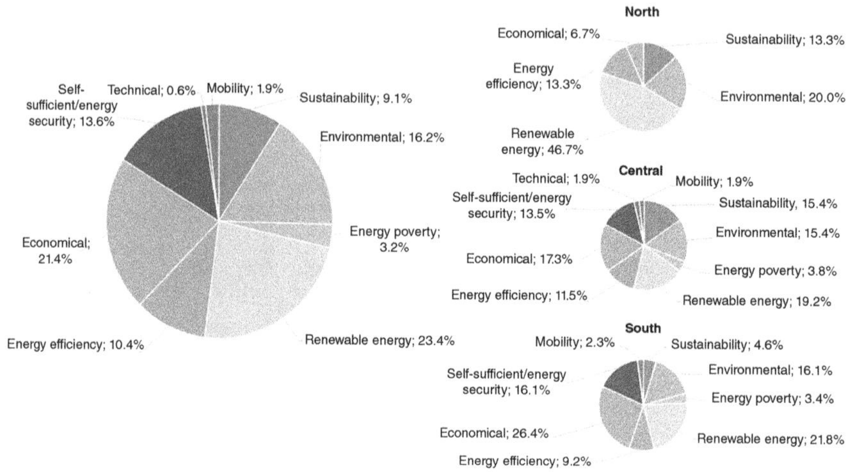

Figure 3.4 Motivations for establishment of the ECs (total on the left in the figure, divided by regions on the right in the figure).

3.2.6 Energy Carriers

Energy carriers identified in the reviewed ECs are presented in Figure 3.5. The figure shows that the most common energy carrier is "electricity" (69%), followed by "Bio" (16%) and "thermal" (12%). A more detailed overview of energy carriers in a regional perspective is presented on the right in the figure. The share of "electricity" differs from 54% in the Central region to 80% in the south region. Other identified energy carriers are "Bio" (north: 27%; central: 20%; south: 11%) and "thermal" (north: 13%;

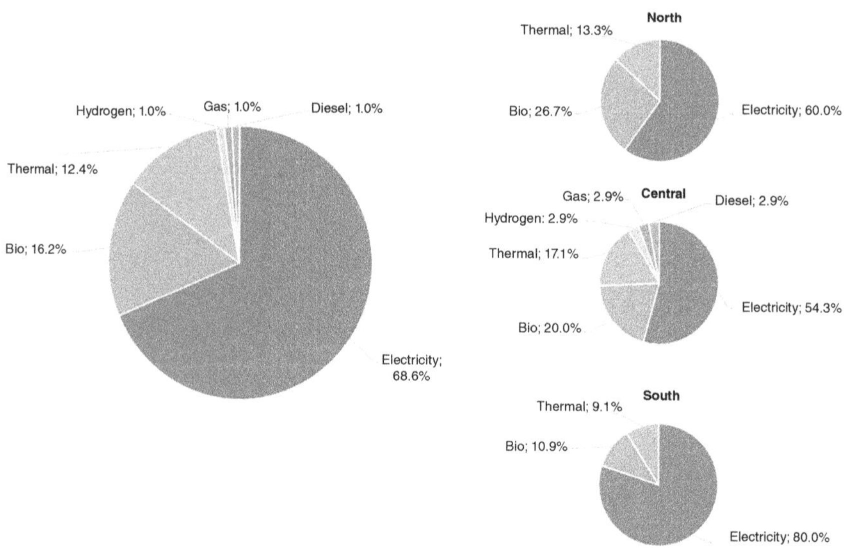

Figure 3.5 Energy carriers identified in reviewed ECs (total on the left in the figure, divided by regions on the right in the figure).

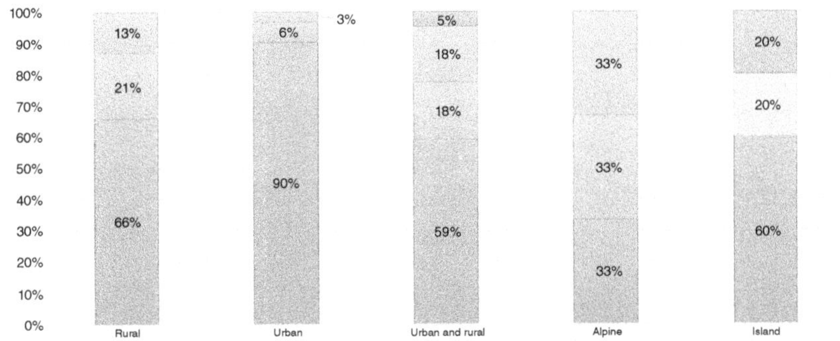

	Rural	Urban	Urban and rural	Alpine	Island
Diesel	0%	0%	0%	0%	20%
Gas	0%	0%	5%	0%	0%
Hydrogen	0%	0%	0%	0%	20%
Thermal	13%	3%	18%	33%	0%
Bio	21%	6%	18%	33%	0%
Electricity	66%	90%	59%	33%	60%

Figure 3.6 Energy carriers according to geographic areas.

central: 17%; south: 9%). Energy carriers such as "hydrogen," "gas," and "diesel" are only specified for the ECs in the central region.

The analysis of reviewed ECs has also studied how energy carriers are spread in different geographic areas such as rural, urban, alpine, and island areas. Some ECs have been identified to cover both urban and rural areas. The results are presented in Figure 3.6. The representative shares for each geographic area are presented in the data table below the bar chart.

The share of "electricity" is largest in urban areas (90%) but varies from 33% to 90% in the different geographic areas. The share of "Bio" varies from 6% to 33% with the largest value in the alpine areas (a limited share of the reviewed ECs is located in alpine areas).

3.2.7 Energy Technologies

Figure 3.7 shows the energy technologies installed in the reviewed ECs. Seven different technologies are included in the review, and the most common is solar panel – photovoltaic (PV), identified in 47% of the reviewed ECs.

"PV" is the most common energy technology in all regions (North: 35%; Central: 31%; South: 63%). In the North region, "storage" (22%), "electric vehicle (EV)/charging station" (17%), and "heat pump" (13%) are also often in use. In the central region, "Wind" (21.6%) and "CHP/Bio" (21.6%) are common technologies. "Storage" is the most common energy technology (11%) after "PV" (63%) in the South region.

Figure 3.8 shows how different energy technologies are spread in different geographic areas such as rural, urban, alpine, and island areas. Some ECs have been identified to cover both urban and rural areas. The representative shares for each geographic area are presented in the data table below the bar chart.

For all geographic areas, "PV" is the most common energy technology in use, with a share varying from 32% to 54%. Additional important energy technologies are "Wind" (varying from 0% to 21%) and "CHP/Bio" (varying from 0% to 29%).

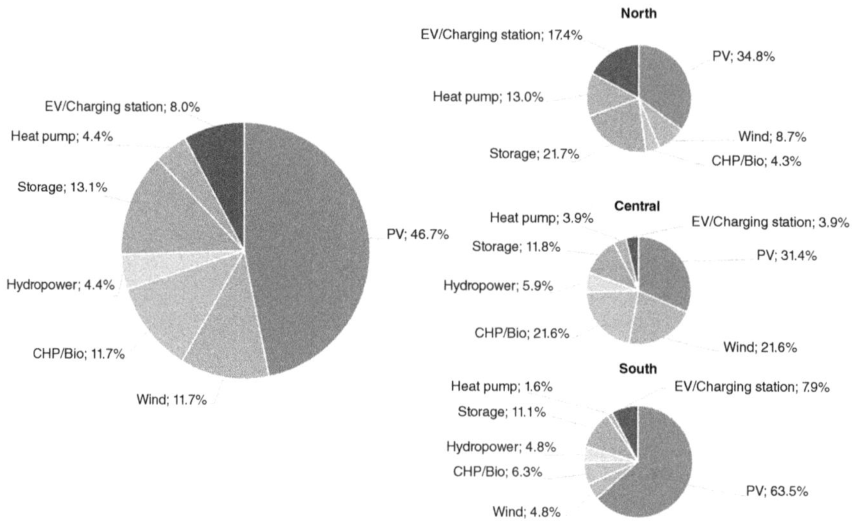

Figure 3.7 Energy technologies installed in reviewed ECs (total on the left in the figure, divided by regions on the right in the figure).

	Rural	Urban	Urban and rural	Alpine	Island
EV/Charging station	3%	10%	9%	14%	13%
Heat pump	3%	6%	3%	0%	13%
Storage	10%	17%	12%	14%	13%
Hydropower	5%	4%	3%	0%	13%
CHP/Bio	10%	6%	21%	29%	0%
Wind	18%	2%	21%	0%	13%
PV	58%	54%	32%	43%	38%

Figure 3.8 Energy technologies according to geographic areas.

3.3 Enabling Technologies

3.3.1 Demand-Side Flexibility Technologies

3.3.1.1 Energy Conversion Technologies

This section deals with energy conversion technologies deployed to enhance the demand-side flexibility in the ILEC. In particular, MES [11] and distributed multienergy systems (DMESs) are taken into account. DMES refers to the presence of MES in distributed locations [12], whose coordinated operation is in line with the scope of effectively developing flexible solutions for the ILEC.

Table 3.1 Types and characteristics of energy conversion technologies.

Adoption of the technology in the ILEC (household level)
Heat pumps, boilers, electric underfloor heaters
Adoption of the technology in the ILEC (community level)
Cogeneration systems, heat pumps, electric chillers, absorption chillers, auxiliary boilers
Energy carriers involved
Electricity, gas, hydrogen, and other types of fuel
Energy demand satisfied
Electricity, heat, cooling, humidification/dehumidification, desiccant cooling, desalination, and production of specific outputs such as hydrogen, methane, water, ammonia, and synthetic fuels
Range size in the market
From kW to MW, depending on the solution. Availability of modular solutions to be used in clusters
Contribution to the demand-side flexibility
Energy shifting, participation in demand-response programs, and in the provision of grid services
Barriers to the demand-side flexibility and potential recommendation to overcome them
Barriers: investment costs of the technologies, return of the investments, inaccurate identification of the demand, need for effective operational scheduling, and operational limits of equipment and energy networks.
Recommendations: design DMES by using technologies most suitable to the user needs, use smart energy metering, incentivize the development of demand-side flexibility, and improve the scheduling algorithms.

The demand considered in a MES can be of different types. Besides electricity, heat, and cooling, other demands can be taken into account, such as humidification/dehumidification, desiccant cooling, desalination, and production of specific outputs such as hydrogen and methane (Table 3.1). However, some of these demands can be supplied through the same energy carrier and differ only for the equipment used in the last stage of the process. For energy efficiency and environmental impact assessment, the demand for a specific use can be considered individually only when it is possible to define the specific power pattern and the efficiency and emission factor for separate production of the useful energy output [13]. Otherwise, the demand for that use must be included in the overall demand supplied by the energy carrier.

DMES flexibility has been defined in Ref. [14] as "the technical ability of a system to regulate multi-energy supply, demand and power flows subject to steady-state and dynamic constraints and while operating within predefined/desired boundary regions for certain energy vectors." When the interactions among DMES technologies and systems with RES are not addressed, the operation of DMES technologies is typically *scheduled* without considering the effects of supply uncertainty for the variable RES, while the uncertainty on the generation and demand of the different energy carriers, in particular the ones depending on temperature, could be considered. Otherwise, the approach for the analysis of the DMES must be structured in an uncertainty-based framework (Section 3.3.2).

On the electrical side, the demand-side flexibility can be characterized by considering various P2X solutions, where "X" means "everything." The P2X conversion is aimed at creating a new energy carrier (different from the energy carriers used as inputs) or supplying the demand of a final user for a different energy carrier.

In the P2X definitions, the first term, "power," refers to the electricity produced by using a primary source (e.g. sun, wind, water, and fuel), with particular interest on using the excess of power production from variable RES. Considering demand-side flexibility for DMES, electricity is the energy carrier that allows the development of the most flexible solutions. In fact, electricity can be distributed through the grid at longer distances with respect to heat or cooling, and the dynamics of the electricity variations are faster than for other energy carriers. In particular, with respect to the grid, demand-side flexibility can be achieved by either increasing or reducing the power taken from the grid in a given node. This could happen for example when the grid operator predicts the possible occurrence of dangerous conditions for the grid (e.g. overloads in high-temperature periods) and offers incentives to the users for changing the power taken from the grid. A user can decide to take the incentive, especially if the user has internal resources (from storage or energy shifting from/to different energy carriers) to avoid or reduce the variation in the local demand served for the various types of energy (electricity, heat, cooling, etc.). The user, also, determines the maximum variation of power available during time (indicated as electricity shifting potential in Ref. [15]) and the profitability of making the power variation starting from the operating condition that the system would have in the absence of the incentive. This leads to the concepts of *upward* and *downward* flexibility, considering a *reduction* or *increase* of the electricity input from the grid, respectively, depending on local generation increase or local demand reduction [14]. Similar concepts may be applied to other energy carriers.

Different solutions can emerge for P2X, such as

- **Power-to-gas:** depending on the gas considered as the output, the energy conversion process can be set up with one or more stages. In the power-to-hydrogen case, the single-stage conversion is carried out by using an electrolyzer. In the power-to-methane case, the process includes two stages, where the first stage produces hydrogen with an electrolyzer and the second stage produces methane through a methanation process. The efficiency of the two-stage cascaded process is the product of the efficiencies of the individual stages, so that the overall efficiency becomes relatively low. An advantage is the availability of the product of the first stage, which can be stored to create an energy buffer, so that the second stage can be activated later. The presence of the energy buffer provides some internal flexibility to the energy conversion process.
- **Power-to-liquid:** the energy conversion generates liquid fuels and chemicals, e.g. ammonia, methanol, and synthetic hydrocarbons [16]. In particular, liquids

with high energy density are of interest, since they can be used in ships and aircrafts or can be stored and transported over long distances. The power-to-liquid process includes two stages. In the first stage, an intermediate carrier is produced starting from electricity, e.g. hydrogen through an electrolyzer, or syngas – a mixture of hydrogen and carbon monoxide – through a reverse water shift process. In the second stage, the intermediate product is converted into the liquid product, e.g. by using the Fischer–Tropsch synthesis and product separation. The liquid product can then be refined to obtain the liquid needed for the specific application. Power-to-Ammonia is a specific case of *power-to-chemicals* with a liquid output. To produce ammonia, electricity is used to electrolyze water, and then hydrogen obtained reacts with nitrogen (obtained through air separation) to produce ammonia. Ammonia is easy to store and transport and has an energy density higher than hydrogen. Further conversion includes *power-to-fuel*, e.g. for producing liquid synthetic fuel.

- **Power-to-heat:** electricity is directly used for ambient heating or water heating. Electric heating where the heater is supplied by the grid could have low efficiency compared with heat production from boilers. However, electric heating may be of interest to provide more flexibility inside the MES. More relevant is electric heating when the electricity comes from RES or electrical storage.
- **Power-to-power:** the usual energy conversions consider alternating current (AC) and direct current (DC) in the different modes (AC/DC, DC/AC, DC/DC, and AC/AC). Moreover, some solutions are indicated by considering the scope of the energy conversion, e.g. *power-to-mobility* to point out the use of electricity to supply EVs.

Considering the technologies used to provide flexibility in an ILEC with dispatchable energy sources:

- The *cogeneration* (or combined heat and power, CHP) system is typically the core of a MES, because of its intrinsic possibility of combined production of electricity and heat from the same input fuel, with higher energy efficiency with respect to the separate production of electricity (taken from the grid) and heat (through boilers supplied by fuel). In the CHP system, the production of electricity and heat is not totally independent, as it is limited inside an operating region (a bounded area, which in some cases is reduced to an operating line with proportional production of electricity and heat given by a cogeneration ratio). The presence of the operating region limits the flexibility that can be provided through the variation of the electricity and heat outputs. The CHP efficiency decreases at partial load operation. The CHP output has a minimum, below which the CHP is switched off to avoid excessively low efficiency. This aspect impacts the CHP availability to play a role in the management of the MES to provide demand-side flexibility and requires careful consideration in the scheduling of the operation of the equipment in the MES and in the definition of coordinated energy shifting strategies.
- The *boiler* is supplied by fuel and produces heat. The boiler is always needed when there is a heat demand, to avoid wasting the ambient excess heat produced

in a MES by a CHP if there is no district heating or thermal energy storage (TES).

- The *electric heat pump* (EHP) is one of the most promising technologies for demand-side flexibility purposes. The EHP converts the electrical input into heat or cooling, depending on the operation mode. The energy produced by the EHP can be directly used or stored in TES systems or in the thermal mass of buildings [17]. A remarkable benefit of using an EHP comes from the high coefficient of performance (COP), defined as the ratio of the desired output (heat or cooling) to the electricity input. The COP mainly depends on the ambient temperature and EHP output capacity, while the COP variation at partial load operation is generally lower.
- The *electric chiller* produces cooling by using the electricity input that comes from the grid or from different equipment (CHP electrical output, local RES, or electrical storage). The general concepts for flexibility purposes are like the ones indicated for the EHP, with different COP values and dependence on the ambient temperature.
- The *electric heater* is used for direct production of heat through the Joule effect, for ambient heating, or in electric underfloor heaters.
- The *absorption chiller* produces cooling by using the heat input that may come from different heat sources (CHP heat output, auxiliary boiler, or TES) at an appropriate temperature. The absorption chiller is useful for performing energy shifting, by sharing the supply of the cooling side with an electric chiller or an EHP.
- The *electrolyzer* uses electricity to split water into hydrogen and oxygen. It is a useful resource for demand-side flexibility, as the hydrogen formed can be directly used or stored for later use. A relatively limited quantity of hydrogen can be injected into the gas network.

The last resource that can be considered in demand-side flexibility is the demand *curtailment* or demand *increase*. In general, demand curtailment can be imposed by the supplier or be part of a *demand-response* program in which the user accepts to change its demand based on proposed changes in the energy price or specific incentives [18]. Demand response typically refers to the electrical demand, because of the easier variations on the electrical side, while the thermal capacity on the thermal side reduces the possibility of making fast changes.

3.3.1.2 Storage Technologies

The storage technologies and their role in the context of ILECs and their contribution to demand-side flexibility are discussed in the following paragraphs.

3.3.1.2.1 Thermal Energy Storage

TES generally involves a temporary storage of high or low temperature heat for later use and nowadays represents an important element in many energy-saving programs in various sectors, like residential, industrial, commercial, and transportation. Typical applications are about the storage of solar energy for overnight heating or cooling, of summer heat for winter use, and of winter ice for space cooling in summer. Due to the electrification of thermal demand, TES is increasingly adopted in the development of demand-side management programs since it allows reducing the demand of energy by storing heat or cooling energy electrically generated during off-peak hours to be used during on-peak demand hours. The reduction of consumption can be achieved by storing the thermal energy generated in excess with respect to the end-user's needs that would normally be released as waste. Considering this, TES can contribute significantly to meet society's desire for more efficient and sustainable energy use by contributing to increase energy savings and to reduce the use of fossil fuels through efficiency increase and/or fuel substitution allowing for the reduction in emissions of pollutants such as CO_2, SO_2, NO_x, and CFCs [19].

TES systems are used in various forms and specific applications. Some of the most common applications are about:

- **Sensible thermal energy storage (STES):** In STES, thermal energy (heat/cold) is absorbed and released due to the variation of the temperature of a storage medium, such as water, oil, rocks, sand, and soil, without a phase transition. The thermal energy stored is proportional to the difference between the final and initial temperatures of the storage medium, its mass, and its heat capacity. STES systems are typically composed of a cylindrical storage tank filled with the storage medium, but also include lakes and water basins. These ones are typically used for seasonal TES.
- **Latent thermal energy storage (LTES):** In LTES, thermal energy (heat/cold) is absorbed and released when a phase change material (PCM) undergoes a phase transition (typically solid–liquid) at a nearly constant temperature, providing for high energy storage density [20].
- **Thermochemical thermal energy storage (TTES):** In TTES, heat is stored through the products of reversible thermochemical reactions. Heat is stored in the form of chemical compounds created by an endothermic reaction, and it is recovered again by recombining the compounds in an exothermic reaction [21]. Table 3.2 summarizes the main characteristics of thermal energy storage.

3.3.1.2.2 H_2 Storage

In the hydrogen economy, an important role is played by hydrogen storage. In fact, there are different types of technologies for storing hydrogen, primarily:

- Physical storage as compressed gas.
- Physical storage as cryogenic liquid hydrogen.
- Material-based storage or solid-state storage.

The first two kinds are the most mature and generally used, whereas the third one is still under investigation [24]. In fact, hydrogen is most normally transferred in pressurized or liquid form (−253 °C) by lorries or tank vessels.

Pressurized Hydrogen Gas Currently, the most widely used method is highly pressurized hydrogen gas. With this approach, the gas is compressed and stored in containers, cylinders, or sometimes underground caverns.

Table 3.2 Characteristics of thermal energy storage.

Adoption of the technology in the ILEC (household level)
At the household level, TES systems are usually used in decentralized heating and cooling systems, placed at the consumption point [22].
Adoption of the technology in the ILEC (community level)
TES systems at the community level are usually used in centralized heating and cooling systems coupled with heat pumps and CHP/combined cooling, heat, and power (CCHP) for district heating and cooling applications. In such a case, seasonal storage can also be used for long-term storage applications [22].
Energy carriers involved
Hot water; chilled water
Energy demand satisfied
Heating (heat storage); cooling (cold storage); domestic hot water
Range size in the market
Apartments < 16 kWth; buildings < 120 kWth; district > 120 kWth [20]
Contribution to the demand-side flexibility
Thermal storage can contribute to the flexibility in the heating and cooling sector by reducing the mismatch between the thermal energy generation and demand. The electricity system flexibility can also be enhanced by using TES by storing the thermal energy electrically generated by power-to-heat technologies during off-peak periods to meet the thermal loads that occur during high demand periods [23]. Moreover, TES systems can store energy for hours, days, weeks, or even months, helping address seasonal variability in supply and demand.
Barriers to the demand-side flexibility and potential recommendation to overcome them
Potential barriers related to TES are mainly connected to the lack of technologies readiness for some TES technologies, the lack of knowledge and awareness of how TES can provide benefits to society, public sector, and industry, and to the barriers related to the intersectoral coupling [23]. Considering this, incrementing the investments into R&D of TES technologies, introducing incentives for removing fossil-fuel technologies in heating and cooling sector, and creating market mechanism for incentivizing demand-side flexibility could represent potential recommendations to overcome these barriers.

Many technologies are based on the compressed gas storage method: stationary, vehicular, and bulk transport. In hydrogen refueling stations, stationary storage is mainly used which requires large-scale and low-cost effort. In the second case, vehicular application, the storage should be not heavy, with high density and affordable. Finally, to distribute hydrogen from the producer to the end users, the third one, i.e. bulk transportation, is used. This type has a large capacity and is lightweight to decrease delivery costs [24].

Liquefied Hydrogen Storage of liquid hydrogen results in higher density and more energy can be stored per unit volume, in comparison to compressed hydrogen gas storage technology. Regarding the economic aspect, the cost for low-pressure liquid hydrogen storage systems can be relatively low; however, the cost and energy consumption for liquefaction are quite high [24]. Moreover, there are still difficulties in the process of evaporation.

Material-Based Storage When hydrogen atoms/molecules are bound with other elements, it is possible to realize a material-based storage. This kind of storage is the most promising, thanks to the large amounts of hydrogen that can be stored at a strikingly small volume. The bonding process can be done through [24]

- **Chemisorption (i.e. absorption):** hydrogen molecules are dissociated into hydrogen atoms and integrated into the lattice of the materials.
- **Physisorption (i.e. adsorption):** hydrogen atoms or molecules are attached to the surface of the materials.

This type of hydrogen storage presents a very high cost and a low storage density. At the moment, the research activities are focused on the choice of the best material to bond the hydrogen able to satisfy the needed criteria of the storage system.

Table 3.3 summarizes the main characteristics of H_2 storage.

3.3.1.2.3 Electrical Energy Storage

Batteries are the most well-known and widely used energy storage devices. They convert chemical energy into electrical energy and can be recharged for multiple cycles. Examples include lead–acid batteries, lithium-ion batteries, and nickel–metal hydride batteries.

Lead–acid batteries [29]:

- **Chemistry:** Lead–acid batteries use a combination of lead plates and sulfuric acid electrolyte.
- **Energy density:** Lead–acid batteries have a relatively low energy density compared to other battery types.
- **Cycle life:** They offer moderate cycle life, typically ranging from a few hundred to a few thousand cycles.
- **Cost:** Lead–acid batteries are generally more affordable compared to other battery technologies.
- **Maintenance:** These batteries require periodic maintenance, including checking and topping up electrolyte levels.
- **Applications:** Lead–acid batteries are commonly used in automotive applications, uninterruptible power supplies (UPSs), and backup power systems.

Lithium-ion batteries [30]:

- **Chemistry:** Lithium-ion batteries use lithium compounds as their electrolyte and electrodes.
- **Energy density:** They have a high energy density, providing a greater amount of stored energy per unit weight or volume.
- **Cycle life:** Lithium-ion batteries offer a relatively high cycle life, typically ranging from hundreds to thousands of cycles.

Table 3.3 Characteristics of H_2 storage.

Adoption of the technology in the ILEC (household level)

Thanks to the fulfillment of control and management system, hydrogen storage can be employed at the household level. Commercially, the most usable storage systems are hydrogen storage with pressure vessels and metal hydride storage systems [25].

Adoption of the technology in the ILEC (community level)

In terms of a community/macroenergy hub level, the hydrogen storage can be employed for storing the surplus energy in the grid, providing many clients by hydrogen, or a combination of the two.

The hydrogen long-term storage can be used mixed with intraday battery storage, also, inside isolated power systems, where the electricity costs are very high. This type of solution could be much faster and cheaper [26].

Energy carriers involved

Electricity; hydrogen

Energy demand satisfied

Electricity

Range size in the market

The energy in 1 kg of hydrogen is about 120 MJ (= 33.33 kWh) [27]

Type of hydrogen	Energy contents (MJ kg^{-1})	
	Lower heating value	Higher heating value
Gaseous hydrogen	119.96	141.88
Liquid hydrogen	120.04	141.77

Contribution to the demand-side flexibility

H2 storage can contribute indirectly to demand-side flexibility by coupling it with an electrolyzer. The stored hydrogen can contribute to grid balancing services and to make available a flexible load.

The hydrogen, in fact, can be stored on a large scale, inside storage facilities over the long term, to collect the renewable energy surplus and to allow this excess for interseasonal storage [28].

Moreover, hydrogen storage does not self-discharge, in addition, presents a longer life and higher temperature tolerance, in comparison with batteries. These features are very useful in extreme weather scenarios [26].

Barriers to the demand-side flexibility and potential recommendation to overcome them

The hydrogen storage systems developed must be efficient, reliable, effective, and surely safe. Moreover, there are several barriers when it comes to the application of the hydrogen storage in demand-side flexibility:

- **Economic aspects:** most low-carbon hydrogen technologies are not cost-effective without government incentives.
- **Safety aspects:** hydrogen is a flammable gas and should be handled with high care.
- **Regulatory aspects:** lack of common EU framework for installing this technology.

- **Cost:** the cost of lithium-ion batteries has been decreasing over the years due to technological advancements and increased demand.
- **Maintenance:** Lithium-ion batteries are generally maintenance-free.
- **Applications:** Lithium-ion batteries are widely used in portable electronics, EVs, renewable energy systems, and various consumer applications.

Nickel–metal hydride (NiMH) batteries [31]:

- **Chemistry:** NiMH batteries use a hydrogen-absorbing alloy for the negative electrode and a nickel-based compound for the positive electrode.
- **Energy density:** NiMH batteries have a lower energy density compared to lithium-ion batteries but higher than lead–acid batteries.
- **Cycle life:** They offer a moderate to high cycle life, typically ranging from hundreds to a few thousand cycles.
- **Cost:** NiMH batteries are generally more affordable compared to lithium-ion batteries.
- **Maintenance:** NiMH batteries are generally maintenance-free.
- **Applications:** NiMH batteries are commonly used in hybrid vehicles, portable electronics, and other applications where high energy density is required without the need for lithium-ion technology.

It is important to note that these characteristics can vary depending on the specific design, manufacturer, and advancements in battery technology.

Although batteries are a well-matured technology, their upfront cost can be a barrier to their widespread adoption especially at the mEH level. Batteries, especially those with large storage capacity, can be expensive. Also, batteries have a limited lifetime, and their performance can degrade over time. This degradation affects their capacity to store and deliver energy efficiently making their participation in demand-side flexibility less efficient. To overcome this barrier, it is important to continue research and development efforts to drive down costs and increase lifetime through technological advancements and economies of scale. Additionally, implementing supportive policies such as incentives, grants, and subsidies can make battery installations more financially viable for energy users. Also, existing regulatory frameworks and market structures may not fully support demand-side flexibility and the participation of batteries. These frameworks often focus on traditional generation and consumption models, making it challenging for batteries to participate in energy markets and provide grid services. Updating regulations to include specific provisions for demand-side flexibility and creating market mechanisms that incentivize battery participation can help remove regulatory barriers and promote their integration [32].

Table 3.4 summarizes the main characteristics of electrical energy storage.

3.3.1.2.4 Electric Vehicle Charging Stations

The EV charging station or EV supply equipment (EVSE) is the main set of elements required to connect the EV to an electrical installation for charging purposes. Physically, the EVSEs can be found in two main configurations: wall box and charging

Table 3.4 Characteristics of electrical energy storage.

Adoption of the technology in the ILEC (household level)
Batteries can be installed at the household level, i.e. at the premises of the consumers. Batteries within an ILEC enable energy storage, load shifting, grid stabilization, demand response, and EV integration.
Adoption of the technology in the ILEC (community level)
Batteries can be installed at the community level as a common ILEC asset. Batteries within an ILEC enable energy storage, load shifting, grid stabilization, demand response, and EV integration.
Energy carriers involved
Electricity
Energy demand satisfied
Electricity
Range size in the market
The two first technologies are the most relevant for the ILECs because of their scope: - Lead–acid batteries in ILECs as mentioned are commonly used in automotive applications so they generally range from 30 to 100 Ah, depending on the vehicle size and power requirements. - Lithium-ion batteries are commonly used in • **EVs:** EV battery sizes typically range from tens of kilowatt-hours (kWh) to a few hundred kWh, with larger batteries enabling longer driving ranges. • **Stationary storage:** In grid-scale applications, i.e. where the batteries are used as common assets of the community, lithium-ion batteries can range from several hundred kWh to several megawatt-hours (MWh). These are used for energy storage in utility-scale renewable projects within ILECs.
Contribution to the demand-side flexibility
Batteries contribute to demand-side flexibility in ILECs through peak shaving, load shifting, providing ancillary services, integrating variable renewable energy, participating in demand-response programs, and enhancing microgrid resilience. By optimizing energy consumption and providing flexibility in response to grid conditions, batteries help create a more balanced, efficient, and resilient energy system.
Barriers to the demand-side flexibility and potential recommendation to overcome them
The upfront cost of batteries is a barrier to widespread adoption. To overcome this, research and development efforts are needed to reduce costs and increase lifetime through technological advancements and economies of scale. Supportive policies like incentives and subsidies can enhance financial viability for energy users. Existing regulations and market structures require updates to include specific provisions and market mechanisms that incentivize flexibility originated from batteries [32].

post. The interface between the EVSE and the grid is directly hard-wired to a control device or a plug and protection box. For most daily charging activities, the energy demand from EVs is satisfied by the overnight slow charging using wall boxes mainly installed at home or at work [33]. But for sporadic long-distance trips, EVs will require to be charged as conventional internal combustion engine (ICE) vehicles at gas stations, achieving 80–90% state of charge (SoC) in 15–30 minutes depending on the nominal power of the charging point and the EV battery capacity [34]. The level of power that an EVSE can deliver to an EV varies in each country depending on the distribution network and supply characteristics: voltage, current, frequency, and the number of phases in the case of AC supply. This is defined mostly by national regulation and utilities' procedures. The IEC 61851 is an international standard dealing with EV conductive charging system, which has become a reference for EV infrastructure [33, 35].

Part 1 of the previous standard categorizes four charging modes:

- **Mode 1** represents the slowest charging level, where the EV is directly connected to standard wall electricity outlet and, therefore, it does not allow any control.
- **Mode 2** is referred to the connection of an EV to a standard socket-outlet, but the connection cable needs to have a control pilot function and a safety system against the electric shock.
- **Mode 3** offers a slow and accelerated charging type in AC through a specific EVSE permanently connected to the AC network.
- **Mode 4** is a DC fast charging method, which uses an external charger that converts AC into DC.

Although most of the chargers installed worldwide are slow chargers (<22 kW), in recent years fast charging (FC) stations (50 kW) and ultrafast charging (UFC) stations (<400 kW) are being installed to meet the growing demand for EVs that require covering long distances, reducing range anxiety when traveling [34, 36]. At the end of 2019, there were installed around 9000 FC and 640 UFC across Europe in the main transport network corridors [36]. According to the reference standard IEC 61851-1, DC power is only used for fast charging in mode 4, mainly using three different types of standardized sockets in the vehicle side: CHAdeMO (usually 50 kW DC output in CHAdeMO 1.0 and 400 kW in CHAdeMO 2.0), combined charging system (CCS) (200 kW DC output for the American CCS1 type and over 300 kW DC output for the European CCS2 option), and the Chinese option, GB/T (125 kW DC output). Additionally, Tesla has developed an additional socket for their models (over 300 kW in V3 superchargers) [33, 37, 38].

Due to the high power involved during the charging process, conventional on-board chargers are not a feasible solution due to cost, size, and weight issues, and for this reason, DC power is delivered to the EV battery by an isolated power converter located outside the vehicle, which includes additional control and protection functions. EV charging stations usually have several charging units that allow charging 10–12 EVs at the same time. For this reason, FC-UFC charging points are directly connected to the medium voltage (MV) grid. From the MV/LV transformer, these charging units can share a common LV-AC link, using independent rectifiers for each charging unit, or use a centralized rectifier, using a common DC link within the charging station. The power delivered from the charging station to the EV is limited by the charge acceptance of the batteries of each specific EV, the nominal ratings of the charger, the connector (the maximum connector rate is defined by the standard), and the cable between the vehicle and the charger. High charging currents require larger cable diameters to avoid overheating, and this issue increases the cable weight (9 kg for a 50 kW-400 V fast charger) [33].

EVs have the potential to play a significant role in enabling demand-side flexibility within an ILEC. As the number of EVs on the road continues to grow, their charging patterns can be strategically managed to provide various benefits to the energy system. By intelligently managing the charging and discharging of EVs, they can actively participate in demand-response programs and support the optimization of energy supply and consumption within an ILEC. One of the keyways EVs can contribute to demand-side flexibility is through vehicle-to-grid (V2G) technology. V2G enables bidirectional energy flow between EVs and the grid, allowing EVs to not only draw energy from the grid for charging but also inject surplus energy back into the grid during peak demand periods [39]. This bidirectional energy flow capability turns EVs into valuable energy storage assets that can be utilized to balance the local energy system. By aggregating a fleet of EVs, their combined storage capacity can provide substantial flexibility to the grid operators [40].

Through V2G technology, EVs can support demand-response initiatives by responding to price signals or grid requests to increase or decrease their charging or discharging rate. During times of high electricity demand, EVs can reduce their charging power or even temporarily halt charging to alleviate the strain on the grid. Conversely, when there is excess renewable energy generation or low electricity demand, EVs can increase their charging rate to absorb the surplus energy, effectively serving as a sink for the intermittent RES [41].

Furthermore, EVs can be integrated into smart energy management systems within an ILEC. By leveraging real-time data and advanced algorithms, EV charging patterns can be optimized to align with the availability of renewable energy and grid conditions. This optimization can ensure that EV charging is coordinated with local generation, energy storage systems, and other energy loads, thereby minimizing peak demand and maximizing the utilization of renewable energy [33].

The integration of EVs into demand-side flexibility strategies also brings economic benefits. EV owners can participate in energy markets and receive incentives for providing flexibility services [42]. By offering their EVs' storage capacity and charging flexibility to the grid, EV owners can potentially earn additional income or reduce their charging costs through dynamic pricing mechanisms.

In summary, EVs have the potential to significantly contribute to the demand-side flexibility of an ILEC. Through V2G technology, optimized charging strategies, and participation in demand-response programs, EVs can actively support the balancing of energy supply and demand, enhance grid stability, and maximize the utilization of RES. The integration of EVs into the energy system can bring both environmental and economic benefits, paving the way for a more sustainable and resilient energy future.

One of the key barriers is the inadequate availability of charging infrastructure, which can limit the widespread adoption of EVs. To overcome this, ILECs should invest in the expansion of public and private charging stations, ensuring convenient access for EV owners and proper incentives. Collaboration with charging network operators and utilities can help streamline the deployment of charging infrastructure.

Many consumers may not fully understand the benefits and potential of demand-side flexibility from EVs. To address this, ILECs can implement awareness campaigns and educational programs to inform consumers about the advantages of demand-response programs, time-of-use tariffs, and other flexible charging options. Providing user-friendly information about the integration of EVs into the energy system can encourage participation and engagement. Some EV owners may be hesitant to participate in demand-side flexibility programs due to concerns about battery degradation caused by frequent charging and discharging cycles. To alleviate these concerns, ILECs can offer incentives such as battery health monitoring, warranty extensions, or financial compensation for potential battery wear. Providing accurate information on the battery management systems of EVs and their ability to handle frequent charging events can also help dispel misconceptions.

A lack of consistent regulations and policies can impede the integration of demand-side flexibility from EVs. ILECs should work closely with regulatory bodies to develop favorable policies that incentivize and facilitate demand-side management programs. These policies could include financial incentives, supportive tariff structures, and streamlined procedures for participation in energy markets.

Concerns around data privacy and security can be a barrier to the implementation of demand-side flexibility programs. ILECs should prioritize the establishment of robust data protection measures, ensuring that EV owners' personal data is securely managed. Implementing transparent data handling practices and obtaining explicit consent from participants can help build trust and encourage broader participation. Last but not least, the lack of interoperability and standardization among EV charging systems can complicate the implementation of demand-side flexibility. ILECs should promote the adoption of common protocols and standards for charging infrastructure, allowing for seamless integration and communication between various charging networks and grid operators. Collaboration with industry stakeholders, standardization organizations, and technology providers can facilitate the development of interoperable solutions.

Table 3.5 summarizes the main characteristics of electric vehicle charging stations.

3.3.2 Distributed Renewable Energy Technologies

3.3.2.1 Photovoltaic Systems

PV systems are the main RES considered for decarbonizing an ILEC. The main solutions for household applications are the installation of PV systems on rooftops and on the façade of buildings. In more general applications, large PV systems can be installed in further available locations, e.g. areas not dedicated to other purposes and viable from the ecological point of view or following the *agrivoltaic* paradigm [43] to coexist with the agricultural production.

The effectiveness of these installations heavily depends on the orientation of the PV modules (south orientation is better) and on the absence of obstacles in the landscape, to avoid operation in shaded conditions. In terms of size, the PV systems

Table 3.5 Characteristics of electric vehicle charging stations.

Adoption of the technology in the ILEC (household level)

The EV charging can be performed at the household level meaning in individual houses participating in the ILEC.

Adoption of the technology in the ILEC (community level)

The EV charging can be performed at the community level meaning the charging spots are publicly available within the ILEC.

Energy carriers involved

Electricity

Energy demand satisfied

Electricity

Range size in the market

According to IEC 61851-1:2017

Mode Charging type Max. Voltage (V) Max. Current (A) Connection
1. AC (slow) 250 (1 ph), 480 (3-ph) 16 Standard socket
2. AC (slow) 250 (1 ph), 480 (3-ph) 32 Via special cable
3. AC (accelerated) 1000[a] – [a]Special cable through the EVSE
4. DC (fast) 1500[a] – [a]Dedicated socket through EVSE

[a]In accordance with the EVSE specification.

Typical values for modes 3 and 4 in Europe are the following: Mode 3: 230/400 V at 16, 32, or 64 A (between 3.7 and 44.3 kW), Mode 4: 400 V at 125 A (between 50 kW). Other supply powers currently exist. It is expected that this power will increase up to 150 kW for lighter vehicles and 500 kW for buses and lorries.

Range size:
- 50 kW (CSS Type 1, 2 – CHAdeMO 1.0)
- 120 kW (Tesla Supercharger)
- 250 kW (Tesla V3 Supercharger)
- 350 kW (CSS)
- 400 kW (CHAdeMO 2.0)
- 900 kW (ChaoJi: CHAdeMO 3.0 – GB/T 2.0)

Contribution to the demand-side flexibility

Through vehicle-to-grid (V2G) technology, EVs can bidirectionally exchange energy with the grid, acting as valuable energy storage assets to balance the local energy system. They can respond to price signals or grid requests, reducing or increasing their charging rate to alleviate strain during peak demand or absorb surplus renewable energy. Optimized charging strategies can minimize peak demand, and EV owners can benefit economically.

Barriers to the demand-side flexibility and potential recommendation to overcome them

Insufficient charging infrastructure poses a challenge, but ILECs can invest in expanding public and private charging stations through collaborations with network operators and utilities. Consumer awareness is crucial, and ILECs can conduct educational campaigns on the benefits of demand-response programs and flexible charging options. Concerns about battery degradation can be alleviated by offering incentives and accurate information on battery management. Clear and favorable regulations should be developed in collaboration with regulatory bodies. Data privacy and security measures should be prioritized to build trust, and promote interoperability and standardization among EV charging systems.

have the advantage of being modular solutions, depending on the space available. During operation of the PV systems, the main issues are the production for a limited time period during the day, also with variable power production even in clear sky conditions, and the variable power output in cloudy conditions. On the technical side, the presence of electrical storage with a capacity coordinated with the PV system size and characteristics would enhance the effectiveness of using the PV energy production to satisfy the local demand. On the economic side, the cost of the PV systems and especially of the batteries for electrical storage is a limiting factor.

With respect to the grid, if the PV production is concentrated in a few sites, it must be verified whether the net power seen at every interface with the grid does not violate the grid limits. Otherwise, PV power curtailment would be needed to avoid violations [44]. With deep diffusion of PV systems, at the planning stage it should be verified that the size of the PV systems does not exceed the limits to avoid PV power curtailment during operation – an issue known as the *hosting capacity* problem [45].

In an ILEC, the benefits of the availability of PV systems depend on various aspects that refer to the local generation, demand, and storage at the sites in which the PV systems are installed and connected to the grid, and on their coordinated management. In particular, the main points are

- **Diversity of the local demand:** the production from PV systems installed in a given site could be available to supply the demand in periods of time in which the local demand at the PV production site is limited. For example, the PV production of a PV system installed on the rooftop of public buildings could be available to supply the demand of households in the ILEC during weekends, when the local demand of the public building is reduced.
- **Correlations among the PV production from different sites:** the location of the PV plants in the ILEC is expected to be relatively close in terms of geographical distance. As such, on a bright day, the correlations[5] among the solar irradiance at different sites are close to unity, so that, in the absence of grid constraints, the whole PV production within the ILEC can be seen as a unique generation for the ILEC. However, on cloudy days, the correlations among the PV production patterns could become lower, and the effects of having multiple PV plants in operation would be favorable, as the local demand at some locations with lower irradiance could be served through the grid by the PV production from other locations in the ILEC.

3.3.2.2 Interactions Between Photovoltaic Supply and Demand

The main concepts to consider are self-sufficiency (SS), self-consumption (SC), and collective self-consumption (CSC). Let us consider, for example, a system with a local demand that can be served by the local electricity production or by the grid, and with the possibility of selling the excess of local electricity production to the grid. In this

5 The correlations among the PV power production patterns should be calculated by removing the night period, otherwise higher correlation would occur because of the contribution of the full correlation among the null PV generation during the night.

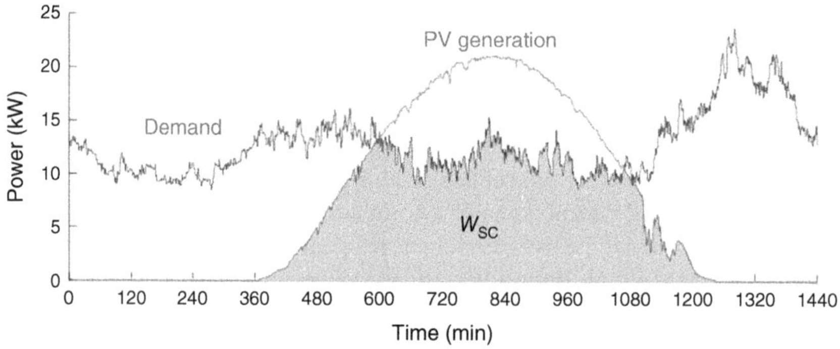

Figure 3.9 Power patterns and indication of the area corresponding to the self-consumed PV electricity.

basic example, the local electricity production is provided by a PV system and there is no energy storage. The daily power patterns of the electricity demand, local PV production, and input from/output to the grid are represented with given regular time steps (e.g. hour by hour). The relevant values of electrical energy are calculated as the total electricity demand W_D for the whole day and the total *PV electricity production* W_{PV} at the grid connection point of the PV system (i.e. the inverter AC output). When the local PV generation exceeds the demand at some time steps, the excess of electricity production from the PV system is represented with W_{G+}, while the self-consumed PV electricity is calculated as $W_{SC} = W_{PV} - W_{G+}$. The terms calculated are important to define two fundamental indicators [46], which are defined as dimensionless and can be also expressed in percent:

1) *Self-consumption* $SC = W_{SC}/W_{PV}$, which represents the part of the PV production used to supply the local demand,
2) *Self-sufficiency* $SS = W_{SC}/W_D$, which represents the part of the local demand *covered* by the PV production.

Figure 3.9 shows an example of power patterns of demand and PV generation with a one-minute time step. The area corresponding to W_{SC} is highlighted. The indicators corresponding to this example are SC = 0.71 and SS = 0.40.

From the definitions of SC and SS, it is apparent that the indicators SS and SC can be improved by increasing the area corresponding to W_{SC}. When a PV system is used, the period without production from the sunset to the next sunrise heavily affects the possibility of improving the indicators. The presence of local storage may provide further technical benefits, by storing (in part or totally) the excess of electricity that would be injected in the grid without storage, using that stored electricity to supply the local demand in the time periods in which the local generation is not available. In the same way, the presence of storage contributes to smoothing the fluctuations of PV production in partially cloudy days.

In an ILEC, the concept of SC has been extended to the CSC (or jointly acting renewable energy SC [9]) for addressing the effects related to the whole community. The CSC is determined by calculating, at each time step of the power

Table 3.6 Types and characteristics of photovoltaic technologies.

Adoption of the technology in the ILEC (household level)
Rooftop PV systems, façade PV systems
Adoption of the technology in the ILEC (community level)
Large PV systems, agrivoltaic solutions
Energy carriers involved
Electricity
Energy demand satisfied
Electricity to cover the demand or to be stored.
Range size in the market
From kW to MW (modular)
Contribution to the demand-side flexibility
Electricity input to the grid or to a local MES, coordinated exchange of electricity inside the ILEC.
Barriers to the demand-side flexibility and potential recommendation to overcome them
Barriers: production limited to a period of the day, sensibility to weather conditions, fluctuating power generation, lack of local measurements of weather variables, difficult prediction of the power production, grid constraints for sending the excess of electricity production, lack of definition of dedicated tariffs, and legal limits to form energy communities. **Recommendations:** identification of solutions that could maximize SC, combined planning and coordinated operation with electricity storage, local measurements of weather variables, smart metering, enhancement of power prediction methods, and evolution of the regulatory and legal framework.

patterns, the minimum between the electricity injected in the grid by the local production systems in the ILEC and the electricity taken from the grid by the ILEC prosumers.

Table 3.6 summarizes the types and main characteristics of photovoltaic technologies.

3.3.3 Information and Communication Technologies

In this paragraph, the ICT groups are described with their role in the context of ILECs. The groups of ICTs have been selected considering the information flow: measurements come from the field (Internet of Things, IoT), have to be transmitted across several devices using the Communication Technologies, have to be stored and accessed (Data Management Technologies), secured (Cybersecurity), and analyzed (Technologies for Analytics), in order to make decisions (Control and Management Technologies). A lot of these tasks are accomplished through the Computing Technologies. Moreover, a paragraph is devoted to a brief description of blockchain technology. In Figure 3.10, the groups of ICTs described here are shown. For each group, a brief description has been provided together with its role in the

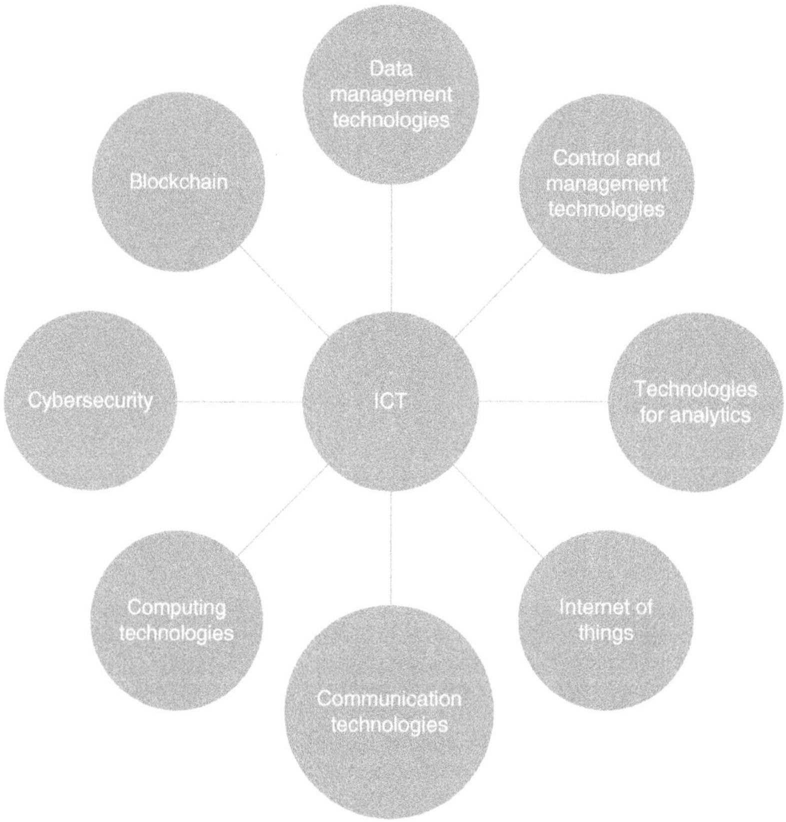

Figure 3.10 ICTs considered in the study.

ILEC (household and community levels), the contribution to the demand-side flexibility and the possible barriers.

3.3.3.1 Data Management Technologies

The data management technologies are the tools and methods that allow the ingestion, processing, securing, and storing of data for the successive decision-making phase [47]. In a smart grid/ILEC context, there are several data sources: sensors, meters, SCADA, etc., that produce a lot of different types of data with high frequency. The volume (high amount of data), velocity (high-frequency generation), and variety (heterogeneous data) of the involved data, often make necessary the adoption of Big Data technologies.

The main components of data management are as follows:

- Data processing
- Data storage
- Data governance
- Data security

In the *data processing* component, the raw data are ingested from several data sources usually following the ETL (extract, transform, load) paradigm: the data are extracted from the data sources, transformed (cleaned, formatted, etc.), and finally loaded in the final structures. The *data storage* is composed of different hardware and software components such as disks, cloud storage, database management systems (DBMSs), and backup and recovery software. In the Big Data context, the relational database management systems (RDBMSs) often show their technological and economic limits. Instead, Big Data platforms, such as Hadoop, can be very helpful given their scalability and distributed architecture. For the management of tabular data, the relational model looks again a suitable solution, but scalable versions should be considered. In the presence of not-relational data (images, videos, texts, etc.), the not-relational database should be considered (e.g. MongoDB for documents, InfluxDB for timeseries, and Neo4J for graph data). With *data governance*, we refer to policies, procedures, and processes for managing data throughout its lifecycle. Data governance ensures that data is accurate, consistent, and available to authorized users when needed. It also helps mitigate risks associated with data breaches or noncompliance. The *data security* is related to the protection of data from unauthorized access, disclosure, or destruction. Data security measures include encryption, access controls, and firewalls.

Table 3.7 summarizes the main characteristics of data management technologies.

3.3.3.2 Control and Management Technologies

The main control and management technologies involved in the energy context are as follows:

- **Energy management systems (EMS, BEMS):** systems that monitor the sensors and, through control algorithms, can apply control actions on several devices in order to obtain the best performance (regards to a particular objective). BEMS are EMS for buildings and their main functions are, hence: monitoring, control, optimization, and reporting related to different types of buildings (e.g. residential and industrial) [51].
- **Supervisory control and data acquisition (SCADA):** system that monitors and controls physical processes such as industrial plants and infrastructure systems. A SCADA system is typically composed of several components such as sensors, actuators, control servers, and communication links. At the field sites, the SCADA monitors continuously the different attributes of involved equipment via sensors and sends control actions to field control devices like remote terminal unit (RTU), programmable logic controller (PLC), or intelligent electronic device (IED) [52].
- **Programmable logic controller (PLC):** industrial computerized device designed to control and automate manufacturing processes and equipment. It uses a programmable memory to store instructions and execute specific functions, which can be customized to meet the specific requirements of a particular industrial application. It acts as the physical interface between devices on the plant and a SCADA. PLCs enable communication, monitoring, and control of intricate automated processes such as robotic cells, temperature regulation, and other industrial machinery. PLCs are usually the main part of automatic systems in industry [53].

Table 3.7 Characteristics of data management technologies.

Adoption of the technology in the ILEC (household level)
Whenever data needs to be archived/accessed/queried, data management technologies come into play, and this is also true in the ILEC at the household level. Big Data technologies may be necessary when dealing with a large volume, variety, or velocity of data. However, small-scale ILECs may not require the use of such technologies.
Adoption of the technology in the ILEC (community level)
Effective management of data generated by sensors and smart meters in multiple microenergy hubs at a macroenergy hub level of an ILEC requires the use of data management technologies. Depending on the number of data sources, their velocity, and variety, it may be necessary to employ Big Data technologies to properly manage the data.
Contribution to the demand-side flexibility
Data management technologies play an important role in enabling demand-side flexibility by providing the tools and infrastructure needed to collect, store, and analyze customer data. For this reason, the data management technologies can be considered enabling technologies for the Demand-Side Flexibility implementation [22].
Barriers to the demand-side flexibility and potential recommendation to overcome them
One of the main barriers is the cost of Data Management technologies that can be expensive to purchase, implement, and maintain. Another potential barrier is the customer's concerns about data privacy and security. The aggregators must ensure that the devices and communication channels used to collect and transmit customer data are secure against cyber-attacks and that personal data are protected. Failure to do so will lead to a lack of trust from customers, which may discourage them from participating in demand-side flexibility schemes [48]. Another barrier is the ambiguity surrounding privacy and data protection legislation, such as the General Data Protection Regulation (GDPR) [49] and the proposed e-Privacy Regulation [50]. There are different interpretations of the restrictions on the use of customer data without their consent, which can vary between Member States. This lack of clarity can hinder the development of demand-side flexibility in the European energy market. Therefore, it is crucial to address these concerns and provide clear guidelines for data protection and privacy, to enable the effective deployment of demand-side flexibility technologies [48].

- **Distributed control system (DCS):** computer-based control system for process or industrial plant. It is composed of several autonomous distributed controllers without a central supervisor. The several intelligent devices cooperate with each other to accomplish a common objective [54].

Table 3.8 summarizes the main characteristics of control and management technologies.

3.3.3.3 Technologies for Analytics

There is a significant amount of data involved in the smart grid/ILEC context coming from multiple sources such as sensors, smart meters, SCADA, and connected vehicles [55]. The process of extracting useful information from this data is called

Table 3.8 Characteristics of control and management technologies.

Adoption of the technology in the ILEC (household level)
At the household level, the EMS/BEMS and PLCs can be adopted.
Adoption of the technology in the ILEC (community level)
For the community level, the SCADA and DCS can be adopted.
Contribution to the demand-side flexibility
Control and Management technologies can contribute to demand-side flexibility by automating the process of controlling and managing energy consumption. This automation can help reduce energy consumption during peak hours and shift it to off-peak hours.
Barriers to the demand-side flexibility and potential recommendation to overcome them
There are no direct barriers for these technologies, except for the possible risks from cyber-attacks and data privacy issues (refer to Sections 3.3.3.1 and 3.3.3.7 for more details).

Data Analytics (or Big Data Analytics in Big Data context). There are four types of Data Analytics: *descriptive analytics* that provides information about the past and present, *diagnostic analytics* that tries to explain why things happened, *predictive analytics* that tries to predict the future, and *prescriptive analytics* that recommends actions to achieve a desired outcome [56].

With the assistance of Artificial Intelligence (AI) techniques, data analytics tools can easily extract insights, recognize patterns and trends, and make predictions based on historical data. There are several definitions for AI, but they can be grouped based on four different points of view that focus on the similarity of the computer programs to the human performance (think humanly/act humanly) or focus on rationality of the computer programs (think rationally/act rationally) [57]. During these years, a specific branch of AI has become the most prominent: machine learning (ML). This refers to computer programs that acquire the ability to perform a task by learning from experience [58].

The increasing availability of data, advancements in hardware technology, and improvements in algorithms and theory have all contributed to the widespread use of a specialized form of ML called deep learning (DL) that use multiple layers to represent data [58]. Three main learning paradigms are supervised, unsupervised, and reinforcement. In Supervised Learning, each input example is linked to output targets, and the aim is to learn the correlation between input and output. Some common examples include image classification and linear regression. On the other hand, Unsupervised Learning does not have associated output targets, and the objective is to learn structure and regularity in the inputs. Various examples of this category are clustering and density estimation. Lastly, reinforcement learning is where the agent learns to interact with the environment to increase the expected cumulative reward. For instance, it is often used for robot navigation and game solving. In the smart grid/ILEC context, the ML is extensively used [59], for

Table 3.9 Characteristics of technologies for analytics.

Adoption of the technology in the ILEC (household level)
In the ILEC at the household level, many tasks are performed to support the decision-making process (e.g. forecasting, anomaly detection, and control algorithms), and in this context the data analytics is crucial.
Adoption of the technology in the ILEC (community level)
In the ILEC at the community level, many tasks are performed to support the decision-making process (e.g. forecasting, anomaly detection, and control algorithms), and in this context the data analytics is crucial.
Contribution to the demand-side flexibility
The implementation of demand-side flexibility can be facilitated by the use of AI and Big Data Analytics that can be seen as enabling technologies [28].
Barriers to the demand-side flexibility and potential recommendation to overcome them
The barriers encountered in data management technologies also apply to technologies for analytics. Furthermore, uncertainties around future AI regulations [63] may pose obstacles to utilizing certain types of analytics for demand-side flexibility. In Europe, for example, even though the EU parliament is establishing guidelines for AI that prioritize safety and transparency [64], the development of specific directives to address AI's impact on energy-related matters will require considerable time and effort.

example, Supervised Learning is useful for predicting load consumption [60], PV production [61], or faults [62]; Unsupervised Learning can be employed in anomaly detection task; and Reinforcement Learning algorithms can aid in solving some optimization problems.

Table 3.9 summarizes the main characteristics of technologies for analytics.

3.3.3.4 Internet of Things (IoT)

IoT is a technology that leverages sensors, actuators, and communication technologies to monitor or control the physical world, represented as a large network of connected things that collect and share data regarding their behavior and the environment around them [65]. Many physical devices related to energy context, from household appliances to industrial machines and from power generation systems to transmission and distribution networks, can communicate with each other to transfer useful data and create an efficient management system.

Smart meters: Smart meters are fundamental for building IoT networks and allow to measure the amount of energy supplied to and withdrawn from the grid. They compose a "smart metering system," enabling demand-response, dynamic pricing, and other energy services. Smart meters are equipped with multiple wired and wireless communication interfaces, including Ethernet, Fiber Optical Communication, RS232, R485, M-Bus, PLC, GSM, Bluetooth, Wi-Fi, and ZigBee [66].

Sensors: Several types of sensors are employed in the energy context to measure different physical quantities (e.g. temperature, humidity, wind speed, current, and voltage) fundamental to optimize energy consumption and cost reduction [67]. When the sensor has additional processing capabilities and can also be programmed to perform specific tasks, such as monitoring a process or controlling a device, it is called a smart sensor [68]. In the energy context, smart sensors are employed in power quality measurement systems, smart building energy management system, smart grid protection systems, etc. [69–71]. The fusion of information coming from different sensors is also crucial to improve the decision-making capability [72].

Actuators: Devices that transform energy into motion (e.g. linear, rotating, and oscillating). Based on the type of involved energy, actuators can be categorized into hydraulic actuators, pneumatic actuators, and electric actuators [65]. The actuators are crucial to apply the control actions in the energy context. For example, connected valves and pumps used to manage the water flow, relays for connecting and disconnecting PV systems, etc.

Sensor network/wireless sensor network: It is a network of sensors that are not directly connected to the internet (as in IoT). Instead, the sensors are connected to a router or to a central point that collects data. There are many different types of data transmission technologies that can be used in sensor networks, such as Ethernet, serial wired connections, GSM (2G, 3G, 4G, LTE), Wi-Fi, and Bluetooth. These technologies use different protocols, which are often created by individual manufacturers. Various technologies and protocols can make the integration of individual systems very difficult [70, 73].

Table 3.10 summarizes the main characteristics of IoT.

3.3.3.5 Communication Technologies

Communication systems are essential to provide effective and secure data transfer among several components (sensors, actuators, meters, etc.). Based on the distance range and speed, the involved network types can be classified as[6]: HAN (Home Area Network), NAN (Neighborhood Area Network) or FAN (Field Area Network) and WAN (Wide Area Network).

- **HAN** is close to the end user (range from 1 to 100 m). It enables the information exchange between household appliances/devices. Data rate: 10–100 kbps; Wireless technologies: ZigBee, Z-wave, Wi-Fi, Bluetooth; Wired technologies: Ethernet, PLC-HomePlug, M-Bus.
- **NAN/FAN** (range from 10 m to 10 km) collects information from HAN and enables communication with the WAN. Data rate: 100 kbps–10 Mbps; Wireless technologies: ZigBee, Wi-Fi, LoRaWAN, WiMAX, Cellular (2G, 3G, 4G, 5G); Wired technologies: Ethernet, PLC.
- **WAN** usually handles communication with utility companies [75, 76] (range from 10 to 100 km). Data rate: 10 Mbps–1 Gbps; Wireless technologies: Z-wave, Cellular (2G, 3G, 4G, 5G); Wired technologies: Ethernet, Fiber Optic.

6 The boundaries between the described typologies are often smooth and depend on the application and network topology.

Table 3.10 Characteristics of IoT.

Adoption of the technology in the ILEC (household level)
Integrating different sensors and actuators is a key part of ILECs at the household level and it can be challenging since there are many different communication protocols and data structures in use. To address this challenge, systems like ILECs need a software platform that, supporting all the communication technologies and protocols involved, can connect individual sensors or networks of sensors into a single system. Improving IoT interoperability is essential for the growth and success of the IoT and several works are focused on this key issue [74].
Adoption of the technology in the ILEC (community level)
In the ILEC at the community level, the use of sensors and actuators will be limited to the integration of existing meter reading and automation control systems. Access to them usually is easier than at the household level because it is limited to create an access channel and not to create measurement/control systems from scratch as in the latter.
Contribution to the demand-side flexibility
The smart meters are essential for demand-side flexibility. An extensive network of sensors and meters expands energy demand management options improving the energy flow control and contributing to the energy savings.
Barriers to the demand-side flexibility and potential recommendation to overcome them
Smart meters are employed widely in Europe, but the relevant entities often do not know, based on the privacy regulation (GDPR and future e-privacy), which data from smart meters can be used and which are the purposes that do not require the explicit consent of the customer. For the development of demand-side flexibility, this problem is crucial [48]. Also, the changes to existing sensor systems and networks can be a problem since this can void the warranty of involved devices.

Both HANs and NANs are involved in the ILECs.

For the data exchange process, the communication protocols are fundamental as well but, given the close relation to the used technology, usually, the separation among communication technology, protocol, or communication system is often impossible. Some used protocols are Modbus, BACnet, KNX, etc. [77]

Table 3.11 summarizes the main characteristics of communication technologies.

3.3.3.6 Computing Technologies

There are different ways to deliver the computing resources:

- **Cloud computing:** storage and computing services provisioned over a network such as Internet. Cloud computing is often used in smart grid context because it provides scalable computing resources and does not require resource provisioning planning [78, 79].
- **On-premise computing**: employment of user's servers/workstation and computing infrastructure. On-premise computing is less flexible than cloud solutions but keeps the data and computation on local servers with full control on them.

Table 3.11 Characteristics of communication technologies.

Adoption of the technology in the ILEC (household level)
The technologies for the data exchange at the household level are crucial since they enable the connection of sensors, actuators, and meters. The huge variety of solutions leads to integration issues. At the household level, the exchange data is often related to sensitive information (e.g. home address, customer identification, and information on energy consumption). For this reason, that information has to be properly secured also through a communication technology that ensures an appropriate security level.
Adoption of the technology in the ILEC (community level)
The selection of technologies to create the communication channels for the ILEC often depends on the solutions already available in the considered area. Creating a new communication network has potential high costs and this can be a limiting factor to ILEC establishment. Also, at the community level, the security of data transmission and storage is crucial.
Contribution to the demand-side flexibility
The communication technologies enable infrastructure management and control. For this reason, they are fundamental to implement any demand-side flexibility solution.
Barriers to the demand-side flexibility and potential recommendation to overcome them
The main concern and, hence, the most limiting factor, is the possible risks from cyber-attacks and data privacy issues (refer to Sections 3.3.3.1 and 3.3.3.7 paragraphs for more details).

- **Edge computing**: the data are processed where they are generated. In edge computing, there is low latency (crucial for real-time solutions), low bandwidth requirements, and the connectivity issues have less impact. In addition, the data privacy and security should be enhanced by the low volume of data transmission. However, the edge devices themselves may be exposed to some risks [80, 81].
- **Fog computing:** the processing is moved nearer to source of data, but the "intelligence" is at the LAN layer. Fog computing can be seen as an intermediate layer between the cloud layer and the edge layer.

Table 3.12 summarizes the main characteristics of computing technologies.

3.3.3.7 Cybersecurity

Cybersecurity involves the implementation of various processes, practices, and technologies to protect critical systems and sensitive information from digital attacks. In addition to protecting valuable information and systems, cybersecurity is also important for maintaining consumer trust.

Three protection goals are defined: confidentiality, integrity, and availability (CIA). Confidentiality refers to restricting access to sensitive information only to authorized individuals. Integrity refers to ensuring the data is not modified during transmission. Availability refers to ensuring the system is responsive and timely in transmitting information from the source to the destination [82].

Table 3.12 Characteristics of computing technologies.

Adoption of the technology in the ILEC (household level)
In the operation of ILEC, several algorithms have to be executed (e.g. load forecasting, anomaly detection, and power control), and hence the computing technologies have a fundamental role. Given the locality of the computation and the low computational requirements, the use of Cloud platforms might not be necessary (the main exception is the training of complex AI model).
Adoption of the technology in the ILEC (community level)
Also, in the ILEC at the community level, the use of Computing technologies is fundamental. Based on the dimension of data to process, the computational requirements can be high, and the Cloud platforms might be a preferable choice.
Contribution to the demand-side flexibility
AI, Big Data Analytics, control algorithms, etc. are enabling technologies for the implementation of the demand-side flexibility [28] and the computing technologies are the necessary infrastructure to execute them.
Barriers to the demand-side flexibility and potential recommendation to overcome them
The barrier aspects are the same as the data management technologies.

In the energy sector, the industrial-specific application impacts on definition of the priority among the three goals. For example, in the context of generation and transmission, ensuring availability and integrity of data is crucial. Failure to do so could result in device misconfiguration and ultimately impact the reliability of the overall system. In the advanced metering infrastructure, guaranteeing the confidentiality of customer personal information has the highest priority [83].

Table 3.13 summarizes the main characteristics of cybersecurity.

3.3.3.8 Blockchain

A blockchain is a decentralized and unalterable ledger that enables the recording of transactions among members of a community. Each participant in the blockchain has an exact replica of the same file, which is regularly updated with new information. In the blockchain is not possible to remove data from the common file, and all parties involved in creating the blockchain collectively and independently verify its accuracy [84].

The blockchain consists of blocks that contain the block header, the block size, a transaction counter, and a list of transactions. Each block header includes important data such as the Previous Block Hash, which is the cryptographic imprint of the preceding block in the chain, the Markle Tree, which is the result of applying a hash function to all transactions in the block, the timestamp, and the block's nonce [85].

Table 3.14 summarizes the main characteristics of blockchain.

Table 3.13 Characteristics of cybersecurity.

Adoption of the technology in the ILEC (household level)
It is the same as at the community hub level, but on a smaller scale.
Adoption of the technology in the ILEC (community level)
The assessment of the attack's impact should be based on cross-cutting criteria, which takes into account the consequences that result from dependencies across different sectors of infrastructure. The security of data in one energy carrier can have an impact on the security of other energy carriers that work together in the same EC by sharing data with each other. This means that ensuring cybersecurity in ILECs is crucial to mitigate the risk of potential disruptive cyber-attacks [82].
Contribution to the demand-side flexibility
Modern energy systems are composed of information and communication infrastructures which are vulnerable to security breaches. This poses a serious threat not only to the security of users' personal data but also to the overall operation and reliability of the system. In order to maintain the integrity of data during transmission, it is essential to implement effective cybersecurity measures. Additionally, confidentiality of data is a critical aspect, particularly in the context of demand-side flexibility where access to certain information must be restricted to authorized personnel only.
Barriers to the demand-side flexibility and potential recommendation to overcome them
The cybersecurity can be wrongly considered only as an additional cost and some stakeholders may prioritize cost and speed of implementation over security measures. This can hinder customer from participating in demand-side flexibility programs due to concerns about the privacy and protection of their data, which they may perceive as vulnerable to cyber threats. It is fundamental to ensure that cybersecurity measures are transparent and built on a foundation of trust between the energy provider and the end users.

3.4 Summary of Main Barriers to the Use of Enabling Technologies in the ILEC

This section summarizes the broad categories of barriers for the enabling technologies in the ILEC as discussed previously:

Cost-related barriers: Implementing data management technologies or DMES technologies in general can be expensive, including the purchase, implementation, and maintenance of the necessary tools and infrastructure.

Data privacy and security barriers: Customers may have concerns about the privacy and security of their data. It is important for aggregators to ensure that the devices and communication channels used to collect and transmit customer data are secure against cyber-attacks. Building trust with customers is crucial for their participation in demand-side flexibility schemes.

Regulatory and compliance barriers: Uncertainties around future AI regulations and blockchain may pose obstacles to utilizing certain types of analytics for demand-side flexibility. Clarifying and establishing regulations related to AI can help overcome these barriers. EU parliament is already negotiating first-ever

Table 3.14 Characteristics of blockchain.

Adoption of the technology in the ILEC (household level)
Not used
Adoption of the technology in the ILEC (community level)
At the community/macroenergy hub level, blockchain technology can be adopted to gather data on community members' energy production and consumption through smart meters that are enabled by blockchain technology.
Furthermore, the control of energy networks through smart contracts enabled by blockchain technology can play a significant role in the implementation of ILECs [86, 87].
Contribution to the demand-side flexibility
One of the advantages that come with integrating blockchain technology into smart grids is the ability to exchange and trade energy in a peer-to-peer (P2P) manner, without relying on a third-party intermediary. This also enables secure energy transactions of installed RES.
In the residential sector, blockchain can aid in managing energy and facilitate households to address energy surplus or deficit with their neighbors. The interaction of smart homes is made secure by leveraging blockchain technologies [88].
Barriers to the demand-side flexibility and potential recommendation to overcome them
The main barrier to the demand-side flexibility is the inability to delete transactions and data stored in the blockchain at a later stage. One way to mitigate this issue is to refrain from storing any data in the blockchain that might require deletion later.
The slowness and energy consumption of blockchain systems, as well as their limited scalability, can be a barrier but it can be overcome through proper system design [87].

rules for safe and transparent AI (European Parliament Press Room, 2023) but of course there is a long way ahead before a related directive can be issued specifically addressing energy-related issues that AI touches upon. Similarly, the regulation of AI in the United States is still in its early stages, and there is no comprehensive federal legislation dedicated [63].

Analysis of the demand-side flexibility barriers: There is a lack of comprehensive analysis of demand-side flexibility, which can limit the effective utilization of enabling technologies. As an example, when the interactions between DMES technologies and RES are not addressed, the operation of DMES technologies is typically scheduled without considering the effects of supply uncertainty for variable RES. This can lead to suboptimal operation and limited flexibility.

Technology limitations: CHP systems, which are typically at the core of a MES, have inherent limitations on flexibility due to the interdependence of electricity and heat production. The presence of operating regions and reduced efficiency at partial load operation restricts the flexibility that can be provided through the variation of electricity and heat outputs. This needs to be rectified through the appropriate planning of different DMES technologies.

Also in many cases, the absence of district heating or TES systems necessitates the use of boilers to avoid wasting excess heat produced by CHP systems. This indicates a lack of sufficient TES infrastructure, which limits the potential for demand-side flexibility and efficient utilization of heat energy. Nevertheless, demand-response programs and incentives for demand curtailment or increase are easier to implement on the electrical side compared to the thermal side. The thermal capacity limitations reduce the possibility of making fast changes, hindering the effective implementation of demand-side flexibility.

Enabling technologies like EHP and electric chillers depend on ambient temperature and have varying COP values. COP variation at partial load operation is generally lower, which affects their flexibility and efficiency. Electrolyzers, which use electricity to split water into hydrogen and oxygen, provide a resource for demand-side flexibility. However, the quantity of hydrogen that can be injected into the gas network is relatively limited, which restricts its utilization.

PVs' main limitations are connected with their stochastic nature. To overcome, batteries are attractive, but they also have a limited cycle of life and high costs that can make this investment nonsustainable for the ILEC concept.

Acknowledgments

Part of the results presented in the chapter has been obtained in eNeuron project. The project eNeuron has received funding from the European Union's Horizon 2020 Research and Innovation Programme under Grant Agreement No. 957779.

List of Abbreviations

AC	alternating current
AI	artificial intelligence
CCHP	combined cooling, heat, and power
CCS	combined charging system
CEC	citizen energy community
CHP	combined heat and power
COP	coefficient of performance
CSC	collective self-consumption
DC	direct current
DMESs	distributed multienergy systems
DL	deep learning
EC	energy community
EHP	electric heat pump
ETL	extract, transform, load
EV	electric vehicle
EVSE	electric vehicle supply equipment
FC	fast charging
GDPR	General Data Protection Regulation
ICE	internal combustion engine
ICT	information and communication technology

IED	intelligent electronic device
ILEC	integrated local energy community
LEC	local energy community
LTES	latent thermal energy storage
MESs	multienergy systems
ML	machine learning
P2X	power-to-X
PCM	phase change material
PLC	programmable logic controller
PV	photovoltaic
REC	renewable energy community
RESs	renewable energy sources
RTU	remote terminal unit
SC	self-consumption
SoC	state of charge
SS	self-sufficiency
STES	sensible thermal energy storage
TES	thermal energy storage
TTES	thermochemical thermal energy storage
UFC	ultrafast charging
V2G	vehicle-to-grid

References

1 Di Somma, M. and Graditi, G. (2022). Integrated energy systems: the engine for energy transition. In: *Technologies for Integrated Energy Systems and Networks* (ed. G. Graditi and M. Di Somma), 15–40. Wiley.

2 Jin, L., Rossi, M., Ciabattoni, L. et al. (2023). Environmental constrained medium-term energy planning: the case study of an Italian university campus as a multi-carrier local energy community. *Energy Conversion and Management* 278: 116701.

3 Europa.eu (2022). REPowerEU: a plan to rapidly reduce dependence on Russian fossil fuels and fast forward the green transition. https://ec.europa.eu/commission/presscorner/detail/en/IP_22_3131 (accessed 11 July 2023).

4 Caliano, M., Buonanno, A., Di Somma, M. et al. (2022).Enabling conditions for the deployment of integrated local energy communities in Europe. *Workshop on Blockchain for Renewables Integration (BLORIN),* Palermo, Italy (02–03 September 2022). 144–148.

5 Comodi, G., Di Somma, M., Graditi, G., and Spinaci, G. (2022). Transition potential of local energy communities. In: *Technologies for Integrated Energy Systems and Networks* (ed. G. Graditi and M. Di Somma), 275–304. Wiley.

6 Papadimitriou, C., Di Somma, M., Charalambous, C. et al. (2023). A comprehensive review of the design and operation optimization of energy hubs and their interaction with the markets and external networks. *Energies* 16: 4018.

7 Sæle, H., Morch, A., Buonanno, A. et al. (2022). Development of energy communities in Europe. *18th International Conference on the European Energy Market (EEM)*, Ljubljana, Slovenia (13–15 September 2022). 1–5.

8 European Commission (2019). Clean energy for all Europeans package. https://energy.ec.europa.eu/topics/energy-strategy/clean-energy-all-europeans-package_en (accessed 11 July 2023).

9 Directive 2018/2001/EU on the promotion of the use of energy from renewable sources, 2018.

10 Directive 2019/944/EU of the European Parliament and of the Council of 5 June 2019 on common rules for the internal market for electricity and amending Directive 2012/27/EU, 2019.

11 Mancarella, P. (2014). MES (multi-energy systems): an overview of concepts and evaluation models. *Energy* 65 (1): 1–17.

12 Mancarella, P. and Chicco, G. (2009). *Distributed Multi-generation Systems: Energy Models and Analyses*. New York, NY: Nova Science Publishers.

13 Chicco, G. and Mancarella, P. (2008). A unified model for energy and environmental performance assessment of natural gas-fueled poly-generation systems. *Energy Conversion and Management* 49 (8): 2069–2077.

14 Chicco, G., Riaz, S., Mazza, A., and Mancarella, P. (2020). Flexibility from distributed multienergy systems. *Proceedings of the IEEE* 108 (9): 1496–1517.

15 Mancarella, P. and Chicco, G. (2013). Real-time demand response from energy shifting in distributed multi-generation. *IEEE Transactions on Smart Grid, Special Section on Real-Time Demand Response* 4 (4): 1928–1938.

16 Dieterich, V., Buttler, A., Hanel, A. et al. (2020). Power-to-liquid via synthesis of methanol, DME or Fischer–Tropsch-fuels: a review. *Energy & Environmental Science* 13: 3207–3252.

17 Evens, M., Mugnini, A., and Arteconi, A. (2022). Design energy flexibility characterisation of a heat pump and thermal energy storage in a comfort and climate box. *Applied Thermal Engineering* 216: 119154.

18 Albadi, M. and El-Saadany, E. (2008). A summary of demand response in electricity markets. *Electric Power Systems Research* 78: 1989–1996.

19 Sadeghi, G. (2022). Energy storage on demand: thermal energy storage development, materials, design, and integration challenges. *Energy Storage Materials* 46: 192–222.

20 Dinçer, İ. and Rosen, M.A. (2021). *Thermal Energy Storage: Systems and Applications*. Wiley.

21 Bauer, T. (2020). Chapter 1: Fundamentals of high temperature thermal energy storage, transfer and conversion. In: *Ultra-High Temperature Thermal Energy Storage, Transfer and Conversion* (ed. A. Datas), 1–34. Woodhead Publishing Series.

22 IRENA (2019). *Innovation Landscape Brief: Renewable Power-to-Heat*. International Renewable Energy Agency.

23 IRENA (2020). Innovation Outlook: Thermal Energy Storage, International Renewable Energy Agency, Abu Dhabi. ISBN: 978-92-9260-279-6.

24 Zhang, F., Zhao, P., Niu, M., and Maddy, J. (2016). The survey of key technologies in hydrogen energy storage. *International Journal of Hydrogen Energy* 41: 14535–14552.

25 IEA (2021). The cost of capital in clean energy transitions. https://www.iea.org/articles/the-cost-of-capital-in-clean-energy-transitions (accessed 11 July 2023).

26 IRENA (2018). Hydrogen from renewable power: Technology outlook for the energy transition, International Renewable Energy Agency, Abu Dhabi. ISBN: 978-92-9260-077-8.

27 Abe, J., Popoola, A.P.I., Ajenifuja, E., and Popoola, O.M. (2019). Hydrogen energy, economy and storage: review and recommendation. *International Journal of Hydrogen Energy* 44: 15072–15086.

28 IRENA (2019). Demand-side flexibility for power sector transformation, International Renewable Energy Agency, Abu Dhabi. ISBN: 978-92-9260-159-1.

29 Yamaki, J. (2009). Secondary batteries – lithium rechargeable systems – lithium-ion: overview. In: *Encyclopedia of Electrochemical Power Sources* (ed. J. Garche), 183–191. Elsevier.

30 Goodenough, J. and Park, K. (2013). The Li-ion rechargeable battery: a perspective. *Journal of the American Chemical Society* 135 (4): 1167–1176.

31 Taniguchi, A., Fujioka, N., Ikoma, M., and Ohta, A. (2001). Development of nickel/metal-hydride batteries for EVs and HEVs. *Journal of Power Sources* 100 (1–2): 117–124.

32 IEA (2022). *Grid-Scale Storage*. Paris: IEA.

33 Habib, S., Khan, M.M., Abbas, F. et al. (2018). A comprehensive study of implemented international standards, technical challenges, impacts and prospects for electric vehicles. *IEEE Access* 6: 13866–13890.

34 Tu, H., Feng, H., Srdic, S., and Lukic, S. (2019). Extreme fast charging of electric vehicles: a technology overview. *IEEE Transactions on Transportation Electrification* 5 (4): 861–878.

35 IEC (2017). Electric vehicle conductive charging system. International Standard IEC 61851, 19 June 2017.

36 Transport & Environment (2020). Recharge EU: how many charge points will EU countries need by 2030. https://www.transportenvironment.org/publications/recharge-eu-how-many-charge-points-will-eu-countries-need-2030 (accessed 11 July 2023).

37 Wang, L., Qin, Z., Slangen, T. et al. (2021). Grid impact of electric vehicle fast charging stations: trends, standards, issues and mitigation measures – an overview. *IEEE Open Journal of Power Electronics* 2: 56–74.

38 Nicholas, M. (2019). Estimating electric vehicle charging infrastructure costs across major U.S. metropolitan areas. International Council on Clean Transportation.

39 Ravi, S. and Aziz, M. (2022). Utilization of electric vehicles for vehicle-to-grid services: progress and perspectives. *Energies* 15 (2): 589.

40 Loschan, C., Schwabeneder, D., Lettner, G., and Auer, H. (2023). Flexibility potential of aggregated electric vehicle fleets to reduce transmission congestions

and redispatch needs: a case study in Austria. *International Journal of Electrical Power & Energy Systems* 146: 108802.
41 IRENA (2019). Innovation landscape for a renewable-powered future: Solutions to integrate variable renewables. International Renewable Energy Agency, Abu Dhabi. ISBN: 978-92-9260-111.
42 Sevdari, K., Calearo, L., Andersen, P., and Marinelli, M. (2022). Ancillary services and electric vehicles: an overview from charging clusters and chargers technology perspectives. *Renewable and Sustainable Energy Reviews* 167: 112666.
43 Sirnik, I., Sluijsmans, J., Oudes, D., and Stremke, S. (2023). Circularity and landscape experience of agrivoltaics: a systematic review of literature and built systems. *Renewable and Sustainable Energy Reviews* 178: 113250.
44 Ghosh, S., Rahman, S., and Pipattanasomporn, M. (2017). Distribution voltage regulation through active power curtailment with PV inverters and solar generation forecasts. *IEEE Transactions on Sustainable Energy* 8 (1): 13–22.
45 Abad, M. and Ma, J. (2021). Photovoltaic hosting capacity sensitivity to active distribution network management. *IEEE Transactions on Power Systems* 36 (1): 107–117.
46 Gaëtan Masson, Jose Ignacio Briano, Maria Jesus Baez (2016). IEA PVPS TASK1, Review and Analysis of PV Self-Consumption Policies. ISBN 978-3-906042-33-6.
47 IBM (2022). What is data management? https://www.ibm.com/topics/data-management (accessed 3 July 2023).
48 European Smart Grids Task Force Expert Group 3 (2019). Demand side flexibility – perceived barriers and proposed recommendations. https://energy.ec.europa.eu/system/files/2019-05/eg3_final_report_demand_side_flexiblity_2019.04.15_0.pdf.
49 Regulation (EU) 2016/679, 2016.
50 European Commission Proposal for an ePrivacy regulation. https://digital-strategy.ec.europa.eu/en/policies/eprivacy-regulation (Accessed 3 July 2023).
51 Lee, D. and Cheng, C. (2015). Energy savings by energy management systems: a review. *Renewable and Sustainable Energy Reviews* 56: 760–777.
52 Upadhyay, D. and Sampalli, S. (2020). SCADA (supervisory control and data acquisition) systems: vulnerability assessment and security recommendations. *Computers & Security* 89: 101666.
53 Alphonsus, E.R. and Abdullah, M.O. (2016). A review on the applications of programmable logic controllers (PLCs). *Renewable and Sustainable Energy Reviews* 60: 1185–1205.
54 Holecko, P. (2011). Overview of distributed control systems formalisms. *Advances in Electrical and Electronic Engineering* 7 (1–2): 253–256.
55 Zhang, Y., Huang, T., and Bompard, E.F. (2018). Big data analytics in smart grids: a review. *Energy Informatics* 1: 1–24.
56 Michigan State University Online (2019). 4 types of data analytics and how to apply them. https://www.michiganstateuniversityonline.com/resources/business-analytics/types-of-data-analytics-and-how-to-apply-them/ (accessed 3 July 2023).

57 Norvig, P. and Russel, S.J. (2020). *Artificial Intelligence: A Modern Approach*. Pearson.

58 Murphy, K.P. (2022). *Probabilistic Machine Learning: An Introduction*. MIT Press.

59 Graditi, G., Buonanno, A., Caliano, M. et al. (2023). Machine learning applications for renewable-based energy systems. In: *Advances in Artificial Intelligence for Renewable Energy Systems and Energy Autonomy* (ed. M. Manshahia, V. Kharchenko, G. Weber, and P. Vasant), 177–198. Cham, Switzerland: EAI/Springer Innovations in Communication and Computing.

60 Buonanno, A., Caliano, M., Pontecorvo, A. et al. (2022). Global vs. local models for short-term electricity demand prediction in a residential/lodging scenario. *Energies* 15 (6): 2037.

61 Buonanno, A., Caputo, G., Balog, I. et al. (2023). Combined machine learning and weather models for photovoltaic production forecasting in microgrid systems. *2023 International Conference on Clean Electrical Power (ICCEP)*, Terrasini, Palermo, Italy (27–29 June 2023). 216–222. IEEE.

62 Atrigna, M., Buonanno, A., Carli, R. et al. (2023). A machine learning approach to fault prediction of power distribution grids under heatwaves. *IEEE Transactions on Industry Applications* 59 (4): 4835–4845.

63 O'Connel, K., Lim, C., Pallaras, L. et al. (2023). Developments in the regulation of artificial intelligence. https://www.kwm.com/global/en/insights/latest-thinking/developments-in-the-regulation-of-artificial-intelligence.html (accessed 27 July 2023).

64 European Parliament Press Room (2023). MEPs ready to negotiate first-ever rules for safe and transparent AI. https://www.europarl.europa.eu/news/en/press-room/20230609IPR96212/meps-ready-to-negotiate-first-ever-rules-for-safe-and-transparent-ai (accessed 27 July 2023).

65 Motlagh, N., Mohammadrezaei, M., Hunt, J., and Zakeri, B. (2020). Internet of things (IoT) and the energy sector. *Energies* 13: 494.

66 Kochański, M., Korczak, K., and Skoczkowski, T. (2020). Technology innovation system analysis of electricity smart metering in the European union. *Energies* 13: 916.

67 Dileep, G. (2020). A survey on smart grid technologies and applications. *Renewable Energy* 146: 2589–2625.

68 Song, E., Fitzpatrick, G., and Lee, K. (2017). Smart sensors and standard-based interoperability in smart grids. *IEEE Sensors Journal* 17 (23): 7723–7730.

69 Morales-Velazquez, L., Romero-Troncoso, R.d.J., Herrera-Ruiz, G. et al. (2017). Smart sensor network for power quality monitoring in electrical installations. *Measurement* 103: 133–142.

70 Naji, N., Riduan Abid, M., Benhaddou, D., and Krami, N. (2020). Context-aware wireless sensor networks for smart building energy management system. *Information* 11: 530.

71 Alonso, M., Amaris, H., Alcala, D., and Florez, R. (2020). Smart sensors for smart grid reliability. *Sensors* 20: 2187.

72 Buonanno, A., Di Gennaro, G., Graditi, G. et al. (2023). Fusion of energy sensors with missing values. *Applied Intelligence* 53: 23613–23627.

73 Alablani, I. and Alenazi, M. (2020). EDTD-SC: an IoT sensor deployment strategy for smart cities. *Sensors* 20: 7191.

74 Atiquzzaman, M., Gaedke, M., and Noura, M. (2019). Interoperability in internet of things: taxonomies and open challenges. *Mobile Networks and Applications* 24: 796–809.

75 Fan, Z., Kulkarni, P., Gormus, S. et al. (2013). Smart grid communications: overview of research challenges, solutions, and standardization activities. *IEEE Communications Surveys & Tutorials* 15 (1): 21–38.

76 Kuzlu, M., Pipattanasomporn, M., and Rahman, S. (2014). Communication network requirements for major smart grid applications in HAN, NAN and WAN. *Computer Networks* 67: 74–88.

77 Domingues, P., Carreira, P., Vieira, R., and Kastner, W. (2016). Building automation systems: concepts and technology review. *Computer Standards & Interfaces* 45: 1–12.

78 Iqbal, A. and Olariu, S. (2021). A survey of enabling technologies for smart communities. *Smart Cities* 4 (1): 54–77.

79 Mishra, N., Kumar, V., and Bhardwaj, G. (2019). Role of cloud computing in smart grid. *International Conference on Automation, Computational and Technology Management (ICACTM)*, London, UK (24–26 April 2019). 252–255. IEEE.

80 Hamilton, E. (2018). What is edge computing: the network edge explained. https://www.cloudwards.net/what-is-edge-computing/ (accessed 18 April 2023).

81 Alwakeel, A.M. (2021). An overview of fog computing and edge computing security. *Sensors* 21 (24): 8226.

82 Blueprint Energy Solutions GmbH (2019). Study on cybersecurity in the energy sector of the energy community. https://www.energy-community.org/dam/jcr:db8e479d-b423-40c9-9ff9-998c7d9045ef/Blueprint_cyber_122019.pdf (accessed 3 July 2023).

83 EECSP (2017). Cyber security in the energy sector. https://energy.ec.europa.eu/document/download/723c8ada-3280-43e5-9d13-db131cd057df_en?filename=eecsp_report_final.pdf (accessed 03 July 2023).

84 Hayes, A. (2023). Blockchain facts: what is it, how it works, and how it can be used. https://www.investopedia.com/terms/b/blockchain.asp (accessed 3 July 2023).

85 Antonopoulos, A.M. (2017). *Mastering Bitcoin*. O'Reilly.

86 Franza, F.M. (2019). The blockchain as enabling technology for energy communities. http://www.foroeuropa.it/documenti/rivista/TesiFrancescoMariaFranza.pdf (accessed 03 July 2023).

87 Andoni, M., Robu, V., Flynn, D. et al. (2019). Blockchain technology in the energy sector: a systematic review of challenges and opportunities. *Renewable and Sustainable Energy Reviews* 143–174.

88 Pop, C., Cioara, T., Antal, M. et al. (2018). Blockchain based decentralized management of demand response programs in smart energy grids. *Sensors* 18: 162.

4

Actors, Business Models, and Key Issues for the Implementation of Integrated Local Energy Communities

Martina Caliano[1], Alberto Borghetti[2], Amedeo Buonanno[1], Marialaura Di Somma[1], Salvatore Fabozzi[1], Giorgio Graditi[1], Carlo Alberto Nucci[2], Christina Papadimitriou[3], and Peter Richardson[4]

[1] *Italian National Agency for New Technologies, Energy and Sustainable Economic Development, Lungotevere Thaon di Revel, 76, 00196, Rome, Italy*
[2] *University of Bologna, Department of Electrical, Electronic and Information Engineering, Viale Risorgimento, 2, 40136, Bologna, Italy*
[3] *Eindhoven Technical University, Department of Electrical Engineering, Groene Loper 3, 5612 AE, Eindhoven, the Netherlands*
[4] *EPRI Europe DAC, Belfield Office Park, Beech Hill Road, D04 V2N9, Dublin, Ireland*

4.1 Introduction

The concept of integrated local energy community (ILEC) refers to energy users deciding to make common choices in terms of satisfying their energy needs in order to maximize the benefits deriving from this collegial approach, thanks to the implementation of multiple energy carrier-related technologies, energy storage, and the optimized management of energy flows [1]. Representing an integrated energy system at the local level, the ILEC inherently requires the coordinated actions and synergetic efforts from multiple actors in the energy value chain for an effective deployment. The identification of their roles and responsibilities as well as their main objectives related to ILEC, is as challenging as it is crucial to identify the needed interactions and, in some cases, the conflicting interests. Each should be properly addressed for fostering the transition from the current energy system paradigm to the future system where ILECs play a central role. These actors can be classified into four main categories, which are (i) *actors in the distributed system* such as end-users or aggregators; (ii) *operators and utilities*, such as distribution system operators (DSOs) or balance responsible parties (BRPs); (iii) *framework setters* such as policymakers, national governments, and local authorities; and (iv) *suppliers and supporters* such as energy and technology suppliers or banks, private investors, and energy cooperatives. In the newly established ILECs context, several actors' interests may vary or substantially differ with respect to their interests, as they appear in the context of the current energy system. The evolution of the roles, interests, and responsibilities that potentially affect the key actors because of their involvement in this new energy

Integrated Local Energy Communities: From Concepts and Enabling Conditions to Optimal Planning and Operation, First Edition. Edited by Marialaura Di Somma, Christina Papadimitriou, Giorgio Graditi, and Koen Kok.

paradigm represented by ILECs, in some cases, may lead to conflicts between the interests of individual actors that were not in place in the current energy system. This may entail the need for the actors to adjust their roles and objectives to make them compatible with the new energy structure and operation proposed by the ILECs.

Enabling conditions for deployment of ILECs are related to technological, environmental, and socioeconomic aspects, energy policy, and regulation framework, which are constantly evolving, affected by institutional changes, energy prices, and the energy development of the local territories. In detail, in implementing ILECs, each of the above aspects should be considered. They can act indeed either as limitations in favor of conventional centralized energy supply systems or as boosters for encouraging ILEC establishment. For instance, key issues related to site conditions, existing support mechanisms, grid connection status, initial investment costs, trust and motivation, or conflicting interest of involved parties are crucial and may represent a barrier to ILECs development. On the other hand, ILECs can play an active role in increasing energy efficiency from both the supply and demand sides, reducing carbon emissions through increasing the use of renewable energy available *in loco*, and putting in practice concepts such as energy democracy and alleviation of energy poverty that are within the Sustainable Development Goal (SDG) 7 of the "2030 Agenda for sustainable development" of United Nations [2], aiming at "*ensuring access to affordable, reliable, sustainable, and modern energy for all*" [3].

Creating sustainable business models (BMs) focusing on the value proposition of these initiatives is also a key for guaranteeing a cost-effective implementation of ILECs and the different actors involved. It is worth mentioning that the literature on energy community BMs is still weak [4] since energy communities were formally defined only recently in European regulation through the Clean Energy for All Europeans Package [5] and the two EU Directives: the revised Renewable Energy Directive (RED II) [6] and the Internal Electricity Market Directive (EMD II) [7], where the renewable energy community (REC) and the citizen energy community (CEC), respectively, were defined for the first time. Moreover, the few BMs present in the literature mainly focus on electricity as the main energy carrier, since it is a common element in both EU directives, whereas the multicarrier aspect is missing [8]. Therefore, to foster this innovative concept of energy community based on the integration and interplay of multiple energy carriers, new community-centered BMs, accounting for the multicarrier aspects, need to be established. These BMs should demonstrate the achievement of tangible economic benefits through the ILEC for a specific actor that in existing traditional schemes would never be achieved.

The aim of this chapter is to discuss all the topics above for analyzing the key enabling conditions for the establishment of this innovative energy paradigm. A detailed mapping of the main actors is presented in Section 4.2, with a critical analysis of their roles and interactions in the context ILECs. The key issues for the implementation of ILECs are categorized as technological, socioeconomic, environmental, and institutional and are discussed in Section 4.3. Innovative BMs for ILECs are outlined in Section 4.4, and a summary of barriers, benefits, and policy implications for ILECs implementation is presented in Section 4.5.

4.2 Actors' Roles and Interactions Within ILECs

Actors represent a key element in the establishment, operation, and success of an ILEC, with their roles and interactions depending on the local conditions and crucial for achieving the goals of an ILEC. Therefore, the mission of an ILEC can be defined by means of the goals and interests of its members which, although variable from case to case, in general include achieving certain environmental, economic, or social goals. Considering this, actors' roles and interactions should be clearly defined to avoid, or at least manage, possible conflicting interests that can arise and that could introduce numerous challenges in terms of the design and operation of an ILEC. The actors formally constituting an ILEC can be members within the community, such as a natural person, a local authority, a school or university, and a nongovernmental organization, or external actors outside the community, such as technology companies, operators and utilities, and governments. Figure 4.1 presents the main key actors that may be involved in ILECs, grouped based on the stakeholder characterization.

The type of actors involved in an ILEC depends on several factors connected to the specific needs and objectives of the ILEC, such as the energy vectors involved (electricity, heat, gas, etc.), territoriality, and the presence of storages and EVs. In the context of an ILEC, an actor may play one or more roles, and the ILEC itself can be seen as a new energy actor assuming one or more roles depending on the needs of the ILEC itself, such as acting as aggregator, local heat/cold distribution system operator

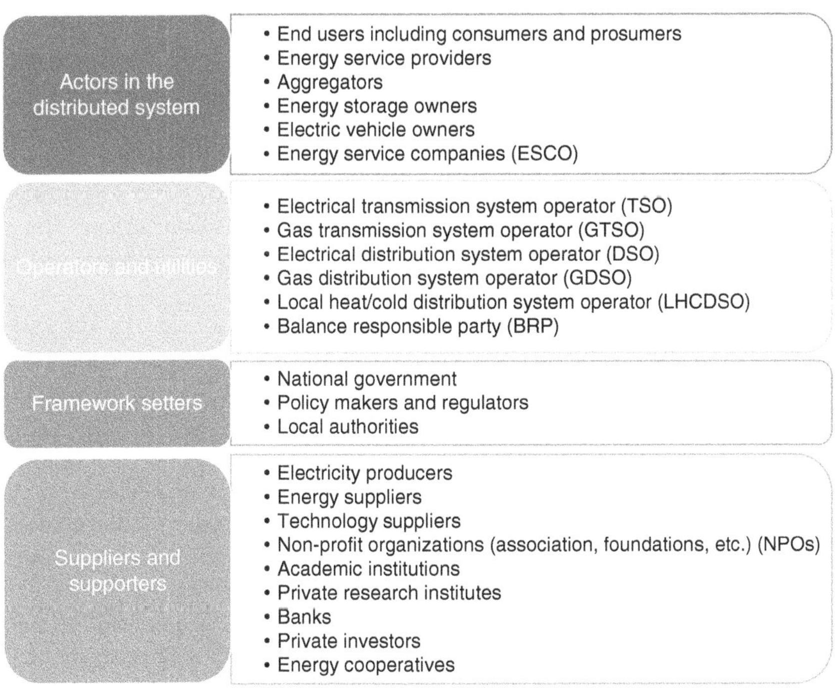

Figure 4.1 Main key actors grouping based on stakeholder characterization.

(LHCDSO), and local gas transmission system operator (GTSO). Therefore, it may happen that within the new energy context represented by the ILEC, the role and responsibilities of a specific actor undergo a change compared to those covered in the context of the current energy system.

The aim of this section is to highlight the evolution of roles covered by actors from the current energy system to an ILEC framework.

The **end-user** has a central role in an ILEC, acting as a simple consumer being the beneficiary of energy commodities or services provided by other actors, or as an active customer, or prosumer, owning its own energy generation units. When acting as a prosumer, it can also assume the role of an energy service provider by sharing or trading energy exceeding its own needs with other members of the ILEC through a peer-to-peer trading platform [9].

The role of the **energy service provider** can be played by any actor within the ILEC dealing with the generation, distribution, storage, and supply of energy or with other energy services like installation and maintenance of equipment or aggregation. The ILEC itself can act as an energy service provider.

In the context of ILECs, the role of the **aggregator** can be covered by an external energy service provider or by the ILEC itself that can aggregate the energy capacity resources, consisting of electricity, heat, gas, etc., from the local community to the grids and seek to maximize the value of collected flexibility in the markets.

End-users can also own energy storage assets and electric vehicles (EVs), acting in these cases as **energy storage owners** and **EV owners**, respectively, potentially contributing to the demand-side management strategies within the ILEC. The ILEC itself can act as an energy storage owner at the community level managing the mismatch between supply and demand of energy (electric, thermal, hydrogen, etc.) within the ILEC.

Within the context of ILECs, additional roles can be identified for **ESCOs** with respect to the current energy system. ESCOs can act by integrating the role of aggregators by aggregating flexible customers and buildings to include the valorization of flexibility through demand response services in the energy efficiency projects.

The main interests for the *actors in the distributed system* are summarized in Table 4.1.

Operators and utilities are legal persons responsible for transmission (transmission system operator [**TSO**], **GTSO**), distribution (DSO, gas distribution system operator [GDSO], LHCDSO), and balancing injections and withdrawals to and from the energy networks (BRP).

DSOs are recognized as having a key role in enabling ILECs proper development [10–12], potentially also acting to support the ILECs' mission to provide environmental, economic, or social community benefits to their members [10]. In detail, according to the RED II [6] and EMD II [7] directives, DSOs are specifically called upon to cooperate with ILECs to facilitate electricity transfers between their members. In the context of ILECs, the ILEC itself can have a similar role as a **DSO/GDSO** by operating its own local networks and can play the role of **LHCDSO** managing the district heating network by interacting with the thermal prosumers within the ILEC, who can feed their thermal generation into the local thermal network.

Table 4.1 Main interests for the *actors in the distributed system* in the context of ILECs.

End-users including consumers and prosumers

- At the private level, the main interests include
  - Use of **local** energy sources, and **clean** and **not expensive energy**
  - **Reduction of energy costs**
  - Providing and **sharing** local energy
  - Possibility to sell the excess of energy generated **reducing the bills** (active customer)
- At the system level, the main interests include
  - **Sale of excess** energy and purchase of missing energy
  - **Increasing RES** and consequent reduction of emissions
  - **Energy independence**
  - **Security of energy supply**
  - **Improve resilience and affordability**
  - **Better quality and reliability of supply up** to a most coordinated integration of renewable sources
  - **Reducing vulnerability and energy poverty**, thus reducing energy supply costs and consumption by encouraging efficiency.

Energy service provider

- The main interest is to provide energy services to make **profit** or **increase benefits** for the ILEC.

Aggregators

- The main interest deals with the creation of business models for generating **profit and maximizing the value of flexibility** in the markets (both with capacity and with energy).

Energy storage owners

- The main interest is to **benefit community users** by increasing self-consumption of the active customers and improving the flexibility.

Electric vehicle owners

- One of the main interests concerns the possibility to **take economic advantages** deriving from the optimized charging/discharging strategies, thanks to a proper energy management system.

ESCO

- At the private level, the main interests are mostly related to the possibility of **profit** deriving from the implementation of energy efficiency measures, including flexibility measures, and **optimized management of local generation**. At the system level, they are mostly related to the **increasing role in the activities of energy efficiency improvement** and in the **management of local generation**.

BRPs provide financial security and identify balance accountability with the market operator entitling the end-users within the ILEC to operate in the market. In this way, the BRP can access the aggregated flexibility that the ILEC is able to offer coming both from its flexible consumption and generation [13, 14]. The ILEC itself can act as a BPR.

The main interests for the *operators and utilities* are summarized in Table 4.2.

Government, policymakers, and regulators are responsible for transposing the RED II [6] and EMD II [7] directives into national contexts. They have the role of setting out the actions needed to engage and enable citizens in energy projects.

Table 4.2 Main interests for *operators and utilities* in the context of ILECs.

Electrical Transmission System Operator (TSO) – Gas Transmission System Operator (GTSO)
• Increase in the share of nontariff revenues • Improve the accuracy of tariffs and budgets • Optimize investment expenditures • Maintain the continuity of the investment process • Reduce the operating costs
Electrical Distribution System Operator (DSO) – Gas Distribution System Operator (GDSO)
• The **development of distributed generation** has and will have a growing impact in the functioning of the DSO/GDSO. The emergence and development of generating units leads to a greater fluctuation in power and gas flow. The growing number of distributed energy sources causes two-way energy flows, which affect the management and operation of the networks. The core responsibilities of DSOs/GDSOs, security of supply and quality of service, will remain the same in the future. To ensure them, DSOs/GDSOs must evolve, thus becoming increasingly active network managers and including the use of flexibility services provided by users of the distribution network.
Local Heat/Cold Distribution System Operator (LHCDSO)
• The main interest concerns with the **management of the thermal flows** within the ILEC to benefit the community users.
Balance Responsible Party (BRP)
• The BRP is interested in being able to count on the aggregated flexibility offered by the ILEC for balancing the energy production and consumption within the networks.

This provides ILECs with technical and financial support by establishing the relevant support schemes and setting the criteria for the regulation of economic aspects (i.e. energy costs, energy prices, taxes, and duties) related to collective self-consumption or energy sharing within the ILECs.

Local authorities play a fundamental role in the context of ILECs being able to simultaneously act as one or more of founders, funders, promoters, end-users, and/or authorization. In some European countries, the local authorities are crucial for the establishment and operation of ILECs, acting as a real catalyst of all processes [4, 15, 16]. They can support the ILECs by provisioning loans, establishing their own financial incentives, and helping ILECs to secure third-party funding. Moreover, local authorities can target and support groups, share knowledge, nurture ideas, and keep stakeholders informed of relevant policy and funding opportunities [15, 16].

The main interests for the *framework setters* are summarized in Table 4.3.

Supporters and suppliers include all the actors that, with their specific characteristics, act by supporting the several phases of an ILEC, from design to operation, supporting them from social (such as **nonprofit organizations** [NPOs] and **energy cooperatives**), economic (such as **banks** and **private investors**) and technical points of view (such as **academic institutions** and **private research institutes**), and by proving energy (such as **electricity producers** and **energy suppliers**), or technologies (**technologies suppliers**) to the ILEC. The role of

Table 4.3 Main interests for *framework setters* in the context of ILECs.

National government, and policymakers and regulators

- The main private interest consists in ensuring **affordable energy supply** for all the users.
- At the system level, this interest translates into the following:
  - Ensure **affordable energy supply** for all the users
  - Promotion of a **sustainable supply system**
  - Transition to a **low carbon energy system**
  - High-**energy security**

Local authorities

- Local authorities have as the main objective the public benefit. Therefore, its main private interests and main interests at the system level are coincident and are as follows:
  - Obtain benefits from community services
  - **Reduction of cost** of energy carriers and of the energy expense through the usage of local, clean, and not expensive energy
  - **Exploit the local and clean energy sources** to empower the image of territory and favor the development of the tourism sector
  - Possibility to sell the excess of electricity **reducing the bills for public buildings** and facilities
  - Reduce the emissions
  - Increase the energy independence
  - Security of energy supply
  - Improve the resilience
  - Alleviate the energy poverty
  - Energy education and awareness raising

supporter can be covered by other actors, especially local authorities and end-users, for example, through the establishment of energy cooperatives.

The main interests for the *supporters and suppliers* are summarized in Table 4.4.

Interaction and cooperation among actors play a crucial role in achieving the objectives of an ILEC. Therefore, when forming an ILEC, individual actors should maintain their rights, goals, and obligations as individual members, but at the same time, they should also be able to act as a collective entity. The main interactions in the context of ILECs among the actors presented above are discussed in Table 4.5.

4.3 Key Issues for the Implementation of ILECs

During their implementation and adaptation, ILECs must deal with several technological, socioeconomic, environmental, and institutional issues, such as site conditions, support mechanisms, issues related to the grid connection, initial costs and financing, trust and motivation of citizens, and spatial issues. These key issues can act either as enabling factors encouraging such systems establishment and growth or as barriers limiting their deployment in favor of centralized systems [1]. In general, they are constantly changing according to several factors, such as institutional changes and energy prices and strongly depend on the energy development of the local territories.

4.3.2.3 Increasing Community Engagement and Willingness to Pay

The role of consumers is evolving from a passive to an active involvement since now they can own generation and flexibility assets that, with appropriate incentives, can serve as valuable resources within the ILECs. Relying solely on economic incentives for creating a community may not be enough since the users could lack appropriate levels of engagement or awareness. Therefore, the social and behavioral factors, such as the sense of community, awareness of benefits, trust levels, and initiatives to increase user commitment and awareness, are fundamental. These factors should be considered when implementing renewable energy technologies within an ILEC, and different technologies may require different approaches.

Moreover, there is a real and growing interest among many consumers for an environmentally friendly energy production. Many people are willing to pay more for renewable energy than for fossil fuel energy, if they know that the funds obtained through the increases will be invested in green energy and will finance infrastructure improving local energy security or if the activities will contribute to the reduction of environmental pollution [32–35].

Investments in renewable energy and new technologies have long-term social benefits but typically have high costs. To implement these projects effectively, social awareness, behavior, and acceptability must be considered. Research should be conducted among potential stakeholders to determine acceptable investments. Raising energy awareness and providing adequate information can turn passive consumers into responsible and active citizens who will participate in shaping the energy and environmental future. Informed end-users are more willing to pay a higher price for energy that brings additional benefits and environmental improvements.

4.3.3 Key Environmental Issues

Environmental key issues in the context of ILECs refer to some specific challenges and concerns related to energy production, distribution, and consumption within the community. The key environmental issues related to the creation of an ILEC are one of the most typical representations of how key issues, in some cases, can act both as a barrier and as a driving force to the implementation of this type of new energy system. If, on the one hand, the ILECs reduce emissions thanks to the higher use of local RES and allow for a better management of the energy sources involved, on the other hand, the local territories development that may derive can lead to an increase in waste and local emissions, especially in the construction and maintenance phases of the ILECs. Therefore, environmental aspects must be constantly monitored to ensure compliance with local targets and ensure good air quality.

A typical example of a measure to reduce some of the environmental concerns in the context of ILECs is the creation of a district heating network serving the community to replace individual boilers located at individual housing units and/or to have less dependence on imported gas. These networks can be fed both with traditional systems (e.g. gas-fired boilers) or with energy conversion technologies from renewable sources (e.g. heat pumps for fourth-generation district heating systems, and solar collectors), or cogeneration systems.

Table 4.4 Main interests for *supporters and suppliers* in the context of ILECs.

Electricity producers
• The main interest consists in **investing** in new power generation systems, including both conventional and RES systems. They could be interested in investing in good interaction with other energy carriers as well to optimize their plants operation.
Energy suppliers
• At the private level, the energy suppliers are interested in **maximizing their profit.**
• At the system level, there is an interest for the energy suppliers to acquire energy from RES and include these into their portfolio.
Technology suppliers
• The main interest consists in **selling innovative technologies**.
Nonprofit organizations (NPOs)
• The main interests are connected to **interventions having social value** acting to limit the social inequality that results from energy poverty.
Academic institutions and private research institutes
• The main interests consist in **providing knowledge and supporting ILEC** in the implementation of innovative solutions.
Banks and private investors
• The main objective is to find the **maximum economic revenues** and to have an active role (or visibility) in sustainable investments.
Energy cooperative
• Main interests in the ILEC context level are mainly related to the **benefits of the local community**. Energy cooperatives exploit local energy sources that would generate incomes (after an initial investment) and then reinvested for the improvement of the community. In addition, energy cooperatives offer education, training, and information programs for the citizens.

In this section, the main barriers and driving forces for the deployment of ILECs, as listed in Figure 4.2, are presented and discussed.

4.3.1 Key Technological Issues

ILECs are localized initiatives where individuals, businesses, or communities come together to generate, consume, and share energy. While ILECs offer numerous benefits, there are several technological key issues associated with their implementation. As detailed in Sections 4.3.1.1–4.3.1.7, these issues include integration with the grid, the need for an energy management system (EMS), mainly for the optimal scheduling of the available energy storage systems, and the need for specific smart metering infrastructure. The initial technological choices in the implementation of ILECs should consider the aspects of scalability (for the development of the community in time), the possibility of changes in the regulatory and policy frameworks, and the opportunity to provide ancillary services to the power distribution system. In any case, addressing the key technological issues requires collaboration among various stakeholders, including technology providers, policymakers, researchers, and local communities. Continuous innovation, investment in research and development,

Table 4.5 Main interactions among the actors in the context of ILECs.

Main interactions among *Actors in the Distributed System*	Main interactions among *Actors in the Distributed System* and *Operators and Utilities*	Main interactions among *Actors in the Distributed System* and *Framework Setters*	Main interactions among *Actors in the Distributed System* and *Suppliers and Suppliers and Supporters*
➢ **End-user**: ● receive energy services from **energy service providers;** ● can interact with the **aggregator** (external or the ILEC itself) that could remotely control the consumer energy use and remunerate users for it; ● acting as an energy community could procure services from **energy storage owners** (external or the ILEC itself) to increase their flexibility; ● acting as active customers, interacting with **EV owners** for selling, and purchasing electricity to/from the local network within the ILEC; ● can be beneficiaries of **ESCO** activities. ➢ **Aggregators** can interact with **end-users** and **ESCOs** to provide the best bidding strategies to customers. ➢ **ESCOs:** ● can provide **end-users** with energy contracts for energy efficiency measures; ● can interact with **aggregators**.	➢ **End-user**: ● acting as active customers, interact with **LHCDSO** (ILEC itself) to manage thermal energy; ● could interact with the **BRP** (external or the ILEC itself) to receive support for balancing services between the production and consumption of the community in the local markets. ➢ **Aggregators** interact with distribution operators (**DSO** and **GDSO**) being the intermediary between the network operators and end-users. ➢ **Energy storage owners:** ● interact with the **TSO** or **DSO**, depending on which point of the network they are connected to provide ancillary services to the system operators (electricity only). In the case of gas network, the storage owner coincides with the **GTSO;** ● can have bilateral agreements with **DSO** and **LHCDSO** to buy the excess of electricity or thermal energy; ● can interact with **BRP** by supporting and providing the balance of the system within the ILEC. ➢ **ESCOs** can interact with **DSO** (**GDSO**) for grid connection.	➢ **End-user** in some countries could require authorizations by the **local authorities** for the realization of new plants that have an impact on the territory. ➢ **Aggregators** can receive regulations from **government, policymakers, and regulators**. ➢ **ESCOs. Local authorities** can be a customer of an ESCO to receive interventions of energy efficiency and integrated energy services.	➢ **End-user**: ● receive energy from the **energy suppliers** who manage the administrative and commercial aspects, such as the final price and sending the bill home; ● can buy or rent from **technology suppliers** energy technologies, and ICT systems or applications; ● can be supported by **nonprofit organizations, academic institutions and research actors, and banks and private investors** for social, technical, and economic issues, respectively; ● can take part in **energy cooperatives**. ➢ **Energy Storage Owners** can have bilateral agreements with **electricity producers** and **suppliers** and support producers from RES to provide uninterrupted power supply to the consumers. ➢ **ESCOs:** ● interact with **energy suppliers;** ● can receive financing from **banks and/or private investors**.

(Continued)

Table 4.5 (Continued)

Main interactions among *Operators and Utilities*	Main interactions among *Operators and Utilities* and *Framework Setters*	Main interactions among *Framework Setters*	Main interactions among *Framework Setters* and Suppliers and Supporters	Main interactions among *Suppliers and Supporters*
➢ **TSO and GTSO** interact with **DSO (GDSO)** for: * developing rules adopted by the distribution system operator for the procurement of services necessary for the efficient, reliable, and secure operations of the distribution system * establishing the technical requirements of all qualified market participants in the markets * the effective participation of market participants connected to their grid in retail, wholesale, and balancing markets * exchanging all necessary information to guarantee the optimal use of resources, to ensure the secure and efficient operation of the system, and to facilitate market development.	➢ **TSO and GTSO** interact with **government, policymakers, and regulators** for: * approval established by TSO specifications for the nonfrequency ancillary services, ensuring nondiscriminatory participation of all the market participants (e.g. RES, DR, energy storage facilities, service aggregators) * approval established by TSO procedures for nondiscriminatory connection of new generating installations and energy storage facilities to the transmission system * approval network development plan based on existing and forecast supply. ➢ **DSO** and **GDSO** can interact with **government, policymakers, and regulators** to establish 1) responsibility: * on network use (cost sharing, e.g. distribution fees) 2) * to ensure the secure and reliable operation of distribution network (e.g. participation in the ancillary services market).	➢ **Local authorities** enforce the laws and regulations defined by **governments, policymakers, and regulators**.	➢ **Local authorities:** ● give authorizations to **NPOs** for building new RES plants; ● can interact with **research actors** for the promotion of ILECs; ● can collaborate with **energy cooperatives** serving the citizens.	➢ In general, *suppliers and supporters* can interact with each other, bringing together and exploiting their specific peculiarities, to support the establishment, development, and all phases of an ILEC.

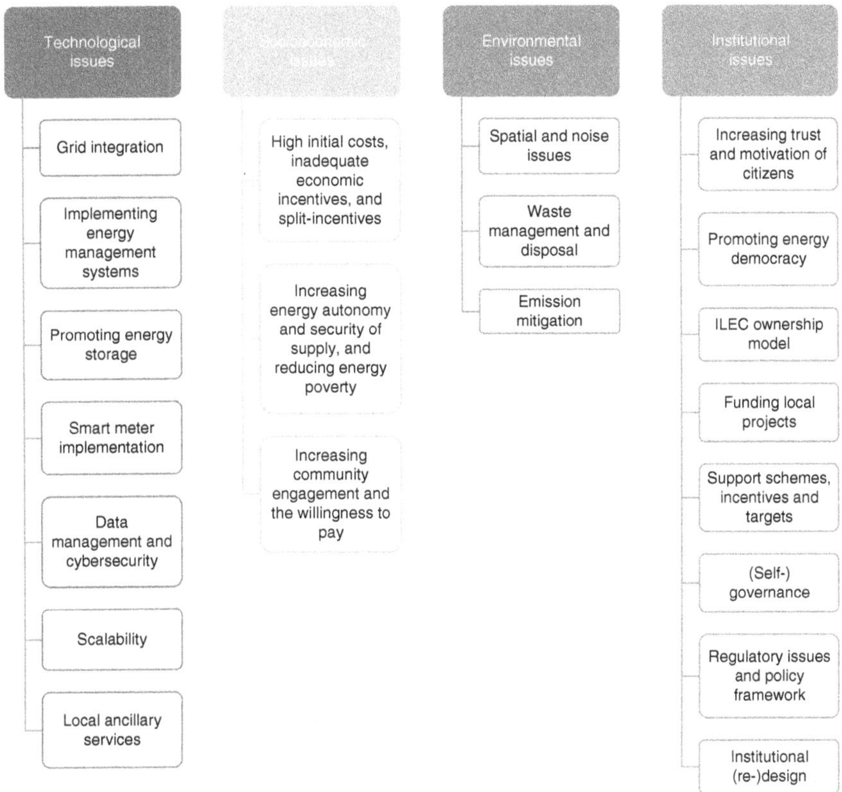

Figure 4.2 Main key issues affecting ILECs deployment.

and supportive policy environments are essential for the successful implementation and growth of ILECs.

4.3.1.1 Grid Integration

ILECs face challenges when integrating their renewable energy generation into the existing power grid. The random/intermittent nature of certain renewable energy source (RES), such as solar and wind, requires careful planning and operation to ensure a stable and reliable electricity supply. Moreover, significant uncertainty also affects the load profiles of the ILEC participants. The specific characteristics and limitations of the distribution network, its topology, and the locations where the ILEC participants are connected can hinder the integration and the expected advantages for the community. Often, the main limitations of the distribution network are associated with bus voltage limits and maximum branch current limits (e.g. [17]).

4.3.1.2 Implementing Energy Management Systems

The optimal operation of ILECs requires the implementation of an EMS to ensure the proper management of the generation and load resources, including storage systems.

The optimal scheduling of the controllable resources of an energy community usually pursues the objective function of minimizing the overall energy procurement cost of the set of users belonging to the community on a daily time horizon, as in [18, 19]. In distributed approaches (e.g. [18]), the information that each prosumer must communicate is limited. The uncertainties in load forecasts can be managed by means of stochastic approaches and intraday corrections to the day-ahead scheduler decisions [20]. Robust optimizations [19, 21] as well as three stages of scheduling are also proposed to cope with uncertainties, namely, day-ahead, intraday, and real-time stage [22]. Short-term load forecasting is used to forecast ILECs' members load profiles, despite the difficulties due to their small size and highly random nature, as shown in [23] concerning microgrids. Robust optimization, stochastic models, and complex load forecast methods, like artificial neural networks [24], may require a computational burden that is not compliant with real-time schedulers.

4.3.1.3 Promoting Energy Storage

Storage technologies play a crucial role in addressing the intermittency of RESs, also improving the ability to manage the energy flow inside the ILEC. Storage devices can be installed at the household level, at the community level, or both, influencing ILECs operation and helping ILECs to balance peaks in demand as well as to achieve power balance. However, the cost and availability of suitable energy storage solutions, such as batteries, can be a significant technological hurdle for ILECs.

4.3.1.4 Implementation of Smart Metering

The use of smart meters (SMs) to monitor the generation and consumption of the ILEC members allows for the creation of a set of historical data which opens several opportunities such as

- helping the consumer to keep track of its consumption and, if possible, reduce it or shift it in hours in which the electricity price is lower or when it can contribute to shared energy increasing;
- providing information to the prosumer to improve the self-consumed energy quota;
- refining the estimation of shared energy and increased self-consumption deriving from the installation of new assets, i.e. renewable plants and storage systems, to perform techno-economic assessments;
- performing load short-term forecasting for scheduling algorithms.

The main advantage of the use of SMs is that the measurement acquisition process is automatic without the need of human intervention.

4.3.1.5 Data Management and Cybersecurity

ILECs rely on data collection, analysis, and sharing to optimize energy usage. However, managing the large volume of data generated and ensuring its security and privacy can be challenging. Cybersecurity measures must be in place to protect against potential cyber threats and unauthorized access to ILEC systems and data.

4.3.1.6 Scalability

As ILECs grow and more participants join, scalability becomes a critical issue. The technical infrastructure must be capable of accommodating increased energy generation, distribution, and storage capacity. Upgrading and expanding the ILEC's technological systems to meet the growing demand can be a complex task.

4.3.1.7 Local Ancillary Services

ILECs are foreseen to be capable of supporting the operation of distribution networks through local ancillary services. These services involve qualified generation and storage units, reactive power compensation devices, and demand response techniques. In some cases, ILECs may also act as DSOs, providing services to connected users and the TSO.

4.3.2 Key Socioeconomic Issues

Social issues can have a very high impact on the deployment of ILECs. Without the acceptance of local communities and citizens, the high technological, economic, and environmental potential of ILECs is not sufficient to guarantee the deployment of this new energy paradigm. Lack of social acceptance can lead to the failure of an ILEC project before it is even implemented. It is therefore important to implement strategies aimed at involving citizens and local communities. One of the first steps in this direction is to consider the main socioeconomic issues, such as the high initial costs, inadequate economic incentives, and split-incentives; increasing energy autonomy and security of supply, and reducing energy poverty; and the paradigm shift through community engagement and willingness to pay.

4.3.2.1 High Initial Costs, Inadequate Economic Incentives, and Split-Incentives

The main economic barriers to the use of RES are inadequate financial support, the high initial costs, and the cheap fossil fuels. Many governments also subsidize fossil fuels to boost economic growth and reduce income inequality [25]. Despite large efforts to promote RES, many countries worldwide, including the United States, continue to extensively use fossil fuels since there is a strong consumers' attachment to gasoline and car travel [26]. These subsidies make it harder to implement clean projects, as they increase the demand for fossil fuels and decrease the production of renewable energy. Clearly, high costs of energy investments can discourage ILEC's progress. To overcome these barriers, fossil fuel subsidies should be eliminated and replaced with incentives for RES (e.g. R&D incentives, fiscal and tax, policy with standard tariffs, grid connection, and tariff incentives) [27]. The economic incentives can create new investment opportunities, disseminate good practices and innovative technological solutions, and support the economic accessibility of different categories of end-users (e.g. private citizens, public entities, and small and medium-sized enterprises (SMEs) [28], by supporting the economic development of local territories. In Europe, financial incentives for the development of distributed energy resources are being implemented at both national and local levels, including rebates, tax credits, financing programs, and solutions that encourage investments in local energy systems. Hence, economic incentives such as financial aid and tax relief or specific programs that support energy communities are needed to foster the ILEC's development.

If the investor in the energy infrastructure does not pay the energy costs, a split incentives effect may take place. The owner may not spend to improve the energy quality and efficiency if another entity gets the benefits (e.g. energy efficiency of rented property, where the owner invests but the tenants are the beneficiary). Coordination and split-incentives issues can arise when costs and benefits of ILECs are not shared by the same actor. Transparency and fairness are therefore needed among market parties and community members to mitigate potential unfair cost–benefit allocation [29]. In this context, ILEC has the potential to serve as a clear and open platform for constructing a business strategy that considers the concerns of all community members, establishes effective communication and coordination of efforts, and offers suitable incentives for investing in adaptable solutions that enhance energy efficiency and boost the share of renewable energy.

4.3.2.2 Increasing Energy Autonomy and Security of Supply and Reducing Energy Poverty

Energy autonomy requires that an energy system meets the three following criteria: it must fulfill the energy needs of its users using energy produced from local resources; it must resolve any discrepancies between the energy supply and demand by using energy storage; and it must function independently, regardless of whether the system is connected to the grid or not [30]. ILEC's structure promotes the use of local resources, leading to increased energy autonomy and decreased dependence on external factors. However, meeting the off-grid requirement and matching supply and demand are not necessary in ILECs: it may not be cost-effective to fulfill the criteria for energy autonomy, and some ILECs may aim for energy neutrality or exporting energy instead. Nevertheless, introducing RES and utilizing demand flexibility and energy storage can make energy autonomy more attractive for ILECs. Energy autonomy can be seen as one of the main drivers if it aligns with the community's needs. It also encourages active participation from actors within ILEC and reduces energy imports.

Due to low-income, high-energy costs, and low energy efficiency of buildings, some households can have difficulty in meeting their basic energy needs [31]. The use of DER improves energy efficiency and security of energy supply by reducing the distance of energy transmission and distribution. This leads to lower costs and expenses on energy carriers and infrastructure as well as a reduction of network losses. It also increases energy independence and ensures continuity of energy supply. Local development and modernization of the grid guarantees quality parameters of electricity distribution by improving the standard of living for consumers. In this context, ILEC development contributes to reduce energy poverty and injustice. ILEC can contribute to decentralization, democratization, and sustainable local economic and social development as well as job creation.

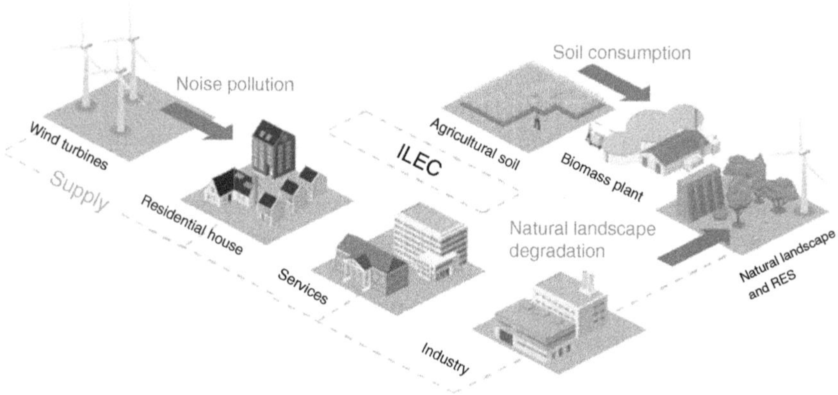

Figure 4.3 Spatial and noise key issues.

Some of the most significant environmental key issues consist of spatial and noise issues, waste management and disposal, and emission mitigation, as discussed below.

4.3.3.1 Spatial and Noise Issues

Spatial and noise issues in ILECs refer to challenges and concerns related to the physical layout and acoustic impacts of community-based energy projects. While these projects offer numerous benefits, such as increased energy independence and reduced carbon emissions, they can also give rise to spatial and noise-related problems.

Figure 4.3 depicts an overview of an ILEC and portrays the conflicts in the use of space for RES technologies that may require extensive land use. Conflicts in this regard may arise, such as in the case of wind energy technologies, in which common complaints relate to the possible negative impact on the natural landscape, local fauna, or noise pollution [36].

The main spatial issues regard the land availability, the land use conflicts, and infrastructure requirements, as discussed below.

- **Land availability:** Finding suitable land for ILEC projects can be a significant challenge. Large-scale renewable energy installations, such as wind farms or solar parks, require substantial land areas. Identifying and securing appropriate sites that balance energy production potential with environmental and social considerations can be complex, especially in densely populated areas [37].
- **Land use conflicts:** ILEC projects may encounter conflicts over land use. ILECs might face opposition from residents or stakeholders who have concerns about the visual impact, altered landscapes, or potential conflicts with existing land uses, such as agriculture or wildlife habitats. Balancing the interests of various parties and ensuring community engagement in decision-making processes is crucial [36].
- **Infrastructure requirements:** ILEC projects often require infrastructure development, such as transmission lines, substations, or energy storage facilities. These

infrastructure requirements can have spatial implications, including the need for land acquisition and potential disruption during construction. Effective planning and coordination are necessary to minimize the impact on local communities [38].

In terms of noise, the main issues are related to specific technologies, such as wind turbines, electrical equipment, and the construction and maintenance phases of the ILECs.

- **Wind turbines:** The noise generated by wind turbines is a common concern in ILECs equipped with wind farms. While modern wind turbines are designed to minimize noise, the rotating blades can still produce audible sounds, particularly at higher wind speeds. Noise pollution can affect the quality of life for nearby residents, especially if turbines are near to homes or sensitive areas. Proper setback distances and noise mitigation measures are essential to address these issues [39].
- **Electrical equipment:** Energy infrastructures, such as inverters or transformers, can generate noise during operation. While these noise levels are typically lower than those from wind turbines, they can still be perceptible in nearby residential areas. Proper design, insulation, and placement of equipment can help minimize noise disturbances [40].
- **Construction and maintenance:** Construction activities associated with ILEC projects, including erecting wind turbines or installing solar panels, can generate temporary noise during the construction phase [41]. Additionally, ongoing maintenance activities might require occasional repairs or equipment servicing, which can result in noise disturbances. Minimizing construction-related noise through efficient construction practices and scheduling maintenance activities during appropriate times can help mitigate these concerns.

Addressing spatial and noise issues in ILECs requires comprehensive planning, careful site selection, and effective engagement with local communities. Prioritizing environmental impact assessments, adopting best practices in infrastructure design and operation, and implementing appropriate noise mitigation measures are crucial to ensure the success and acceptance of energy community projects.

4.3.3.2 Waste Management and Disposal

The issue of waste management and disposal in an ILEC is a complex and critical challenge that arises from various energy production and consumption processes. It involves the proper handling, disposal, and reduction of waste generated throughout the energy sector, including the extraction, generation, transmission, and consumption of energy resources. The main waste management issues in ILECs may include the following items:

- **Renewable energy waste:** Although RES like solar and wind have lower waste production during their operational phase, there are waste management concerns during their manufacturing, installation, and decommissioning. This includes electronic waste from solar panels, turbine blades, and other components.
- **E-waste:** The energy sector heavily relies on electronic devices, equipment, and infrastructure. The disposal and recycling of electronic waste, including batteries

and obsolete technologies, is crucial to prevent environmental contamination from hazardous materials [42].

- **Waste-to-energy conversion:** Waste-to-energy technologies aim to convert waste materials into usable energy sources, reducing waste volume and producing electricity or heat. Proper waste segregation, treatment, and monitoring are essential to ensure effective waste-to-energy conversion while minimizing emissions and pollutants [43].

Efficient waste management practices are vital for minimizing environmental impacts, protecting human health, and promoting sustainable energy development. Collaboration among governments, industries, and communities is necessary to tackle these waste management challenges effectively in ILEC projects. Additionally, investing in research and technological advancements can typically help developing innovative solutions for waste reduction, recycling, and safe disposal in the community. Developing and enforcing comprehensive waste management policies and regulations is crucial to address waste-related challenges in the ILEC. This includes setting waste reduction targets, implementing proper waste disposal practices as well as encouraging research and innovation in waste management technologies. An example of an incentive for waste reduction can be to implement additional cost charges for waste above a certain weight limit, as suggested by life cycle assessment (LCA) studies [44]. Implementing a circular economy approach in the ILEC involves reducing waste generation, promoting resource efficiency, and maximizing recycling and reuse, all of which are key considerations in the LCA framework [45]. This approach aims to minimize waste, optimize material flows, and create a more sustainable and resilient energy system by assessing the environmental impact across the entire life cycle of energy production and consumption [46].

4.3.3.3 Emission Mitigation

Fossil fuel combustion for electricity and thermal generation and transportation releases pollutants such as carbon sulfur dioxide, nitrogen oxides, and particulate matter methane, which can contribute to air pollution [47]. These pollutants have adverse health effects on the community and contribute to smog formation and respiratory diseases. Shifting to cleaner energy sources like wind, solar, and hydropower can mitigate air pollution.

Figure 4.4 presents a traditional configuration of a city district and its possible evolution toward an ILEC. In the figure, there are four sections: related to power plants, the residential, the tertiary, and the industrial sectors.

Mitigating emissions in an ILEC is crucial for addressing key environmental issues such as climate change, air quality degradation, and resource depletion. Possible strategies related to emissions mitigation in energy communities are as follows:

Renewable energy integration: Integrate a higher share of RESs like solar, wind, hydro, and geothermal into the energy mix. This reduces reliance on fossil fuels and lowers greenhouse gas (GHG) emissions.

Energy efficiency measures: Implement energy-efficient technologies and practices to reduce energy consumption and emissions. This includes efficient

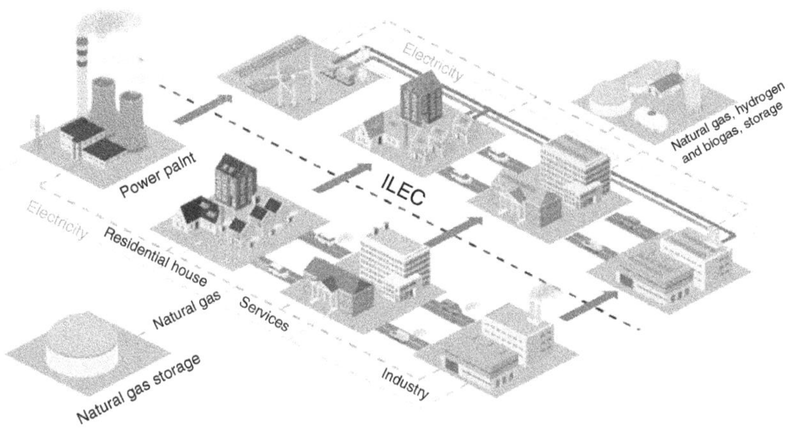

Figure 4.4 Emission mitigation from traditional system to ILEC.

building design, industrial process improvements, energy-efficient appliances, and high-efficiency energy technologies, i.e. CHP systems.

Demand response: Foster a smart grid context that optimizes energy distribution and allows for efficient demand response mechanisms. This reduces the need for peak power generation from fossil fuels and improves overall energy efficiency.

Electrification of transportation: Promote the adoption of EVs and develop adequate charging infrastructure. This reduces emissions from the transportation sector by replacing internal combustion engine vehicles.

Carbon capture and storage (CCS): Implement technologies that capture CO_2 emissions from industrial processes and power plants, followed by safe storage underground to prevent their release into the atmosphere.

Energy storage solutions: Invest in energy storage technologies like batteries and pumped hydro storage to store excess renewable energy and provide a stable energy supply when intermittent sources are not generating.

4.3.4 Key Institutional Issues

The institutional issues deal with regulative, normative, and cultural-cognitive elements that, all together with their activities and resources, provide stability to ILECs from the design phase to the operation phase. Institutional issues are strictly related to social norms, economic and financial aspects, and regulatory framework, as discussed in the following.

4.3.4.1 Increasing Trust and Motivation of Citizens

Trust is linked to citizen engagement, and it is the strongest factor influencing willingness to participate before environmental aspects and higher income [48]. In an ILEC, a high trust level gives a competitive advantage. A shared vision, the sense of the identity in ILEC, is crucial for effective strategic decision-making. The common vision fosters trust and helps to coordinate the necessary actions for the community's creation, operation, and evolution. A clear, tangible, and measurable vision enhances stakeholder alignment and attracts new members. The dissemination of results via EU and RES projects can aid in attracting new members. Moreover, monitoring and improving members' energy behavior can have a positive impact on the local community and increase awareness about energy efficiency and environmental protection.

4.3.4.2 Promoting Energy Democracy

There is no commonly recognized definition of energy democracy. This theme is related to social movements that focus on the advancement in the renewable energy transition and asking for the democratic restructuring of energy "regimes" [49]. By moving toward social ownership, decentralization, and public participation, it is possible to create an energy system that is clean, affordable, and accessible to all. ILECs promote decentralized structures and democratization of energy production and supply and hence can lead to new opportunities for participation and contributions as well as a more equitable distribution of economic and political power [29]. However, it is important to note that decentralization of energy generation does not necessarily lead to decentralization of power.

4.3.4.3 ILEC Ownership Model

The ownership model of ILEC projects can significantly affect trust and engagement. The projects can be entirely community owned or co-owned with private or public sectors. As described in [50], the most common models are *cooperatives* where the members can be participants in the ILEC or, alternatively, finance projects through its shares. Another model is the community NPO that provides facilities to the ILEC (e.g. NPO can manage the district heating). Other ownership models include *development trusts* and *shares owned by a local community organization*.

There are several incentives for various actors to get involved in a community. These benefits include generating income and opportunities for the local community through investments in community-owned projects, having simpler approval and favorable treatment for permits related to these investments, ensuring local control, reducing energy costs, and guaranteeing a reliable energy supply, showcasing environmental commitment, and managing energy loads [51].

4.3.4.4 Funding Local Projects

There are several factors that contribute to the energy needs of a region, including the type, size, and structure of local economic activities, the habits of the people, the climate, and the wealth of the region and country. This required energy is used for various daily activities, such as heating, lighting, and industry [29].

When developing a clean energy transition plan, it is important to study the local energy system thoroughly. This involves categorizing and describing the energy system in terms of electricity production and consumption, transportation, heating, cooling, etc. Additionally, it is crucial to study the area's natural and environmental profile to ensure the appropriate spatial planning and weather conditions for sustainable renewable energy projects. It is fundamental to make decision responsibly in order to make people feel safe. The local context of the communities where these projects are planned has a significant impact on their success. There are two models regarding the role of local government in relation to energy communities in general. The first model views the municipality's role as supportive, where they provide the necessary infrastructure for renewable projects. The second model sees the municipality as the initiator of the process, developing energy projects with their companies, implementing virtual power plants, and working toward reducing municipal bills and combating energy poverty. ILECs are more efficient in providing services to local energy customers and are accountable to their consumer base. They can play a crucial role in funding local renewable energy projects and establishing local energy associations.

4.3.4.5 Support Schemes, Incentives, and Targets

Support schemes can aid in achieving the new renewable energy targets by 2030 and increasing the share of renewables in the energy sector. These schemes can be classified based on criteria such as whether they are generation-based or capacity-based, price-driven or volume-driven, and provide partial or overall financial support. The main support schemes used are feed-in tariffs (FITs), feed-in premium (FIP), auctions and tendering schemes, investment grants, tax incentives or exemptions, and low-interest loans. FITs offer a fixed payment to power plant operators for each unit of electricity generated, while FIP involves power plant operators participating in the market and receiving an additional payment to the electricity market price. Long-term contracts may be considered to manage risks. Auctions and tendering schemes, managed by public authorities, stimulate competition among participants but may pose cost and risk barriers. Tax incentives or exemptions are the main support instruments for RES heating and cooling, while low-interest loans help reduce the initial investment cost as a barrier for renewable energy technologies. The RED II directive acknowledges and values the contribution of RECs in achieving the targets set. Member states are encouraged to focus on creating energy communities and providing them with fair competition in the market. In addition, the directive RED II requires member states to establish frameworks that will facilitate the progress of RECs. However, there are currently limited dedicated support schemes for these communities.

The EU suggests that support schemes should be designed with a specific focus to make renewable sources competitive in the market. Additionally, support for technologies that have reached maturity should be removed, while remaining flexible to unanticipated changes in production costs. Consequently, Flexible Investment Promotion schemes should be preferred over FIT schemes that fail to account for market fluctuations. To improve the investor confidence, it is crucial to avoid retroactive or unpredictable changes. Implementing cooperation mechanisms between EU nations to exploit external energy potential may help to reduce costs for consumers and boost investor confidence [52].

4.3.4.6 (Self-)governance

The main challenge for governance in ILECs is the lack of knowledge on how to involve local community members in energy governance policies and structures [53]. Self-organizing ILECs face four major categories of challenges [54], including economic and financial issues, legal barriers, sociocultural conditions, and micro-political struggles and conflicts. For economic and financial issues, there is a complex contractual arrangement between various stakeholders, and some ILECs struggle to establish BMs and obtain funding. Legal barriers include actual legal restrictions and perceived limitations that prevent ILEC initiators from developing. Sociocultural conditions require intrinsic motivation and strong community relationships to succeed. Additionally, knowledge and skills sharing depend heavily on the social context. Finally, ILECs struggle with micropolitical struggles, conflicts, and free-riding behavior, leading to unequal cost-sharing and potential failure. To mitigate this issue, a crucial measure is to establish clear roles for all potential actors within an ILEC and develop governance models that can assist them, such as housing cooperatives. The concepts of "multiactor perspective" [54] and "meta-governance" [55] have already been introduced in this regard. The multiactor perspective considers four sectors: the state, the market, the community, and the Third Sector, which acts as a connecting element. These sectors, which build upon their traditional definitions, provide flexibility in defining the role of each actor in ILECs. These distinguished actors can be seen as "meta-governance" centers [55] as they coordinate various governance processes.

4.3.4.7 Regulatory Issues and Policy Framework

In recent years, at the European level, there has been progress in regulating energy communities by defining RECs with the RED II (2018/2001/EU) and extending it to CECs with the EMD Directive 2019/944 [50]. However, ILECs, as systems that operate with an integrated approach, currently lack a defined regulatory framework. By implementing legislation on a national level, different BMs could be created to fit emerging ILECs. A regulatory framework should be established that clearly defines the actors, geographical scope, activities, technologies involved, and membership rules, considering national technical, financial, and social contexts. A similar approach is already used in the United States, where energy regulation is largely in the hands of individual states. The positive side of this is that the system is already decentralized offering the possibility for individual cities or counties to decide who to rely on for energy supplies. Creating different BMs that align with these regulatory frameworks will support emerging ILECs and provide a path for growth. A well-defined regulatory framework will help ILECs to achieve their goal of providing clean and sustainable energy, contributing to a more sustainable and cleaner future.

4.3.4.8 Institutional (Re-)design

Several steps, involving energy production, conversion, transformation, and distribution, as well as public and private sector contributors, are required for traditional networks to deliver energy to end consumers. Current research on ILECs places significant emphasis on evaluating pre-existing energy system structures to determine their adequacy in achieving an efficient and dependable energy system. As indicated in [56], institutional planning should shift focus from designing organizations for diffusion to designing relevant institutions that can accommodate ILEC operations. This design should be flexible and adaptable and should incorporate lessons learned from past experiences as well as feedback and monitoring from members and their systems. Institutional traits such as the spirit of community, a cooperative culture, and locality and responsibility regulations have an impact on citizen involvement decisions as well as the scale and position of facilities. The changing institutional structures within the energy sector significantly affect the institutional space that local energy systems can develop into [29, 57].

4.4 Business Models for ILECs

As business and regulatory landscapes evolve, together with advancements in technology and changing customer expectations, ILEC stakeholders must continuously assess and adapt BMs to enable effective ILEC use cases and maximize revenues. Through analysis and refinement of BMs, stakeholders can optimize performance, promote innovation, and achieve sustainable operation in an increasingly dynamic and evolving sector. This section presents a summary of BMs developed with a particular focus on the value propositions of multienergy carrier initiatives, with the aim of enhancing the cost-effectiveness of their implementation in ILECs. Each BM is defined from the perspective of a specific actor and outlines how that actor can take advantage of the tangible economic benefits via ILECs. Then, a Business Model Canvas (BMC) is developed and presented for two of the identified BMs, providing, for them, a full description of each of the BM components.

4.4.1 Overview of Selected Business Models

Sixteen BMs are identified and developed. Short descriptions for each of the BMs are provided as follows:

A. Flexibility for local balancing and increase of self-consumption for multicarrier systems
   This BM, developed from the perspective of prosumers, enables them to participate in local energy balancing schemes by incentivizing the operation of their RES in a flexible manner to maximize the use of renewable energy in their ILEC and contribute to a reduction in local carbon emissions. In doing so, prosumers can avail of financial payments for their participation.
B. Flexibility harvesting at the community level for services to the upper grid
   Through this BM, prosumers within an ILEC can be incentivized to operate their energy resources in a flexible manner to provide coordinated energy-related services to the higher levels of the energy delivery systems, e.g. electricity transmission systems. The incentive mechanisms can be offered through a peer-to-peer market approach to provide services to higher grid levels, such as load shifting and congestion management.
C. Co-operative ownership of district heating network
   A co-operative ownership BM for a district heating network within an ILEC could be utilized to coordinate the overall heating demand in response to energy

price signals. Such a BM is developed from the perspective of the end-user to allow them to collectively optimize their heat-energy usage while taking advantage of lower heat energy prices and additional income from participation in flexibility provision services.

D. Energy service provider (e.g. retailer and aggregator) optimal portfolio management and participation in the local market
This BM relates to energy brokers representing ILEC prosumers within wholesale energy, balancing, and/or local flexibility markets and managing their associated energy flexibility and imbalances through appropriate energy contracts. The BM has been developed from the perspective of the energy service provider, which would design the energy contract schemes with the ILEC prosumers and hold responsibility for the energy and flexibility billing.

E. Flexible energy tariffs for flexible assets optimal utilization to avoid peak power
Through this BM, prosumers within an ILEC can be incentivized, via varying energy price signals, to operate their flexible energy assets in a coordinated manner to maximize local energy use and balancing as well as avoiding energy demand peaks at times when energy prices can be high thereby reducing energy costs for the prosumers.

F. ESCO minimizing customers' energy bills and optimizing energy usage
The BM for ESCOs is based on providing energy savings for ILECs without sacrificing the comfort of productivity of the ILEC members. ESCOs are companies that specialize in providing services such as energy audits, retrofits, and maintenance of energy-efficient equipment for ILEC prosumers. Through this BM, they would benefit by providing these services in exchange for a portion of the energy cost savings to the prosumers.

G. Heat recovery and storage for effective integration of district heating and cooling
This BM involves a LHCDSO operating as an ILEC, selling excess thermal energy from heat recovery and storage systems to customers that are external to ILEC. Specific energy supply contracts can be set up between the end-users and the LHCDSO allowing the LHCDSO to manage the excess thermal energy and exploit this opportunity on behalf of the ILEC. The benefits of such a model include a revenue stream for the LHCDSO and free thermal energy to the internal ILEC customers.

H. Prosumer selling self-produced energy
Prosumers with sufficient levels of DER installed capacity could have the option to sell their excess electricity back to the grid. This BM would involve prosumers within an ILEC selling this excess electricity directly to other electricity customers within the ILEC through a peer-to-peer market mechanism. The sale of excess electricity to customers outside of the ILEC could also be possible and would require a third-party aggregator to manage the interactions.

I. Software development companies providing forecasting and resource coordinating services
The development and sale of software products providing forecasting and resource coordinating, scheduling, and management services are crucial factors in the successful implementation and operation of ILECs. This BM, developed

from the perspective of technology suppliers, outlines the opportunity for such entities to supply these required services and technologies to ILECs. These services are crucial to the successful coordination and operation of energy resources within an ILEC and could be supplied as a "Software-as-a-Service" offering.

J. Combined heat and power, absorption heat pumps, fuel cells manufacturers taking advantage of multienergy coupling and district heating networks
The implementation of ILECs promotes the use of technologies that allow for energy self-consumption by the ILEC members. Such technologies can include combined heat and power, absorption heat pump, and fuel cells that allow producing electrical and thermal (heating and cooling) energy, leading to higher energy efficiencies and reducing GHG emissions. Manufacturers could exploit this sector coupling capability by producing and supplying appropriate technologies to ILEC-based markets.
K. SMs and sensor manufacturers for optimal energy management and innovative devices manufacturers (ICT or hardware) for control, measuring, and monitoring in general
With participation in ILECs, prosumers will have the ability to optimize their load and/or generation according to certain use cases. A key component of this capability will be the monitoring of energy use/production throughout the ILECs. As such, there will be an opportunity for manufacturers of advanced metering infrastructure technologies to produce and supply appropriate metering and sensory equipment to facilitate ILEC operation.
L. Blockchain platform providers for local energy markets
Within ILECs, new market opportunities are becoming available for prosumers to actively provide grid services via flexible operation of distributed resources. Local energy markets can allow prosumers to trade energy on market platforms with predefined price mechanisms. Blockchain technologies can enable LEMs to facilitate the trading of energy within and outside of ILECs. This BM relates to opportunity for blockchain technology providers to supply this enabling service to ILECs.
M. Reliability services to energy users in case of islanding
This BM relates to the opportunity for ESCOs to operate ILECs as microgrids to continue supplying customers during contingency scenarios. DSOs would request and pay for this service from ESCOs to continue to supply their customers during abnormal grid conditions.
N. Sharing excess PV generation in the scope of energy communities
The implementation of this BM would allow for prosumers within ILECs to offer excess energy generated from RES to other customers in return for payment. The ILEC itself or an aggregator would be responsible for managing the delivery of the excess energy within the ILEC and establishing the appropriate contracting mechanisms.
O. Reducing RES curtailment through the exploitation of synergies among energy carriers
This BM relates to the management of energy resources within ILECs to avoid unnecessary curtailment of RES by actively coordinating the operation of other

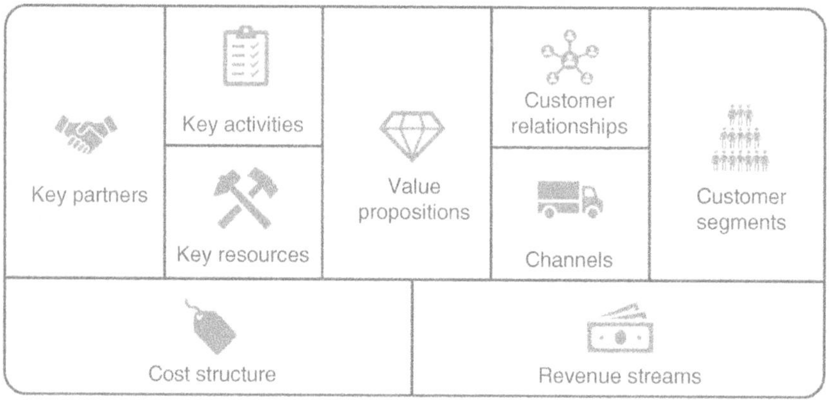

Figure 4.5 Overview of business model canvas. Source: Adapted from Leclercq et al. [58].

technologies such as energy storage and demand response. ILEC operators or aggregators would have the opportunity to offer this technology coordination as a service to prosumers and would receive payment for energy that would otherwise have to be wasted.

P. Use of hydrogen-based nonconventional storage for PV power
This BM involves the use of electrolyzer technologies to convert excess RES energy generation into hydrogen, which can be stored for use at a later time, rather than curtailing the RES output. ILEC prosumers would receive a payment for the energy that would otherwise have been wasted, and the stored hydrogen could then be used for a variety of end-use applications providing multiple business opportunities for the service providers.

4.4.2 Development of Business Models

Each of the identified BMs can be developed using the BMC template reported in Figure 4.5, which describes the main components of the BM in detail and provides a visual overview of their relationships [58].

A detailed description is presented for each of the BMC components for *BM A. (Flexibility for local balancing and increase of self-consumption for multicarrier systems)* and for *BM G. (Heat recovery and storage for effective integration of district heating and cooling)* to demonstrate the information captured as part of the BM development process.

4.4.2.1 BM A – Flexibility for Local Balancing and Increase of Self-consumption for Multicarrier Systems

♦ **Description**
The advent of smart grids and decentralization of energy systems has radically changed the conventional view of the energy sector. Flexibility for local balancing and the increase of self-consumption is a key consideration for multicarrier

energy systems that combine different energy sources, storage systems, and loads. One of the sectors that will realize several benefits as a result of this innovative context is the residential sector. Optimizing the use of energy systems (e.g. EV, heat pumps, photovoltaics, and storage systems) can potentially provide flexibility to the main grid, while prioritizing the self-consumption of end-users.

In the context of ILECs, a cooperative energy management approach for multi-carrier networked energy hubs is fundamental. Several studies have been previously examined, which show that by considering a cooperative EMS based on multienergy resources, operating costs, and GHG emissions can be reduced, and self-reliability of the networked system can be improved [59].

A BM on flexibility for local balancing and increasing self-consumption for multicarrier systems can be based on offering energy management services to customers (e.g. commercial and industrial loads and local energy communities). These services can be based on the provision of energy efficiency programs and control and optimization devices to enable local load balancing and maximization of self-consumption from RES. The main revenue streams will come from (i) the service fees to be paid by the customer for access to the EMS including hardware and software installation and maintenance, (ii) excess renewable energy sold by the involved customers, (iii) receiving payments for demand response programs by shifting energy consumption from periods of peak demand, and (iv) the provision of ancillary services to the grid (e.g. voltage regulation). The connection to the grid is essential for the participation of demand response programs and provision of ancillary services by customers.

- **Key partners**

  (Stakeholders crucial to the implementation and delivery of the service)
  - End-users (acting as prosumers) and EV owners that can provide demand response are the main actors in this BM.
  - Energy suppliers and/or DSOs are involved in the establishment and operation of demand response programs and ancillary service requests.
  - ESCOs establish and offer energy-saving/renewable energy projects.
  - Technology suppliers provide hardware and software devices to facilitate energy management.
  - Aggregators combine multiple customer loads and/or energy system devices for the sale, purchase, or auction of energy or services in one or more electricity markets.
- **Key activities**

  (Actions needed to develop and deliver the service)
  - Development of EMS devices to enable flexible operation of actor load and/or energy systems.
  - Installation and maintenance of EMS devices.
  - Provision of demand response programs and participation in ancillary services.
- **Key resources**

  (Resources required to develop and deliver value propositions)
  - EMS devices (hardware and software) that can provide optimized operations for local balancing (physical resources).

- RES systems for energy generation (physical resources).
- Technicians that can install and maintain all the devices involved (human resources).
- Initial investment from the end-users (financial resources).

♦ **Customer value proposition**
(Added value to the customer or the problem that the service is solving)
- Increasing the self-consumption from RES, providing flexibility services to the main grid, and reducing GHG emissions.
- Participation in flexibility markets through the provision of flexibility services provided by the main actors.
- Optimized energy management, increased energy efficiency, and participation in flexibility markets for the main actors.
- Increased energy autonomy as a result of the increase in self-consumption.
- Exploitation of local energy resources from local RES generation.

♦ **Customer relationship**
(Customer perception of the service)
- Participation in energy management schemes is perceived to be positive due to a likely reduction in energy costs.
- Relatively low effort is required by customers to participate due to the relative ease of installation of required devices and the availability of support on device operation and maintenance.
- Customer minimum comfort conditions should be met at all times.

♦ **Channels**
(Channels through which service is delivered and the customer is engaged)
- Services can be advertised through online marketing, etc., and delivered through direct sales to customers (e.g. industrial loads or to end-users by energy suppliers).
- The customer engagement can be procured by
  - **Awareness:** Raised through online marketing.
  - **Evaluation:** Proposal of the contract and providing information on potential energy cost reductions.
  - **Purchase:** Customer agreement to participate.
  - **Delivery:** Provision and installation of required devices and services.
  - **After sales support:** Through direct contact (email, phone, etc.) with customers for the provision of regular maintenance.

♦ **Customer segments**
(Identification of customer types and their respective needs)
- Commercial and industrial loads, ILECs, and end-users are involved in assisting with the exploitation of local RES, optimal energy system management, reduced energy bills, and reduction of GHG emissions.

♦ **Cost structure**
(Costs related to resources and activities)
- Installation and maintenance costs for required devices including RES systems, EMS devices, etc.
- Development costs for all involved devices.

- EMS operation and maintenance costs.
- Marketing and advertising costs.

♦ **Revenue streams**

(Sources and methods of payment for the service)

- Provision of demand response and ancillary services.
- Sale of excess energy from the customer to the grid.
- Technology supplier device sales.
- Revenues to ESCOs for promoting the schemes.

4.4.2.2 BM G – Heat Recovery and Storage for Effective Integration of District Heating and Cooling

♦ **Description**

Waste heat recovery, its storage, and the integration within district heating and district cooling networks represent a fundamental step toward the decarbonization of energy systems in Europe. Typical systems allowing for heat recovery are cogeneration and trigeneration systems, but many others are also under consideration and studied such as the heat recovery from refrigeration systems, data centers, and wastewater.

Several studies highlight how heat storage systems associated with waste heat and the district heating and cooling networks can bring numerous advantages from an energy, environmental, and economic point of view [60, 61]. The presence of heat storage has shown a high potential for integrating intermittent RES, such as wind and solar, in the heating or cooling sector. This can be achieved using power-to-heat devices, such as electric heat pumps or electrical boilers, thus increasing the cross-vector integration of the energy systems and becoming effective tools for electricity demand-side management programs [62]. Considering this, the benefits associated with heat recovery and storage can generally be measured by the decrease in primary energy consumption for thermal demand, but also for the electricity demand, if power-to-heat technologies are involved. For example, the presence of thermal energy storage can be used as a reserve by the heat pumps to increase the coefficient of performance (in both cooling and heating modes) and, consequently, decrease electricity demand. Finally, it should be emphasized that the advantages that can be exploited using recovered heat and thermal storage are often associated with a fourth-generation district heating and cooling network [63] which, unlike more traditional versions, has lower operating temperatures (50–70 °C), allows the integration of decentralized resources, and assures higher efficiencies [64].

With ILECs, the possibilities of recovering waste heat and the installation of thermal storage can be exploited, both in industrial districts and in residential districts equipped with a district heating network operating at low temperatures (50–70 °C). Industrial waste heat generally has higher temperature levels and lends itself better to being used within the same districts because producers are located far from city centers and the losses along the network would usually not justify the economic return. Concerning the residential sector, the heat recovered from data centers, from the refrigeration systems of retail food stores and cogeneration and trigeneration systems, represents the major sources currently analyzed in the literature [65].

In terms of BM development, the LHCDSO, acting as an ILEC, can sell the excess thermal energy (from recovery and storage) to commercial, industrial, and residential customers that are external to the ILEC. The service can be offered through specific energy supply contracts between users and the LHCDSO. In this way, the LHCDSO can obtain a revenue stream but even ILEC users can benefit from receiving free thermal energy when supplied from the LHCDSO. Other related benefits can include a reduction of GHG emissions using the heat recovered and an increase in the energy autonomy by deploying higher generation from local RES and fighting energy poverty reducing the dependency on energy imports.

- **Key partners**
  (Stakeholders crucial to the implementation and delivery of the service)
  - LHCDSO, acting as an ILEC, is the main actor.
  - End-users, acting as both consumers and prosumers of thermal energy.
  - Energy storage owners (thermal energy storage) manage the mismatch between the supply and demand for thermal energy.
  - **Energy suppliers:** thermal energy suppliers provide thermal energy to distribution heating network (DHN); electricity suppliers provide electricity to power-to-heat technologies.
  - ESCOs provide energy-saving projects or renewable energy projects.
  - Technology suppliers provide hardware and software systems.
- **Key activities**
  (Actions needed to develop and deliver the service)
  - Management of the thermal flows within the ILEC.
  - Thermal storage management.
  - Heat excess production.
  - Development of multi-EMS software and related sensors by technology suppliers.
- **Key resources**
  (Resources required to develop and deliver value propositions)
  - Multienergy converters, sensors, DHN infrastructure, storage, energy technology, EMS, communication infrastructure (physical resources).
  - Databases with information shared among the energy community (intellectual resource).
  - Operators of the ILEC skilled in the energy and economy sectors (human resources).
  - Possible investments for the implementation (financial resources).
- **Customer value proposition**
  (Added value to the customer or the problem that the service is solving)
  - Reduction of GHG emissions. Using heat recovered, GHG emissions can be reduced.
  - Improvement of local economy. Local energy generation (exploiting local resources) may reduce the need to purchase similar energy quantities from

commercial providers. The income received may also be invested in additional projects bringing social benefits for the whole community.
- Energy savings and optimal energy management. The purchase of prosumer self-generated energy can be cheaper than buying energy at retail prices.
- Improving energy autonomy. Higher production from local RES and heat recovery means less dependency on energy imports.
- Exploitation of local energy resources. Heat recovery can be considered a local resource.
- Fighting Energy Poverty. Higher production from local RES and heat recovery means less dependency on energy imports and help to fight energy poverty.

♦ **Customer relationship**
(Customer perception of the service)
- The service is perceived as a good practice by the end-user because he/she will see a reduction in bills related to gas consumption.
- A low effort is required in establishing and keeping the relationship, since the end-user is guaranteed to have benefits by signing a contract with LHCDSO.
- The needed trust level is related to the guarantee of the comfort condition in the built environment and of the reduction in end-users' energy bills.
- The service is important in the measure of reduction in energy bills.

♦ **Channels**
(Channels through which service is delivered and the customer is engaged)
- The service is delivered through specific energy supply contracts between end-users external to ILEC and LHCDSO.
- A cost-effective channel can be an energy supply contract with time-varying prices (time-of-use tariff).
- The customer engagement can be assured by the following:
  - Awareness raising social campaign.
  - **Evaluation:** proposal of the contract and informing about the reduction in the energy bill.
  - **Purchase:** signature of contract.
  - **Delivery:** provision of thermal energy from DHN based on the contract conditions.
  - **After sales:** monitoring and customer satisfaction, and quality of service.
  - Telemarketing and customer satisfaction questionnaires can be used for marketing the service and post-implementation support.

♦ **Customer segments**
(Identification of customer types and their respective needs)
- External commercial, industrial, and residential loads are involved.
- ILEC participants can reduce energy bills and increase/maintain the satisfaction of comfort conditions.
- ILEC participants' concerns related to potential service interruptions they might have compared to the already established services.

♦ **Cost structure**
(Costs related to resources and activities)
- Installation and/or maintenance costs for RES systems, DHN, thermal storage, and others.

- Development and maintenance costs for all involved devices (e.g. EMS).
- Marketing and advertising costs.

♦ **Revenue streams**

(Sources and methods of payment for the service)

- Revenues for selling the excess of thermal energy to customers external to ILEC.
- ILEC participants receive free thermal energy when distributed from the LHCDSO (acting as an ILEC).

4.5 Conclusion and Lessons Learned on Barriers, Benefits, and Policy Implications for ILECs Implementation

ILECs involve various actors with distinct roles and interactions. The previous sections of this chapter provided a bird's eye view of the actors, their interaction, and multifaceted concerns in implementing ILECs. Understanding these actors and their dynamics is essential for successful implementation in the future. This section summarizes the barriers, benefits, and policy implications associated with ILECs and the actors involved.

4.5.1 Barriers

Conflicting interests among actors within ILECs can lead to potential disruptions, emphasizing the importance of managing these conflicts effectively. Regulatory challenges also arise, requiring the overcoming of barriers and compliance with national directives, which necessitates the involvement of policymakers and regulators to create a supportive framework.

The technical complexity of integrating diverse energy technologies within ILECs poses another challenge that requires expertise and support from academic institutions and private research institutes. Additionally, financial support is crucial for the establishment and operation of ILECs. Local authorities can provide loans and incentives, while banks and private investors can contribute additional funding.

However, the lack of financial support and the presence of cheap fossil fuels present significant barriers to the implementation of ILECs. Overcoming these barriers requires eliminating fossil fuel subsidies and implementing incentives for RES, such as R&D incentives, fiscal policies, and grid connection and tariff incentives.

Another obstacle is the split-incentives effect, where the investor may not invest in improving energy efficiency if another entity benefits from it. To mitigate this, transparency and fairness among market parties and community members are necessary, with ILECs serving as platforms for effective communication and fair cost–benefit allocation.

Implementing ILECs also faces regulatory challenges due to their unique energy systems, which may not fit within existing regulations. Technological integration poses additional technical challenges, including interoperability issues and the need for advanced control systems.

Financial constraints are prevalent as the upfront costs of establishing ILECs and deploying infrastructure can be substantial, hindering implementation. Furthermore, stakeholder engagement is crucial but challenging due to differing interests and priorities among prosumers, energy suppliers, technology providers, and regulators.

Addressing these barriers and challenges is essential to fully realize the benefits of ILECs, including increased renewable energy adoption, energy efficiency, and community empowerment. Policymakers, regulators, industry stakeholders, and research institutions must collaborate to develop effective policies and strategies to overcome these barriers and promote the successful implementation of ILECs.

4.5.2 Benefits

From an economic standpoint, ILECs bring significant advantages. They create local job opportunities, support local businesses, and drive economic growth. The decentralized nature of ILECs fosters local resilience and promotes community-based economic development, ensuring that the benefits of the energy transition are distributed widely.

Social coherence is enhanced through ILECs as they actively engage and empower local communities. Community members can become energy prosumers, participating in shared energy resources and peer-to-peer trading platforms. This not only fosters social cohesion but also enables individuals to actively contribute to the energy transition.

ILECs also maximize the value of flexibility in energy markets through aggregation and coordination. This allows for the efficient integration of diverse energy resources into the grid, contributing to grid stability and resilience. Energy autonomy and security of supply are promoted by ILECs through the utilization of local resources and reduced dependence on external factors. By generating energy locally, resolving supply-demand discrepancies through energy storage, and functioning independently from the grid, ILECs improve security of supply and reduce energy imports. One of the critical benefits of ILECs is their contribution to reducing energy poverty. Through improved energy efficiency and security of energy supply, ILECs help alleviate energy poverty. By utilizing distributed energy resources and modernizing the grid, costs are lowered, energy independence is increased, and a continuous supply of energy is ensured. Furthermore, ILECs support local economic and social development and job creation.

Community engagement and willingness to pay are crucial aspects of ILECs. By providing appropriate incentives and raising awareness about the benefits of renewable energy, ILECs encourage active engagement from users. Many consumers are willing to pay more for renewable energy if it contributes to green energy infrastructure or reduces environmental pollution.

ILECs enable the efficient integration and utilization of RES, facilitating the transition to a more sustainable energy system. They also improve cost-effectiveness by optimizing energy use, balancing supply and demand, and leveraging flexibility, ultimately reducing energy costs for prosumers and potentially decreasing grid infrastructure investments. Furthermore, ILECs enhance the resilience of energy systems by promoting local energy generation, storage, and distribution. By reducing dependence on centralized grids and minimizing the impact of grid outages, they contribute to local energy resilience.

ILECs play a crucial role in carbon emission reduction by facilitating the integration of multiple energy carriers. This integration leads to the reduction of local carbon emissions, making a significant contribution to global climate change mitigation efforts.

Lastly, ILECs create new business opportunities for various actors, such as energy brokers, software developers, technology manufacturers, and service providers. These actors can offer innovative products and services tailored to the needs of integrated energy systems, further driving economic growth and technological advancement.

4.5.3 Policy Implications

Policymakers and regulators need to define criteria for collective self-consumption, establish support schemes, and address economic regulations related to energy costs, prices, taxes, and duties. By creating these frameworks, governments can enable the growth of ILECs and provide a clear path for their implementation.

Government support is crucial in transposing European directives into national contexts and providing technical and financial assistance to ILECs. Governments can facilitate citizen engagement, establish funding opportunities, and break down regulatory barriers that hinder the progress of ILECs.

Local authorities also play a fundamental role in the development of ILECs. They can act as founders, funders, promoters, and end-users, providing essential support through loans, financial incentives, and knowledge sharing. Engaging local authorities is vital too for the success and sustainability of ILECs.

Collaboration and knowledge sharing among various actors are vital for achieving the objectives of ILECs. The active involvement and cooperation of end-users, aggregators, energy storage owners, EV owners, ESCOs, TSOs/GTSOs, DSOs/GDSOs, BRPs, framework setters, and supporters/suppliers are necessary to unlock the full potential of ILECs.

To facilitate the transition to clean energy and accelerate the growth of ILECs, governments should consider implementing specific policies and measures. First, eliminating subsidies for fossil fuels and redirecting financial support towards RES will create a level playing field for clean energy projects and incentivize the adoption of ILECs.

Additionally, governments should introduce economic incentives such as financial aid, tax relief, and dedicated programs to support energy communities. These incentives can create investment opportunities, disseminate best practices, and improve the economic accessibility of different categories of end-users.

Policies should also focus on fostering community engagement through shared visions and trust-building. Disseminating results, monitoring energy behavior, and raising awareness about energy efficiency and environmental protection will help align stakeholders and promote active community involvement.

Spatial issues related to land use for renewable energy technologies should be addressed in policies. Engaging stakeholders and considering the community's access to land can help mitigate conflicts and opposition to the introduction of RES. Exploring alternative technologies like rooftop solar installations and district heating networks may be more attractive in certain contexts.

Energy democracy should be promoted through policy measures that empower social movements advocating for renewable energy transition. Involving citizens in decision-making processes and providing access to renewable energy resources can foster energy democracy and ensure a fair and equitable energy transition.

Mitigating environmental impacts is also crucial. Policies should prioritize emission reduction and waste management in ILECs. Shifting to cleaner energy sources like wind, solar, and hydropower can help reduce air pollution. It is worth noting that effective waste management strategies should be developed to address any environmental concerns.

In summary, to support the establishment and operation of ILECs, policymakers should adapt regulatory frameworks, provide incentive mechanisms, foster standardization and interoperability, encourage stakeholder engagement and collaboration, and facilitate knowledge sharing and the replication of best practices.

Acknowledgments

Part of the results presented in the chapter have been obtained in eNeuron project. The eNeuron project has received funding from the European Union's Horizon 2020 Research and Innovation Programme under Grant Agreement No. 957779.

List of Abbreviations

BM	business model
BMC	business model canvas
BRP	balance responsible party
CCS	carbon capture and storage
CEC	citizen energy community
DERs	distributed energy resources
DHN	district heating network
DSO	distribution system operator
EMS	energy management system
ESCO	energy service company
FIP	feed-in-premium
FIT	feed-in-tariff

GDSO	gas distribution system operator
GHG	greenhouse gas
GTSO	gas transmission system operator
ICTs	information and communication technologies
ILEC	integrated local energy community
LCA	life cycle assessment
LHCDSO	local heat/cold distribution system operator
NPO	nonprofit organization
REC	renewable energy community
RESs	renewable energy sources
SM	smart meter
TSO	transmission system operator

References

1 Caliano, M., Buonanno, A., Di Somma, M. et al. (2022). Enabling conditions for the deployment of integrated local energy communities in Europe. *Workshop on Blockchain for Renewables Integration (BLORIN)*, Palermo (Italy) (2-3 September 2022).

2 Comodi, G., Spinaci, G., Di Somma, M., and Graditi, G. (2022). Transition potential of local energy communities. In: *Technologies for Integrated Energy Systems and Networks* (ed. G. Graditi and M. Di Somma), 275–304. Wiley.

3 IEA, IRENA, UNSD, WB, WHO (2019), Tracking SDG 7: The Energy Progress Report 2019, Washington DC. https://sdgs.un.org/sites/default/files/2021-05/Report%20-%202019%20Tracking%20SDG7%20Report.pdf (accessed 25 September 2023).

4 Reis, I., Gonçalves, I., Lopes, M., and Antunes, C. (2021). Business models for energy communities: a review of key issues and trends. *Renewable and Sustainable Energy Reviews* 144: 111013.

5 European Commission (2019). Clean energy for all Europeans package. European Commission.

6 European Commission (2018). European Commission Directive (EU) 2018/2001 of the European Parliament and of the Council of 11 December 2018 on the promotion of the use of energy from renewable sources. European Commission.

7 European Commission (2019). Directive (EU) 2019/944 of the European Parliament and of the Council of 5 June 2019 on common rules for the internal market for electricity and amending Directive 2012/27/EU. European Commission.

8 Cielo, A., Margiaria, P., Lazzeroni, P. et al. (2021). Renewable energy communities business models under the 2020 Italian regulation. *Journal of Cleaner Production* 316: 128217.

9 Gjorgievski, V.Z., Cundeva, S., and Georghiou, G.E. (2021). Social arrangements, technical designs and impacts of energy communities: a review. *Renewable Energy* 169: 1138–1156.

10 EDSO (2021). DSOs as facilitators of energy communities. www.edsoforsmartgrids.eu/wp-content/uploads/E.DSO-Position-Paper-on-Energy-communities-FINAL.pdf (accessed 13 September 2023).

11 Morch, A.Z., Sæle, H., Merino, J. et al. (2021). Local multi-vector energy systems within the European political and regulatory landscape: scope and key priorities for the study. eNeuron project D2.1.

12 Sæle, H., Merino, J., Di Somma, M. et al. (2021). Introduction and development of local energy communities in Europe. eNeuron project D3.1.

13 Pressmair, G., Amann, C., and Leutgob, K. (2021). Business models for demand response: exploring the economic limits for small- and medium-sized prosumers. MDPI.

14 Leutgöb, K., Amann, C., Tzovaras, D., and Dimosthenis, I. (2019). New business models enabling higher flexibility on energy markets. European Council for an Energy Efficient Economy Summer Study.

15 Schönberger, P. (2013). Municipalities as key actors of German renewable energy governance.

16 Community Energy (2017). Community energy England, community energy: a local authority perspective.

17 Zhao, J., Arefi, A., Borghetti, A., and Ledwich, G. (2022). Indices of congested areas and contributions of customers to congestions in radial distribution networks. *Journal of Modern Power Systems and Clean Energy* 10: 656–666.

18 Lilla, S., Orozco, C., Borghetti, A. et al. (2020). Day-ahead scheduling of a local energy community: an alternating direction method of multipliers approach. *IEEE Transactions on Power Systems* 35 (2): 1132–1142.

19 Tostado-Véliz, M., Rezaee Jordehi, A., Icaza, D. et al. (2023). Optimal participation of prosumers in energy communities through a novel stochastic-robust day-ahead scheduling model. *International Journal of Electrical Power & Energy Systems* 147: 108854.

20 Orozco, C., Borghetti, A., De Schutter, B. et al. (2022). Intra-day scheduling of a local energy community coordinated with day-ahead multistage decisions. *Sustainable Energy, Grids and Networks* 29: 100573.

21 Zhou, Y., Wei, Z., Sun, G. et al. (2018). A robust optimization approach for integrated community energy system in energy and ancillary service markets. *Energy* 148: 1–15.

22 Rana, M.J., Zaman, F., Ray, T., and Sarker, R. (2021). Real-time scheduling of community microgrid. *Journal of Cleaner Production* 286: 125419.

23 Liu, N., Tang, Q., Zhang, J. et al. (2014). A hybrid forecasting model with parameter optimization for short-term load forecasting of micro-grids. *Applied Energy* 129: 336–345.

24 Hippert, H.S., Pedreira, C.E., and Souza, R.C. (2001). Neural networks for short-term load forecasting: a review and evaluation. *IEEE Transactions on Power Systems* 16 (1): 44–55.

25 IEA (2023). Fossil fuels consumption subsidies 2022. https://www.iea.org/reports/fossil-fuels-consumption-subsidies-2022 (accessed 11 June 2023).

26 Toni Allen (2022). US States fossil fuel consumption: these 15 states are most dependent. https://commodity.com/blog/states-fossil-fuels/ (accessed 05 June 2023).

27 Nelson, D. and Shrimali, G. (2014). Finance mechanisms for lowering the cost of renewable energy in rapidly developing countries. Climate Policy Initiative.

28 Mutani, G., Santantonio, S., Lowitzsch, J. et al. (2020). Economic incentives for energy efficiency measures and low-emissions technologies. *International Conference and Workshop in Óbuda on Electrical and Power Engineering*, Budapest, Hungary, IEEE Publication (18-19 November 2020).

29 Koirala, B.P., Koliou, E., Friege, J. et al. (2016). Energetic communities for community energy: a review of key issues and trends shaping integrated community energy systems. *Renewable and Sustainable Energy Reviews* 56: 722–744.

30 Rae, C. and Bradley, F. (2012). Energy autonomy in sustainable communities: a review of key issue. *Renewable and Sustainable Energy Reviews* 16 (9): 6497–6506.

31 European Commission (2022). Energy poverty in the EU. https://energy.ec.europa.eu/topics/markets-and-consumers/energy-consumer-rights/energy-poverty-eu_en (accessed 5 May 2023).

32 Mamica, L. (2021). Willingness to pay for the renewable energy sources of the residents of Kraków. *Energy Policy Journal* 24 (2): 117–136.

33 Danne, M., Meier-Sauthoff, S., and Musshoff, O. (2021). Analyzing German consumers' willingness to pay for green electricity tariff attributes – a discrete choice experiment. *Energy, Sustainability and Society* 11: 15.

34 Kowalska-Pyzalska, A. (2019). Do consumers want to pay for green electricity? A case study from Poland. *Sustainability* 11 (5): 1310.

35 Ntanos, S., Kyriakopoulos, G., Chalikias, M. et al. (2018). Public perceptions and willingness to pay for renewable energy: a case study from Greece. *Sustainability* 10: 687.

36 Bauwens, T. (2013). What roles for energy cooperatives in the diffusion of distributed generation technologies? *SSRN Electronic Journal* https://doi.org/10.2139/SSRN.2382596.

37 Schunder, T., Yin, D., Bagchi-Sen, S., and Rajan, K. (2020). A spatial analysis of the development potential of rooftop and community solar energy. *Remote Sensing Applications* 19: 100355. https://doi.org/10.1016/J.RSASE.2020.100355.

38 Wächter, P., Ornetzeder, M., Rohracher, H. et al. (2012). Towards a sustainable spatial organization of the energy system: backcasting experiences from Austria. *Sustainability* 4 (2): 193–209.

39 Karasmanaki, E. (2022). Is it safe to live near wind turbines? Reviewing the impacts of wind turbine noise. *Energy for Sustainable Development* 69: 87–102. https://doi.org/10.1016/J.ESD.2022.05.012.

40 Wang, C., Li, Y.G., Li, G.M., and Li, H.L. (2022). Equipment noise evaluation based on auditory saliency map. *Applied Acoustics* 201: 109125. https://doi.org/10.1016/J.APACOUST.2022.109125\stquote\stquote.

41 Mir, M., Nasirzadeh, F., Lee, S.H. et al. (2022). Construction noise management: a systematic review and directions for future research. *Applied Acoustics* 197: 108936. https://doi.org/10.1016/J.APACOUST.2022.108936.

42 Jain, M., Kumar, D., Chaudhary, J. et al. (2023). Review on E-waste management and its impact on the environment and society. *Waste Management Bulletin* https://doi.org/10.1016/J.WMB.2023.06.004.

43 Foster, W., Azimov, U., Gauthier-Maradei, P. et al. (2021). Waste-to-energy conversion technologies in the UK: processes and barriers – a review. *Renewable and Sustainable Energy Reviews* 135: 110226. https://doi.org/10.1016/J.RSER.2020.110226.

44 Maldet, M. (2023). Integration of sustainable water-and waste management into energy communities. Das Energiesystem nach Corona: Irreversible Strukturänderungen – Wie? p. 18. https://repositum.tuwien.at/handle/20.5 (accessed 6 July 2023).

45 Smith, J.R., Johnson, A.B., and Davis, C.E. (2022). Life cycle assessment of energy systems in the ILEC. *Environmental Sustainability Journal* 5 (2): 123–145.

46 Mishra, R., Naik, B.K.R., Raut, R.D., and Paul, S.K. (2022). Circular economy principles in community energy initiatives through stakeholder perspectives. *Sustainable Production and Consumption* 33: 256–270. https://doi.org/10.1016/J.SPC.2022.07.001.

47 Emission Energy Education (n.d.). Emissions. https://energyeducation.ca/encyclopedia/Emissions (accessed 6 July 2023).

48 Declerck, C.H., Boone, C., and Emonds, G. (2016). Citizens' willingness to participate in local renewable energy projects: the role of community and trust in Germany. *Energy Research & Social Science* 13: 60–70.

49 Burke, M.J. and Stephens, J.C. (2017). Energy democracy: goals and policy instruments for sociotechnical transitions. *Energy Research and Social Science* 33: 35–48.

50 Sæle, H., Morch, A., Buonanno, A. et al. (2022). Development of energy communities in Europe. *18th International Conference on the European Energy Market (EEM)*, Ljubljana, Slovenia, IEEE (13-15 September 2022).

51 Gordon, W. (2008). What are the barriers and incentives for community-owned means of energy. *Energy Policy* 36 (12): 4401–4405.

52 European Commission (2023). Support schemes for renewable energy. https://energy.ec.europa.eu/topics/renewable-energy/financing/support-schemes-renewable-energy_en (accessed 8 May 2023).

53 Parag, Y., Hamilton, J., White, V., and Hogan, B. (2013). Network approach for local and community governance of energy: the case of Oxfordshire. *Energy Policy* 62: 1064–1077.

54 Avelino, F., Bosman, R., Paradies, G. et al. (2014). *The (Self-)Governance of Community Energy: Challenges & Prospects*. Rotterdam: Dutch Research Institute For Transitions (DRIFT).

55 Frantzeskaki, N., Wittmayer, J., and Loorbach, D. (2014). The role of partnerships in 'realising' urban sustainability in Rotterdam's City Ports Area, The Netherlands. *Journal of Cleaner Production* 65: 406–417.

56 Frantzeskaki, N., Avelino, F., and Loorbach, D. (2013). Outliers or frontrunners? Exploring the (self-) governance of community-owned sustainable energy in Scotland and the Netherlands. *Renewable Energy Governance* 29: 101–116.

57 Oteman, M., Wiering, M., and Helderman, J.-K. (2014). The institutional space of community initiatives for renewable energy: a comparative case study of the Netherlands, Germany and Denmark. *Energy, Sustainability and Society* 4 (1): 1–17.

58 Leclercq, L., Tagliapietra, M., and Eijgelaar, M. (2020). D9.1 Business model development for the deployment of the IPT and other technological solutions. WP9 – INSULAE business cases and exploitation strategies.

59 Tiwari, S. and Singh, J.G. (2022). Optimal energy management of multi-carrier networked energy hubs considering efficient integration of demand response and electrical vehicles: A cooperative energy management framework. *Journal of Energy Storage* 51: 104479.

60 Carotenuto, A., DeLuca, G., Fabozzi, S. et al. (2016). Energy analysis of a small geothermal district heating system in Southern Italy. *International Journal of Heat and Technology* 34 (Special Issue 2): S519–S527.

61 DallaRosa, A. and Christensen, J. (2011). Low-energy district heating in energy-efficient building areas. *Energy* 36 (12): 6890–6899.

62 Tosatto, A., Dahash, A., and Ochs, F. (2023). Simulation-based performance evaluation of large-scale thermal energy storage coupled with heat pump in district heating systems. *Journal of Energy Storage* 61. https://doi.org/10.1016/j.est.2023.106721.

63 Lund, H., Werner, S., Wiltshire, R. et al. (2014). 4th generation district heating (4GDH): integrating smart thermal grids into future sustainable energy systems. *Energy* 68: 1–11.

64 Fabozzi, S., DeLuca, G., and Vanoli, L. (2022). Fourth generation district heating and cooling. In: *Polygeneration Systems* (ed. F. Calise, L. Vanoli, M.D. D'Accadia, and M. Vicidomini), 323–350. Academic Press https://doi.org/10.1016/B978-0-12-820625-6.00003-7.

65 Kauko, H., Kvalsvik, K., Rohde, D. et al. (2018). Dynamic modeling of local district heating grids with prosumers: a case study for Norway. *Energy* 151: 261–271. https://doi.org/10.1016/j.energy.2018.03.

5

Comprehensive Analysis and Future Outlook of Planning and Operation Approaches for Multicarrier Energy Systems Under the Integrated Local Energy Community Concept

Christina Papadimitriou[1], Marialaura Di Somma[2], Dimitrios Tzelepis[3,4], Koen Kok[1], and Giorgio Graditi[2]

[1] *Eindhoven University of Technology, Electrical Engineering Department, Groene Loper 3, 5612 AE, Eindhoven, the Netherlands*
[2] *Italian National Agency for New Technologies, Energy and Sustainable Economic Development, Lungotevere Thaon di Revel, 76, 00196, Rome, Italy*
[3] *Smart Power Networks Ltd., 65 More Close, London W14 9BN, UK*
[4] *University of Strathclyde, EEE Department, Royal College Building, 204 George St, Glasgow G1 1XW, UK*

5.1 Introduction

The new geopolitical and energy market reality requires drastically accelerating the clean energy transition and increasing Europe's energy independence from unreliable suppliers and volatile fossil fuels. Therefore, the need for a rapid clean energy transition has never been stronger and clearer. In the context of phasing out the European Union (EU's) dependence before 2030, the EC proposed the REPowerEU plan [1] that will increase the resilience of the EU-wide energy system based on two key pillars, with the second pillar being "reducing faster the use of fossil fuels in our homes, buildings, industry, and power system, by boosting energy efficiency, increasing renewables and electrification, and addressing infrastructure bottlenecks." Therefore, key answers addressing the question of how the EU energy transition should be shaped, include: (i) **self-sufficiency** coming from local resources, i.e. meaning *diversification* and (ii) **energy-independency** achieved through high penetration of renewable energy sources (RES) in all energy sectors, i.e. meaning *decarbonization.*

Both key aspects suggest that the bottom-up approach in the energy transition is the way forward. This means that local energy communities are the effective means of decarbonization and especially by optimally planning and operating them, which can truthfully contribute to the EU energy vision. On top of that, the local energy communities are seen as the technical hubs where the different energy carriers can collaborate and integrate forming the integrated local energy community (ILEC) vision as analyzed in Chapters 2 and 3. This suggests that the ILEC is a complex system whose optimal design and operation is multifaceted and thus getting more challenging.

Integrated Local Energy Communities: From Concepts and Enabling Conditions to Optimal Planning and Operation, First Edition. Edited by Marialaura Di Somma, Christina Papadimitriou, Giorgio Graditi, and Koen Kok.

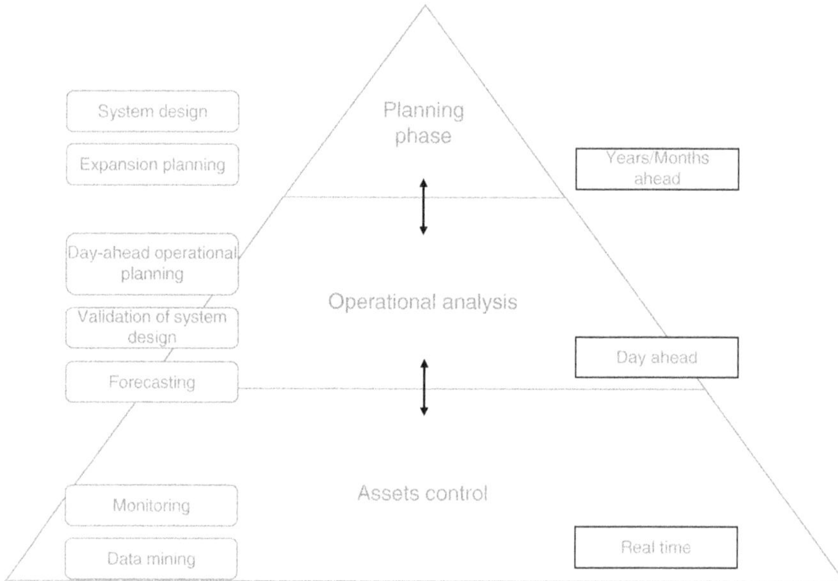

Figure 5.1 Overview of the planning and operation processes in an ILEC.

Different approaches have been presented in the past and the aim of this chapter is to analyze them and highlight the main challenges giving a holistic approach and covering different timelines and hierarchical levels of decision-making within the ILEC framework. Figure 5.1 shows schematically the levels of hierarchy, timelines, and the interaction of processes within the ILEC and will be analyzed in the next.

In more detail, the chapter will highlight the most important aspects when designing and operating multicarrier energy systems by considering uncertainties, operational constraints, and interaction with the markets. Therefore, the chapter will review the long-term planning approaches of an ILEC and will also cover the optimal operation planning in both day-ahead and real-time focusing on the ILEC interaction with the markets and demand-side management programs. It will also focus on the actual management and control of the ILEC assets while shedding light on data architectures for better visibility within the system.

The structure of the chapter is as follows. Section 5.2 presents the optimal planning and decision-making of the ILEC for the long-term horizon. Section 5.3 discusses the decision-making in operational planning phase (day-ahead) and how uncertainties affect this layer. Section 5.4 focuses on optimal operation in real time and Section 5.5 focuses on real-time control and related technical aspects. Section 5.6 shares the holistic vision of planning and operating an ILEC while Section 5.7 concludes the chapter.

5.2 Optimal Planning of Multicarrier Energy Systems

5.2.1 Introduction to Optimal Planning and Technical Aspects

Optimal planning – including expansion and reinforcement – of multicarrier energy systems is a complex and interdisciplinary endeavor that necessitates expertise in engineering, economics, environmental science, and policy analysis. The overarching objective is to develop an integrated, flexible, and sustainable energy system that can facilitate the transition to a low-carbon future [2]. Traditionally, energy system planning has been conducted on a carrier-by-carrier basis, with each vector planned in isolation without considering its interconnection with other carriers [3]. However, over the years, energy systems have encountered numerous techno-economic challenges such as excessive investments in network assets, network outages, variability in renewable resources, and limitations of natural resources. Simultaneously, advancements in energy systems research have revealed that adopting a synergistic approach among different energy carriers can yield substantial benefits for both operational efficiency and planning of energy systems [4].

Emerging as areas of growing importance, several key technical aspects are gradually receiving attention but have yet to be fully integrated into energy systems planning. These aspects encompass various domains and play a crucial role in optimized and resilient energy systems planning.

Modeling and interactions: The advent of multicarrier energy systems has brought about a heightened level of complexity, necessitating a comprehensive approach for their modeling and planning. These systems involve intricate interactions and interdependencies among different energy carriers, requiring planners to account for various factors. Ensuring sector coupling, where different energy sectors such as transportation, industry, and buildings are interconnected, poses an additional challenge. It requires a holistic approach that considers the unique requirements and interactions of each sector. Addressing these complexities and challenges necessitates the development of accurate models and sophisticated algorithms. These tools play a crucial role in capturing and optimizing the complex interactions among different carriers, ensuring efficient system operation and maximizing energy utilization.

Electrification of transport: The electrification of transport has witnessed a remarkable surge, leading to a substantial impact on the necessary infrastructure for charging. However, alongside the widespread adoption of electric vehicles (EVs), the momentum behind alternative fuel options, such as hydrogen-powered EVs, is also gaining traction. Consequently, the analysis required for developing infrastructure extends beyond the mere coupling of energy carriers. It now necessitates a comprehensive understanding of various factors, including travel patterns, diverse road networks, and vehicle specifications. To effectively cater to the growing demand, infrastructure planning must consider the geographic distribution of charging stations, ensuring convenient accessibility for EV owners across different regions. Understanding travel patterns becomes crucial in determining optimal charging locations along frequently traversed routes, enabling seamless long-distance travel. Furthermore, incorporating different road networks into infrastructure planning guarantees that charging facilities are strategically placed, considering urban areas, highways, and rural regions. Moreover, the analysis must consider the diverse range of vehicles utilizing charging infrastructure. Tailoring solutions to accommodate different vehicle types, such as cars, buses, and trucks, ensures that charging stations can cater to the specific requirements and charging capabilities of various EV models [5, 6].

Electrification of heat: Planning energy systems for the electrification of heat, coupled with the integration of alternative fuels, presents several challenges that must be addressed to ensure efficient and sustainable solutions. One crucial aspect is the consideration of heat comfort levels, as different areas and buildings have varying requirements for heating. A comprehensive analysis of heat demand patterns, along with the identification of suitable energy sources, is essential for designing an optimized network. Furthermore, the electrification of heat extends beyond residential heating. Industrial processes, such as manufacturing and large-scale heating applications, need to be considered. Planning networks should incorporate the diverse energy needs of industries, including the integration of alternative fuels like hydrogen or biomass, to meet their specific requirements effectively [7, 8].

5.2.2 Decision-making Process for Multicarrier Energy Systems Planning

The planning of a multicarrier energy system is a complex decision-making process – as already established – aiming at the system design, namely the identification of the optimal configuration in terms of choice of technologies, and their numbers and sizes by considering one or multiple objective functions.

The complexity of this decision-making process is linked to (i) the presence of multiple generation, conversion technologies, and thermal storage systems which convert and store, respectively, (ii) a series of energy carriers with complex interactions between them, and (iii) the multienergy loads of a user or group of users varying in dynamic conditions [9]. The complexity is also linked to the presence of various stakeholders interested in the development and consequent diffusion of multicarrier energy systems serving districts. A purely economic optimization aimed, for example, at minimizing costs, is not sufficient to ensure the long-term sustainability of the energy supply system. The latter is considered increasingly important by European energy legislation [10].

In general, in the last century, energy planning efforts have mainly been directed toward energy models aimed at exploring the relationship between energy and the economy and only one carrier as already mentioned. Therefore, a single-objective approach was used to identify the cheapest energy supply options [11]. Since 1980, with the growing awareness of the need to ensure a sustainable development of energy supply characterized by low environmental impacts, the decision-making framework related to energy planning has slightly changed [12].

The need to integrate considerations especially of environmental nature into energy planning has, in fact, led to the growing use of multiobjective approaches. These methods provide a better understanding of the intrinsic characteristics of the decision problem, promoting the role of stakeholders in the decision-making process, and facilitating collective compromised decisions. Furthermore, they contribute to improving the quality of the decision, making it explicit, and therefore rational and more efficient, facilitating not only negotiations between the various stakeholders but also the quantification and determination of priorities [13].

5.2.3 Features and Modeling Approaches of Optimization Problems for the Multicarrier Energy Systems Planning

An optimization problem formulated for planning a multicarrier energy system can be formally expressed as follows:

$$\begin{gathered} \min F_{\text{obj}}(x,y,\theta) = \min\left([f_1(x,y,\theta), f_2(x,y,\theta), \dots, f_m(x,y,\theta)]\right) \\ \text{subject to} \\ G(x,y,\theta) = 0 \\ H(x,y,\theta) < 0 \\ x \in \{0,1\} \\ y \in \Omega \end{gathered} \tag{5.1}$$

where f_i represents the ith objective function; m represents the number of objective functions; x is the vector of binary variables that are usually used to represent logic relations or choices, e.g. whether to install a technology or not; y is the vector of continuous decision variables, e.g. size of technologies; Ω is the domain of the continuous decision variables; θ is the vector representing parameters involved in the optimization problem, such as energy price or user load; $G(x,y,\theta)$ represent the equality constraints (e.g. energy balances to meet user loads); and $H(x,y,\theta)$ represent the inequality constraints (e.g. capacity constraints of technologies or storage systems).

In most cases, optimization problems formulated for multicarrier energy systems planning are based on the mixed-integer programming (MIP) approach, which can contain both integer and continuous decision variables. When the integer variables are limited to the values 0 and 1, then they are referred to as binary decision variables and can, for example, be used to model decisions such as the on/off type of a certain technology within the system. Continuous decision variables, on the other hand, can assume any value between a lower and an upper limit, and can be representative, for example, of the thermal or electrical power supplied by a certain technology in a certain instant of time. Furthermore, these types of optimization problems can be linear or nonlinear.

A linear programming problem is characterized by objective functions that are linear in the decision variables, as well as by constraints that are equalities or inequalities, which are also linear in the decision variables [14]. Optimization problems of mixedinteger linear programming (MILP) type are usually difficult to solve, since a set of decision variables is limited to integer values. The branch-and-cut algorithm is the most used deterministic method to solve these types of problems, since it is already implemented in several commercial optimization software [13]. This type of method is an exact resolution method, i.e. it allows solving a combinatorial optimization problem in an exact way, finding a feasible solution that corresponds to the optimum of the objective function among all the feasible solutions.

On the other hand, a nonlinear programming problem includes at least one nonlinear function in the decision variables, which could be an objective function or some or all of the constraints. Problems of mixed-integer nonlinear programming (MINLP) type combine the difficulties of solving a combinatorial problem with the complexity of the nonlinear (and often nonconvex) character of objective functions and/or constraints [15]. Usually, there are two approaches used to solve this type of programming problem. The first approach consists of applying some simplifying hypotheses during the formulation of the problem itself, such as the linearization of the objective function and/or of the constraints. In this way, it is possible to solve the optimization problem, transforming it into a linear programming problem, for which efficient exact solution methodologies are available. The application of exact methods is not always possible, essentially for two concomitant reasons, such as the intrinsic complexity of the problem and the time available to generate the solution. In general, the renunciation of the use of exact methods must be a weighted choice, based not only on the theoretical complexity of the problem but also on the effective practicability of exact methods linked to the processing time available and the size of the problem to be solved. However, sometimes the complexity of the problem could make it practically impossible to arrive at a sufficiently accurate and manageable formulation [16]. When the problem does not allow applying exact solution techniques, it becomes necessary to provide *good* admissible solutions at *reasonable* times. Therefore, the second approach is based on the use of heuristic methods, such as genetic algorithms and evolutionary algorithms [13].

5.2.4 Exact Modeling/Exact Optimization Dilemma

In the context of multicarrier energy systems planning, the choice of the solution methodology represents a real dilemma between exact modeling and exact optimization [15]. When a real optimization problem is simplified in the modeling phase, i.e. in the problem formulation phase, the optimal solutions found are *real solutions to a nonproblem*, as defined by [17]. For example, a simplified linear model of the network components, optimized with a very accurate solution method, will produce solutions which, even if accurate, are not optimal for the real problem, given the simplifying assumptions adopted. Similarly, a realistic and complex model of the network, optimized with an inaccurate solution method, generates *nonsolutions to a real problem* [15], i.e. solutions that are not optimal, despite the formulated realistic model. These examples illustrate the dilemma between exact modeling and exact optimization, depicted in Figure 5.2, faced in solving real-world optimization problems.

The formulation of the optimization problem should be as close as possible to the upper right corner of the figure. This requires a good compromise between the

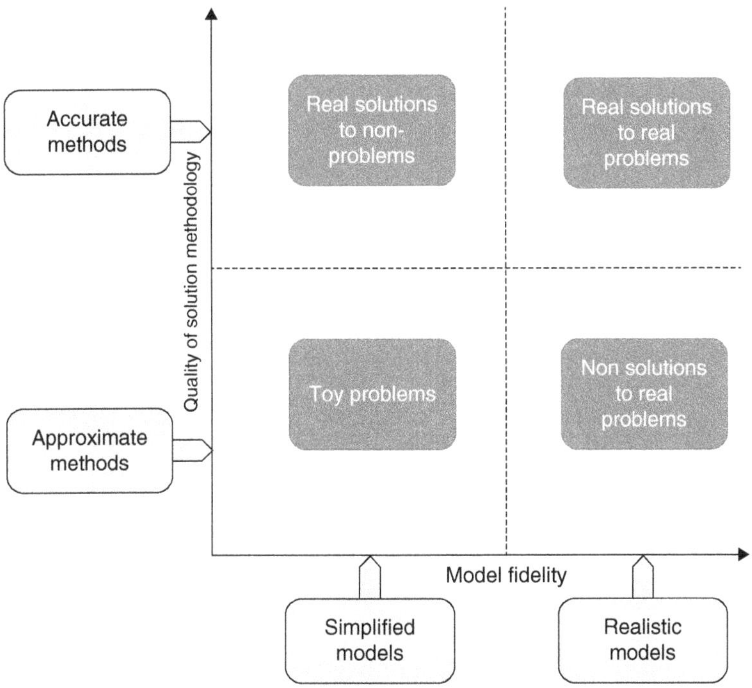

Figure 5.2 Representation of the exact modeling/exact optimization dilemma in the planning of multicarrier energy systems. Source: Adapted from Alarcon-Rodriguez et al. [15].

accuracy of the optimization method and the complexity, with the relative level of detail, of the formulated multicarrier energy system model.

5.2.5 Need for a Multiobjective Approach

For planning a multicarrier energy system, a single-objective approach aimed at minimizing costs is the practical one for planners, whose greatest interest is linked to the economic factor. Through the economic optimization of the system, the planner can in fact obtain information on the system technologies which allow to minimize the total costs for investments, operation, and maintenance, for all energy carriers.

However, in a realistic scenario, the planning of a multicarrier energy system sees the interest of different stakeholders, with objectives formulated from different perspectives. If, on the one hand, there are the planners, whose greatest interest is certainly represented by the economic factor, on the other hand, there are the energy legislations, including the European ones, which are instead particularly interested in reducing the environmental impacts linked to the energetic supply. These types of goals are conflicting in nature, and as a result, there is no single solution that can satisfy all stakeholders. Hence, the need to move to a multiobjective approach that allows compromised solutions to be found for the benefit of the various stakeholders is a one-way direction. By assigning objective functions with inherently conflicting objectives, such as economic/environmental ones, the quality

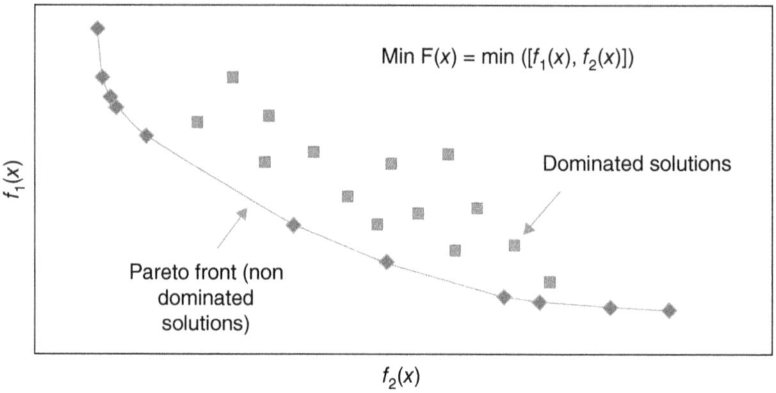

Figure 5.3 Graphical representation of the Pareto front in the form of a trade-off curve for two objective functions.

of the decision-making process is improved, making it rational, and therefore efficient. Furthermore, the use of multiobjective approaches in energy network planning can encourage the use of policy tools aimed at favoring the integration of these systems, showing their economic and environmental benefits as well as their impacts. When an optimization problem includes only one objective function, there is only one *best solution*. Conversely, a multi-objective problem with conflicting objectives does not have a single best solution, but a set of best solutions. In this case, the multidimensional concept of *dominance* is used to evaluate whether one solution is better than the others. A solution *a* is dominant over *b* if the following two conditions are true [18]:

- *a* is not worse than *b* in all objectives.
- *a* is better than *b* in at least one objective.

In that case, *b* is said to be *dominated* by *a*, or that *a* is *nondominated* by *b*. A dominated solution is also called *sub-optimum*. Thus, the solution of a multi-objective problem is the nondominated set of solutions, known as the Pareto set. With reference to its objective functions, it becomes a Pareto front, whose graphical representation is generally a surface, but if there are only two objective functions, it degenerates into a trade-off curve as shown in Figure 5.3. A solution belongs to the Pareto front, if it is not possible to improve one goal without making all the other goals worse.

The search for a single solution to a multiobjective optimization problem includes two steps: optimization and decision-making. Depending on the order in which these two operations are performed, there are two different approaches to obtain a unique solution from a multiobjective optimization problem, namely:

1. For the first approach, the preference information of the stakeholders is known *a priori* and single-objective optimization techniques are used. The decision-making process precedes the optimization process. This procedure is very useful when precise information on stakeholders' preferences is available.

2. For the second approach, the optimization process is performed before the decision-making process, and it is preferable in case of unavailability of information related to the interests of the stakeholders. In fact, the multiobjective optimization process becomes crucial and is based on the analysis of the Pareto front. The latter, in fact, makes compromise solutions visible between the various stakeholders or between incommensurable objectives. Pareto front analysis facilitates the decision-making process and allows understanding of the intrinsic characteristics of the problem. Therefore, once the Pareto front is obtained, the preference is expressed as *a posteriori*. This second approach is considered the most representative of reality since it is less subjective and more transparent than the first method and allows obtaining a more realistic representation of the planning problem.

5.2.6 Analysis of Works in the Literature Dealing with Multicarrier Energy Systems Planning

The optimal planning of multicarrier energy systems is widely investigated in the literature by using both single-objective functions aimed at minimizing total costs and multiobjective functions aiming at reducing costs and environmental emissions (mostly CO_2 reduction). Table 5.1 summarizes the main features of the most significant works analyzed in the literature on the planning of such systems. They focus both on single poly-generation technologies, such as cogeneration and trigeneration, and on complex multicarrier energy systems with multiple technologies such as cogeneration and trigeneration systems, gas-fired boilers, electric heat pumps, electrolyzes, fuel cells, electric, thermal, and hydrogen storage systems.

The main goal of these optimization problems is to find the mix of technologies, their number with the relative sizes, and the operational strategies in representative days for a given user or group of users. Among these problems, the single-objective ones have as their objective function the minimization of total costs or the maximization of the Net Present Value. Multiobjective optimization problems instead use economic/environmental criteria intended either as minimization of total annual costs and CO_2 emissions or as maximization of energy/exergy and economic savings. The economic objective function is typically formulated as the sum of various cost functions such as the investment costs of technologies, maintenance costs, costs of energy carriers, and costs related to the carbon tax, where present, to minimize. The most used solution methodology is that typical for linear programming problems.

5.3 Operational Planning of Multicarrier Energy Systems for Day-Ahead Optimization and Decision-Making Under Uncertainties

Once the system configuration is set, the operational planning of multicarrier energy systems becomes the next step to address for achieving daily crucial objectives as minimization of daily energy costs and/or environmental impacts.

Table 5.1 Main features of the most significant works analyzed in the literature on multicarrier energy systems planning.

References	Energy system under study	Objective(s)	Modeling approach of the optimization problem
[19–22]	Cogeneration	Min {total annual costs}	Linear programming
[23, 24]	Cogeneration	Min {total annual costs}	Nonlinear programming
[25]	Trigeneration	Max {net present value}	Linear programming
[26]	Trigeneration	Multiobjective: Min {total annual costs, CO_2 emissions}	Linear programming
[27]	Trigeneration	Multiobjective: Min {total annual costs, CO_2 emissions}	Evolutive algorithm
[28–34]	Multicarrier energy system	Min {total annual costs}	Linear programming
[35]	Multicarrier energy system	Max {net present value}	Genetic algorithm
[36]	Multicarrier energy system	Multiobjective: Min {total annual costs, CO_2 emissions}	Evolutive algorithm/Linear programming
[9, 37–40]	Multicarrier energy system	Multiobjective: Min {total annual costs, CO_2 emissions}	Linear programming
[41, 42]	Multicarrier energy system	Multiobjective: Min {total annual costs, fossil primary energy consumption}	Linear programming
[43]	Multicarrier energy system	Multiobjective: Min {total annual costs, total exergy efficiency}	Linear programming

Challenges related to daily operation of multicarrier energy systems are multiple. First, considering the interaction between energy carriers (e.g. electricity, heat, cooling, gas, hydrogen, and mobility), and related technologies is not a trivial task. The imbalance between supply and demand sides, mainly related to the typical instantaneous variation of user demand and the limited operation flexibility of certain technologies, is another challenge that needs to be properly managed in the operation planning phase [44–46].

The considerations presented in Section 5.2 related to the features and modeling approaches of optimization problems, the exact modeling/exact optimization dilemma, and the need of a multiobjective approach are valid also for the operational planning of multicarrier energy systems. However, in this case, the goal of the optimization problem is to find the daily operation strategies (typically with one hour as time step) of the multicarrier energy system that minimize or maximize one or multiple objective functions under specific problem constraints as, for instance,

Table 5.2 Main features of the most significant works analyzed in the literature on multicarrier energy systems operational planning.

References	Energy system under study	Objective(s)	Modeling approach of the optimization problem
[47, 48]	Cogeneration	Min {energy costs}	Linear programming
[49]	Trigeneration	Min {energy costs}	Genetic algorithm
[50–53]	Multicarrier energy system	Min {energy costs}	Linear programming
[54]	Multicarrier energy system	Min {energy and emission costs}	Linear programming
[55]	Multicarrier energy system	Max {ILEC's profit}	Linear programming
[56, 57]	Multicarrier energy system	Multiobjective: Min {energy costs, exergy of input energy carriers}	Linear programming
[58]	Multicarrier energy system	Multiobjective: Min {energy costs, exergy losses}	Nonlinear programming
[10, 59–64]	Multicarrier energy system	Multiobjective: Min {energy costs, CO_2 emissions}	Linear programming
[65]	Multicarrier energy system	Multiobjective: Min {OPEX, exergy of input energy carriers, stress indexes}	Nonlinear programming
[66]	Multicarrier energy system	Multiobjective: Min {energy costs, primary energy}	Linear programming

energy balances (electricity and thermal energy balances) to meet user loads and capacity constraints of technologies or storage systems.

The topic of operational planning of multicarrier energy systems has been widely investigated in the literature, as summarized in Table 5.2 that presents the main features of the most related significant works. They focus both on single poly-generation technologies, such as cogeneration and trigeneration, and on complex multicarrier energy systems with multiple energy technologies and different types of user loads. Moreover, the optimization problems are formulated with a deterministic or a stochastic approach. In detail, in deterministic optimization models, parameters θ defined in Eq. (5.1) are assumed to be known. In relation to the latter aspect, a crucial issue to be considered in the operational planning process is related to uncertain parameters that can be associated with both supply and demand sides such as renewables, energy prices (e.g. electricity and gas), and energy demand uncertainties [44]. Considering parameters known through a deterministic approach, thus failing to consider uncertainties in the decision-making process, may lead to suboptimal or even infeasible solutions. In practice, the feasible operation strategies of multicarrier energy systems may deviate from the optimal ones [44–46].

Day-ahead operational planning of multicarrier energy systems may need support of forecasting tools for renewables, energy prices, and users' energy demand related to electricity, heating, and cooling, which are input data for the operation optimization problem. This way, it can be treated with a deterministic approach by using the forecasted data as assigned input data and the optimization model can be run on the daily basis for providing the day-ahead optimized operation strategies of the multicarrier energy system under study.

Another way to achieve day-ahead optimized operation strategies of multicarrier energy systems is to model uncertainties and consider them in the optimization problem formulation. The assessment of uncertainties in modeling of multicarrier energy systems is a widely investigated topic in the literature, with most of the works focusing on assessment of variable renewables as photovoltaic (PV) or wind turbines (WTs) [63, 67], and the uncertainties related to electrical and thermal loads [68–70]. Sometimes, uncertainties related to energy prices such as electricity and gas prices and EVs are also modeled in the operation optimization problems.

Different methodologies can be used for handling uncertainties in operation optimization of multicarrier energy systems.

- The *stochastic optimization* or *stochastic programming* generates and uses random variables related to uncertain variables. For stochastic problems, the random variables appear in the formulation of the optimization problem itself, which involves random objective functions or random constraints.
- The *robust optimization* defines the optimal solution considering more adverse scenarios regardless of the probability of occurring.
- The *chance-constrained optimization* uses probabilistic constraints to find compromises between the optimal value and the robustness of solution.

In more detail, the difference between the three approaches consists of the degrees of risk aversion and ways of uncertainties characterization. Stochastic programming is a risk-neutral approach, which aims to optimize the expected outcome over a known probability distribution. Robust optimization problems typically involve min–max type of operators. Chance-constrained optimization can be seen as solving a stochastic program with probabilistic constraints, which specify that some constraints with uncertain parameters are satisfied with a given level of probability. The main differences of the three methods are summarized in Table 5.3 [71].

There are also additional mathematical frameworks used in the literature to model decision-making under uncertainty such as the Markov-based model as in [54, 72], where PV uncertainties are modeled by a Markovian process and, for an effective coordination, other devices are modeled as Markov processes with states depending on PV states in the operation optimization problem for an ILEC and a microgrid, respectively.

Among the analyzed references, the stochastic optimization is the most used methodology. In this case, the probability distributions of the uncertain parameters

Table 5.3 Features of stochastic, robust, and chance-constrained optimization methods.

Stochastic optimization
• **Degree of risk aversion:** Risk neutral
• **Characterization of uncertainty:** Probability distribution
Robust optimization
• **Degree of risk aversion:** Worst case scenario
• **Characterization of uncertainty:** Uncertainty set
Chance-constrained optimization
• **Degree of risk aversion:** Risk averse
• **Characterization of uncertainty:** Probability distribution, risk level

are assumed known *a priori*. The uncertainties are characterized by some discrete realizations of the uncertain parameters as an approximation to the real probability distribution [67, 71]. Each of these realizations represents a *scenario*, and the aim of the stochastic optimization is to optimize the expected value of the objective function(s) over all the scenarios. The generic expected objective function can be thus formally expressed as

$$\mathbb{E}[f(x, y, \xi)] = \sum_{k=1}^{K} \pi_k f(x, y, \xi_k) \tag{5.2}$$

where ξ is the vector of scenarios and π_k is the probability of occurrence of scenario ξ_k, with k being the index of each scenario.

Generation of scenarios is a key for using stochastic programming and the high number of scenarios may affect the computational effort of the stochastic optimization phase. It is thus important also to reduce the number of scenarios for preserving the most representative ones. For instance, in [67, 73], a comprehensive tool is presented for the scenario generation of solar irradiance profiles by using historical data for a specific location, being particularly useful for creating scenarios in the context of the stochastic operation optimization of multicarrier energy systems. By using the Roulette Wheel mechanism for generating an initial set of scenarios, the tool applies a reduction process based on the Fast-Forward method, which allows the preservation of the most representative ones while reducing the computational efforts in the next stochastic optimization phase.

A particular case of stochastic optimization is represented by the two-stage stochastic programming shown in Figure 5.4, where the first decision is made *here and now* and cannot depend on future realizations. The second decision is *wait and see* which can be made after some of the random parameters are observed [71, 74]. The generated scenario tree is then used to solve the stochastic optimization problem and the quality of the solution can be evaluated by measures related to the statistical optimality gap [75].

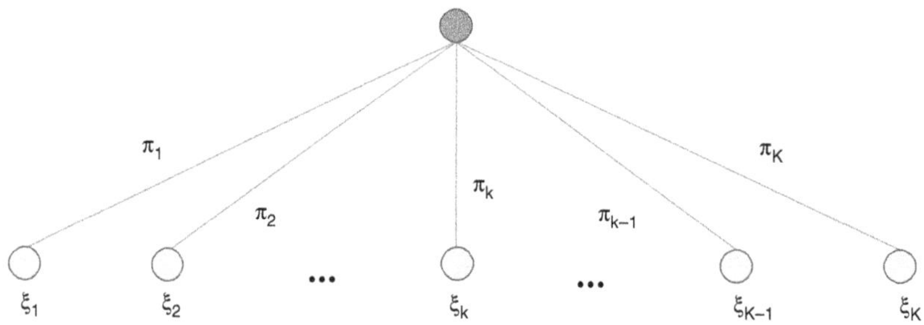

Figure 5.4 Graphical representation of a two-stage scenario tree.

5.4 Optimal Operation of Multicarrier Energy Systems in Real Time Under Multiobjective Approaches Considering Demand-Response Programs and Market Interaction

So far, the optimal planning in long term and operational day-ahead have been presented. Multicarrier energy systems' optimal operation in real time is also a challenging problem due to issues such as complexity of the system and the presence of uncertainty that will be analyzed in Section 5.4.1 and are in close resemblance with the ones discussed in Section 5.2.

Several studies have investigated the optimal operation of multicarrier systems under multiobjective approaches in the past. Less of them are also considering the interaction with demand-response programs (DRPs) while very few deal with the market interaction as well. Even less research studies deal with the interaction of multiple markets, e.g. other than electricity markets. The scientific resources are briefly described below.

5.4.1 Real-Time Operation of Multicarrier Energy Systems: Drivers and Challenges

One of the primary challenges in real-time operation is the dynamic nature of multicarrier energy systems, which consist of diverse energy sources, storage devices, and loads. The integration of these different components requires careful coordination and control to ensure reliable and cost-effective operation. Moreover, the variability and uncertainty associated with renewable energy generation and the dynamic nature of electricity market prices further complicate the real-time operation.

To address these challenges, advanced control and optimization algorithms play a crucial role. These algorithms leverage real-time data from various sources, including sensors, smart meters, and communication networks, to make informed decisions in real time. They enable effective coordination and control of energy generation, storage, and consumption, of different carriers considering multiple objectives such as cost minimization, energy efficiency, and/or environmental impact as will be explained in Section 5.4.2.

Forecasting techniques play a vital role in real-time operation by providing accurate predictions of load demand, renewable energy generation, and market prices. Load forecasting helps anticipate the electricity consumption patterns, enabling optimal scheduling and dispatch of energy resources. Renewable energy forecasting assists in incorporating renewable generation into the system while managing its intermittent nature. Market price forecasting aids in decision-making, allowing peers to optimize energy trading and participate in real-time markets.

Real-time market interaction mechanisms are essential for efficient operation and market participation. Real-time pricing provides dynamic prices based on the current supply-demand conditions, encouraging consumers to adjust their consumption patterns accordingly. Demand response activation, triggered by market signals, enables consumers to modify their energy usage in response to price fluctuations or grid conditions, contributing to demand-side management.

Together, these aspects facilitate the seamless operation of multicarrier energy systems in real time, promoting energy efficiency, cost-effectiveness, and grid stability.

5.4.2 Multiobjective Approaches for Optimal Real-Time Operation

The real-time management of multicarrier energy systems must consider the diverse requirements and preferences of different stakeholders. This is ultimately true when it comes to ILEC concept. For example, an economic optimization that satisfies aggregators' needs may not guarantee an optimal operation strategy from an environmental or social standpoint, which is essential for the energy community and its peers in real time. This inherent conflict between objectives necessitates the adoption of a multiobjective approach when addressing the real-time optimal operation of integrated systems.

In the literature, there are only a few of research papers that deal with the multiobjective approach, and they are mostly focusing on the management systems of houses or buildings, i.e. potential peers in ILEC fostering technologies from different carriers. The objectives can range from cost-related factors such as operational costs to environmental or social concerns such as minimizing CO_2 emissions and maximizing comfort, respectively.

A weighted sum multiobjective home energy management system (HEMS) is proposed in [76] to minimize the energy procurement cost and peak load demand using the demand-side flexibility provided by the storage system. Alıç and Filik [77] introduces five types of discomfort zones resulting from the HEMS's operation for detailed cost–comfort analysis using the weighted sum method. The energy management of a multicarrier structured smart home is carried out in [78] by considering the uncertainties involved in PV generation using a two-point estimate method. Kazemi et al. [79] presents a HEMS whose primary objective is to achieve cost-effectiveness and the secondary to minimize the carbon footprint. The system considers the electricity and thermal energy carriers, with bidirectional energy exchange capability of electricity and heating as well as gas carrier.

5.4.3 Demand-Response Programs and Market Interaction

As already mentioned, DRPs and market interactions are two important schemes that can be considered in the optimal operation of such systems. DRPs allow ILEC peers to adjust their energy consumption in response to real-time pricing signals or other incentives, which can help balance the supply and demand. Market interactions, on the other hand, can help generate revenue and reduce the overall cost of energy.

5.4.3.1 Demand-Response Programs in Multicarrier Energy Systems

Most existing literature for DRP approaches focuses on the electrical networks. When it comes to multicarrier systems, DRP and its impact are often considered for the configuration or/and day-ahead scheduling of the system. So, there is a few existing literature in the context of real-time scheduling as discussed in the next.

Lu et al. [80] considers a robust optimization approach for optimal load dispatch and integrates electrical and thermal DRPs for shiftable loads. Dan et al. [81] designs a Predictive Control model for scheduling devices to perform price-based demand response and reduce operating costs in a multicarrier system building. Certain models focus on shifting loads from high-price to low-price time slots to reduce operational costs [82], or transferring loads from peak demand to off-peak periods [83] under energy hubs and multicarrier systems respectively. Furthermore, time of use (TOU) tariff prices are presented as an incentive scheme for end-users to reduce operational costs by shifting loads from high- to low-price periods [84]. Deferrable and critical loads are also considered in specific cases for an energy hub [85]. The latter constitutes a possible restriction of converting between the different energy carriers across time and for different consumers. For instance, these critical loads could be only electrical related. Therefore, it is not possible to participate in the DRPs [13, 86]. In addition, the execution of real-time price (RTP) is integrated into the model subject to minimize the operation cost and to maximize the consumer's welfare index for a residential energy hub [87].

It is worth noting that while demand response (DR) mechanisms are predominantly incorporated to facilitate flexible multicarrier energy systems operation, these studies do not model the DR mechanisms from a market or network operation perspective. It is also remarkable that the research so far is dealing only with flexible loads and not flexible resources as well. Table 5.4 summarizes the related research papers.

5.4.3.2 Market Interaction with Multicarrier Energy Systems

In this section, the market approach is investigated, exploring various aspects such as wholesale electricity markets, grid services, and gas market/trading [55, 63, 64, 88–90].

Regarding the reviewed literature, multicarrier energy systems actively participate in the *day-ahead market (DAM)*, considering variable costs and different timeframes. Typically, a single entity seen as an ILEC is connected to the utility grid at the point of common coupling (PCC), assuming the presence of a wholesale market. The interaction with the DAM can be asymmetric, focusing exclusively on energy imports or

Table 5.4 Multicarrier energy systems' approach to demand-response programs.

References	Approach to demand response in multicarrier energy systems
[80, 85, 86]	Electrical and thermal demand-response programs for shiftable loads
[81–84]	Load shifting from high-price to low-price times to reduce operational costs
[87]	Real-time price for minimization of operational costs and/or maximizing comfort

the ILEC can both import and export excess energy to the wholesale market. In the latter case, energy exports may arise from surplus generation in a combined heat and power (CHP) system, the available electrical flexibility of distributed energy resources (DERs), or EVs operating in vehicle-to-grid (V2G) mode, etc.

After the DAM, *the intraday market (IDM)* serves as the second stage, allowing for the correction of deviations from the DAM schedule and operation. Most research models deal with their optimization operation models for an ILEC, i.e. buy or sell electricity to meet its own electricity and heat demand while minimizing costs and considering the market interaction using day-ahead and real-time prices. But they do not specifically address wholesale electricity markets for each timeframe as the uncertainties of "real-time prices" are highly correlated with DAM prices. In [91], a multiple-stage optimization is considered with three temporal horizons: day-ahead dispatch with 1-hour resolution, and intraday operation with 15-minutes and 5-minutes beforehand scheduling intervals. This market design representation is closer to the realistic market dispatch.

In general, when participating in the day-ahead electricity spot market, the ILEC or a proxy acting on behalf of the ILEC, such as a market agent, must submit selling or buying bids before the market gate closure, usually the day before the electricity delivery or withdrawal. Between the gate closure and the delivery of energy cleared in the day-ahead market, several intraday electricity markets take place. During these IDMs, the market agent representing the ILEC can submit bids to modify the generation schedule determined in the day-ahead market or the previous IDM.

Recently, Coelho et al. [92] combined day-ahead and real-time optimization strategies to address this gap, focusing on an aggregator representing multienergy assets. Nevertheless, further research is needed to investigate additional actors under various realistic business models and different market settings such as balancing and peer-to-peer (P2P) transactions.

Apart from the wholesale electricity markets, other ancillary services should be considered as well in favor of network operators, such as grid congestion, reserve, balancing markets, frequency control, or voltage regulation. Nevertheless, ancillary or grid services are hardly considered in the literature under the scope of multicarrier energy systems.

Another point that needs to be highlighted – when it comes to ILECs – is the negotiation of natural gas supply and consumption that also occurs in day-ahead and IDMs or sessions. However, the time horizon and temporal resolution of the day-ahead and intraday gas market sessions differ from those of the electricity markets. Gas imbalances are computed in the gas market, with the gas sold or purchased in each market or session settled at the market clearing price (usually marginal). Excess or deficit gas imbalances are cleared at the imbalance price, which typically differs for excess and deficit imbalances. Although Coelho et al. [92] considers the day-ahead gas market through multienergy bidding, the intraday gas market is not addressed, meaning that the entire process has not been fully considered thus far from the research community. In most papers where the gas price is considered, a single gas price is used, presumably corresponding to the day-ahead gas market price. Finally, the uncertainty of gas prices is not considered so far.

5.5 Data Architectures, Control Technologies, and the Scaling of Energy Systems

At the heart of the energy transition paradigm lies the intricate interplay between data architectures, control technologies, and the scaling of energy systems. Data Architectures serve as the strategic backbone of modern energy systems, enabling the efficient collection, storage, and analysis of vast amounts of data. By establishing robust frameworks for data integration, these architectures empower energy stakeholders to gain valuable insights into energy generation, distribution, and consumption patterns. This data-driven approach forms the basis for informed decision-making, enabling the optimization of resources, identification of inefficiencies, and the development of innovative strategies for sustainable energy management. Complementing the role of data architectures, control technologies provide the strategic framework for orchestrating the operations of energy systems. By integrating intelligent control algorithms and real-time data, these technologies enable precise monitoring, regulation, and coordination of energy resources. Control technologies play a pivotal role in a resilient and sustainable energy ecosystem.

5.5.1 The Importance of Visibility and Controllability in Managing Multicarrier Energy Systems

Effective management of multicarrier energy systems requires a deep understanding of the system's behavior, real-time visibility into its components, and the ability to control and optimize its operation. Visibility and controllability are crucial aspects of managing these complex systems as they provide operators of such systems with the necessary information and tools to make informed decisions, ensure system stability, and maximize efficiency [93, 94]. Some of the most important benefits of system monitoring are shown in Table 5.5 [95–97].

Controllability also plays a pivotal role in the effective management of multicarrier energy systems, offering relevant operators the ability to actively influence and optimize the system's operation. An overview of the importance of controllability and its

Table 5.5 Benefits of system monitoring in multicarrier energy systems.

	Key aspects	Benefits
Monitoring	**System understanding**	• Extended visibility among carriers and their interdependencies • Early bottlenecks identification • Informative decision-making • Optimization of planning and operational strategies
	Real-time monitoring	• Up-to-date information on system conditions, performance, and behavior (including assets) • Early detection of anomalies, grid disturbances, or equipment failures • Prompt intervention and minimizing potential disruptions
	Load balancing and demand	• Visibility into consumer demand patterns, load profiles, and energy usage to support load balancing and demand response strategies • Identification of peak demand periods • Easier load shifting techniques implementation • Informative incentivization of consumers • Improved grid stability
	Renewable energy integration	• Renewable energy generation forecasts, for optimized integration • Effective management of energy storage systems and curtailment strategies • Effective balancing of energy supply and demand

impact on various aspects of multicarrier energy system management is presented in Table 5.6 [98–103].

In conclusion, visibility and controllability in multicarrier energy systems is crucial for optimizing the cross-vector nature of these systems bringing along important benefits. However, achieving a high degree of controllability and visibility over the multicarrier energy system comes with the following overarching challenges as shown in Table 5.7 [104–108].

Of course, on top of the technical challenges, there are also challenges related to stakeholder engagement and collaboration. Multicarrier energy systems involve various stakeholders, including energy providers, grid operators, consumers, and regulatory bodies. Engaging and coordinating these diverse stakeholders is a challenge, as it requires alignment of interests, collaboration in decision-making, and agreement on control strategies. For example, optimizing electricity generation may conflict with the control of gas network pressure or thermal energy demand. Coordinating and balancing these conflicting objectives requires sophisticated control strategies, optimization algorithms, and coordination mechanisms that can address cross-vector interactions and find optimal solutions. Establishing effective governance structures, communication channels, and collaborative platforms is essential to foster stakeholder engagement and achieve coordinated visibility and control in a nonconflicting way.

Table 5.6 Benefits of system controllability in multicarrier energy systems.

	Key aspects	Benefits
Controllability	**Cross-carrier energy Integration and optimization**	• Efficient energy exchange and coordination between different energy carriers • Optimized cross-carrier energy flows • Integrated control strategies • Energy efficiency maximization • Better grid balancing and overall system performance
	Stability and resilience	• Better response to grid disturbances, voltage fluctuations, or equipment failures • Better frequency stability, voltage regulation • Rapid fault detection • Reliable energy supply
	Flexibility and flexibility trading	• Identification and optimization of the flexible resources • Better coordination of flexible assets among carriers • Cost-effective and sustainable energy operation
	Synergistic resource utilization	• Optimal coupling of different types of energy resources • Maximizing combined benefits and minimizing waste • Enhanced system efficiency

5.5.2 Outlook on the Complexities and Challenges of Data and Data Architectures

As the demand for energy continues to rise and the global transition toward sustainable sources accelerates, the need to scale up energy systems has become paramount. Achieving scalability in energy systems requires more than just expanding infrastructure; it necessitates a comprehensive understanding of data architecture and its pivotal role in enabling effective management and optimization.

As depicted in Figure 5.5, data and data architectures have the potential to span across various aspects, extending beyond control functions to influence strategy, planning, and organizational and social dimensions of energy systems [109–111]. At the operational level, which this section focuses on, data management plays a critical role in enabling efficient control strategies, optimizing energy flows, and facilitating real-time decision-making. Robust data architectures provide a foundation for seamless data integration, enabling effective monitoring, forecasting, and control of energy systems. However, the impact of data architecture goes beyond operational aspects. It extends to organizational aspects by fostering data-driven cultures within energy organizations, promoting collaboration and knowledge sharing, and facilitating evidence-based decision-making. Moreover, data architectures also have educational implications, as they support the development of data analytics skills and enhance the understanding of complex energy systems among professionals

Table 5.7 Challenges for extended controllability and visibility over multicarrier energy systems.

Challenges	Description
Transitioning from traditional systems	• Existing infrastructure, control mechanisms, and operational practices may not be suitable for multicarrier systems • Requirement for careful planning, coordination, and investment
Data integration	• Diverse data sources spanning different energy carriers • Integrating data complexity • Lack of data standardization, interoperability, and integration frameworks
Communication and interoperability	• Need for development and adoption of communication protocols, interfaces, and interoperable technologies • Interoperability challenges include data synchronization, real-time data sharing, and ensuring compatibility between control systems of different carriers
System complexity and scalability	• Need for advanced algorithms, modeling techniques, and control strategies • Scalable control architectures and algorithms that can handle the complexity and growth of such systems
Uncertainty and variability	• Real-time forecasting techniques, adaptive control strategies, and decision-making frameworks • Robust monitoring and predictive capabilities are required
Cybersecurity and privacy	• Ensuring the security of critical infrastructure • Protecting data integrity • Safeguarding consumer privacy required
Dynamic market conditions and regulatory challenges	• Adaptation to changing market conditions • Optimizations of energy trading and pricing strategies • Efficient response to market signals • Appropriate regulatory framework

and researchers. By leveraging data and well-designed architectures, energy systems can foster a culture of continuous learning, innovation, and adaptation. Ultimately, data architectures have the potential to shape the overall energy landscape, driving transformative changes, fostering sustainability, and enabling a future powered by scalable and efficient energy systems.

Especially for multicarrier energy systems, data and data architectures play a crucial role in their efficient management. Figure 5.6 provides an overview of the main components and data exchanges involved in monitoring, optimization, and control of multicarrier energy systems. While the pictorial representation is simplified, it is important to acknowledge that the complexities associated with these systems are significant. These latter arise from potential interdependencies, the multitude of monitored and controlled nodes, the intricate network routes, and the wide range of timeframes involved, spanning from days to milliseconds [112–115].

Ensuring the widespread adoption and risk-free operation of such systems requires data exchanges and data architectures that support scalability and replicability in a plug-and-play fashion, minimizing the need for manual configurations.

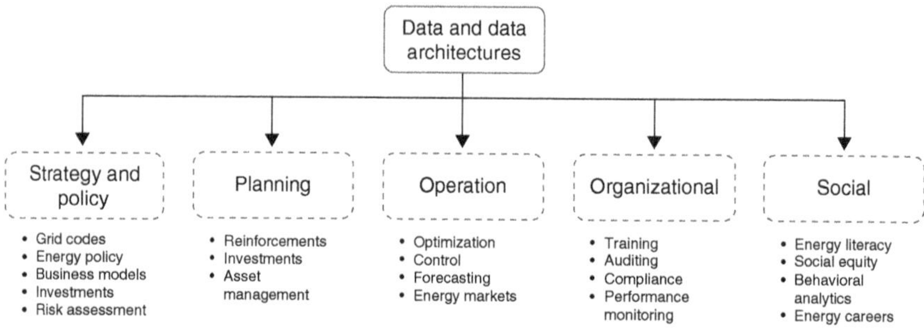

Figure 5.5 Overview of multiple dimensions of data and data architectures in energy systems.

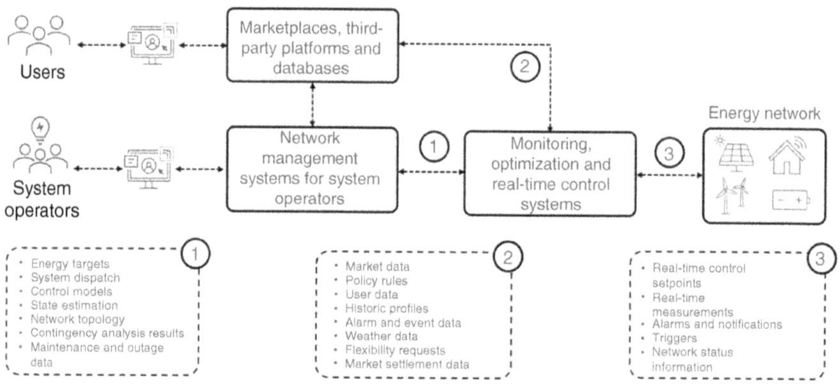

Figure 5.6 Overview of data exchanges for monitoring, optimization, and control of multicarrier energy systems.

In this context, several key considerations are analyzed in Table 5.8 to highlight the importance of robust data architectures for multicarrier energy systems.

5.5.3 Developments and Trends on System Architectures and Control Technologies

The architecture of energy systems is undergoing a significant transformation, shifting from a traditional centralized model to a more decentralized framework. As a result, data and control technologies and relevant subsystems must adapt and expand to accommodate this new landscape. The transition to decentralized energy systems necessitates the development of robust data management and control systems that can handle the increased volume, velocity, and variety of data generated by distributed energy assets. Furthermore, control technologies need to be deployed at multiple scales, including local distribution networks and individual energy assets, enabling real-time monitoring, control, and optimization. Over the past decade, there has been a significant leap in the development of system architecture, subsystems, and control technologies, as the power sector embraces this evolutionary shift.

Table 5.8 Key considerations for data architectures in multicarrier energy systems.

Key considerations	Robust data architecture qualities in multicarrier energy systems
Interoperability	Interoperability ensures that data from various sources can be integrated and shared, enabling effective coordination between energy vectors, demand-response programs, and grid management systems [112, 116].
Data integration	This integration of data – from diverse sources, including sensors, meters, and control systems – enables a holistic view of the system, facilitating effective monitoring, optimization, and control of energy flows.
Scalability	The architecture should be able to handle large volumes of data generated from various sources and support future expansion as the system evolves. Scalable data architectures ensure that the system can accommodate growing data requirements without compromising performance [117].
Real-time data processing	Data architectures should support efficient processing and analysis of real-time data streams, allowing operators to monitor system performance, detect anomalies, and respond quickly to changing conditions. Real-time data processing is crucial for multicarrier energy systems, as it enables timely decision-making and control.
Cyber security	Data security and privacy are paramount in multicarrier energy systems, where sensitive information about energy consumption, infrastructure, and customers is involved. Data architectures should incorporate robust security measures to protect against unauthorized access, data breaches, and cyber threats. Privacy considerations should also be addressed to comply with data protection regulations and maintain trust among stakeholders [118].
Advanced analytics and decision support	Data architectures should support advanced analytics techniques, such as machine learning, optimization algorithms, and predictive modeling. Advanced analytics enhance system efficiency, resilience, and enable proactive management of multicarrier energy systems.
Visualization and reporting	Effective data visualization enables operators, stakeholders, and decision-makers to gain insights, monitor performance, and make informed decisions.
Flexibility and adaptability	Data architectures should be flexible and adaptable to accommodate these changes such as integration of new energy sources and regulatory updates to evolving customer demands. They should support modular designs, easy integration of new data sources, and the ability to incorporate emerging technologies and standards.
Data quality and validation	Data architectures should include mechanisms for data validation, quality control, and error detection. Ensuring data accuracy and reliability through automated validation processes enhances the overall effectiveness of multicarrier energy systems.

5.5.3.1 Energy Management Systems

Energy management systems, including supervisory control and data acquisition (SCADA), advanced distribution management system (ADMS), distributed energy resource management system (DERMS), and distribution management system (DMS), have played a crucial role in the management of energy systems [119–121]. However, with the energy systems landscape rapidly evolving, these systems are gradually becoming outdated and are no longer considered an optimal technological option. Specifically, they face challenges such as limited scalability, a lack of real-time analytics, and interoperability issues among diverse DERs. Additionally, data management and integration pose challenges, along with regulatory frameworks that can hinder their deployment and effectiveness. Implementing these systems can be costly, and ensuring data quality remains a challenge. Moreover, the operation and maintenance of these systems require skilled personnel. Acknowledging these limitations, ongoing efforts are being made to address these issues and enhance the capabilities of energy management systems to meet the evolving needs of the energy sector. Advancements in energy system control have been made in key areas, showing promise for addressing some of the limitations in energy management, including in the context of multicarrier energy systems.

One area of progress is adaptive control algorithms, which have been developed to handle the dynamic nature of multicarrier energy systems. These algorithms continuously adjust control parameters based on real-time measurements and system changes across multiple energy carriers, ensuring optimal performance and stability while coordinating the interactions between different energy carriers [122–124]. Coordinated control becomes even more critical in multicarrier energy systems, as it involves managing the certain parameters – e.g. coordinated voltage control in electricity grid and pressure in other energy carriers like gas and heat networks.

Wide-area monitoring and control (WAMC) systems have also emerged as valuable tools in multicarrier energy systems [125, 126]. By utilizing synchronized phasor measurement units (PMUs) and advanced control algorithms, WAMC systems can monitor and control the interactions and flows between different energy carriers over a wide geographic area. Consider a scenario where an integrated energy comprises an electricity grid, a natural gas network, and a district heating system. A WAMC system can be deployed to monitor and control the interactions between these energy carriers. In this example, the WAMC system utilizes synchronized PMUs and advanced control algorithms to gather real-time data from different points in the electricity grid, gas network, and district heating system. The PMUs provide synchronized measurements of voltage, current, and frequency, enabling a holistic view of the entire energy system. Using the gathered data, the WAMC system performs real-time analysis and coordination to optimize the operation of the multicarrier energy system. For instance, the WAMC system can monitor electricity demand, renewable energy generation, and energy storage levels. Based on this information, it can determine the optimal allocation of electricity and gas resources to meet the demand while considering system constraints and cost factors. Moreover, the WAMC system can enable dynamic interactions between the energy carriers. For example, it can regulate power-to-gas (P2G) conversion processes, where excess electricity is used to produce hydrogen or synthetic natural gas, which can then be stored and utilized in the gas network when needed. The WAMC system can monitor the gas network's pressure and demand, and based on the real-time data, it can control the P2G conversion rates to maintain system stability and efficient energy utilization.

To manage the complexities of multicarrier energy systems, hierarchical control architectures have been also developed. These architectures organize control functions into different levels, allowing for efficient coordination and control across multiple energy carriers and system components [127, 128]. Hierarchical control architectures enable effective management of interactions between electricity, gas, heat, and transportation networks, ensuring smooth integration and optimal utilization of diverse energy resources. For example, a hierarchical control architecture enables the seamless integration of electricity and gas networks while ensuring reliable and efficient operation. The high-level control optimizes the system performance by making decisions on energy flow coordination, while the lower-level control ensures the smooth operation of individual network components.

5.5.3.2 Energy Systems Modeling

Energy systems modeling plays a critical role in the efficient and reliable operation of the energy sector as briefly discussed in Section 5.2.1. Two notable modeling standards are the Common Information Model (CIM) and the Common Smart Grid Model (CSGM) [129–131]. CIM is a widely adopted data exchange standard that offers a unified representation of energy system components. It encompasses various aspects such as generation, transmission, distribution, and consumption. CIM enables seamless data exchange and interoperability between different energy management systems, promoting efficient grid operation and the integration of DERs. By utilizing CIM, utilities and grid operators can efficiently exchange information, streamline their operations, and make informed decisions for optimal energy management. Similarly, CSGM provides a standardized model for data exchange and interoperability specifically tailored for smart grids. It offers a consistent representation of grid assets, network topology, and operational parameters. This standardization allows for efficient integration of advanced grid management systems, real-time monitoring, and control applications. By leveraging CSGM, utilities can enhance their grid optimization efforts, improve the reliability of their operations, and enable effective control mechanisms for grid stability. Both CIM and CSGM contribute to the usability and effectiveness of energy system models. They enable seamless data exchange, promote interoperability among diverse systems, and facilitate the integration of advanced management and control technologies. It should be noted that energy systems modeling is crucial for the efficient operation of multicarrier energy systems. By using standards like CIM and CSGM, which offer unified representations of various energy system components across all carriers, multicarrier systems can achieve seamless data exchange and interoperability. Practically, an energy asset can be modeled and integrated into an optimization and control framework, regardless of the carrier it belongs to (e.g. gas, electricity, and transport). Consequently, such frameworks can be used to help utilities across carriers optimize the control and operation of their multienergy systems effectively.

5.5.3.3 Virtualization

The traditional approach to substation control system deployments involves time-consuming physical visits for equipment connection, inspection, and configuration. This hinders the rapid modernization of the power grid and limits the scalability of advanced control solutions. However, a new solution leveraging virtualization and remote connectivity technologies provides a faster and more cost-effective alternative. By utilizing a central server and operating system called a hypervisor and creating virtual machines (VMs), physical devices at substations can be hosted as software apps, reducing hardware costs and deployment delays [132]. This approach enables quick integration of software apps, real-time measurements, communication with edge intelligence devices, and generation of control actions. Virtualization offers numerous benefits to utilities in managing their substations and potentially, wider energy system operations. It simplifies management by centralizing system monitoring and resource allocation, leading to efficient maintenance and troubleshooting. It improves operational efficiency by reducing the number of physical servers needed, resulting in energy savings and cost reductions. Virtualization enhances flexibility by enabling easy resource scaling without physical server additions or removals. It strengthens security by isolating applications on separate VMs, minimizing the impact of system compromises. Additionally, virtualization enhances reliability by facilitating seamless migration of VMs to different servers during failures, ensuring uninterrupted service. Lastly, it enables safe testing and development of new applications and systems in isolated environments without affecting production systems. Virtualization plays a pivotal role, particularly for the control and operation of multicarrier energy systems. Imagine that control systems of different carriers (e.g. a gas flow controller and a voltage controller) can be hosted as software apps on the same hardware device within VMs. This can facilitate secure and very fast data exchange among such controllers and, most significantly, minimize the costs related to interconnecting large physical devices (such interconnections shall cost time, capital and introduce operational disruptions). This level of flexibility can subsequently increase confidence and reduce the complexity of utilities' operation when it comes to bringing everything together in the physical world.

5.5.3.4 Cloud Computing

Cloud computing has revolutionized the optimization, control, and monitoring of energy systems through its advanced features. The scalability of cloud platforms enables energy systems to dynamically allocate computing resources, ensuring efficient operation during varying workloads [133–135]. High-performance computing capabilities offered by the cloud empower energy systems to execute complex optimization algorithms and simulations, enabling accurate modeling and efficient decision-making. Cloud-based data storage and processing capabilities handle the massive influx of data generated by energy systems, facilitating real-time monitoring, predictive analytics, and anomaly detection. This enables proactive maintenance, precise fault detection, and optimized asset performance. This collaborative approach enhances grid resilience, enables demand response initiatives, and supports the integration of renewable energy resources. Cloud-based data storage and processing are essential for multicarrier systems to handle the massive influx of data generated by different energy carriers. This enables real-time monitoring, predictive analytics, and anomaly detection, which are crucial for efficiently managing the diverse energy sources and maintaining grid stability in multicarrier setups. Most importantly, the collaborative nature of cloud platforms is vital for devising multicarrier systems as it will be the instrument to allow grid operators, utilities, and stakeholders to work together on a shared platform, exchanging data, models, insights, and operations, and therefore jointly optimize the energy systems. Practically, without the advancement and evolution of cloud computing, the realization of truly multicarrier systems will be in question.

5.5.3.5 Digital Twins

Digital twins, virtual replicas of physical assets or systems, are being leveraged in the energy sector for asset optimization, predictive maintenance, and simulation-based control strategies. Digital twins have played a transformative role in the energy sector by offering virtual replicas of physical assets or systems [136, 137]. By creating a real-time digital representation of energy infrastructure, digital twins facilitate accurate monitoring, performance analysis, and predictive modeling. This helps optimize operational efficiency, improve asset utilization, and reduce downtime. Digital twins enable energy companies to make data-driven decisions, simulate various scenarios, and proactively address maintenance needs, ultimately leading to improved reliability, cost savings, and overall performance of energy systems.

5.5.3.6 Cybersecurity

Advancements in cybersecurity for the power sector have been crucial in protecting critical infrastructure from evolving cyber threats. These advancements have focused on strengthening the resilience and defense capabilities of energy systems against potential attacks [130, 138–140]. One significant development is the adoption of robust cybersecurity protocols, such as the IEC 62351 series of standards and NIST cybersecurity framework. These developments provide guidelines for secure communication and data exchange between different components of the power system, including substations, control centers, and intelligent electronic devices (IEDs). Implementing these protocols ensures authentication, access control, data integrity, and confidentiality, mitigating the risk of unauthorized access or tampering. Additionally, the power sector has embraced advanced threat detection and prevention mechanisms. Intrusion detection and prevention systems (IDPS) are deployed to monitor network traffic and detect any suspicious activities or anomalies [141]. These systems use techniques such as signature-based detection, anomaly detection, and behavioral analysis to identify potential cyber threats. Furthermore, the sector has adopted Security Information and Event Management (SIEM) solutions that aggregate and analyze security logs and events from various sources to provide real-time visibility into potential security incidents [142]. These technologies enable proactive threat management, timely incident response, and continuous monitoring of the power infrastructure's security posture. Overall, these advancements in cybersecurity for the power sector have strengthened the protection of critical infrastructure by implementing robust protocols, deploying advanced threat detection mechanisms, and adopting proactive security management practices. These measures ensure the integrity, availability, and confidentiality of power systems, safeguarding against cyberattacks and helping in maintaining the reliable and secure operation of the power grid. Multicarrier energy systems require a highly tailored approach to cybersecurity. The adoption of standards like the IEC 62351 series and the NIST cybersecurity framework is crucial because they provide guidance on securing the interconnected components of multicarrier systems. In a multicarrier context, this ensures that data shared between these carriers remains confidential and unaltered. For example, it helps protect the integrity of data exchanged between electric grids and heating systems, which is essential for seamless coordination between carriers. Moreover, multicarrier energy systems can deploy IDPS that are tailored to the specific energy carriers involved. These intrusion detection and prevention solutions can monitor network traffic within each carrier, identifying and mitigating threats that may target particular aspects of the system. For instance, they can detect anomalies in the data communication between transportation fuel infrastructure and electricity grids. In summary, the relevance ensures a holistic view of the cybersecurity posture of multicarrier systems, enabling a coordinated response to potential threats or breaches, regardless of the energy carrier involved.

5.6 Holistic Approach in Planning and Operating an ILEC

What has been established so far is that the complex system of an ILEC needs a holistic approach for optimal planning, operation, and control. The holistic approach needs to consider important technical aspects, concerns, and interactions from within the ILEC but also from exogenous systems such as the upper grid or the markers.

From the literature analyzed in Sections 5.2–5.4, it is seen that most works focus either on the planning part or on the real-time operation and control and in most cases the exogenous systems are neglected. This section will give an overview of how these processes need to come together in the different hierarchical layers of an ILEC as have been shared earlier in Figure 5.1. Main challenges are also analyzed.

5.6.1 Higher Hierarchical Level: Planning Phase

This phase is the higher hierarchical one in the sense that the managerial or administrative instrument of the ILEC needs to take informative decisions on the system design, expanding scenarios or altering structure scenarios. The long-term planning is designated by considering the multiple carriers with a multiobjective framework to pick up not only techno-economical causes but also other motivation that is in line with the ILEC as explained in Section 5.2. In that sense, it handles the process of determining the optimal system design including optimal sizing, placement, number and type of technologies, operation strategies, and discrete investment alternative scenarios.

This stage can provide the investors with optimal investment strategies for developing the local energy system. Future developments of the system constitute the projections for energy demand, energy prices, and the different investment possibilities for energy supply, conversion between energy carriers, distribution, storage, end-use measures, and restrictions on CO_2 emissions.

To do so, the planning phase needs to have inputs from the system and on-site data that are related to the local specificities. Furthermore, this stage needs to be in collaboration with the operational analysis phase in the next where the operational scenarios and day-ahead scheduling will be based on the updated architecture layout and/or expansion strategies.

In Table 5.9, the inputs and outputs of the planning phase are summarized.

5.6.2 Lower Hierarchical Layer: Operational Analysis

Based on the system design, technologies, and their characteristics in the planning phase, this layer optimizes the dispatch of multicarrier energy systems by preferably

Table 5.9 Inputs and outputs involved in the planning phase of an ILEC.

Characterization	Functionality	Data handled
Input	System design data	• Time series data, e.g. RES output, market prices, and load profiles
Input	Data processing	• Based on clustering techniques applied to time series data, representative time series, e.g. RES output, market prices, and load profiles, are generated so that representative days can be evaluated under the planning
Output	Optimal system design	• Optimal configuration of the multienergy system in terms of types of energy technologies • Optimal sizing of energy technologies • Optimal operation strategies in representative days • Net-present value of costs, and cost data for different categories and periods • Cost-optimal investment strategies • Operation for all representative days in all years considered by the OPEX optimization
Interaction with the lower level	Operational analysis	• Information exchange about the system architecture and structural data • Information exchange about any infeasible and/or undesirable system configurations that should be avoided

pursuing multiple objectives when analyzing the various operational scenarios or day-ahead dispatch. This phase can have two different approaches, i.e. stochastic approach or determinist approach.

So, either through scenario-generation approach or using forecasting tools as input, the optimal expected hourly operation strategies of the technologies in the multicarrier systems are extracted by considering e.g. total daily costs and CO_2 emissions combining more than one operational objective. Further input data are needed such as user demand, local climate data, energy prices, and technical information of the energy technologies before this layer determines the optimal operation scheduling of the multienergy system by considering the uncertainties related to input data.

Two approaches are included to determine RES generation, load profiles, and market prices as already mentioned:

- **Forecasting:** Functionality that forecasts data for the next day, required for scheduling the day-ahead operation of the system. This can apply to PV production, day-ahead market prices, and/or load profiles consumption.
- **Scenario generation:** Functionality that generates scenarios by employing historical data to be analyzed for the purpose of verifying a given system design from the planning phase under various operational conditions. The generated scenarios can foster stochastic parameters by using artificial intelligence-based methods and by providing the probability of their occurrence. Such scenarios will help assess the system's operation under various conditions to assess the feasibility of potential system designs and then close the loop of interaction with the upper layer.

This layer is in close collaboration with the lowest layer of the real-time operation and assets' control. Based on the business models of the ILEC in serve, the interaction may differ. Of course, data from the field are used as inputs for the operational analysis and in order that the loop is closed the operational analysis may channel its output in different ways. In case that the ILEC has a direct control access to the assets, the relevant set points are imposed as per the operational strategies. In case that the ILEC has a local market operating, it provides the power needs per time blocks with the energy prices bids to the P2P market.

In Table 5.10, the inputs and outputs of the operational analysis phase are summarized.

5.6.3 Lowest Hierarchical Layer: Real-Time Operation and Assets' Control

This layer is directly connected to the ILEC prosumers where the controllable assets are expected to be in their premises. Based on Section 5.5 and the business concept in serve, it can be assumed that there is a centrally controlled system where the previous layer of operational analysis can directly impose the decided operational strategies. In that case, the functionalities of this layer are limited to data extraction, monitoring, and real-time control.

Table 5.10 Inputs and outputs involved in the operational analysis phase of an ILEC.

Characterization	Functionality	Data handled
Input	Forecasting	• 24 hours ahead PV production forecasting • Day-ahead load forecast per outgoing circuit (Street) and phase. Self-consumption (Demand drop), per outgoing circuit (street), per phase • Day-ahead market price data
Input	Scenario generation	• Based on clustering techniques applied to time series data, representative time series for RES output, market prices, load profiles, and so on is generated with the probability of occurrence
Output	Operational analysis	• Pareto frontier of solutions' pairs • Optimal hourly operation strategies of energy technologies • Aggregated hourly energy generation profiles for ILEC
Interaction with the lower level	Assets control	• Energy market prices for the local market • Set points for the ILEC assets

But if we assume that the centrally controlled system is relaxed then each consumer has control of operation of its individual energy-related assets. In this case, each consumer acts as a peer and responds to the local market price which is co-shaped by the operational analysis' layer bidding and the interaction of the peers. So, in this case, the local market – whose operational time domain is real time – is the way to settle energy transactions among peers in a decentralized manner under the higher-level energy profiles asked. It is expected that the global objective set through the day-ahead optimal dispatch in the operational analysis is matched as closely as possible within an operating window of the market in real time.

The integration of dynamics and real-time control is revolutionizing the conventional approach to operational planning tools. Traditionally, static operational tools were used for scenario analysis over extended time horizons, with simulation time steps ranging from minutes to days. However, recognizing the vital role of operational strategies and real-time control systems in multicarrier energy systems planning, it has become increasingly crucial to incorporate them into the planning process [141, 143]. This integration ensures that the resulting data aligns more closely with reality, leading to desired outcomes in terms of flexibility, efficiency, grid stability, DERs integration, and system resilience. A recent plan outlined by the UK's Department for Energy Security and Net Zero, Ofgem, and Department for Business, Energy and Industrial Strategy highlights the benefits of integrating smart control and flexibility into the UK energy system. It reveals that such integration could reduce system costs by up to £10bn per year by 2050 by minimizing the need for additional generation and network infrastructure to meet peak demand [144]. By leveraging real-time control and dynamic capabilities in planning, we can optimize resource utilization, enhance grid performance, and achieve substantial cost savings while ensuring a sustainable and resilient energy future.

Table 5.11 Inputs and outputs involved in the real-time operation of the peers through the energy management unit.

Characterization	Functionality	Data handled
Input	Forecasting	• RES generation profile • Load profile
Input	Assets	• Real-time status feedback of control devices • System configuration
Input	Local market	• Electricity prices
Output	Energy management unit	• Allocation of active powers for all assets in peer's portfolio (real-time horizon) • Set points for the assets' control

So, in the latter case, the peers need to have their own energy management unit that would allow them to interact with the market in real time. This unit handles the real-time scheduling of each peer by considering the energy prices broadcasted from the local market and considering as inputs the forecasted load and RES production profiles. The energy management unit can have different objectives served according to the peer aims, e.g. consumption costs minimization or/and environmental footprint minimization. The appliances allocated for such a multicarrier system are based on the schemes and scenarios determined in the planning phase. It can include WT, PV, BESS, CHP, EV with V2G capability, air conditioning (AC), heat pump, fuel cell, electric boiler, and thermal storage systems and the energy management unit is expected to optimally allocate their flexibility.

The description of inputs and outputs of different functionalities at that layer is provided in Table 5.11.

5.6.4 Interaction with External Entities

The ILEC is serving its objectives, i.e. to be self-sufficient and/or cost-efficient through the holistic planning, operation, and control as explained earlier, but it is expected that the interaction with external entities such as the grids and the markets is also fostered. In this section, a general overview of these interactions is given.

According to the literature, two external interactions are mainly considered: the interaction with the markets and the network perspective. The market approach refers to the monetary/financial energy transactions between the ILEC operator and the surrounding energy system or market of interest, while the network approach is limited to the physical/technical interaction and constraints regarding the amount of energy exchange between those two main systems. These interactions reflect also on the operational analysis of the ILEC. In Figure 5.7, a high-level schematic capturing the interactions is shown. The ILEC operator can directly collaborate with the operators of different grids while he can participate in both local and external markets directly as a market participator or through brokers, i.e. aggregator or retailer. In more detail:

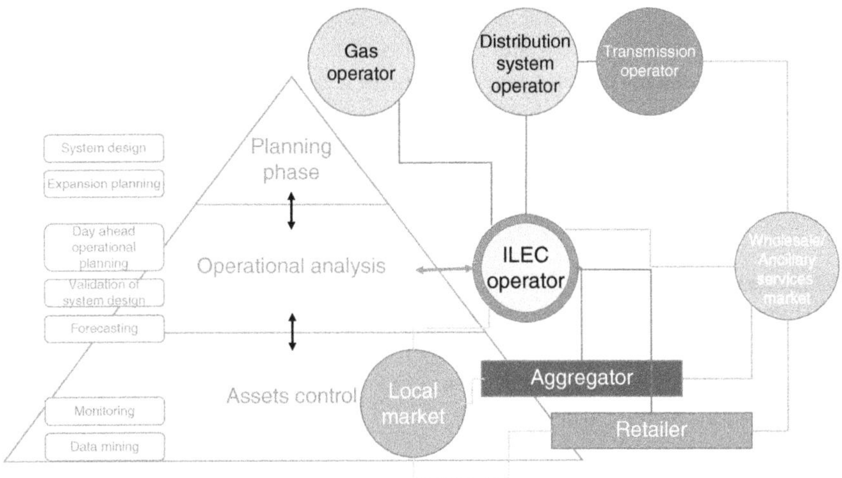

Figure 5.7 Interaction of an ILEC with the external entities.

5.6.4.1 Network Interaction

ILEC operator is in direct communication with the Distribution System Operator and/or Gas operator, e.g. through bilateral contract, so that his requests, e.g. flexibility requests, are treated as inputs or constraints in the optimization models of the operational analysis layer. These can be also technical constraints associated with the electrical and gas networks, such as the ones seen in the literature. These constraints can include power and energy balance at the electrical PCC, an admissible voltage range, transmission line capacity, and operational limits related to the gas sub-network. For example, in Ref. [143], costs such as generation cost, start-up/shutdown cost, upward/downward reserve cost, and real-time operation cost (regulation cost, curtailment/shedding penalties) have been considered, where a co-optimization between the power system and the natural gas system is presented. Similarly, in Ref. [145], the electric power system, natural gas system, and district heating system are coordinated to achieve the optimal economical operation of the whole system (energy and reserve scheduling) with minimum wind curtailment, in both the day-ahead and real-time stage.

The interaction of the ILEC operator with the electrical distribution grids can be also considered in terms of grid stability or performance. For example, in Ref. [144], where a strong interaction via importing/exporting electrical power in large quantities may result in a destabilization and performance degradation of the grid, and it can reduce the economic performance of the system. In Ref. [146], the renewable penetration level and optimal scheduling of P2G and gas-fired generating (GfG) are decided by the grid operator considering the cost that comes with it.

Other works evaluate and solve these grid issues inside the multicarrier energy systems and not under the upstream utility network perspective. In Ref. [143], the power output and up/down reserve capacity of gas-fired units are determined in the day-ahead scheduling stage in the multienergy coupled system for handling wind power uncertainty, which influences the operation feasibility of the gas system and the availability of reserve of gas-fired units in real time. In Ref. [145], simultaneous energy and reserve scheduling is performed to minimize the operation costs, while accommodation of wind power production is achieved through reserves in the second stage (real time). In both works, the reserve availability is used to manage uncertainties inside ILECs, but not as a network service to the grid. Ref. [147] estimates the ancillary service fees and other charges depending on the energy schedule but does not apply the issue upstream in the utility grid.

5.6.4.2 Local Market Interaction

The ILEC operator acts as a crucial intermediary in the local flexibility market. It identifies and mobilizes the flexibility resources within the community, establishes a digital marketplace for trading these resources, and ensures smooth and efficient interactions among the participants.

5.6.4.3 Wholesale Market Interaction

The ILEC operator acts as an intermediary between the local energy community peers and wholesale markets. It engages in energy trading, participates in ancillary service markets, and monitors the wholesale market conditions to optimize the community's energy transactions either directly or through appropriate intermediaries as seen in Figure 5.7.

5.6.4.4 Interaction with Aggregators/Retailers

The ILEC operator acts as a facilitator and coordinator between the community, retailers, and aggregators. It collaborates with retailers to meet the energy needs of the community members and shares information with aggregators to integrate flexible resources into the wider energy market. By facilitating these interactions and optimizing energy transactions, the operator promotes efficient energy use, fosters economic benefits, and enhances the overall resilience of the community energy ecosystem.

5.7 Conclusion

This chapter provides an outlook on the challenging issue of planning and operating an ILEC. As already established, this is a multifacet issue that needs to be addressed holistically and so far, only a few publications deal with this aspect in full. In specific:

Optimal planning in multicarrier energy systems is a complex task that requires expertise in various disciplines, including engineering, economics, environmental science, and policy analysis. Some key challenges include modeling and interactions among different energy carriers, ensuring sector coupling, and addressing the electrification of transport and heat. These challenges require sophisticated algorithms, and a comprehensive understanding of factors such as integrated planning, road networks, and integration of alternative fuels. As such, the decision-making process for multicarrier energy systems planning is demanding and this is the main reason for use of multiobjective approaches that helps in addressing conflicting objectives and finding compromised solutions.

In the *operational planning* of multicarrier energy systems, challenges include managing the interaction between energy carriers, addressing supply-demand imbalances, and handling uncertainties related to renewables, energy prices, and energy demand. Stochastic optimization is commonly used to handle these uncertainties.

When it comes to *real-time operation,* it involves advanced control and optimization algorithms, multiobjective approaches, integration of DRPs, and market interactions. The latest trend is the interactions with the operational planning through real-time data and info. The challenges include system complexity, renewable energy variability, uncertainty, and the need to balance conflicting objectives and stakeholder preferences in real time.

Finally, data architectures that allow the efficient control and thus implementation of previous stages have many challenges that relate to the standardized process of communication, privacy, and security issues that are more complex when it comes to multicarrier systems.

List of Abbreviations

AC	air conditioning
ADMS	advanced distribution management system
CHP	combined heat and power
CSGM	common smart grid model
DAM	day-ahead market
DERMSs	distributed energy resource management systems
DERs	distributed energy resources
DMSs	distribution management systems
DRPs	demand-response programs
EU	European Union
EVs	electric vehicles
GfG	gas-fired generating
HEMS	home energy management system
IDM	intraday market
IDPSs	intrusion detection and prevention systems
IEC	International Electrotechnical Commission
IED	intelligent electronic device
ILECs	integrated local energy communities
P2G	power-to-gas
PCC	point of common coupling
PMUs	phasor measurement units
PVs	photovoltaics

RES	renewable energy sources
SCADA	supervisory control and data acquisition
SIEM	security information and event management
TOU	time of use
V2G	vehicle-to-grid
VMs	virtual machines
WAMC	wide-area monitoring and control
WT	wind turbine

References

1 EU (2022). Communication from the Commission to the European Parliament, the European Council, the Council, the European Economic and Social Committee and the Committee of the Regions REpowerEU Plan. https://eur-lex.europa.eu/legal-content/EN/TXT/?uri=COM%3A2022%3A230%3AFIN&qid=1653033742483.

2 Rousis, A.O., Konstantelos, I., and Strbac, G. (2020). A planning model for a hybrid AC–DC microgrid using a novel GA/AC OPF algorithm. *IEEE Transactions on Power Systems* 35 (1): 227–237. https://doi.org/10.1109/TPWRS.2019.2924137.

3 Liu, X. and Liu, Y. (2019). Optimal planning of AC-DC hybrid transmission and distributed energy resource system: review and prospects. *CSEE Journal of Power and Energy Systems*. https://doi.org/10.17775/CSEEJPES.2019.00540.

4 Nozarian, M., Fereidunian, A., and Barati, M. (2023). Reliability-oriented planning framework for smart cities: from interconnected micro energy hubs to macro energy hub scale. *IEEE Systems Journal* 1–12. https://doi.org/10.1109/JSYST.2023.3244498.

5 Shao, C., Li, K., Hu, Z., and Shahidehpour, M. (2022). Coordinated planning of electric power and natural gas distribution systems with refueling stations for alternative fuel vehicles in transportation system. *IEEE Transactions on Smart Grid* 13 (5): 3558–3569. https://doi.org/10.1109/TSG.2022.3166965.

6 Gan, W., Shahidehpour, M., Yan, M. et al. (2020). Coordinated planning of transportation and electric power networks with the proliferation of electric vehicles. *IEEE Transactions on Smart Grid* 11 (5): 4005–4016. https://doi.org/10.1109/TSG.2020.2989751.

7 Hartel, P. and Ghosh, D. (2022). Modelling heat pump systems in low-carbon energy systems with significant cross-sectoral integration. *IEEE Transactions on Power Systems* 37 (4): 3259–3273. https://doi.org/10.1109/TPWRS.2020.3023474.

8 Klyapovskiy, S., You, S., Cai, H., and Bindner, H.W. (2019). Integrated planning of a large-scale heat pump in view of heat and power networks. *IEEE Transactions on Industry Applications* 55 (1): 5–15. https://doi.org/10.1109/TIA.2018.2864114.

9 Di Somma, M., Graditi, G., and Yan, B. (2021). Cost-sustainability trade-off solutions for the optimal planning of local integrated energy systems from

nanogrids to communities. In: *Handbook of Smart Energy Systems* (ed. M. Fathi, E. Zio, and P.M. Pardalos), 1–34. Cham: Springer International Publishing https://doi.org/10.1007/978-3-030-72322-4_80-1.

10 Yan, B., Di Somma, M., and Graditi, G. (2021). Chapter 4: Multiobjective operation optimization of DER for short- and long-run sustainability of local integrated energy systems. In: *Distributed Energy Resources in Local Integrated Energy Systems* (ed. G. Graditi and M. Di Somma), 89–123. Elsevier https://doi.org/10.1016/B978-0-12-823899-8.00001-7.

11 Meier, P. and Mubayi, V. (1983). Modelling energy-economic interactions in developing countries: a linear programming approach. *European Journal of Operational Research* 13 (1): 41–59. https://doi.org/10.1016/0377-2217(83)90006-1.

12 Nijkamp, P. and Volwahsen, A. (1990). New directions in integrated regional energy planning. *Energy Policy* 18 (8): 764–773. https://doi.org/10.1016/0301-4215(90)90029-4.

13 Papadimitriou, C.N., Anastasiadis, A., Psomopoulos, C.S., and Vokas, G. (2019). Demand response schemes in energy hubs: a comparison study. *Energy Procedia* 157 (2018): 939–944. https://doi.org/10.1016/j.egypro.2018.11.260.

14 Luenberger, D.G. and Ye, Y. (2008). *Linear and Nonlinear Programming*, vol. 116. New York, NY: Springer US https://doi.org/10.1007/978-0-387-74503-9.

15 Alarcon-Rodriguez, A., Ault, G., and Galloway, S. (2010). Multi-objective planning of distributed energy resources: a review of the state-of-the-art. *Renewable and Sustainable Energy Reviews* 14 (5): 1353–1366. https://doi.org/10.1016/j.rser.2010.01.006.

16 Korte, B. and Vygen, J. (2011). *Ottimizzazione Combinatoria*. Milano: Springer Milan https://doi.org/10.1007/978-88-470-1523-4.

17 Irving, M.R. and Song, Y.-H. (2000). Optimisation techniques for electrical power systems. Part 1: Mathematical optimisation methods. *Power Engineering Journal* 14 (5): 245–254. https://doi.org/10.1049/pe:20000509.

18 Deb, K. (2011). Multi-objective optimisation using evolutionary algorithms: an introduction. In: *Multi-objective Evolutionary Optimisation for Product Design and Manufacturing* (ed. L. Wang, A.H.C. Ng, K. Deb, et al.), 3–34. https://doi.org/10.1007/978-0-85729-652-8_1. London: Springer London.

19 Oh, S., Lee, H., Jung, J., and Kwak, H. (2007). Optimal planning and economic evaluation of cogeneration system. *Energy* 32 (5): 760–771. https://doi.org/10.1016/j.energy.2006.05.007.

20 Ren, H., Gao, W., and Ruan, Y. (2008). Optimal sizing for residential CHP system. *Applied Thermal Engineering* 28 (5–6): 514–523. https://doi.org/10.1016/j.applthermaleng.2007.05.001.

21 Yokoyama, R. and Ito, K. (2006). Optimal design of gas turbine cogeneration plants in consideration of discreteness of equipment capabilities. *Journal of Engineering for Gas Turbines and Power* 128 (2): 336–343. https://doi.org/10.1115/1.2131889.

22 Horii, S., Ito, K., Pak, P.S., and Suzuki, Y. (1987). Optimal planning of gas turbine co-generation plants based on mixed-integer linear programming.

International Journal of Energy Research 11 (4): 507–518. https://doi.org/10.1002/er.4440110407.

23 Beihong, Z. and Weiding, L. (2006). An optimal sizing method for cogeneration plants. *Energy and Buildings* 38 (3): 189–195. https://doi.org/10.1016/j.enbuild.2005.05.009.

24 Cui, C., Yang, X., Tian, F. et al. (2015). Optimization of a combined heat and power generation system with ice thermal storage. In: *Low-carbon City and New-type Urbanization: Proceedings of Chinese Low-carbon City Development International Conference* (ed. S. Feng, W. Huang, J. Wang, et al.), 13–23. https://doi.org/10.1007/978-3-662-45969-0_2. Springer Berlin Heidelberg.

25 Arcuri, P., Florio, G., and Fragiacomo, P. (2007). A mixed integer programming model for optimal design of trigeneration in a hospital complex. *Energy* 32 (8): 1430–1447. https://doi.org/10.1016/j.energy.2006.10.023.

26 Cardona, E. and Piacentino, A. (2007). Optimal design of CHCP plants in the civil sector by thermoeconomics. *Applied Energy* 84 (7–8): 729–748. https://doi.org/10.1016/j.apenergy.2007.01.005.

27 Burer, M., Tanaka, K., Favrat, D., and Yamada, K. (2003). Multi-criteria optimization of a district cogeneration plant integrating a solid oxide fuel cell–gas turbine combined cycle, heat pumps and chillers. *Energy* 28 (6): 497–518. https://doi.org/10.1016/S0360-5442(02)00161-5.

28 Zhou, Z., Liu, P., Li, Z., and Ni, W. (2013). An engineering approach to the optimal design of distributed energy systems in China. *Applied Thermal Engineering* 53 (2): 387–396. https://doi.org/10.1016/j.applthermaleng.2012.01.067.

29 Omu, A., Choudhary, R., and Boies, A. (2013). Distributed energy resource system optimisation using mixed integer linear programming. *Energy Policy* 61: 249–266. https://doi.org/10.1016/j.enpol.2013.05.009.

30 Hawkes, A.D. and Leach, M.A. (2009). Modelling high level system design and unit commitment for a microgrid. *Applied Energy* 86 (7–8): 1253–1265. https://doi.org/10.1016/j.apenergy.2008.09.006.

31 Mehleri, E.D., Sarimveis, H., Markatos, N.C., and Papageorgiou, L.G. (2013). Optimal design and operation of distributed energy systems: application to Greek residential sector. *Renewable Energy* 51: 331–342. https://doi.org/10.1016/j.renene.2012.09.009.

32 Mehleri, E.D., Sarimveis, H., Markatos, N.C., and Papageorgiou, L.G. (2012). A mathematical programming approach for optimal design of distributed energy systems at the neighbourhood level. *Energy* 44 (1): 96–104. https://doi.org/10.1016/j.energy.2012.02.009.

33 Ren, H. and Gao, W. (2010). A MILP model for integrated plan and evaluation of distributed energy systems. *Applied Energy* 87 (3): 1001–1014. https://doi.org/10.1016/j.apenergy.2009.09.023.

34 Weber, C. and Shah, N. (2011). Optimisation based design of a district energy system for an eco-town in the United Kingdom. *Energy* 36 (2): 1292–1308. https://doi.org/10.1016/j.energy.2010.11.014.

35 Li, H., Nalim, R., and Haldi, P.-A. (2006). Thermal-economic optimization of a distributed multi-generation energy system—a case study of Beijing. *Applied*

Thermal Engineering 26 (7): 709–719. https://doi.org/10.1016/j.applthermaleng.2005.09.005.

36 Weber, C., Maréchal, F., Favrat, D., and Kraines, S. (2006). Optimization of an SOFC-based decentralized polygeneration system for providing energy services in an office-building in Tōkyō. *Applied Thermal Engineering* 26 (13): 1409–1419. https://doi.org/10.1016/j.applthermaleng.2005.05.031.

37 Buoro, D., Casisi, M., De Nardi, A. et al. (2013). Multicriteria optimization of a distributed energy supply system for an industrial area. *Energy* 58: 128–137. https://doi.org/10.1016/j.energy.2012.12.003.

38 Jin, L., Rossi, M., Ciabattoni, L. et al. (2023). Environmental constrained medium-term energy planning: the case study of an Italian university campus as a multi-carrier local energy community. *Energy Conversion and Management* 278: 116701. https://doi.org/10.1016/j.enconman.2023.116701.

39 Piazza, G., Bracco, S., Delfino, F. et al. (2023). Impact of electric mobility on the design of renewable energy collective self-consumers. *Sustainable Energy, Grids and Networks* 33: 100963. https://doi.org/10.1016/j.segan.2022.100963.

40 Foiadelli, F., Nocerino, S., Di Somma, M., and Graditi, G. (2018). Optimal design of DER for economic/environmental sustainability of local energy communities. *2018 IEEE International Conference on Environment and Electrical Engineering and 2018 IEEE Industrial and Commercial Power Systems Europe (EEEIC/I&CPS Europe)* (12–15 June 2018). IEEE: Palermo, Italy. pp. 1–7. https://doi.org/10.1109/EEEIC.2018.8493898.

41 Di Somma, M., Caliano, M., Cigolotti, V., and Graditi, G. (2021). Investigating hydrogen-based non-conventional storage for PV power in eco-energetic optimization of a multi-energy system. *Energies (Basel)* 14 (23): 8096. https://doi.org/10.3390/en14238096.

42 Di Somma, M., Caliano, M., Graditi, G. et al. (2020). Designing of cost-effective and low-carbon multi-energy nanogrids for residential applications. *Inventions* 5 (1): 7. https://doi.org/10.3390/inventions5010007.

43 Di Somma, M., Yan, B., Bianco, N. et al. (2017). Multi-objective design optimization of distributed energy systems through cost and exergy assessments. *Applied Energy* 204: 1299–1316. https://doi.org/10.1016/j.apenergy.2017.03.105.

44 Di Somma, M., Graditi, G., Heydarian-Forushani, E. et al. (2018). Stochastic optimal scheduling of distributed energy resources with renewables considering economic and environmental aspects. *Renewable Energy* 116: 272–287. https://doi.org/10.1016/j.renene.2017.09.074.

45 Zhou, Z., Zhang, J., Liu, P. et al. (2013). A two-stage stochastic programming model for the optimal design of distributed energy systems. *Applied Energy* 103: 135–144. https://doi.org/10.1016/j.apenergy.2012.09.019.

46 Lasemi, M.A., Arabkoohsar, A., Hajizadeh, A., and Mohammadi-ivatloo, B. (2022). A comprehensive review on optimization challenges of smart energy hubs under uncertainty factors. *Renewable and Sustainable Energy Reviews* 160: 112320. https://doi.org/10.1016/j.rser.2022.112320.

47 Shaneb, O.A., Taylor, P.C., and Coates, G. (2012). Optimal online operation of residential μCHP systems using linear programming. *Energy and Buildings* 44: 17–25. https://doi.org/10.1016/j.enbuild.2011.10.003.

48 Gustafsson, S.-I. and Karlsson, B.G. (1991). Linear programming optimization in CHP networks. *Heat Recovery Systems and CHP* 11 (4): 231–238. https://doi.org/10.1016/0890-4332(91)90068-F.

49 Sakawa, M., Kato, K., and Ushiro, S. (2001). Operation planning of district heating and cooling plants through genetic algorithms for nonlinear 0–1 programming. *Computers & Mathematics with Applications* 42 (10–11): 1365–1378. https://doi.org/10.1016/S0898-1221(01)00246-2.

50 Yan, B., Luh, P.B., Sun, B. et al. (2013). Energy-efficient management of eco-communities. *2013 IEEE International Conference on Automation Science and Engineering (CASE)* (17–20 August 2013). IEEE: Madison, WI. pp. 106–111. https://doi.org/10.1109/CoASE.2013.6654057.

51 Guan, X., Xu, Z., and Jia, Q. (2010). Energy efficient buildings facilitated by micro grid. *IEEE PES General Meeting* (25-29 July 2010). Minneapolis, MN: IEEE. pp. 1–1. https://doi.org/10.1109/PES.2010.5590040.

52 Shafie-khah, M., Vahid-Ghavidel, M., Di Somma, M. et al. (2020). Management of renewable-based multi-energy microgrids in the presence of electric vehicles. *IET Renewable Power Generation* 14 (3): 417–426. https://doi.org/10.1049/iet-rpg.2019.0124.

53 Yan, B., Di Somma, M., Luh, P.B., and Graditi, G. (2018). Operation optimization of multiple distributed energy systems in an energy community. *2018 IEEE International Conference on Environment and Electrical Engineering and 2018 IEEE Industrial and Commercial Power Systems Europe (EEEIC/I&CPS Europe)* (June 2018). IEEE: Palermo, Italy. pp. 1–6. https://doi.org/10.1109/EEEIC.2018.8494476.

54 Yan, B., Di Somma, M., Graditi, G., and Luh, P.B. (2020). Markovian-based stochastic operation optimization of multiple distributed energy systems with renewables in a local energy community. *Electric Power Systems Research* 186: 106364. https://doi.org/10.1016/j.epsr.2020.106364.

55 Di Somma, M., Graditi, G., and Siano, P. (2019). Optimal bidding strategy for a DER aggregator in the day-ahead market in the presence of demand flexibility. *IEEE Transactions on Industrial Electronics* 66 (2): 1509–1519. https://doi.org/10.1109/TIE.2018.2829677.

56 Di Somma, M., Yan, B., Bianco, N. et al. (2015). Operation optimization of a distributed energy system considering energy costs and exergy efficiency. *Energy Conversion and Management* 103: 739–751. https://doi.org/10.1016/j.enconman.2015.07.009.

57 Di Somma, M., Bing, Y., Bianco, N. et al. (2015). Influence of energy quality management on CO_2; emissions in operation optimization of a distributed energy system. *2015 International Conference on Clean Electrical Power (ICCEP)* (June 2015). IEEE: Taormina, Italy. pp. 297–304. https://doi.org/10.1109/ICCEP.2015.7177639.

58 Yan, B., Di Somma, M., Bianco, N. et al. (2016). Exergy-based operation optimization of a distributed energy system through the energy-supply chain. *Applied Thermal Engineering* 101: 741–751. https://doi.org/10.1016/j.applthermaleng.2016.02.029.

59 Roberto, R., De Iulio, R., Di Somma, M. et al. (2019). A multi-objective optimization analysis to assess the potential economic and environmental benefits of distributed storage in district heating networks: a case study. *International Journal of Sustainable Energy Planning and Management* 20: https://doi.org/10.5278/ijsepm.2019.20.2.

60 Di Somma, M., Yan, B., Bianco, N. et al. (2016). Multi-objective operation optimization of a distributed energy system for a large-scale utility customer. *Applied Thermal Engineering* 101: 752–761. https://doi.org/10.1016/j.applthermaleng.2016.02.027.

61 Di Somma, M., Ciabattoni, L., Comodi, G., and Graditi, G. (2020). Managing plug-in electric vehicles in eco-environmental operation optimization of local multi-energy systems. *Sustainable Energy, Grids and Networks* 23: 100376. https://doi.org/10.1016/j.segan.2020.100376.

62 Di Somma, M., Graditi, G., Mongibello, L. et al. (2018). Trade-off solutions between economy and CO_2 emissions for the daily operation of a distributed energy system: a real case study in Italy. *2018 IEEE International Conference on Environment and Electrical Engineering and 2018 IEEE Industrial and Commercial Power Systems Europe (EEEIC/I&CPS Europe)* (June 2018). IEEE: Palermo, Italy. pp. 1–6. https://doi.org/10.1109/EEEIC.2018.8494360.

63 Di Somma, M., Buonanno, A., Caliano, M. et al. (2022). Stochastic operation optimization of the smart savona campus as an integrated local energy community considering energy costs and carbon emissions. *Energies (Basel)* 15 (22): 8418. https://doi.org/10.3390/en15228418.

64 Di Somma, M., Buonanno, A., Caliano, M. et al. (2023). Stochastic energy management for the Italian UNIVPM campus as a multi-carrier energy hub participating in the day-ahead market. *IEEE EUROCON 2023 – 20th International Conference on Smart Technologies* (6-8 July 2023). IEEE: Torino, Italy. pp. 251–256. https://doi.org/10.1109/EUROCON56442.2023.10199000.

65 Caliano, M., Delfino, F., Di Somma, M. et al. (2022). An energy management system for microgrids including costs, exergy, and stress indexes. *Sustainable Energy, Grids and Networks* 32: 100915. https://doi.org/10.1016/j.segan.2022.100915.

66 Di Somma, M., Di Dio, V., Favuzza, S. et al. (2023). Performance optimization of a residential microgrid balancing economic and energy issues. *IEEE EUROCON 2023 – 20th International Conference on Smart Technologies* (July 2023). IEEE: Torino, Italy. pp. 752–757. https://doi.org/10.1109/EUROCON56442.2023.10198931.

67 Buonanno, A., Caliano, M., Di Somma, M. et al. (2022). A comprehensive tool for scenario generation of solar irradiance profiles. *Energies (Basel)* 15 (23): 8830. https://doi.org/10.3390/en15238830.

68 Majidi, M. and Zare, K. (2019). Integration of smart energy hubs in distribution networks under uncertainties and demand response concept. *IEEE Transactions on Power Systems* 34 (1): 566–574. https://doi.org/10.1109/TPWRS.2018.2867648.

69 Khani, H., Sawas, A., and Farag, H.E.Z. (2021). An estimation–based optimal scheduling model for settable renewable penetration level in energy hubs. *Electric Power Systems Research* 196: 107230. https://doi.org/10.1016/j.epsr.2021.107230.

70 Good, N., Karangelos, E., Navarro-Espinosa, A., and Mancarella, P. (2015). Optimization under uncertainty of thermal storage-based flexible demand response with quantification of residential users' discomfort. *IEEE Transactions on Smart Grid* 6 (5): 2333–2342. https://doi.org/10.1109/TSG.2015.2399974.

71 Li, C. and Grossmann, I.E. (2021). A review of stochastic programming methods for optimization of process systems under uncertainty. *Frontiers in Chemical Engineering* 2: https://doi.org/10.3389/fceng.2020.622241.

72 Yan, B., Luh, P.B., Warner, G., and Zhang, P. (2017). Operation and design optimization of microgrids with renewables. *IEEE Transactions on Automation Science and Engineering* 14 (2): 573–585. https://doi.org/10.1109/TASE.2016.2645761.

73 Buonanno, A., Caliano, M., Di Somma, M. et al. (2021). Comprehensive method for modeling uncertainties of solar irradiance for PV power generation in smart grids. *2021 International Conference on Smart Energy Systems and Technologies (SEST)* (6-8 September 2021). IEEE: Vaasa, Finland. pp. 1–6. https://doi.org/10.1109/SEST50973.2021.9543245.

74 Chen, Y.-Y. and Fan, H.-Y. (2015). An application of stochastic programming in solving capacity allocation and migration planning problem under uncertainty. *Mathematical Problems in Engineering* 2015: 1–16. https://doi.org/10.1155/2015/741329.

75 Bounitsis, G.L., Papageorgiou, L.G., and Charitopoulos, V.M. (2022). Data-driven scenario generation for two-stage stochastic programming. *Chemical Engineering Research and Design* 187: 206–224. https://doi.org/10.1016/j.cherd.2022.08.014.

76 Huy, T.H.B., Dinh, H.T., and Kim, D. (2023). Multi-objective framework for a home energy management system with the integration of solar energy and an electric vehicle using an augmented ε-constraint method and lexicographic optimization. *Sustainable Cities and Society* 88: 104289. https://doi.org/10.1016/j.scs.2022.104289.

77 Alıç, O. and Filik, Ü.B. (2021). A multi-objective home energy management system for explicit cost-comfort analysis considering appliance category-based discomfort models and demand response programs. *Energy and Buildings* 240: 110868. https://doi.org/10.1016/j.enbuild.2021.110868.

78 Rastegar, M., Fotuhi-Firuzabad, M., Zareipour, H., and Moeini-Aghtaieh, M. (2017). A probabilistic energy management scheme for renewable-based residential energy hubs. *IEEE Transactions on Smart Grid* 8 (5): 2217–2227. https://doi.org/10.1109/TSG.2016.2518920.

79 Kazemi, M., Papadimitriou, C., Paterakis, N. et al. (2023). Optimal design of multi-carrier and -objective home energy management system. IEEE SEST: Mugla, Turkey. pp. 1–6.

80 Lu, X., Liu, Z., Ma, L. et al. (2020). A robust optimization approach for optimal load dispatch of community energy hub. *Applied Energy* 259: https://doi.org/10.1016/j.apenergy.2019.114195.

81 Dan, M., Srinivasan, S., Sundaram, S. et al. (2020). A scenario-based branch-and-bound approach for MES scheduling in urban buildings. *IEEE Transactions on Industrial Informatics* 16 (12): 7510–7520. https://doi.org/10.1109/TII.2020.2978870.

82 Bollenbacher, J. and Rhein, B. (2017). Optimal configuration and control strategy in a multi-carrier-energy system using reinforcement learning methods. *IESC 2017 – International Energy and Sustainability Conference* (December 2017). Farmingdale, NY, USA. pp. 1–6. https://doi.org/10.1109/IESC.2017.8167476.

83 Ren, H., Zhou, W., Nakagami, K. et al. (2010). Multi-objective optimization for the operation of distributed energy systems considering economic and environmental aspects. *Applied Energy* 87 (12): 3642–3651. https://doi.org/10.1016/j.apenergy.2010.06.013.

84 Zhou, Y., Wu, J., and Long, C. (2018). Evaluation of peer-to-peer energy sharing mechanisms based on a multiagent simulation framework. *Applied Energy* 222: 993–1022. https://doi.org/10.1016/j.apenergy.2018.02.089.

85 Liu, T., Zhang, D., Wang, S., and Wu, T. (2019). Standardized modelling and economic optimization of multi-carrier energy systems considering energy storage and demand response. *Energy Conversion and Management* 182: 126–142. https://doi.org/10.1016/j.enconman.2018.12.073.

86 Vahid-Ghavidel, M., Sadegh Javadi, M., Gough, M. et al. (2020). Demand response programs in multi-energy systems: a review. *Energies (Basel)* 13 (17): 1–17. https://doi.org/10.3390/en13174332.

87 Mirhedayati, A., Shahinzadeh, H., Nafisi, H., et al. (2021). CHPs and EHPs effectiveness evaluation in a residential multi-carrier energy hub. *2021 25th Electrical Power Distribution Conference (EPDC)* (August 2021). IEEE: Karaj, Iran. pp. 42–47. https://doi.org/10.1109/EPDC53015.2021.9610816.

88 Papadimitriou, C., Di Somma, M., Charalambous, C. et al. (2023). A comprehensive review of the design and operation optimization of energy hubs and their interaction with the markets and external networks. *Energies (Basel)* 16 (10): 4018. https://doi.org/10.3390/en16104018.

89 Sousa, T., Soares, T., Pinson, P. et al. (2019). Peer-to-peer and community-based markets: a comprehensive review. *Renewable and Sustainable Energy Reviews* 104: 367–378. https://doi.org/10.1016/j.rser.2019.01.036.

90 Bjarghov, S., Löschenbrand, M., Saif, A.I. et al. (2021). Developments and challenges in local electricity markets: a comprehensive review. *IEEE Access* 9: 58910–58943. https://doi.org/10.1109/ACCESS.2021.3071830.

91 Li, P., Guo, T., Abeysekera, M. et al. (2021). Intraday multi-objective hierarchical coordinated operation of a multi-energy system. *Energy* 228: https://doi.org/10.1016/j.energy.2021.120528.

92 Coelho, A., Iria, J., and Soares, F. (2021). Network-secure bidding optimization of aggregators of multi-energy systems in electricity, gas, and carbon markets. *Applied Energy* 301: 117460. https://doi.org/10.1016/j.apenergy.2021.117460.

93 Miao, A., Yuan, Y., Wu, H., and Zhu, J. (2023). Novel planning method and coordinated operation strategy of multi-station integration system. *CSEE Journal of Power and Energy Systems* 1–15. https://doi.org/10.17775/CSEEJPES.2022.02920.

94 Habibi, M., Vahidinasab, V., Mohammadi-Ivatloo, B. et al. (2022). Exploring potential gains of mobile sector-coupling energy systems in heavily constrained networks. *IEEE Transactions on Sustainable Energy* 13 (4): 2092–2105. https://doi.org/10.1109/TSTE.2022.3182871.

95 Tzelepis, D., Tsotsopoulou, E., Hong, Q. et al. (2023). Measuring technologies for future power grids. In: *Encyclopedia of Electrical and Electronic Powe Engineering*, 310–319. Elsevier https://doi.org/10.1016/B978-0-12-821204-2.00147-1.

96 Ullah, M., Narayanan, A., Wolff, A., and Nardelli, P.H.J. (2022). Industrial energy management system: design of a conceptual framework using IoT and big data. *IEEE Access* 10: 110557–110567. https://doi.org/10.1109/ACCESS.2022.3215167.

97 Al-Ghaili, A.M., Kasim, H., Al-Hada, N.M. et al. (2021). Energy management systems and strategies in buildings sector: a scoping review. *IEEE Access* 9: 63790–63813. https://doi.org/10.1109/ACCESS.2021.3075485.

98 Ma, R. and He, P. (2020). Dynamic modeling gas-electricity combined system with wind power and its stability control method. *CSEE Journal of Power and Energy Systems* https://doi.org/10.17775/CSEEJPES.2020.02670.

99 Huang, Y., Sun, Q., Zhang, N., and Wang, R. (2021). A multi-slack bus model for bi-directional energy flow analysis of integrated power-gas systems. *CSEE Journal of Power and Energy Systems* https://doi.org/10.17775/CSEEJPES.2020.04190.

100 Liu, C., Shahidehpour, M., Fu, Y., and Li, Z. (2009). Security-constrained unit commitment with natural gas transmission constraints. *IEEE Transactions on Power Systems* 24 (3): 1523–1536. https://doi.org/10.1109/TPWRS.2009.2023262.

101 Martinez-Mares, A. and Fuerte-Esquivel, C.R. (2012). A unified gas and power flow analysis in natural gas and electricity coupled networks. *IEEE Transactions on Power Systems* 27 (4): 2156–2166. https://doi.org/10.1109/TPWRS.2012.2191984.

102 Clegg, S. and Mancarella, P. (2016). Integrated electrical and gas network flexibility assessment in low-carbon multi-energy systems. *IEEE Transactions on Sustainable Energy* 7 (2): 718–731. https://doi.org/10.1109/TSTE.2015.2497329.

103 Kroposki, B., Bernstein, A., King, J. et al. (2020). Autonomous energy grids: controlling the future grid with large amounts of distributed energy resources.

IEEE Power and Energy Magazine 18 (6): 37–46. https://doi.org/10.1109/MPE.2020.3014540.

104 Mancarella, P. (2014). MES (multi-energy systems): an overview of concepts and evaluation models. *Energy* 65: 1–17. https://doi.org/10.1016/j.energy.2013.10.041.

105 Vrba, P., Mařík, V., Siano, P. et al. (2014). A review of agent and service-oriented concepts applied to intelligent energy systems. *IEEE Transactions on Industrial Informatics* 10 (3): 1890–1903. https://doi.org/10.1109/TII.2014.2326411.

106 McArthur, S.D.J., Davidson, E.M., Catterson, V.M. et al. (2007). Multi-agent systems for power engineering applications—part II: technologies, standards, and tools for building multi-agent systems. *IEEE Transactions on Power Systems* 22 (4): 1753–1759. https://doi.org/10.1109/TPWRS.2007.908472.

107 McArthur, S.D.J., Davidson, E.M., Catterson, V.M. et al. (2007). Multi-agent systems for power engineering applications—part I: concepts, approaches, and technical challenges. *IEEE Transactions on Power Systems* 22 (4): 1743–1752. https://doi.org/10.1109/TPWRS.2007.908471.

108 Zhao, Z. and Chen, G. (2018). An overview of cyber security for smart grid. *2018 IEEE 27th International Symposium on Industrial Electronics (ISIE)* (13-15 June 2018). IEEE: Cairns, QLD, Australia. pp. 1127–1131. https://doi.org/10.1109/ISIE.2018.8433816.

109 Riekstin, A.C., Rodrigues, B.B., Nascimento, V.T. et al. (2016). Sustainability information model for energy efficiency policies. *IEEE Communications Magazine* 54 (11): 176–184. https://doi.org/10.1109/MCOM.2016.1600222CM.

110 Tao, Y., Qiu, J., and Lai, S. (2022). A supervised-learning assisted computation method for power system planning. *IEEE Transactions on Artificial Intelligence* 3 (5): 819–832. https://doi.org/10.1109/TAI.2021.3133821.

111 Subbiah, R., Pal, A., Nordberg, E.K. et al. (2017). Energy demand model for residential sector: a first principles approach. *IEEE Transactions on Sustainable Energy* 8 (3): 1215–1224. https://doi.org/10.1109/TSTE.2017.2669990.

112 Degefa, M.Z., Lundkvist, H., Sanchez-Acevedo, S., and Gregertsen, K.N. (2023). Challenges of TSO-DSO voltage regulation under real-time data exchange paradigm. *IEEE Open Journal of the Industrial Electronics Society* 4: 75–84. https://doi.org/10.1109/OJIES.2023.3239946.

113 Bush, S.F., Goel, S., and Simard, G. (2013). IEEE vision for smart grid communications: 2030 and beyond roadmap. *IEEE Vision for Smart Grid Communications: 2030 and Beyond Roadmap*. pp. 1–19. https://doi.org/10.1109/IEEESTD.2013.6690098.

114 Chen, H., Xuan, P., Wang, Y. et al. (2016). Key technologies for integration of multitype renewable energy sources—research on multi-timeframe robust scheduling/dispatch. *IEEE Transactions on Smart Grid* 7 (1): 471–480. https://doi.org/10.1109/TSG.2015.2388756.

115 Cheng, X., Zhang, R., and Yang, L. (2016). Consumer-centered energy system for electric vehicles and the smart grid. *IEEE Intelligent Systems* 31 (3): 97–101. https://doi.org/10.1109/MIS.2016.52.

116 Rousis, A.O., Tzelepis, D., Pipelzadeh, Y. et al. (2021). Provision of voltage ancillary services through enhanced TSO-DSO interaction and aggregated distributed energy resources. *IEEE Transactions on Sustainable Energy* 12 (2): 897–908. https://doi.org/10.1109/TSTE.2020.3024278.

117 Chard, K., Tuecke, S., and Foster, I. (2014). Efficient and secure transfer, synchronization, and sharing of big data. *IEEE Cloud Computing* 1 (3): 46–55. https://doi.org/10.1109/MCC.2014.52.

118 Humayed, A., Lin, J., Li, F., and Luo, B. (2017). Cyber-physical systems security—a survey. *IEEE Internet of Things Journal* 4 (6): 1802–1831. https://doi.org/10.1109/JIOT.2017.2703172.

119 IEEE guide for distributed energy resources management systems (DERMS) functional specification. *IEEE Std 2030.11-2021*. pp. 1–61. https://doi.org/10.1109/IEEESTD.2021.9447316.

120 Poudel, S., Sharma, P., Dubey, A., and Schneider, K.P. (2020). Advanced FLISR with intentional islanding operations in an ADMS environment using GridAPPS-D. *IEEE Access* 8: 113766–113778. https://doi.org/10.1109/ACCESS.2020.3003325.

121 Babayigit, B. and Abubaker, M. (2023). Industrial internet of things: a review of improvements over traditional SCADA systems for industrial automation. *IEEE Systems Journal* 1–14. https://doi.org/10.1109/JSYST.2023.3270620.

122 Takaghaj, S.M., Macnab, C.J.B., Westwick, D., and Boiko, I. (2014). Neural-adaptive control of waste-to-energy steam generators. *IEEE Transactions on Control Systems Technology* 22 (5): 1920–1926. https://doi.org/10.1109/TCST.2013.2292818.

123 Jan, M.U., Xin, A., Rehman, H.U. et al. (2021). Frequency regulation of an isolated microgrid with electric vehicles and energy storage system integration using adaptive and model predictive controllers. *IEEE Access* 9: 14958–14970. https://doi.org/10.1109/ACCESS.2021.3052797.

124 Davidson, J., Genest, R., and Ringwood, J.V. (2018). Adaptive control of a wave energy converter. *IEEE Transactions on Sustainable Energy* 9 (4): 1588–1595. https://doi.org/10.1109/TSTE.2018.2798921.

125 Chenine, M., Ullberg, J., Nordstrom, L. et al. (2014). A framework for wide-area monitoring and control systems interoperability and cybersecurity analysis. *IEEE Transactions on Power Delivery* 29 (2): 633–641. https://doi.org/10.1109/TPWRD.2013.2279182.

126 Chenine, M. and Nordstrom, L. (2011). Modeling and simulation of wide-area communication for centralized PMU-based applications. *IEEE Transactions on Power Delivery* 26 (3): 1372–1380. https://doi.org/10.1109/TPWRD.2011.2106805.

127 Skarvelis-Kazakos, S., Papadopoulos, P., and Unda, I.G. (2014). Agent-based control of multiple energy carriers and energy hubs. *IEEE PES Innovative Smart Grid Technologies,* Europe (12-15 October 2014). IEEE: Istanbul, Turkey. pp. 1–6. https://doi.org/10.1109/ISGTEurope.2014.7028912.

128 Xiang, Y., Fang, M., and Liu, J. (2023). Distributed dispatch of multiple energy systems considering carbon trading. *CSEE Journal of Power and Energy Systems* https://doi.org/10.17775/CSEEJPES.2021.09050.

129 Bytyqi, A., Gandhi, S., Lambert, E., and Petrovic, N. (2022). A review on TSO-DSO data exchange, CIM extensions and interoperability aspects. *Journal of Modern Power Systems and Clean Energy* 10 (2): 309–315. https://doi.org/10.35833/MPCE.2021.000770.

130 Karie, N.M., Sahri, N.M., Yang, W. et al. (2021). A review of security standards and frameworks for IoT-based smart environments. *IEEE Access* 9: 121975–121995. https://doi.org/10.1109/ACCESS.2021.3109886.

131 Shahid, K., Nainar, K., Olsen, R.L. et al. (2021). On the use of common information model for smart grid applications—a conceptual approach. *IEEE Transactions on Smart Grid* 12 (6): 5060–5072. https://doi.org/10.1109/TSG.2021.3095896.

132 Wojtowicz, R., Kowalik, R., Rasolomampionona, D.D., and Kurek, K. (2022). Virtualization of protection systems – tests performed on a large environment based on data center solutions. *IEEE Transactions on Power Delivery* 37 (4): 3401–3411. https://doi.org/10.1109/TPWRD.2021.3128993.

133 Bera, S., Misra, S., and Rodrigues, J.J.P.C. (2015). Cloud computing applications for smart grid: a survey. *IEEE Transactions on Parallel and Distributed Systems* 26 (5): 1477–1494. https://doi.org/10.1109/TPDS.2014.2321378.

134 Li, J., Gu, C., Xiang, Y., and Li, F. (2022). Edge-cloud computing systems for smart grid: state-of-the-art, architecture, and applications. *Journal of Modern Power Systems and Clean Energy* 10 (4): 805–817. https://doi.org/10.35833/MPCE.2021.000161.

135 Moghaddam, M.M., Manshaei, M.H., Soorki, M.N. et al. (2021). On coordination of smart grid and cooperative cloud providers. *IEEE Systems Journal* 15 (1): 672–683. https://doi.org/10.1109/JSYST.2020.2987017.

136 Jafari, M., Kavousi-Fard, A., Chen, T., and Karimi, M. (2023). A review on digital twin technology in smart grid, transportation system and smart city: challenges and future. *IEEE Access* 11: 17471–17484. https://doi.org/10.1109/ACCESS.2023.3241588.

137 Bazmohammadi, N., Madary, A., Vasquez, J.C. et al. (2022). Microgrid digital twins: concepts, applications, and future trends. *IEEE Access* 10: 2284–2302. https://doi.org/10.1109/ACCESS.2021.3138990.

138 Tuyen, N.D., Quan, N.S., Linh, V.B. et al. (2022). A comprehensive review of cybersecurity in inverter-based smart power system amid the boom of renewable energy. *IEEE Access* 10: 35846–35875. https://doi.org/10.1109/ACCESS.2022.3163551.

139 Abdelmalak, M., Venkataramanan, V., and Macwan, R. (2022). A survey of cyber-physical power system modeling methods for future energy systems. *IEEE Access* 10: 99875–99896. https://doi.org/10.1109/ACCESS.2022.3206830.

140 Elbez, G., Nahrstedt, K., and Hagenmeyer, V. (2023). Early attack detection for securing GOOSE network traffic. *IEEE Transactions on Smart Grid* 1–1. https://doi.org/10.1109/TSG.2023.3272749.

141 Radoglou-Grammatikis, P.I. and Sarigiannidis, P.G. (2019). Securing the smart grid: a comprehensive compilation of intrusion detection and prevention systems. *IEEE Access* 7: 46595–46620. https://doi.org/10.1109/ACCESS.2019.2909807.

142 Bhatt, S., Manadhata, P.K., and Zomlot, L. (2014). The operational role of security information and event management systems. *IEEE Security and Privacy* 12 (5): 35–41. https://doi.org/10.1109/MSP.2014.103.

143 Huang, W., Zhang, N., Yang, J. et al. (2019). Optimal configuration planning of multi-energy systems considering distributed renewable energy. *IEEE Transactions on Smart Grid* 10 (2): 1452–1464. https://doi.org/10.1109/TSG.2017.2767860.

144 BEIS, OFGEM (2021). *Transitioning to a Net Zero Energy System Smart Systems and Flexibility Plan 2021*. London: Crown Copyright.

145 Zhang, L., Zou, Y., Yu, J. et al. (2020). Simultaneous optimization of renewable energy and energy storage capacity with hierarchical control. *CSEE Journal of Power and Energy Systems* https://doi.org/10.17775/CSEEJPES.2019.01470.

146 Zhang, Y., Liu, Y., Shu, S. et al. (2021). A data-driven distributionally robust optimization model for multi-energy coupled system considering the temporal-spatial correlation and distribution uncertainty of renewable energy sources. *Energy* 216: 119171. https://doi.org/10.1016/j.energy.2020.119171.

147 Cedillos Alvarado, D., Acha, S., Shah, N., and Markides, C.N. (2016). A technology selection and operation (TSO) optimisation model for distributed energy systems: mathematical formulation and case study. *Applied Energy* 180: 491–503. https://doi.org/10.1016/j.apenergy.2016.08.013.

6

Analytical Framework for Coordinated Planning and Operation of Multicarrier Energy Systems

Marialaura Di Somma[1], Christina Papadimitriou[2], Anastasios Oulis Rousis[3], Angelos Patsidis[3], Miadreza Shafie-Khah[4], Vahid Shahbazbegian[4], and Magnus Askeland[5]

[1] *Italian National Agency for New Technologies, Energy and Sustainable Economic Development, Department of Energy Technologies and Renewable Sources, Lungotevere Thaon di Revel, 76, 00196, Rome, Italy*
[2] *Eindhoven Technical University, Electrical Engineering Department, Groene Loper 3, 5612 AE, Eindhoven, the Netherlands*
[3] *Smart Power Networks (SMPnet) Ltd., 65 More Close, London W14 9BN, UK*
[4] *School of Technology and Innovations, University of Vaasa, Wolffintie 32, 65200, Vaasa, Finland*
[5] *SINTEF Energy Research, Department of Energy Systems, Sem Sælands vei 11, 7034, Trondheim, Norway*

6.1 Introduction

6.1.1 Benefits of Local Multicarrier Energy Systems

Sector coupling can increase flexibility of the energy system to cope with the fluctuations in energy demand and variable renewable energy supply, while also avoiding renewable energy sources (RESs) curtailment and reducing the need to reinforce existing network infrastructure [1, 2]. The so-called power-to-X technologies, that allow for the decoupling of power from the electricity sector for use in other sectors such as heating, cooling, transport, and chemicals, introduce circularity in the energy system. In fact, excess from RES can be converted into fuel as hydrogen or synthetic methane that can be used on site or stored for later use as well as reconverted again into electricity when renewable electricity supply is not sufficient to satisfy the loads. Another example of power-to-X is represented by power-to-heat technologies, such as heat pumps (HPs), that can lead to cost-effective and efficient heating and cooling of buildings and industrial process heat production, allowing for lower primary energy consumption. If combined with thermal storage, they can allow thermal energy production to be shifted when renewable electricity is in excess, thus closing the loop and enabling a circular approach.

Although the several benefits associated with sector coupling solutions, it is also clear that implementing it at a large-scale regional or national level is extremely challenging, mainly due to the involvement of too many stakeholders with different and often conflicting goals [3]. Therefore, a bottom-up approach is preferable focusing on local multicarrier energy systems (MCESs) offering the unique opportunity to implement sector coupling solutions and energy communities at the

Integrated Local Energy Communities: From Concepts and Enabling Conditions to Optimal Planning and Operation, First Edition. Edited by Marialaura Di Somma, Christina Papadimitriou, Giorgio Graditi, and Koen Kok.

local level, thereby exploiting synergies among energy carriers available locally and increasing the efficiency in the energy resources use. The MCES paradigm is also fully in line with the ongoing paradigm shift of the energy system from centralized energy generation (largely dependent on fossil-fueled power plants) to decentralized generation. The latter relies on small-scale RES and combined heat and power (CHP) units not connected to the high-voltage or gas grid, namely usually small-scale plants that supply electricity and heat to a building, industrial site, or community, potentially selling back surplus electricity. MCESs may offer several benefits deriving from the linking of sector coupling and decentralized generation that are discussed below.

- Producing energy close to where it is used allows **exploiting RES available locally**, such as wind, solar, biomass, geothermal, hydro, and waste.
- An **increased energy efficiency** is achievable through (i) the use of waste heat from power generation processes or reduced transmission losses; (ii) the increased use of renewable, carbon-neutral, or low-carbon sources of fuel; and (iii) reduced exchange of energy on large distances that also contribute to delay the need for new infrastructures or their reinforcement.
- **Higher flexibility** for generation matching local demand patterns for electricity and heat is achievable.
- **End-user engagement** is promoted including households, small communities, or businesses that own and operate energy generation and storage assets available locally, including solar photovoltaic (PV) panels, wind turbines (WTs), CHP, battery storage, and electric mobility (e.g. in electric vehicles [EVs], e-bikes, and e-scooters). Moreover, end-user engagement is also fostered through the inclusion of demand-response (DR) programs. On the other hand, development of local energy generation contributes to achieving energy self-sufficiency.
- **Addressing energy system vulnerability** since local energy production and consumption increases energy security in terms of optimal use of local resources, reduction of losses, congestion, and reduction or delay in infrastructural investments [4].
- **Addressing energy poverty** as the energy generated locally allows profits remaining in the local system as a community helping bring down the cost of energy in the long run, while also inducing the emergence of local value chains addressing energy poverty in low-income, inhabited, or marginal areas.

To achieve the aforementioned benefits and make MCES sustainable not only from the energy and environmental perspective but also from the economic one thus having a concrete option to centralized energy systems, coordinated planning and operation on multiple time horizons is extremely important. As extensively discussed in Chapter 5, planning of MCES is a complex decision-making process aiming at their optimal design, namely the identification of the optimal configuration in terms of choice of technologies and sizes by considering one or multiple objective functions. On the other hand, optimal operation planning (day-ahead or real-time) aims at finding the optimized daily or real-time operation strategies of the MCES, also in this case, by considering one or multiple objective functions.

6.1.2 The Optimization Framework for Coordinated Planning and Operation of MCES

The aim of this chapter is to cover all the transversal aspects related to the coordinated planning and operation of MCES, by presenting a comprehensive analytical framework for the optimal design and operation problems, structured to cover different time horizons from long-term system planning to day-ahead and real-time operation. The system planning framework has the goal to come up with the optimal configuration in MCES, by replying to questions such as how to develop the system in terms of types and sizes of technologies. Essential inputs are system predefined superstructure with potential technologies to be installed and energy flows needed by the end-users; characteristics of potential technologies options; users' demand profiles in representative days; and renewables availability. The approach used for the optimal planning can be the economic or the multicriteria optimization, considering for instance economic and environmental criteria. The day-ahead operation planning framework, represented in Figure 6.1a allows achieving the day-ahead scheduling of MCES with the support of forecasting tools for RES generation, energy prices for electricity and gas, and users' multienergy demand or scenario generation tools for the same data aiming at the stochastic optimization of MCES. The short-term operation planning framework represented in Figure 6.1b

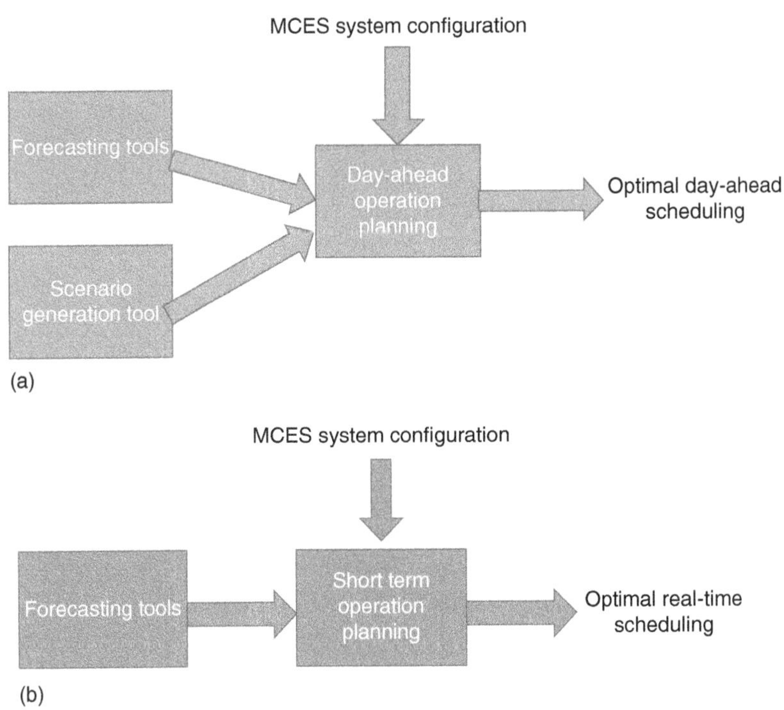

Figure 6.1 Scheme of the optimization framework for MCES for (a) day-ahead scheduling and (b) real-time scheduling.

has the main goal to find the optimal real-time scheduling of the MCES, given its configuration, with the support of forecasting tools for RES generation, energy prices, and user demand.

The optimization of MCESs necessitates the establishment of objective functions that capture the desired outcomes and priorities of the system's stakeholders. These objective functions serve as quantitative metrics to guide the optimization process, enabling the efficient utilization of energy resources and the achievement of specific system goals. The choice of objective function(s) is driven by the specific objectives, system characteristics, and stakeholder preferences. Some examples of objective functions that can be used in the context of MCES are presented as follows:

1. **Cost minimization:** One of the primary objectives is often to minimize the overall cost of the MCES. This includes minimizing the costs associated with energy generation, storage, conversion, and distribution, considering investment and/or operation costs.
2. **Emissions reduction:** Another important objective is to reduce the environmental impact of the energy system by minimizing greenhouse gas emissions or pollutant emissions. This objective promotes the integration of RES and the efficient utilization of low-carbon technologies.
3. **Reliability and resilience:** This objective function focuses on enhancing the reliability and resilience of the MCES. This may involve minimizing the risk of energy shortages or blackouts, ensuring sufficient energy reserves, or optimizing the response to contingencies and disturbances.
4. **DR and flexibility:** This objective function aims to incentivize DR programs or flexible energy usage patterns to achieve a more balanced and efficient system operation. This involves optimizing the coordination between energy supply and demand and maximizing the utilization of flexible loads and energy storage.
5. **Renewable energy integration:** The objective function prioritizes the integration of RES to maximize their utilization and minimize dependence on fossil fuels. This can include maximizing the share of renewable energy in the overall energy mix or optimizing the utilization of available renewable resources.

It is important to note that the selection of the objective function depends on the specific goals, system characteristics, and stakeholder priorities in each case. The combination and weighting of these objective functions may vary to reflect the unique requirements and objectives of the MCES under consideration. By understanding and appropriately formulating the objective function(s), stakeholders can effectively navigate the trade-offs and complexities inherent in MCESs, paving the way for sustainable and efficient system operation.

In the following, the modeling of a wide range of generation, conversion, and storage technologies that can be part of MCESs is presented in Section 6.2. The detailed optimization frameworks for the design and the day-ahead and the real-time operation problems are discussed in Sections 6.3–6.5, respectively, by also discursive representative case studies to demonstrate the effectiveness of the optimization models. A detailed overview of the current commercial tools available for the optimal planning of MCESs is presented in Section 6.6, by mapping them according to several criteria such as type, type of users, coverage of multicarrier aspects, objective functions, functionalities, mathematical approach, temporal resolution, and time horizons as well as pros and cons in their application to MCESs.

6.2 Modeling of Energy Technologies in MCES

In this section, modeling of energy technologies that can be part of MCES is analyzed in detail. In particular, the modeled technologies are subdivided into the following categories: RES, primary energy technologies, secondary energy technologies, and storage systems as represented in Figure 6.2.

6.2.1 Renewable Energy Sources (RESs)

PV systems, solar collectors for domestic hot water production, and WTs are the RES here considered.

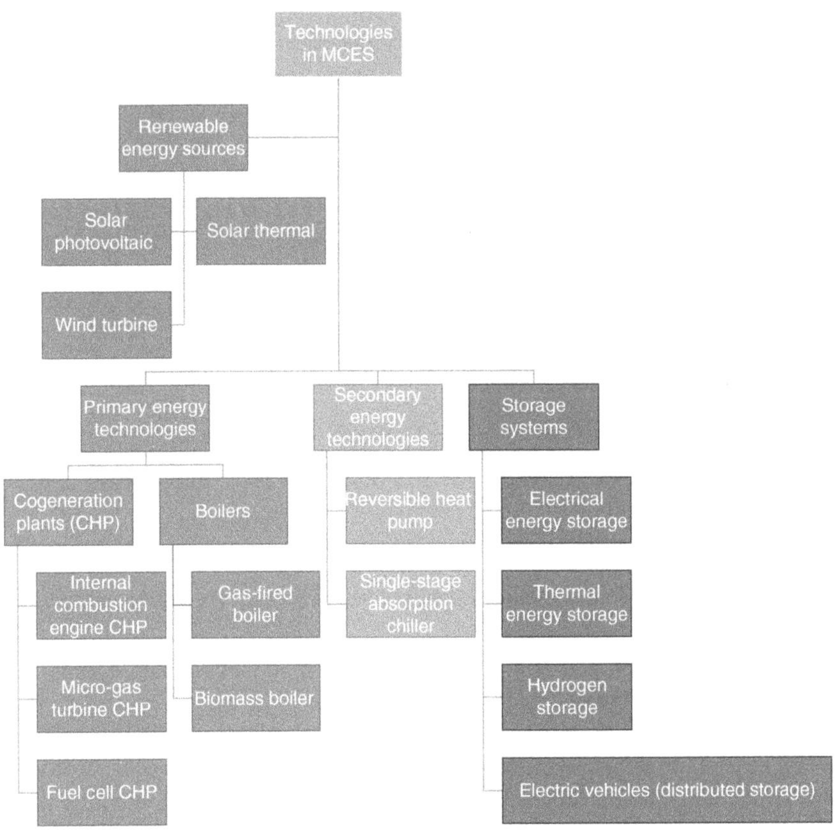

Figure 6.2 Energy technologies modeled in MCES.

6.2.1.1 Solar PV

Solar PV can be used to satisfy the user electricity demand within MCES. The model representing the PV system is expressed below:

$$P_{\mathrm{PV},d,t} = A_{\mathrm{PV}} \eta^{e}_{\mathrm{PV}} I_{d,t}, \forall d, t \tag{6.1}$$

where $P_{\mathrm{PV},d,t}$ is the generated power at time t of day d; depending on the installed PV area, A_{PV}; the electric efficiency of the PV panels, η^{e}_{PV}; and the solar irradiance at time t of day d,$I_{d,t}$.

6.2.1.2 Solar Thermal

Solar thermal (ST) can be used to satisfy the user domestic hot water within MCES. The model representing the ST system is expressed below:

$$H_{\mathrm{ST},d,t} = A_{\mathrm{ST}} \eta^{\mathrm{th}}_{\mathrm{ST}} I_{d,t}, \forall d, t \tag{6.2}$$

where $H_{\mathrm{ST},d,t}$ is the generated thermal power at time t of day d, depending on the installed ST area, A_{ST}; the thermal efficiency of the solar collectors, $\eta^{\mathrm{th}}_{\mathrm{ST}}$; and the solar irradiance at time t of day d, $I_{d,t}$.

6.2.1.3 Wind Turbines

Wind turbines (WTs) can be used to satisfy the user electricity demand within MCES. For a typical WT, it is assumed that it starts generating at the cut-in wind speed v_{c}, the generated power increases linearly as the wind speed increases from v_{c} to the rated wind speed v_{R}, the rated power is provided when the wind speed varies from v_{R} to the cut-out wind speed v_{F} at which the WT will be shut down for safety considerations (all the variables are assumed known based on weather and type). Then, the model of WT generators is expressed below:

$$P_{\mathrm{WT},d,t} = \begin{cases} P_{\mathrm{WT}} \dfrac{v_{d,t} - v_{\mathrm{c}}}{v_{\mathrm{R}} - v_{\mathrm{c}}} & v_{\mathrm{c}} \le v_{d,t} \le v_{\mathrm{R}} \\ P_{\mathrm{WT}} & v_{\mathrm{R}} \le v_{d,t} \le v_{\mathrm{F}} \\ 0 & v_{d,t} < v_{\mathrm{c}} \quad v_{d,t} > v_{\mathrm{F}} \end{cases} \quad \forall d, t \tag{6.3}$$

where $P_{\mathrm{WT},d,t}$ is the generated power at time t of day d, whereas P_{WT} is the WT-rated power.

6.2.2 Primary Energy Technologies

Primary energy technologies include cogeneration plants (CHPs) and gas-fired and biomass boilers.

6.2.2.1 Cogeneration Plants

CHPs may include internal combustion engines (ICEs), microgas turbines (MGTs), and fuel cells (FCs) as prime movers and can be used to satisfy the user electricity and thermal demand and the cooling demand through an absorption chiller (AC) within MCES. Taking the CHP with an ICE as an example, the amount of natural gas required by the prime mover to produce the power is formulated as follows:

$$G_{\text{CHP ICE},d,t} = \frac{P_{\text{CHP ICE},d,t}}{\eta^{e}_{\text{CHP ICE}}\text{LHV}_{\text{NG}}}, \forall d, t \tag{6.4}$$

where $\eta^{e}_{\text{CHP ICE}}$ is the electrical efficiency of the CHP and LHV_{NG} is the lower heat value of natural gas.

The thermal power recovered from the prime mover is formulated as follows:

$$H_{\text{CHP ICE},d,t} = \frac{P_{\text{CHP ICE},d,t}\eta^{\text{th}}_{\text{CHP ICE}}}{\eta^{e}_{\text{CHP ICE}}}, \forall d, t \tag{6.5}$$

where $\eta^{\text{th}}_{\text{CHP ICE}}$ is the thermal efficiency of the CHP.

The following equations can be considered to model the CHP FC and electrolyzer (EZ) [5].

$$H_{2\text{CHP FC},d,t} = \frac{P_{\text{CHP FC},d,t}}{\left(\eta^{e}_{\text{CHP FC}}\text{LHV}_{\text{H}_2}\right)}, \forall d, t \tag{6.6}$$

$$P^{\text{req}}_{\text{EZ},d,t} = \frac{\text{H}_{2\text{CHP FC},d,t}\text{LHV}_{H_2}}{\eta^{e}_{\text{EZ}}}, \forall d, t \tag{6.7}$$

Equation (6.6) allows calculating the amount of hydrogen needed by the CHP FC to produce power and depends on the electrical efficiency of the FC and the hydrogen lower heat value. Equation (6.7) allows calculating the power required by the EZ to produce hydrogen and depends on the EZ electrical efficiency.

The thermal power recovered from the CHP FC is formulated as in Eq. (6.5).

6.2.2.2 Boilers

Boilers may include gas-fired and biomass boilers and can be used to satisfy the thermal user demand in MCES. Taking the gas-fired boiler (GB) as an example, the amount of natural gas required by the GB to produce the thermal power is formulated as follows:

$$G_{\text{GB},d,t} = \frac{H_{\text{GB},d,t}}{\eta^{\text{th}}_{\text{GB}}\text{LHV}_{\text{NG}}}, \forall d, t \tag{6.8}$$

where $\eta^{\text{th}}_{\text{GB}}$ is the GB thermal efficiency. The model of the biomass boiler can be expressed similarly, by considering the lower heat value of biomass used to feed the boiler.

6.2.3 Secondary Energy Technologies

Secondary energy technologies include reversible heat pumps and single-stage ACs.

6.2.3.1 Reversible Heat Pump

Reversible heat pump (HP) can be used to satisfy the user thermal or cooling demand within MCES in the heating or cooling mode, respectively. Considering the heating mode, the HP model can be expressed as follows:

$$P^{\text{HM,req}}_{\text{HP},d,t} = \frac{H_{\text{HP},d,t}}{\text{COP}^{\text{HM}}_{\text{HP}}}, \forall d, t \tag{6.9}$$

which links the power required by the HP, $P^{\text{HM,req}}_{\text{HP},d,t}$, to the thermal power provided, $H_{\text{HP},d,t}$, through its coefficient of performance $\text{COP}^{\text{HM}}_{\text{HP}}$. The model is similar for the HP operating in cooling mode.

6.2.3.2 Single-Stage Absorption Chiller

Single-stage AC can be used to meet the user cooling demand within MCES and can be powered by the CHPs and boilers in the MCES. The general model for the AC is expressed below:

$$C_{\text{AC},d,t} = \left(\sum_{\text{CHP}} H^{\text{SC}}_{\text{CHP},d,t} + \sum_{\text{GB}} H^{\text{SC}}_{\text{GB},d,t} \right) \text{COP}_{\text{AC}}, \forall d, t \tag{6.10}$$

where $C_{\text{AC},d,t}$ is the cooling power provided by the AC, depending on the thermal power for space cooling purposes provided by all installed CHPs and the thermal power for space cooling purposes provided by all installed boilers.

6.2.4 Storage Systems

Storage systems include electric and thermal storage and EVs acting as distributed storage, based on the bidirectional charging concept, meaning that they can not only absorb power from the grid while charging but also provide vehicle-to-X services.

6.2.4.1 Electrical Energy Storage

Electrical energy storage systems may include battery energy storage systems (BESSs) such as lithium ion. The model is expressed below:

$$\text{SOC}_{\text{BESS},d,t} = \text{SOC}_{\text{BESS},d,t-\Delta t} + \frac{P^{\text{Ch}}_{\text{BESS},d,t} \eta^{\text{Ch}}_{\text{BESS}} \Delta t}{\text{Cap}_{\text{BESS}}} - \frac{P^{\text{Disch}}_{\text{BESS},d,t} \Delta t}{\eta^{\text{Dis}}_{\text{BESS}} \text{Cap}_{\text{BESS}}}, \forall d, t \tag{6.11}$$

that defines the battery state-of-charge (SOC) dynamics, where $P^{\text{Ch}}_{\text{BESS},d,t}$ and $P^{\text{Disch}}_{\text{BESS},d,t}$ are the charging and discharging power, respectively, $\eta^{\text{Ch}}_{\text{BESS}}$ and $\eta^{\text{Dis}}_{\text{BESS}}$ are the charging and discharging efficiencies, respectively; and Cap_{BESS} is the battery capacity.

This model is completed with the additional equations below:

$$\text{SOC}^{\min}_{\text{BESS}} \leq \text{SOC}_{\text{BESS},d,t} \leq \text{SOC}^{\max}_{\text{BESS}}, \forall d, t \tag{6.12}$$

$$0 \leq P^{\text{Ch}}_{\text{BESS},d,t} \leq x^{\text{Ch}}_{\text{BESS},d,t} P^{\text{Ch,max}}_{\text{BESS},d,t}, \forall d, t \tag{6.13}$$

$$0 \leq P^{\text{Disch}}_{\text{BESS},d,t} \leq x^{\text{Disch}}_{\text{BESS},d,t} P^{\text{Disch,max}}_{\text{BESS},d,t}, \forall d, t \tag{6.14}$$

$$x^{\text{Ch}}_{\text{BESS},d,t} + x^{\text{Disch}}_{\text{BESS},d,t} \leq 1, \forall d, t \tag{6.15}$$

The upper and lower limits of SOC are enforced by Eq. (6.12), whereas the battery charging/discharging power limits are enforced in Eqs. (6.13)–(6.15).

6.2.4.2 Thermal Energy Storage

Thermal energy storage (TES) may include thermal storage for heating and cooling purposes. As for the TES for heating, the state dynamic is modeled as follows:

$$H_{\text{TES-Th},d,t} = H_{\text{TES-Th},d,t-\Delta t}(1 - \varphi_{\text{TES-Th}}(\Delta t)) + \left(H^{\text{Ch}}_{\text{TES-Th},d,t} - H^{\text{Disch}}_{\text{TES-Th},d,t} \right) \Delta t, \forall d, t \tag{6.16}$$

meaning that the energy stored at time t of day d, $H_{\text{TES}-\text{Th},d,t}$, depends on the nondissipated energy stored at time $(t - \Delta t)$ of the same day $H_{\text{TES}-\text{Th},d,t-\Delta t}$ (based on the loss fraction of the storage, $\varphi_{\text{TES}-\text{Th}}(\Delta t)$), and on the net energy flow $\left(H^{\text{Ch}}_{\text{TES-Th},d,t} - H^{\text{Disch}}_{\text{TES-Th},d,t}\right)$. The model for TES for cooling is similar.

6.2.4.3 Hydrogen Storage

The state dynamic of the hydrogen storage can be modeled as follows:

$$H_{2\text{Hsto},d,t}^{\text{sto}} = H_{2\text{Hsto},d,t-\Delta t}^{\text{sto}}\eta_{\text{Hsto}} + \left(H_{2\text{Hsto},d,t}^{\text{Ch}} - H_{2\text{Hsto},d,t}^{\text{Disch}}\right)\Delta t, \forall d, t \tag{6.17}$$

that relates the amount of hydrogen stored at time t of day d with the one stored at previous time $(t - \Delta t)$ of the same day that depends on the efficiency of the hydrogen storage, η_{Hsto}.

6.2.4.4 Electric Vehicles

As mentioned earlier, EVs can act as distributed storage with bidirectional charging services, namely grid-to-vehicle (G2V) and vehicle-to-grid (V2G).

The EV modeling is discussed below [6]. The EV is characterized by a specific value for (i) battery capacity, (ii) arrival and departure times at/from the charging stations, (iii) the SOC at the arrival time, and (iv) the SOC desired at the departure time.

Before the arrival and after the departure at/from the charging stations, the EV can be modeled as follows:

$$\begin{cases} P^{\text{Ch}}_{\text{EV},d,t} = 0 \\ P^{\text{Disch}}_{\text{EV},d,t} = 0 \end{cases}, \text{If } t < t^{\text{arrival}}_{\text{EV}}, \text{and } t > t^{\text{departure}}_{\text{EV}}, \forall d \tag{6.18}$$

where $P^{\text{Ch}}_{\text{EV},d,t}$ is the charging power for the EV at time t of day d; $P^{\text{Disch}}_{\text{EV},d,t}$ is the discharging power of EV at time t of day d; and $t^{\text{arrival}}_{\text{EV}}$ and $t^{\text{departure}}_{\text{EV}}$ are the arrival and departure times, respectively, at/from the charging stations for the EV.

During the parking period, the EV can be modeled as follows:

$$\begin{cases} \text{SOC}_{\text{EV},d,t^{\text{arrival}}_{\text{EV}}} = \text{SOC}^{\text{initial}}_{\text{EV},d} \\ \text{SOC}_{\text{EV},d,t} = \text{SOC}_{\text{EV},d,t-\Delta t} + \dfrac{P^{\text{Ch}}_{\text{EV},d,t}\eta^{\text{Ch}}_{\text{EV}}\Delta t}{\text{Cap}_{\text{EV}}} - \dfrac{P^{\text{Disch}}_{\text{EV},d,t}\Delta t}{\eta^{\text{Disch}}_{\text{EV}}\text{Cap}_{\text{EV}}}, \text{if } t^{\text{arrival}}_{\text{EV}} < t < t^{\text{departure}}_{\text{EV}}, \quad \forall d \\ \text{SOC}_{\text{EV},d,t^{\text{departure}}_{\text{EV}}} \geq \text{SOC}^{\text{Desired}}_{\text{EV},d} \end{cases} \tag{6.19}$$

where $\text{SOC}_{\text{EV},d,t}$ is the SOC of the battery of EV at time t of day d; $\text{SOC}^{\text{initial}}_{\text{EV},d}$ is the SOC of the battery at arrival time at charging station in day d; $\eta^{\text{Ch}}_{\text{EV}}$ and $\eta^{\text{Disch}}_{\text{EV}}$ are the charging and discharging efficiencies, respectively; Cap_{EV} is the battery capacity of EV; $\text{SOC}^{\text{Desired}}_{\text{EV},d}$ is the desired SOC at departure time from the charging station in day d.

In addition to the EV operating mode expressed in Eqs. (6.18) and (6.19), additional operation constraints for EV are defined below:

$$\begin{cases} 0 \le P^{\mathrm{Ch}}_{\mathrm{EV},d,t} \le x^{\mathrm{Ch}}_{\mathrm{EV},d,t} P^{\mathrm{Ch,max}}_{\mathrm{EV},d,t} \\ 0 \le P^{\mathrm{Disch}}_{\mathrm{EV},d,t} \le x^{\mathrm{Disch}}_{\mathrm{EV},d,t} P^{\mathrm{Disch,max}}_{\mathrm{EV},d,t} \\ x^{\mathrm{Ch}}_{\mathrm{EV},d,t} + x^{\mathrm{Disch}}_{\mathrm{EV},d,t} \le 1 \\ \mathrm{SOC}^{\min}_{\mathrm{EV}} \le \mathrm{SOC}_{\mathrm{EV},d,t} \le \mathrm{SOC}^{\max}_{\mathrm{EV}} \end{cases}, \text{if } t^{\mathrm{arrival}}_{\mathrm{EV}} < t < t^{\mathrm{departure}}_{\mathrm{EV}}, \forall d \tag{6.20}$$

which allows limiting the charging and discharging power for the EV's battery and the battery SOC.

6.3 The Optimal Design Problem for MCES

Traditionally, optimal design problems are considered an economic objective only. As comprehensively discussed in Chapter 5, as the desired outcome is more complex, the objective formulation of optimal design problems has later been extended to also account for other factors such as environmental outcomes, thereby requiring the use of a multiobjective approach.

6.3.1 Objective Function

The main metric for an economic objective is the total MCES costs, considering both operational costs of the system and investment costs. The economic objective to minimize can be formulated as follows:

$$C^{\mathrm{Tot}} = C^{\mathrm{INV}} + C^{\mathrm{O\&M}} + C^{\mathrm{Energy}} \tag{6.21}$$

The investment cost function C^{INV} may be formulated as follows:

$$C^{\mathrm{INV}} = \sum_i \mathrm{CRF}_i (C_{c,i} S_i) \quad \mathrm{CRF}_i = r(1+r)^{N_i} / \left[(1+r)^{N_i} - 1\right] \tag{6.22}$$

The total annualized investment cost of all technologies is represented in Eq. (6.22). For technology i, the total annualized investment cost is calculated as the product of the capital recovery factor (CRF) (a parameter depending on the lifetime of the technology, N_i, and the interest rate, r), the capital cost (a parameter), and its designed size that is a continuous decision variable of the optimization model.

The total annual operation & maintenance (O&M) cost of the MCES can be formulated as follows:

$$C^{\mathrm{O\&M}} = \Delta_t \sum_i \sum_d \sum_t \mathrm{OM}_i \mathrm{R}_{i,d,t} \tag{6.23}$$

For technology i, the total annual O&M cost is calculated by multiplying its O&M-specific cost (a parameter) by its generation level, R, at time t of day d, and the time step (one hour). For storage systems, the O&M costs can be calculated through their capacities.

The total energy cost can be formulated as follows:

$$C^{\text{Energy}} = \Delta t \left(\sum_i \sum_d \sum_t \text{Pr}_{\text{NG},d} G_{i,d,t} + \sum_d \sum_t \text{Pr}_{\text{PG},d,t} P_{\text{PG},d,t} \right) \tag{6.24}$$

It is the sum of the total cost of gas calculated by multiplying the gas price (a parameter) by the total amount of gas consumed by the gas-fired technologies in the MCES, and the total cost of buying grid power calculated by multiplying the time-varying unit price of grid power (a parameter) and the total amount of power taken from the grid.

As already mentioned, a multiobjective approach is often preferable for long-term planning of MCESs to reduce also the environmental impacts represented by carbon emissions.

The environmental objective can be represented by the minimization of the total annual carbon emissions, consisting of the sum of the following functions:

$$\text{Env}^{\text{Tot}} = \text{Env}^{\text{NG}} + \text{Env}^{\text{PG}} \tag{6.25}$$

$$\text{Env}^{\text{NG}} = \Delta t \sum_i \sum_d \sum_t G_{i,d,t} \text{CI}_{\text{NG}} \text{LHV}_{\text{NG}} \tag{6.26}$$

$$\text{Env}^{\text{PG}} = \Delta t \sum_d \sum_t P_{\text{PG},d,t} \text{CI}_{\text{PG}} \tag{6.27}$$

In Eq. (6.26), the total carbon emission associated with the gas consumed by the gas-fired technologies in the MCES depends on the carbon intensity (CI) of gas, whereas in Eq. (6.27), the total carbon emission associated with the grid power depends on the carbon intensity of the power grid, which the MCES is connected to.

With the economic and environmental objectives, the optimal planning problem has two types of objective functions to be minimized. To solve this multiobjective optimization problem, the weighted-sum method can be used to have a single objective function formulated as follows:

$$\text{Fobj} = c\omega C^{\text{TOT}} + (1 - \omega)\text{Env}^{\text{Tot}} \tag{6.28}$$

where c is a constant scaling factor to keep the two objectives at the same order of magnitude and ω is the weight for the total annual cost varying in the range of 0–1. When $\omega = 1$, it is to find the solution that minimizes the total annual cost of the MCES, and when $\omega = 0$, it is to find the solution that minimizes the total environmental impact of MCES. When varying the weight ω in the range of $[0, 1]$, the Pareto frontier between economic and environmental objectives can be found.

6.3.2 Constraints

The optimization problem constraints mainly consist of three types of constraint, i.e.

Design constraints: ensure that the designed size of each energy technology installed in MCES (a continuous decision variable) is within the minimum and maximum size values available in the market.

Operation constraints: for instance, the common operation constraint for technologies in the MCES is the technology capacity constraint.

System balance constraints: ensure that at each time t, the multienergy user demand (electrical, heating, and cooling demand) is satisfied by the MCES.

6.3.3 Solution Methodology

The problem is formulated through a mixed-integer linear programming (MILP) approach. It must be mentioned that from the analysis carried out in Chapter 5, it emerged that MILP formulations are the most widely used in the literature for the design optimization of MCES, being these a good compromise between model fidelity and complexity of the optimization process, by also taking advantage of powerful MILP solution methods as branch-and-cut. In detail, the optimization problem is linear and involves both discrete (binary) and continuous variables. The binary decision variables are the existence and the operation on/off status of the energy technologies, whereas the continuous decision variables include the size of the energy technologies with the corresponding energy rate provided (power, heating, and cooling), capacity of storage systems, with charging/discharging rate, and power taken from the grid. All the other variables can be classified as dependent variables.

6.3.4 Case Study and Simulation Results

In the case study, the investigated MCES corresponds to an integrated local energy community (ILEC) located in a neighborhood in Turin, Italy. It consists of three multienergy microgrids, each of them associated with an end-user, namely a supermarket, a hotel, and a residential building cluster of five buildings. The entire year is divided into four seasons (i.e. cold, cold mid, hot mid, and hot), each of them with a representative day, and the optimization is conducted on an hourly basis for a representative day per season. It is assumed that each microgrid can be composed of the technologies with related characteristics shown in Table 6.1 [7, 8]. The distances between the end-users are assumed known, namely 50 m between the supermarket and the residential building clusters, 100 m between the hotel and the supermarket, and 70 m between the residential building cluster and the hotel, whereas the distance between the energy hub and its associated user is assumed null.

The multiobjective optimization model is implemented by using IBM ILOG Cplex Optimization Studio, version 12.10.0 and can be solved within 20 hours with a mixed-integer gap lower than 0.15%. The Pareto frontier obtained is shown in Figure 6.3. With the economic optimization, the total annual cost is minimum, equal to 572.19 k€ and the total annual emissions are maximum, i.e. 1581.05 tons of CO_2. With the environmental optimization, the total annual cost is maximum equal to 680.72 k€ and the total annual emissions are minimum, i.e. 1338.30 tons of CO_2.

The optimized configurations obtained for the multienergy microgrids are shown in Table 6.2 for the economic and environmental optimization (extreme points of the Pareto frontier), and they are discussed below.

Table 6.1 Technologies candidates for the optimal planning of the ILEC with related characteristics.

		Efficiency	
Energy technology	**Investment cost**	**El**	**Th**
CHP ICE	770–1380 €/kW	0.30–0.45	0.43–0.65
CHP MGT	1630–2490 €/kW	0.24–0.31	0.44–0.53
GB	100 €/kW		0.80
PV	2000 €/kW_p		0.14
Solar thermal	200 €/m^2		0.60
Heat pump	460 €/kW		3.5/3.0
Absorption chiller	230–510 €/kW		0.80
Battery	400 €/kWh	0.75/0.75	
TES	20 €/kWh		0.95

Source: Refs. [7, 8].

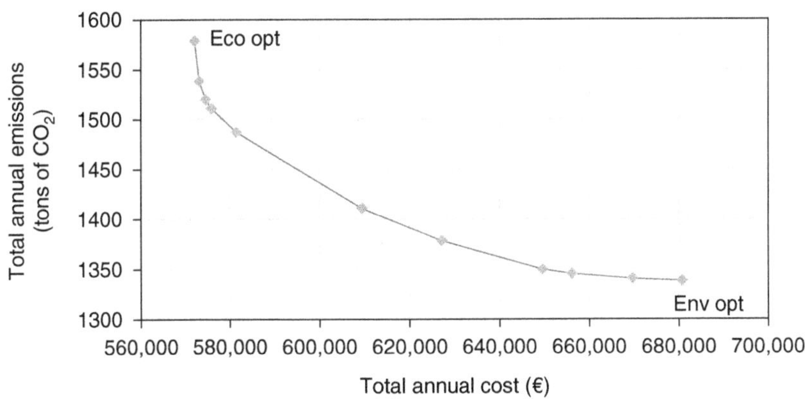

Figure 6.3 Pareto frontier obtained with the multiobjective optimization model. Source: Refs. [7, 8].

Economic optimization: The PV installed area is maximum for all the users, highlighting the economic benefits coming from the low investment costs of this technology. Large storage for space heating demand is preferred over a large heat pump, since it is more convenient to store a larger amount of heat recovered from CHPs, instead of installing larger heat pumps.

Environmental optimization: The sizes of ST and heat pumps for all users are all at their maximum, highlighting the environmental benefits of ST for thermal purposes and of electric heat pumps supplied by the electricity network characterized in Italy by a low carbon intensity.

Table 6.2 Optimal configuration of the ILEC.

	Residential building		Supermarket		Hotel	
Technology	**Eco Opt**	**Env Opt**	**Eco Opt**	**Env Opt**	**Eco Opt**	**Env Opt**
CHP ICE (kW_e)	221	22	153	118	106	21
PV (m^2)	750	433	600	408	242	290
Solar thermal (m^2)	0	317	0	192	0	210
Gas-fired boiler (kW_{th})	0	0	0	0.0	0	0
Heat pump (kW_{th})	270	1074	326	482	50	191
Absorption chiller (kW_{th})	500	0	0	0	100	0
Battery (kWh_{el})	0	0	0	0	0	10
TES (DHW) (kWh_{th})	0	522	0	200	14	242
TES (heat) (kWh_{th})	873	0	160	0	100	0
TES (cooling) (kWh_{th})	387	31	0	0	20	0

Source: Refs. [7, 8].

Under the case-specific circumstances, the calculated total annual cost is 838.95 k€ and increases of about 32% as compared with the cost for the MCES obtained with the economic optimization, whereas the calculated total annual emissions are 2003.78 tons of CO_2 and increase of about 33% as compared with the emissions for the MCES obtained with the environmental optimization.

6.4 Optimal Day-Ahead Scheduling of MCES Under Uncertainties and by Considering DR Programs

In this section, the optimal operation of MCES represented by multienergy microgrids is studied, consisting of the electricity and the natural gas distribution systems. DR programs are also integrated into the operation of the microgrid to effectively manage and optimize energy consumption in response to various conditions, such as peak demand periods when consumption is at the highest or intermittent supply due to the RES integration. The uncertainty in the parameters such as electricity demand and output power of RES is also considered to provide more realistic solutions.

DR programs are initiatives implemented by utility companies and grid operators to adjust energy consumption in response to changes in grid conditions, such as high demand, grid instability, or the need to balance energy supply and demand [9]. These programs aim to incentivize consumers to reduce or shift their electricity usage, contributing to a more reliable and efficient energy system. There are two main types of DR programs: tariff-based and incentive-based. Tariff-based programs involve the use of dynamic pricing, where electricity rates vary based on the time of day or overall grid conditions. In this case, consumers are encouraged to adjust their consumption patterns to take advantage of lower rates during off-peak hours. On the other hand, incentive-based programs offer financial rewards or other benefits to consumers who voluntarily curtail their electricity usage when called upon during peak demand events. These programs encourage consumers to participate actively in DR efforts, fostering a more flexible and responsive electricity grid.

The uncertainty in the problem of optimal operation of a multienergy microgrid can be effectively captured through various approaches [10]. As already discussed in Chapter 5, the choice of the method depends on the specific characteristics of the microgrid and the nature of the uncertainty involved. In the problem of optimal operation of a multienergy microgrid, where historical data is available to characterize the uncertainty in parameters, stochastic programming emerges as a more fitting approach. By modeling these uncertain parameters as random variables with known probability distributions and generating scenarios based on historical data, stochastic programming effectively captures the probabilistic nature of uncertainty.

In the following, the problem of optimal day-ahead operation of a multienergy microgrid under uncertainty in the presence of tariff-based and incentive-based DR programs is discussed.

6.4.1 Objective Function

Equation (6.29) represents the objective function of the optimal operation of a microgrid, considering both the natural gas and electricity distribution systems. The equation comprises two main terms in which the first term deals with the cost of electricity network operation (Eq. (6.30)). It encompasses various factors, such as the cost of purchased electricity from the main grid, the revenue generated by selling excess electricity back to the main grid, the cost associated with load-shedding strategies to manage in emergency cases, and the incentives provided to consumers participating in incentive-based demand response (EDR) programs. The second term focuses on the cost of operating the natural gas systems within the microgrid (Eq. (6.31)). This includes expenses related to purchasing natural gas from the main network, the management cost of stored gas within pipelines, and the cost of gas shedding when necessary.

$$C^{\text{tot}} = C^{\text{elec}} + C^{\text{gas}} \tag{6.29}$$

$$C^{\text{elec}} = \Delta t \left(\sum_t \text{Pr}_{\text{PG},t} \left(P^{\text{buy}}_{\text{PG},t} - P^{\text{sell}}_{\text{PG},t} \right) + \sum_t C^{\text{lsh}} P^{\text{lsh}}_{\text{PG},t} + \sum_t \text{inc} \left(\text{Load}_t - \text{Load}^{\text{DR}}_t \right) \right) \tag{6.30}$$

$$C^{\text{gas}} = \Delta t \left(\sum_t \text{Pr}_{\text{NG}} G_t + \sum_t C^{\text{gsh}} G^{\text{sh}}_t + \sum_t C^{\text{lp}} \Delta G_t \right) \tag{6.31}$$

6.4.2 Constraints

6.4.2.1 Electricity Network Constraints

The microgrid scheduling model needs to adhere to various operational constraints to guarantee secure and reliable demand provision. Among these constraints, *active power balance* holds the utmost significance. It mandates that the total active power generated by distributed energy resources, along with imported power from the main grid, must precisely match the aggregate active power consumed, losses in distribution lines, and exported power to the main grid. By adhering to this principle, the microgrid efficiently fulfills the energy demands of its users, preventing any risk of system overload or underload.

In addition to active power balance, another crucial constraint is *reactive power balance* within the microgrid. The scheduling model must ensure that the reactive power generated by distributed energy resources matches the total reactive power consumed by the loads and the reactive power losses in the distribution lines. This equilibrium of reactive power is essential for upholding the voltage stability of the microgrid.

Furthermore, *Kirchhoff's law* presents another fundamental constraint in microgrid scheduling. According to this law, the voltage and current at every node within the microgrid must comply with the principle of energy conservation. The microgrid scheduling model must diligently guarantee that the voltage and current at each node in the system are balanced and satisfy the requisites of Kirchhoff's law. By adhering to this crucial principle, the microgrid can maintain a stable and efficient electrical network, ensuring the proper distribution and utilization of energy throughout the system.

Additionally, the *capacity of dispatchable units* constitutes a vital constraint that necessitates consideration in microgrid scheduling. Dispatchable units, such as diesel generators, must be operated within their designated capacity to prevent overloading or underloading. Thus, the microgrid scheduling model must meticulously consider the output of each dispatchable unit and ensure that it remains within the specified limits. The *ramp-rate limits* of dispatchable units constitute another crucial constraint in microgrid scheduling. These limits govern the rate at which the dispatchable units can ramp up or down their power output.

6.4.2.2 Gas Network Constraints

To accurately simulate the natural gas network within the microgrid, a set of constraints is considered. These constraints encompass various aspects of the natural gas distribution system, ensuring its safe and efficient operation. Among the constraints, the *natural gas flow balance* ensures that the amount of purchased natural gas from the main grid must be equal to the consumption at each time step within the microgrid. It guarantees that the energy demands are adequately met without shortages or wastage. *Maximum and minimum pressure at each node* is also considered to maintain the integrity of the natural gas network and limits on the pressure at each node are imposed. Furthermore, *maximum and minimum natural gas flow within pipelines' constraints* are added to restrict the maximum and minimum flow rates of natural gas through pipelines and are set to avoid overloading or underutilization of the pipeline infrastructure. Additionally, linepack refers to the amount of natural gas stored in the pipelines. *Constraints related to linepack changes* help manage the storage of natural gas, allowing the system to respond to demand fluctuations effectively.

The *Panhandle equation* is utilized to relate the natural gas flow within pipelines to the pressure at nodes. This equation provides a crucial link between flow rates and pressure, enabling accurate modeling of the natural gas distribution system.

In addition to the previously mentioned constraints, a *coupling constraint* must be taken into consideration to calculate the required amount of natural gas for gas-fired technologies and add it to the gas flow constraint. This coupling constraint establishes the interdependency between the electricity and gas networks within the microgrid.

6.4.2.3 Demand Response Constraints

In Eq. (6.32), the load in the presence of a tariff-based and an incentive-based DR is calculated [11]. The former is a time-of-use (TOU) DR program, and the latter is an EDR program in which consumers decide to change their demand based on prices and incentives (INC_i). It should be noted that the cost of incentive payment for the EDR program has been added to the operating cost of the microgrid. For the TOU DR program, distinct electricity prices are considered for specific periods, including valley periods, off-peak periods, and peak periods ($\mathrm{Pr}^{\mathrm{ini}}_{\mathrm{PG},t'}$ is replaced by $\mathrm{Pr}_{\mathrm{PG},t'}$). In contrast, the EDR program entails a fixed electricity purchase rate, complemented by an extra incentive offered during peak periods. In the formulation, $E_{t,t'}$ is a matrix that indicates the elasticity of demand, and $\mathrm{Load}^{\mathrm{DR}}_{b,t}$ and $\mathrm{Load}_{b,t}$ represent the electric load in the presence and absence of the DR program, respectively.

$$\mathrm{Load}^{\mathrm{DR}}_{b,t} = \mathrm{Load}_{b,t}\left(1 + \sum_{t'=1}^{T} E_{t,t'} \frac{\left(\mathrm{Pr}_{\mathrm{PG},t'} - \mathrm{Pr}^{\mathrm{ini}}_{\mathrm{PG},t'}\right) + \mathrm{INC}_i}{\mathrm{Pr}^{\mathrm{ini}}_{\mathrm{PG},t'}}\right) \tag{6.32}$$

6.4.2.4 Uncertainty Consideration

To capture uncertainty and propose a stochastic model for the problem under study, two-stage stochastic programming is utilized, expressed mathematically in Eq. (6.33). In this formulation, in the first stage, a decision is made based on available information, and in the second stage, after observing uncertain factors, an adjustment decision is made. More precisely, the formulation involves defining decision variables, objective functions, and constraints for both stages, while incorporating probabilities for the uncertain factors. This approach helps in finding optimal solutions that balance initial decisions with adjustments to achieve the best possible outcome considering uncertain scenarios.

$$\begin{aligned} &\mathrm{Min}\, z = cx + \sum \mathrm{prob}^s b^s y^s \\ &\text{Subject to } Bx + D^s y^s = h^s \forall s \in S \\ &x \geq 0, y^s \geq 0, \text{and} \sum \mathrm{prob}^s = 1 \end{aligned} \tag{6.33}$$

Considering the above formulation, the objective function of the stochastic operation of the multienergy microgrid should be rewritten. The changes must be implemented in the objective function of operation of the electricity subsystem (Eq. (6.30)), and constraints should be revised consequently.

$$C^{\text{elec}} = \Delta t \left(\sum_t \text{Pr}_{\text{PG},t} \left(P_{\text{PG},t}^{\text{buy}} - P_{\text{PG},t}^{\text{sell}} \right) + \sum_s \text{prob}^s \left(\sum_t C^{\text{lsh}} P_{\text{PG},t}^{\text{lsh}} + \sum_t \text{inc} \left(\text{Load}_t - \text{Load}_t^{\text{DR}} \right) \right) \right) \tag{6.34}$$

6.4.3 Solution Methodology

The optimization problem developed in the section is in the form of mixed-integer nonlinear programming (MINLP). The panhandle equation in the gas system's operating model and a set of constraints connecting voltage and current in the electricity system's operating model (i.e. the application of Kirchhoff's voltage law) are nonlinear constraints. Moreover, the binary variables represented the status of generating units. To solve this class of problem, DICOPT (DIscrete and Continuous OPTimizer) solver is employed in the GAMS software, which is based on the extensions of the outer-approximation algorithm for the equality relaxation strategy [12]. The MINLP algorithm inside DICOPT solves a series of nonlinear programming (NLP) and mixed-integer programming (MIP) subproblems, which can be solved using any NLP or MIP solver that runs under GAMS. The performance will heavily depend on the choice of the selected subsolvers.

6.4.4 Case Study and Simulation Results

In this case study, the focus lies on examining the optimal operation of a multienergy microgrid with DR under uncertainty located in the United States. The microgrid is composed of two networks: an 11-node gas distribution network and a 33-bus electricity distribution network. Gas distribution involves 11 nodes interconnected by 13 pipelines, connected to the main grid via node one. Similarly, the electricity network consists of 33 buses and 32 lines, connected to the main grid through bus one. Distributed generators, like PV systems and WTs, are linked to specific buses (15, 29, 30, and 32) in the electricity network. Dispatchable units are connected to buses 2 and 9, with gas-powered dispatchable units supplied through nodes 5 and 6 in the gas network. The study aims to optimize the microgrid's operation, accounting for uncertainties, and leverage DR strategies to manage energy resources effectively. More detailed characteristics of the networks can be found in Ref. [13]. To assess DR programs, the baseline scenario without DR assumes a constant grid power price, set at 16 \$ MWh^{-1}. In the case of the TOU DR program, different grid power prices are considered for specific periods: 8 \$ MWh^{-1} during valley periods, 16 \$ MWh^{-1} during off-peak periods, and 21.5 \$ MWh^{-1} during peak periods. On the other hand, the EDR program involves a flat rate of 16 \$ MWh^{-1} for electricity purchases, and an additional incentive of 7 \$ MWh^{-1} is provided during peak periods. More detailed data about the DR programs can be found in Ref. [14].

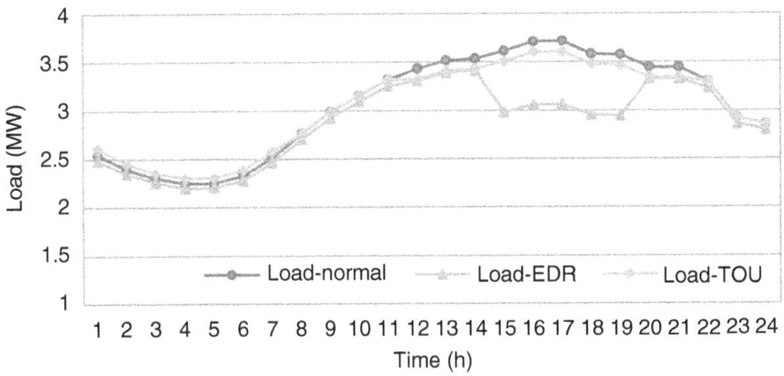

Figure 6.4 Load profile in the presence and absence of the demand-response programs.

The optimization of the stochastic model for the operation of the multi-energy microgrid, considering DR, incorporates five scenarios, including a base case and four additional scenarios where the renewable output power and demands can increase and decrease by 5%, resulting in varying conditions to account for uncertainty. However, in practical applications, when relying on historical data, it is essential to perform scenario reduction to identify a concise yet representative set of scenarios. This approach helps capture the key characteristics of uncertainty while ensuring computational feasibility and simplicity in the analysis [15].

6.4.4.1 Simulation Results

Figure 6.4 illustrates the effects of TOU DR and EDR on consumer electricity demand. During peak hours from 14 : 00 to 20 : 00, consumers participating in the TOU program show a modest reduction in electricity consumption, while those in the EDR program demonstrate a more significant decrease of approximately 0.6 MW. This substantial reduction in electricity demand through the EDR program provides valuable support to the grid operator in ensuring a balanced supply-demand provision during critical periods, enhancing grid stability and reliability.

Figure 6.5 presents a comparative examination of the microgrid dispatch in three cases: without DR, with TOU DR, and with EDR. The graph clearly shows that in the presence of the EDR, there is a substantial decrease in the reliance on gas-fired units during peak operational hours compared to the other cases. This advantage of reduced reliance on gas-fired units through the EDR program offers the potential for emission reduction, making the microgrid's operations more environmentally friendly and efficient.

Figure 6.6 examines the impact of DR programs on the operation of the gas network (i.e. gas supply and linepack), influenced by their effects on the electricity network. The graph illustrates that both the EDR program and the TOU DR program lead to reduced operation of gas-fired units. As a result of this reduction, the consumption of natural gas during peak hours also decreases. The connection between electricity and gas networks through these DR programs creates a favorable

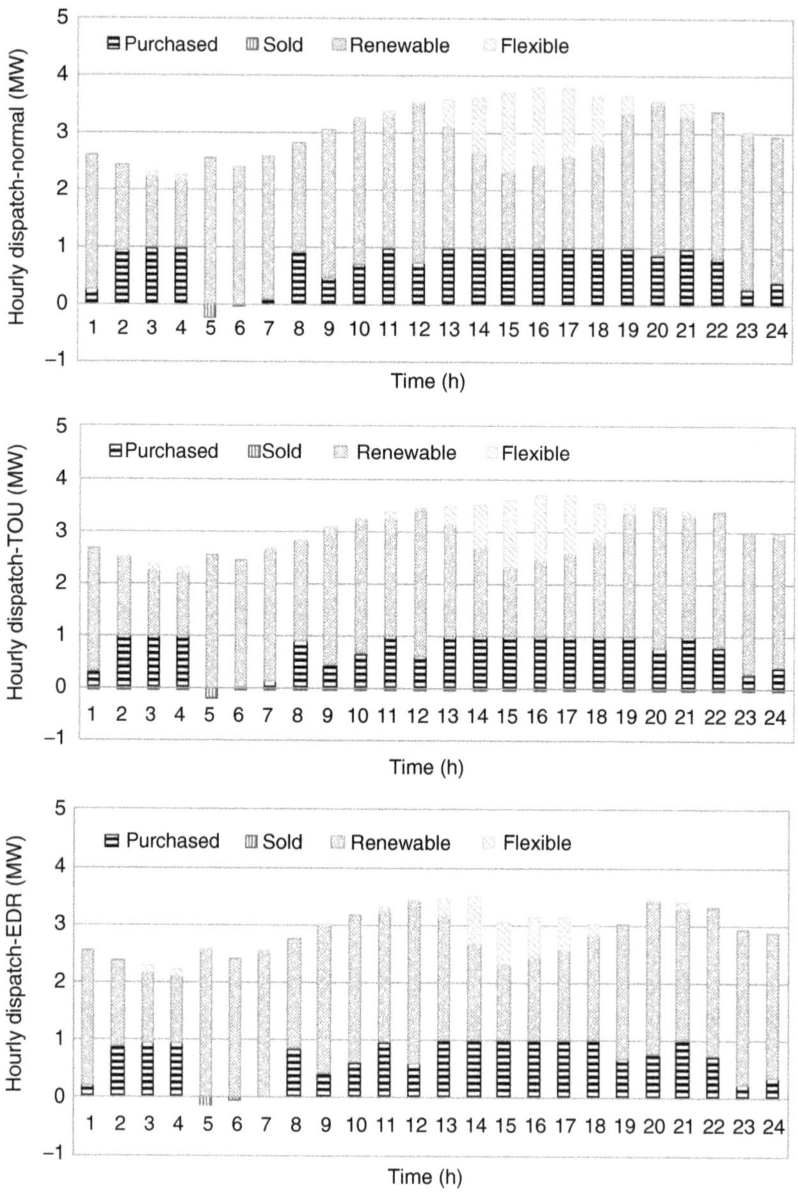

Figure 6.5 Hourly dispatch of the microgrid in the presence of the demand-response programs.

outcome, contributing to the optimization of gas network operations and enabling more efficient energy consumption during critical peak hours. Additionally, the EDR program further decreases fluctuations in stored gas within the pipelines of the gas network. This reduction in linepack fluctuations enhances the gas network's capability to effectively cope with sudden changes in demand. By moderating the variations in stored gas, the EDR program optimizes the gas network's

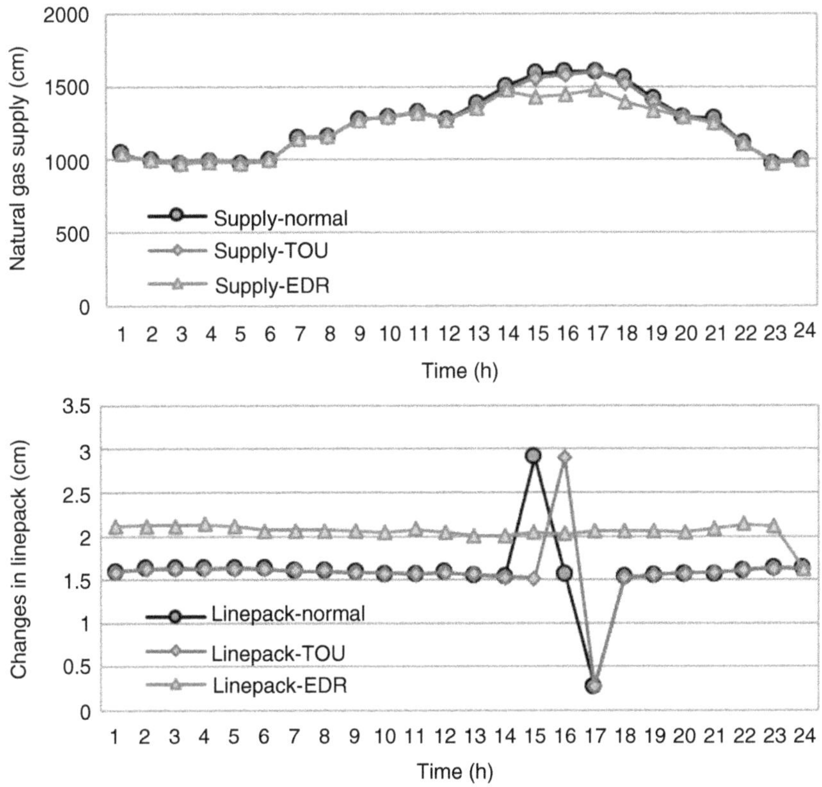

Figure 6.6 Natural gas supply and changes in linepack in the presence and absence of the DR programs.

responsiveness and resilience, enabling it to efficiently adjust to shifts in demand patterns.

Figure 6.7 showcases the microgrid dispatch when a stochastic model is employed. The graph provides insights into how the microgrid's energy resources are allocated and utilized under the stochastic modeling approach. Furthermore, Table 6.3 presents the operating costs of the microgrid in different scenarios. It compares the costs when DR programs are incorporated, and when both stochastic and deterministic considerations are considered. The table provides a comprehensive overview of the various operating cost scenarios, allowing for a detailed analysis of cost efficiency under different conditions. Together, Figure 6.7 and Table 6.3 offer valuable information regarding the microgrid's dispatch strategy and operating costs, considering the integration of DR programs and stochastic modeling techniques. These insights can be crucial for optimizing the microgrid's performance, cost-effectiveness, and overall energy management. In the microgrid's stochastic model, uncertainty is effectively managed by utilizing gas-fired units, allowing for proper handling of the inherent variability in RESs. While the stochastic model's operating cost may be higher compared to the deterministic one, it holds a crucial advantage in addressing increased energy demands during periods of

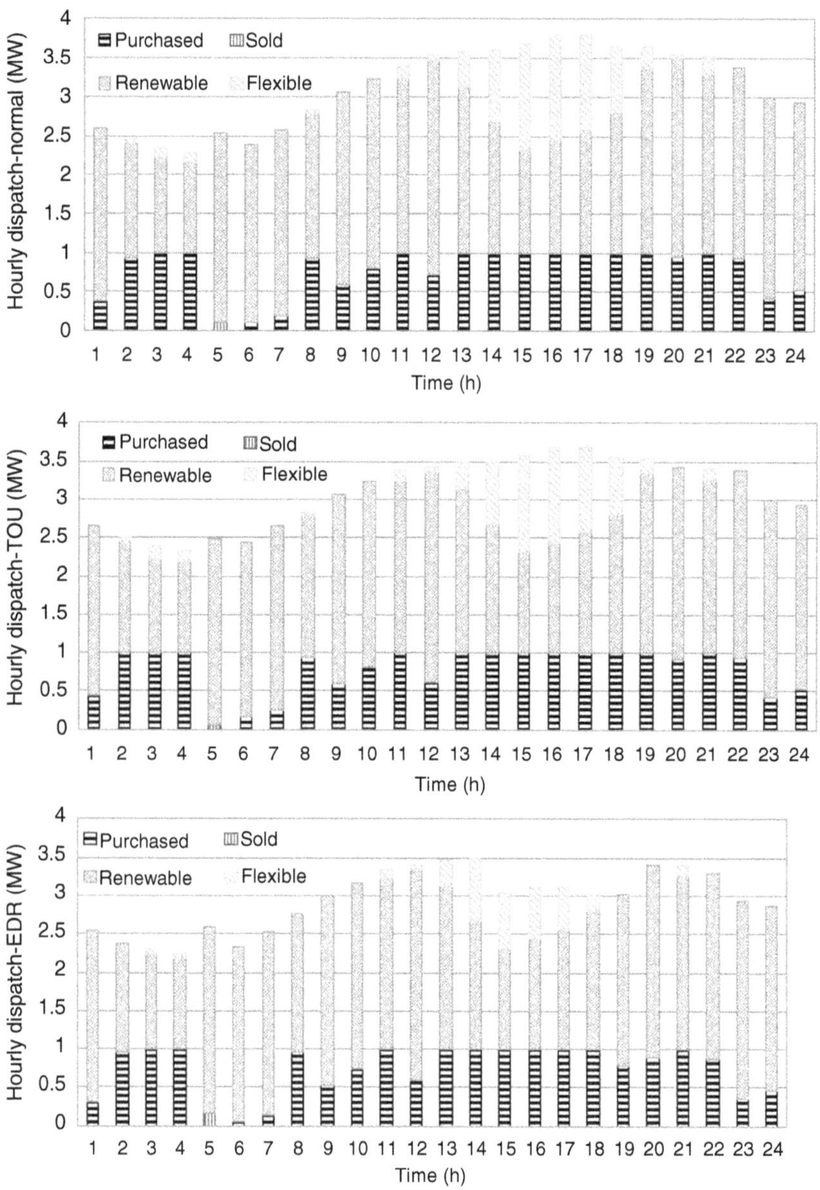

Figure 6.7 Hourly dispatch of the microgrid considering uncertainty.

Table 6.3 Operating cost of the microgrid.

		Cost	
DR program	**Deterministic/Stochastic**	**Electricity network ($)**	**Gas network ($)**
—	Deterministic	2947.96	10 927.12
TOU	Deterministic	2744.84	10 476.27
EDR	Deterministic	2043.42	10 130.32
—	Stochastic	3066.513	11 368.23
TOU	Stochastic	2857.56	10 907.40
EDR	Stochastic	2130.34	10 561.38

limited renewable energy availability. This adaptability to changing conditions and uncertain energy supply equips the stochastic model to handle fluctuating energy demands with greater proficiency. Properly evaluating the expected costs between the stochastic and deterministic models necessitates consideration of uncertainty's value. While the deterministic model may seem more cost-effective, it assumes perfect foresight and it overlooks the dynamic nature of RES and the potential risks associated with varying energy demand. In contrast, the stochastic model provides a more realistic representation of the microgrid's performance and cost-effectiveness, considering the inherent uncertainty in renewable energy generation. In conclusion, the stochastic model's higher operating cost is justified by its ability to effectively manage uncertainties and establish a reliable and resilient energy management strategy. To compare the expected costs of both models accurately, a thorough assessment of uncertainty and the advantages it offers in ensuring flexible and robust microgrid operation is essential.

6.5 Optimal Real-Time Operation of MCES

This section delves into the crucial aspect of formulating the operation of MCESs that include energy assets from three energy carriers, namely electricity, heating, and transportation, focusing on the optimization and real-time control.

Managing MCESs creates a unique opportunity to exploit flexibility provided by carriers that until today are heavily underutilized. MCESs are inherently complex and dynamic, and their optimal operation and management encompasses the integration of various assets. These assets range from traditional electricity assets such as WT, solar PV, and BESS to assets from other energy carriers such as EV, HP, and heating, ventilation, and air-conditioning (HVAC). Although significant work has been conducted with respect to managing networks that are penetrated by RESs and BESSs, assets from other energy carriers introduce a multitude of features, constraints, and social considerations that should be accurately modeled in order to be part of a management system [16]. These characteristics render the optimization of MCES a challenging task. For example, EVs, via the appropriate protocols, can be controlled in order to essentially serve as energy storage units capable of storing and discharging electricity to the grid when instructed, however accurately forecasting their charging patterns can be quite challenging. Similarly, HPs and HVAC systems can be appropriately operated to exploit market conditions and participate in other network services; however, comfort levels of the end-users shall be considered and not violated at any time.

6.5.1 Objective Function

The objective function of the optimization problem is formulated as a maximization problem (expressed as the opposite minimization problem for computational efficiency) of the renewable energy penetration. Specifically, the objective function is calculated based on Eq. (6.35) and aims to maximize the active power output of both PV and WT units in order to reduce reliance on nonrenewable energy sources.

$$\max\left(\sum_{c\in C}\sum_{t\in T}\sum_{k\in N_{\mathrm{WT}}} P_{\mathrm{WT},d,t}(k,c)+\sum_{c\in C}\sum_{t\in T}\sum_{k\in N_{\mathrm{PV}}} P_{\mathrm{PV},d,t}(k,c)\right) \tag{6.35}$$

where c is the contingency associated with that asset. It should be noted that the problem has been formulated as a mathematical optimization model, which is solved 24 hours ahead in time steps of 5 minutes. The 24-hour time horizon allows for the consideration of daily trends and patterns in energy consumption and generation, while the 5-minute intervals provide the necessary granularity to capture the dynamic nature of energy demand and supply.

6.5.2 Constraints

The development of a comprehensive mathematical model that considers the characteristics of each asset, their interactions, and finally their constraints, is a prerequisite for the optimization problem formulation. Decision (control) variables represent the controllable aspects of the system that can be adjusted to achieve the optimization goal. Parameters, on the other hand, are fixed values that define the characteristics of the system, such as the capacity of assets or their minimum operating power. By defining decision variables and parameters, constraints can then be formulated to ensure the optimization problem is feasible.

These constraints mainly refer to *technology operation constraints* such as capacity constraints. The constraints considered in this work model are upper and lower, active and reactive power of conventional synchronous generator (SG), WT, PV, EV, and BESS. With respect to the latter asset, other considered constraints are related to the BESS power and energy capacity, the energy conservation law, and the efficiency during charging and discharging. Finally, the constraints that relate to the EV have been developed with two main operational designs at the epicenter. First, the EV is based on the bidirectional charging concept, meaning that it can absorb power from the grid while charging, but can also provide V2G services as power is drawn from the EV toward the grid. Second, when the EV is charging, it is modeled as an impedance–current–power (ZIP) model. In order to include EV in the operational strategy, the anticipated plug-in period is forecasted and used as an input in the optimization problem. Finally, from the electrified heating energy vector, a grand-source heat pump (GSHP) is modeled, based on a vertical U-pipe ground heat exchanger (GHX) configuration. The modeling of the GSHP unit has been based on heat transfer balance equations, thermal exchange equations between building and external environment, in-building heat transfer equation, and lastly, comfort and capacity limits.

Typical *network constraints* shall also be considered when designing an optimization algorithm for an MCES. These constraints ensure the optimal operation of the system while maintaining safety and stability. The constraints modeled here can be broadly classified into two types: equality constraints and inequality constraints. Equality constraints ensure that the power flow in the system satisfies certain relationships, such as Kirchhoff's laws and power balance equations. Inequality constraints impose limits on the system parameters, such as the voltage magnitude and phase at buses as well as thermal constraints of the lines. These constraints ensure that the system operates within safe limits and prevents the occurrence of voltage instability or line overloads. Modeling of the network constraints follows well-established mathematical formulations which can be found in Ref. [17] and are based upon a fully-detailed AC optimal power flow (ACOPF).

6.5.3 Solution Methodology

Optimal power flow (OPF) has long been recognized as a vital tool employed by network operators in various applications. Over the years, the literature has explored several OPF formulations to achieve optimal operation in distribution networks. These formulations include DC optimal power flow (DCOPF), second-order cone programming-based AC optimal power flow (SOCP-ACOPF), and ACOPF. DCOPF serves as a linearized alternative to detailed ACOPF, often employed for large-scale problems or networks where solving an ACOPF formulation would require significant time. ACOPF introduces a comprehensive mathematical representation of the network, encompassing the classical power flow equations that capture the network's physical characteristics. This meticulous modeling approach ensures an accurate representation of vital network features, including frequency, voltage, and voltage angles, which are essential considerations in deriving feasible solutions. The need for detailed formulations, especially in the context of MCESs, is further amplified by the need to calculate and investigate network metrics that cannot be monitored through alternative optimization modeling (e.g. DCOPF).

The formulation of the constraints and objective functions described previously is implemented using the Python programming language and leverages the Pyomo toolbox [18]. To solve the optimization problem, Pyomo employs the open-source solvers IPOPT and GLPK to solve the resulting MINLP. Pyomo is a powerful, open-source tool that can be used to develop mathematical models of MCES and interfaces directly with a wide range of open-source and commercial solvers.

While temporal aspects that can impede the achievement of set points derived from ACOPF have been explored in the literature, there has been limited investigation on integrating and coordinating the ACOPF considering

- high-level market mechanisms,
- emerging flexibility provision schemes, and
- real-time changes in network and asset status, and short-term weather and demand forecasts.

The strategy proposed in this section is targeting MCES that require constant dispatch considering the dynamic nature of the above points. The framework includes constant calculation and dispatch of set points from the ACOPF, executed within a couple of minutes (e.g. five minutes). The latter builds upon two main operational directives:

1. At the distribution system level, optimization schemes are developed to dispatch control set points to energy assets over relatively large time periods. However, to effectively handle the dynamic nature of the system, it becomes necessary to employ more granular timeframes.
2. Traditional optimization schemes operate in an open-loop framework, disregarding sophisticated market products and changes in the network conditions. To address this issue, feedback-based, adaptive optimization algorithms and architectures offer a practical solution by considering the above and constantly redispatching the network according to the latest available information.

When combined, a feedback-based, adaptive optimization scheme that is based on a granular, rolling-window operation demonstrates significant potential and substantial benefits for the system operator. These algorithms, supported by relevant computing and communication systems, can enable effective control of MCESs.

6.5.4 Case Study and Simulation Results

In order to validate and demonstrate the effectiveness of the proposed analytical framework for the optimal real-time operation of MCESs, a comprehensive case study was conducted. This case study serves as a practical application of the concepts and methodologies discussed thus far, providing valuable insights into the real-world implications of the optimization strategies. Through the use of simulation tools and realistic system data, the performance of the optimized operation strategy was evaluated. By examining the simulation results, the aim is to showcase the practical viability and advantages of the optimized real-time operation approach in MCESs.

To perform the case study, the considered MCES depicted in Figure 6.8 has been modeled in Python using Pyomo. The MCES is composed of various generation and consumption assets including

- An interface with the upstream network
- A 33 kV/11 kV transformer
- SG unit (connected at Bus 1)
- WTs (connected to Bus 1, Bus 3, Bus 4, and Bus 5)

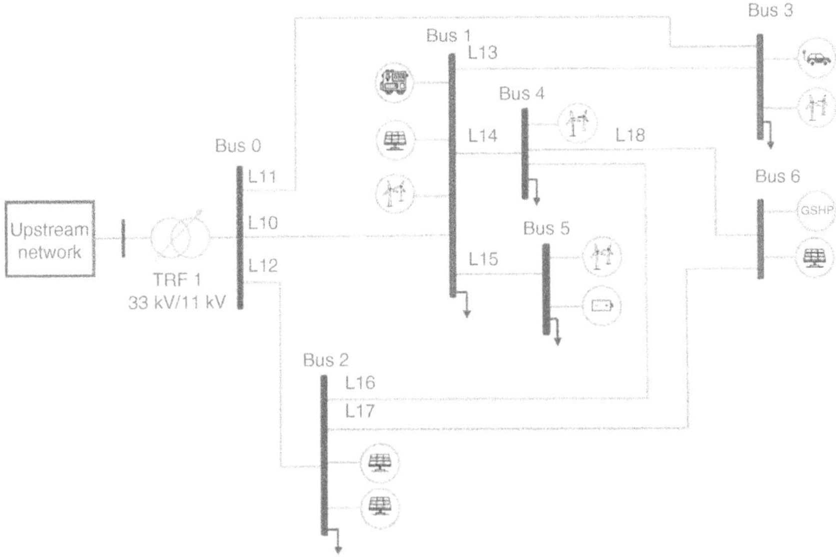

Figure 6.8 MCES network used in the case study.

- PVs (connected to Bus 1, Bus 2, and Bus 6)
- A BESS (connected to Bus 5)
- An EV charger (connected to Bus 3)
- A GSHP (connected at Bus 6)

Furthermore, the MCES contains noncontrollable demand (connected at Bus 1, Bus 2, Bus 3, Bus 4, and Bus 5) representing the typical electricity demand.

The studies presented in this section seek to showcase the outcomes of employing a fully detailed ACOPF tailored to the MCES, utilizing the mathematical formulations discussed previously. To benchmark the results, a classical load flow analysis was conducted using typical real-world input data for all the assets of the MCES (see examples in Figures 6.9 and 6.10).

Based upon the load flow analysis, a power constraint at the substation is expected to be violated (refer to Figure 6.11). This is mainly due to the noncontrollable nature of the RES and the participation of the BESS unit in the electricity market (i.e. predefined power profile based upon an assumed contractual agreement). The constraint

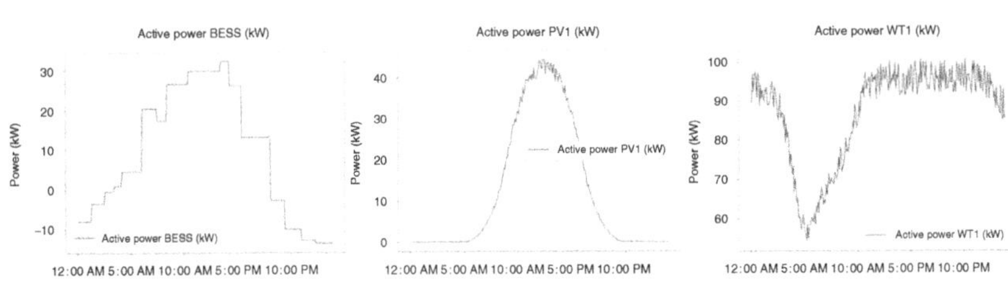

Figure 6.9 BESS, PV, and WT input profiles for load flow analysis.

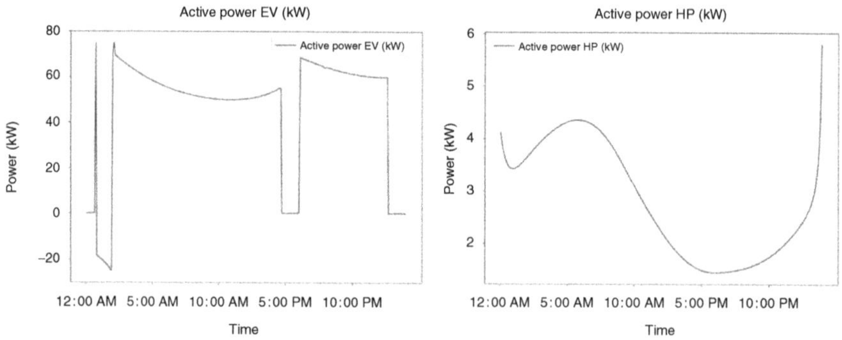

Figure 6.10 EV and GSHP active power input profiles for load flow analysis.

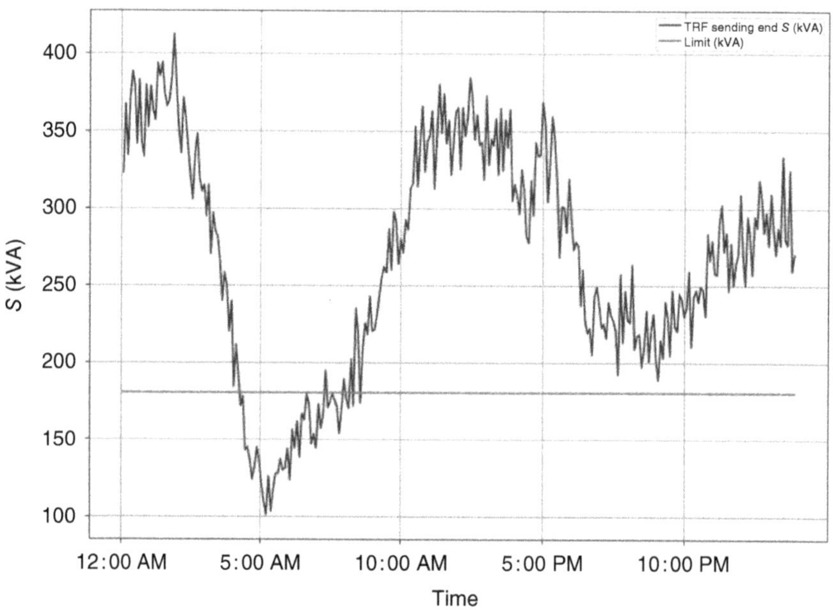

Figure 6.11 Transformer apparent power profile and limit extracted by load flow analysis.

violation, if not mitigated, can lead to congestion issues and render disconnections of units imperative for the operator, meaning that end-users and asset owners' operations are affected by disruptions. This underscores the importance of incorporating congestion management strategies in network management approaches, to ensure reliable and efficient operation of the MCES.

To benchmark the ACOPF-based approach discussed previously, the problem was formulated with characteristics that are discussed hereafter. The objective function used for this case study is to maximize the penetration of renewable generation while mitigating any congestion challenges at the upstream network (when power flow surpasses the maximum capacity threshold of the transformer). The latter was not modeled through the objective function, but rather expressed as a constraint due to

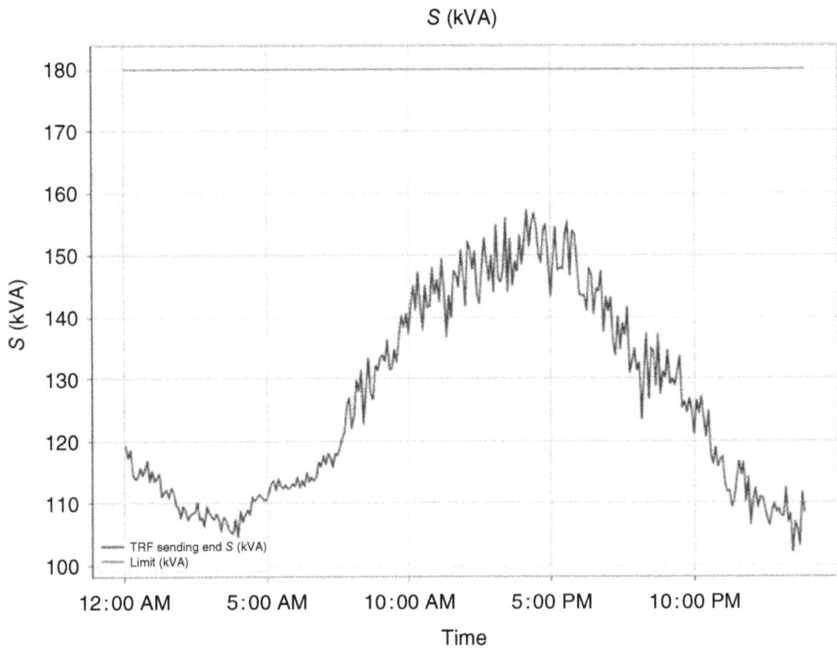

Figure 6.12 Transformer apparent power expected profile and limit following MCES optimization.

its emergent nature (prevent overloading and potential damage, outages, and disconnections).

The resulting apparent power profile at the upstream substation is presented in Figure 6.12. Note that through the MCES optimization objective function and relevant constraints, three direct and indirect objectives have been realized:

1. Upstream substation expected constraint violation is alleviated.
2. Renewable resources are not curtailed and inject all their available power into the network.
3. Disconnections, outages, and possible equipment damage risk are minimized.

6.6 Analysis of Commercial Tools for the Optimal Design and Operation of MCES

Based on the literature review, numerous tools and models exist to tackle the optimal operation and planning of MCESs. Such tools have their distinct advantages and limitations and can be categorized against different criteria. So far, universal categorization is not in place. Instead, energy planning tools can be categorized based on several criteria, including their purpose, scope, level of detail, and modularity. This section provides a comparison among the different commercial tools based on their characteristics and specifications that are briefly described in the next [19, 20].

1. **Purpose**: Energy planning tools can be categorized based on their purpose, such as
   - **Energy demand forecasting tools:** These tools are used to forecast the future energy demand for a region or country based on various economic and demographic factors.
   - **Energy supply planning tools:** These tools are used to determine the optimal mix of energy sources (e.g. coal, gas, and renewable energy) to meet the forecasted energy demand.
   - **Energy efficiency assessment tools:** These tools are used to evaluate the potential energy savings that can be achieved through various energy efficiency measures.
   - **Energy system modeling tools:** These tools are used to simulate the operation of the energy system and evaluate the impact of different policies and scenarios on the system's performance.

2. **Scope:** Energy planning tools can also be categorized based on their scope, which can be regional, national, or global. Some tools are designed for specific regions or countries, while others are used to model the global energy system. So, this specification provides the targeted scale of the system model, which can range from large, e.g. national networks to comparatively small structures such as local distribution grids. The scale that each tool target is related to the intention of use and the included models are developed accordingly. As an example, at a national level, network assets/devices or regions energy modeling is relatable, whereas at the microgrid level single technologies or buildings energy modeling are important. Also, under this specification, the spatial scope is involved. Spatial resolution has a strong impact on which studies a tool can conduct depending on whether component models correspond to individual network assets and devices or aggregated structures like buildings or districts.
3. **Level of detail:** Energy planning tools can also be categorized based on the level of detail they provide. Some tools provide a high level of detail, such as hourly or daily energy consumption data, while others provide a more aggregated view of the energy system, such as annual energy consumption data. So, this category provides a measure for the temporal resolution of the component models used by a tool. The temporal resolution has strong implications for the potential applications a tool can be applied to. For instance, models for control applications need to address time scales in the order of the dynamics of the underlying process. Other models, e.g. the ones focusing on energy balancing or economic studies may require only aggregated or averaged information.
4. **Modeling approach:** Energy planning tools can be categorized based on their modeling approach, such as
   - **Optimization models:** These models are used to determine the optimal mix of energy sources and technologies that minimize the cost of meeting energy

demand while satisfying various constraints. The optimization models may focus on one single objective optimization or more than one.
- **Simulation models:** These models simulate the operation of the energy system over time and can be used to evaluate the impact of different policies and scenarios on the system's performance.
- **Bottom-up models:** These models are used to analyze energy consumption in various sectors (e.g. residential, commercial, and industrial) and can provide insights into the potential for energy efficiency improvements.
- **Top-down models:** These models use aggregated data to estimate energy consumption and can be used to forecast future energy demand.

5. **Modularity:** This is related to the structure of the tool and has an impact on the user-friendliness and expandability of the tool. Three main categories are distinctive:
   a) **Chain the different models together through specialized code:** This approach has flexibility and efficiency for one specific pair of tools but expert users are needed, and thus it has low modularity.
   b) **Monolithic program:** This approach means extending one tool by directly including one or more others. For the user, the resulting tool is easy to learn. However, creating such a tool becomes increasingly difficult as the complexity of the system increases.
   c) **Workflow management:** This approach focuses on creating an interaction framework where different tools can communicate through a shared database by using standardized data structures. Once established, usage is relatively easy for regular users and offers flexibility and scalability options as well.

The authors of Refs. [21, 22] provide an overview of the most popular commercial models and tools. The relevant ones for the multicarrier systems are briefly described below:

Energy system planning and optimization:

- **EnergyPLAN:** A tool for energy system analysis and optimization that evaluates the technical and economic feasibility of various energy supply scenarios. It allows users to model different energy supply technologies and demand-side measures to create optimized energy systems that meet specific objectives [23, 24].
- **Calliope:** A tool for energy system modeling and optimization that analyzes the feasibility and performance of a range of energy supply technologies and demand-side measures. It enables users to simulate and optimize energy systems at a national or regional level [25, 26].

Energy system design and optimization:

- **HOMER Pro:** A tool for energy system design and optimization that is focused on off-grid and grid-connected systems, hybrid renewable energy systems, and microgrids. It allows users to simulate and optimize different energy system configurations and determine the optimal mix of technologies to meet energy demand [27, 28].

- **SPINE Toolbox:** A tool for energy system modeling and analysis that explores different design options and identifies optimal solutions in the early stages of system design. It is focused on building and district-level energy systems [29, 30].

MCES modeling and simulation:

- **Integrate:** A tool for energy system optimization and management that integrates multiple energy carriers and demand sectors to optimize system design and operation based on various objectives. It is designed for district- and regional-level energy systems [31].
- **EnergyExemplar PLEXOS:** A tool for energy system planning and optimization that models and simulates large-scale energy systems at the national or regional level. It enables users to optimize energy system operation based on multiple objectives [32].

Energy management and data analysis:

- **EnergyCAP:** A tool for utility bill tracking, energy management, and sustainability reporting for the building and facility level. It helps users track energy usage and costs and provides basic energy management tools [33].
- **HYSYS:** A tool for process simulation that models and simulates energy systems at a detailed process level, including process design, optimization, and control [34].

The above-mentioned tools are summarized and compared in Table 6.4, against the five main characteristics analyzed in the first part of this section.

The first column describes the main purpose or function of each tool. Most of the tools are designed for energy system planning, optimization, and simulation, but they differ in terms of their specific applications and focus. For example, HOMER Pro is designed for designing and optimizing off-grid and grid-connected energy systems, while EnergyPLAN is designed for evaluating the technical and economic feasibility of different energy supply scenarios. The second column describes the scope or scale of the energy systems that each tool can model or analyze. The tools range from the building and facility level to the national and regional level, with some tools able to model both small and large systems. For example, HOMER Pro and SPINE Toolbox are focused on building- and facility-level energy systems, while EnergyPLAN and Calliope are designed for analyzing energy systems at a national or regional level. Level of detail column describes the level of detail that each tool can provide in modeling energy systems. The tools range from low to very high levels of detail, depending on their intended use and the complexity of the energy system being modeled. For example, EnergyCAP provides a low level of detail and is designed for utility bill tracking and basic energy management, while HYSYS provides a very high level of detail and is designed for process simulation of complex energy systems. The fifth column describes the modeling approach used by each tool to simulate and optimize energy systems. The last column categorizes the tools against the three categories of the modularity characteristic.

Table 6.4 A comparison for commercial tools for energy planning of MCES.

Tool	Purpose	Scope	Level of detail	Modeling approach	Modularity
HOMER Pro	Energy system design and optimization for off-grid and grid-connected systems, hybrid renewable energy systems, and microgrids	Building and facility level	Medium to high	Optimization modeling and simulation	a
EnergyPLAN	Energy system analysis and optimization for evaluating technical and economic feasibility of various energy supply scenarios	National and regional level	Low to high	Mathematical modeling and optimization	a
SPINE Toolbox	Energy system modeling and analysis for exploring different design options and identifying optimal solutions in the early stages of system design	Building and district level	Low to medium	Optimization modeling and simulation	*c*
Integrate	Energy system optimization and management for integrating multiple energy carriers and demand sectors, and optimizing system operation based on various objectives	District to regional level	Medium to high	Optimization modeling	*c*
Calliope	Energy system modeling and optimization for analyzing the feasibility and performance of a range of energy supply technologies and demand-side measures	National and regional level	Low to medium	Mathematical modeling and optimization	*c*
EnergyCAP	Utility bill tracking, energy management, and sustainability reporting for building and facility level	Building and facility level	Low	Data management and analysis	*b*
EnergyExemplar PLEXOS	Energy system planning and optimization for modeling and simulating large-scale energy systems at the national or regional level	National and regional level	High	Optimization modeling and simulation	*a*
HYSYS	Process simulation for modeling and simulating energy systems at a detailed process level, including process design, optimization, and control	Building and facility level to national and regional level	Very high	Process simulation and optimization	b

6.7 Conclusions and Lessons Learned

Formation, development, and operation of MCES is motivated by possible benefits such as producing where energy is consumed, increasing energy efficiency and flexibility, end-user engagement, reducing system vulnerabilities, and addressing energy poverty. Planning and operation of MCES is very complex due to many technical and stakeholder interdependencies in addition to the need for considering several time horizons. As coordination of operation and planning is increasingly important, there is a need for sound methodologies addressing the various aspects of MCES.

The aspects of coordinated planning and operation of MCES are presented both in the form of an analytical framework and with detailed model examples for the planning, operation, and real-time aspects that comprise MCES. These aspects represent different problem formulations and are also interlinked. System configuration needs to consider operational situations when designing the system. Furthermore, when implementing the system that has been designed and evaluated through operational analyses, real-time operation needs to be compatible with the analyses performed.

Traditionally, cost minimization has been the most used approach to optimize MCES, but several other objective functions are relevant to consider. This includes emission reduction, reliability and resilience, DR and flexibility, and renewable energy integration. Choosing what combination of these to employ depends on the priorities of stakeholders and system characteristics on a case-by-case basis. A crucial point is to be aware of the various objective formulations and their impact on the results to effectively navigate the trade-offs between them.

List of Abbreviations

AC	absorption chiller
ACOPF	AC optimal power flow
BESS	battery energy storage system
CHP	combined heat and power
CI	carbon intensity
CRF	capital recovery factor
DCOPF	DC optimal power flow
DICOPT	discrete and continuous optimizer
DR	demand response
EDR	incentive-based demand response
EV	electric vehicle
EZ	electrolyzer
FC	fuel cell
G2V	grid-to-vehicle
GB	gas-fired boiler
GHX	ground heat exchanger
GSHP	grand-source heat pump
HP	heat pump

HVAC	heating, ventilation, and air-conditioning
ICE	internal combustion engine
ILEC	integrated local energy community
MCES	multicarrier energy system
MGT	microgas turbine
MILP	mixed-integer linear programming
MINLP	mixed-integer nonlinear programming
MIP	mixed-integer programming
NLP	nonlinear programming
O&M	operation & maintenance
OPF	optimal power flow
PV	photovoltaic
RESs	renewable energy sources
SG	synchronous generator
SOC	state-of-charge
SOCP-ACOPF	second-order cone programming-based ACOPF
ST	solar thermal
TES	thermal energy storage
TOU	time-of-use
V2G	vehicle-to-grid
WT	wind turbine

References

1 Piazza, G., Delfino, F., Bergero, S. et al. (2023). Economic and environmental optimal design of a multi-vector energy hub feeding a local energy community. *Applied Energy* 347: 121259. https://doi.org/10.1016/j.apenergy.2023.121259.

2 Di Somma, M. and Graditi, G. (2022). Integrated energy systems: the engine for energy transition. In: *Technologies for Integrated Energy Systems and Networks* (ed. D. Somma and G. Graditi), 15–40. https://doi.org/10.1002/9783527833634.ch2. Wiley.

3 Jin, L., Rossi, M., Ciabattoni, L. et al. (2023). Environmental constrained medium-term energy planning: the case study of an Italian university campus as a multi-carrier local energy community. *Energy Conversion and Management* 278: 116701. https://doi.org/10.1016/j.enconman.2023.116701.

4 Ciavarella, R., Di Somma, M., Graditi, G., and Valenti, M. (2019). Congestion management in distribution grid networks through active power control of flexible distributed energy resources. *2019 IEEE Milan PowerTech*, Milan (Italy). IEEE. pp. 1–6. https://doi.org/10.1109/PTC.2019.8810585.

5 Di Somma, M., Caliano, M., Cigolotti, V., and Graditi, G. (2021). Investigating hydrogen-based non-conventional storage for PV power in eco-energetic optimization of a multi-energy system. *Energies (Basel)* 14 (23): 8096. https://doi.org/10.3390/en14238096.

6 Di Somma, M., Ciabattoni, L., Comodi, G., and Graditi, G. (2020). Managing plug-in electric vehicles in eco-environmental operation optimization of local multi-energy systems. *Sustainable Energy, Grids and Networks* 23: 100376. https://doi.org/10.1016/j.segan.2020.100376.

7 Di Somma, M., Graditi, G., and Yan, B. (2023). Cost-sustainability trade-off solutions for the optimal planning of local integrated energy systems from nanogrids to communities. In: *Handbook of Smart Energy Systems* (ed. M. Fathi, E. Zio, and P.M. Pardalos), 2573–2606. https://doi.org/10.1007/978-3-030-97940-9_80. Cham: Springer International Publishing.

8 Foiadelli, F., Nocerino, S., Di Somma, M., and Graditi, G. (2018). Optimal design of DER for economic/environmental sustainability of local energy communities. *2018 IEEE International Conference on Environment and Electrical Engineering and 2018 IEEE Industrial and Commercial Power Systems Europe (EEEIC/I&CPS Europe)*, Palermo (Italy). IEEE. pp. 1–7. https://doi.org/10.1109/EEEIC.2018.8493898.

9 Stanelyte, D., Radziukyniene, N., and Radziukynas, V. (2022). Overview of demand-response services: a review. *Energies (Basel)* 15 (5): 1659. https://doi.org/10.3390/en15051659.

10 Qiu, H., Gu, W., Liu, P. et al. (2022). Application of two-stage robust optimization theory in power system scheduling under uncertainties: a review and perspective. *Energy* 251: 123942. https://doi.org/10.1016/j.energy.2022.123942.

11 Heydarian-Forushani, E., Golshan, M.E.H., Shafie-khah, M., and Catalão, J.P.S. (2020). A comprehensive linear model for demand response optimization problem. *Energy* 209: 118474. https://doi.org/10.1016/j.energy.2020.118474.

12 Henrion, R. and Römisch, W. (2022). Problem-based optimal scenario generation and reduction in stochastic programming. *Mathematical Programming* 191 (1): 183–205. https://doi.org/10.1007/s10107-018-1337-6.

13 Borchers, B. and Mitchell, J.E. (1997). A computational comparison of branch and bound and outer approximation algorithms for 0–1 mixed integer nonlinear programs. *Computers and Operations Research* 24 (8): 699–701. https://doi.org/10.1016/S0305-0548(97)00002-6.

14 Shahbazbegian, V., Dehghani, F., Shafiyi, M.A. et al. (2023). Techno-economic assessment of energy storage systems in multi-energy microgrids utilizing decomposition methodology. *Energy* 283: 128430. https://doi.org/10.1016/j.energy.2023.128430.

15 Henrion, R. and Römisch, W. (2022). Problem-based optimal scenario generation and reduction in stochastic programming. *Math Program* 191(1): 183–205. doi: 10.1007/s10107-018-1337-6.

16 Liu, T., Zhang, D., Wang, S., and Wu, T. (2019). Standardized modelling and economic optimization of multi-carrier energy systems considering energy storage and demand response. *Energy Conversion and Management* 182: 126–142. https://doi.org/10.1016/j.enconman.2018.12.073.

17 Rousis, A.O., Konstantelos, I., and Strbac, G. (2020). A planning model for a hybrid AC–DC microgrid using a novel GA/AC OPF algorithm. *IEEE Transactions on Power Systems* 35 (1): 227–237. https://doi.org/10.1109/TPWRS.2019.2924137.

18 Pyomo software. http://www.pyomo.org/ (accessed June 2023).

19 Askeland, M., Morch, A., Papadimitriou, C. et al. (2023). Workflow-based architecture for optimal planning of integrated local multi-energy systems. *2023 International Conference on Smart Energy Systems and Technologies (SEST)*, Mugla (Turkey) (September 2023). IEEE. pp. 1–6. https://doi.org/10.1109/SEST57387.2023.10257503.

20 Connolly, D., Lund, H., Mathiesen, B.V., and Leahy, M. (2010). A review of computer tools for analysing the integration of renewable energy into various energy systems. *Applied Energy* 87 (4): 1059–1082. https://doi.org/10.1016/j.apenergy.2009.09.026.

21 Fodstad, M., del Granado, P.C., Hellemo, L. et al. (2022). Next frontiers in energy system modelling: a review on challenges and the state of the art. *Renewable and Sustainable Energy Reviews* 160: 112246. https://doi.org/10.1016/j.rser.2022.112246.

22 Ringkjøb, H.-K., Haugan, P.M., and Solbrekke, I.M. (2018). A review of modelling tools for energy and electricity systems with large shares of variable renewables. *Renewable and Sustainable Energy Reviews* 96: 440–459. https://doi.org/10.1016/j.rser.2018.08.002.

23 EnergyPlan tool. https://www.energyplan.eu/ (accessed June 2023).

24 Østergaard, P.A. (2015). Reviewing EnergyPLAN simulations and performance indicator applications in EnergyPLAN simulations. *Applied Energy* 154: 921–933. https://doi.org/10.1016/j.apenergy.2015.05.086.

25 Calliope tool. https://www.callio.pe/ (accessed June 2023).

26 Laha, P. and Chakraborty, B. (2021). Cost optimal combinations of storage technologies for maximizing renewable integration in Indian power system by 2040: multi-region approach. *Renewable Energy* 179: 233–247. https://doi.org/10.1016/j.renene.2021.07.027.

27 HOMER Pro® microgrid software. https://www.homerenergy.com/products/pro/index.html (accessed June 2023).

28 Papaioannou, D.I., Papadimitriou, C.N., Dimeas, A.L. et al. (2014). Optimization & sensitivity analysis of microgrids using HOMER software – a case study. In: *MedPower 2014*, Athens, 1–7. https://doi.org/10.1049/cp.2014.1668.

29 Spine Toolbox. http://www.spine-model.org/spine_toolbox.htm (accessed June 2023).

30 Gonzato, S., Bruninx, K., and Delarue, E. (2021). Long term storage in generation expansion planning models with a reduced temporal scope. *Applied Energy* 298: 117168. https://doi.org/10.1016/j.apenergy.2021.117168.

31 Di Somma, M., Papadimitriou, C., Morch, A. et al. (2023). An innovative toolbox for the optimal design and operation of integrated local energy communities. *27th International Conference on Electricity Distribution (CIRED 2023)*, Rome

(Italy). Institution of Engineering and Technology. pp. 676–680. https://doi.org/10.1049/icp.2023.0460.
32 Energy Exemplar tool. https://www.energyexemplar.com/plexos (accessed June 2023).
33 EnergyCAP software. https://www.energycap.com/ (accessed June 2023).
34 Aspen HYSYS® software. https://www.aspentech.com/en/products/engineering/aspen-hysys (accessed June 2023).

7

Integrated Flexibility Solutions for Effective Congestion Management in Distribution Grids

Bart van der Holst[1], Gijs Verhoeven[1], Milad Kazemi[1], Christina Papadimitriou[1], Marialaura Di Somma[2], and Koen Kok[1]

[1] *Eindhoven University of Technology, Electrical Energy System Group, Electrical Engineering Department, De Rondom, 5612AP, Eindhoven, the Netherlands*
[2] *Italian National Agency for New Technologies, Energy and Sustainable Economic Development, Lungotevere Thaon di Revel, 76, 00196, Rome, Italy*

7.1 Introduction

In recent years, the global energy landscape has witnessed a significant transformation, driven by the urgent need to address climate change, reduce greenhouse gas emissions, and transition toward a sustainable and decentralized energy system. This transformation has great implications for the functioning and operation of the electricity grid as distributed renewable energy sources are rapidly integrated and energy demand is being electrified. As a result, grid components like cables and transformers are experiencing higher loads than before, and often higher than was expected during the grid planning phase. In several countries, this has resulted in component overloading, or congestion, in both the transmission and distribution systems.

Integrated local energy communities (ILECs) present a paradigm shift in the energy sector, fostering a bottom-up approach that empowers local communities to actively participate in energy production, consumption, and management, by making use of multiple energy carriers and sector coupling solutions. By leveraging advanced technologies, including smart grids, energy storage systems, and intelligent demand response, ILECs offer a novel framework to optimize the integration of diverse energy resources. The coordinated control of the multicarrier energy flows within the ILECs gives rise to the flexibility to shift demand/production in space, time, and between the various carriers. This chapter aims to explore how this concept of integrated flexibility in ILECs can be used to mitigate congestion in the electricity grid. We thereby explicitly focus on the distribution grid, and thus the medium- and low-voltage grid levels of the network, because that is where most ILECs operate.

The chapter will thus provide a comprehensive overview of the potential of integrated flexibility from ILECs for congestion management in distribution grids.

Integrated Local Energy Communities: From Concepts and Enabling Conditions to Optimal Planning and Operation, First Edition. Edited by Marialaura Di Somma, Christina Papadimitriou, Giorgio Graditi, and Koen Kok.

Section 7.2 first introduces the topic of congestion management in distribution grids by showing how current developments in distribution grids increasingly give rise to congestion issues. The traditional forms of congestion management are then discussed. In Section 7.3, the concept of integrated flexibilities in ILECs is introduced and several modeling cases are presented. Section 7.4 describes the legal and practical aspects of various instruments by which integrated flexibility can be harnessed to alleviate congestion by an operator of the distribution grid. The discussed topic is supported by a simulation study to show the potential of such instruments in theory. Furthermore, a detailed analysis is provided of the new European Clean Energy package and its view on the role and regulations for using flexibility for congestion management in practice. The chapter is then concluded with Section 7.5, where current challenges and an outlook for the future are presented.

With a solid understanding of ILECs and their dynamics established in Chapters 3 and 4, this chapter delves deeper into the specific area of integrated flexibility for congestion management. It explores how these communities can effectively identify and employ flexibility measures, such as demand response, energy storage, and electric vehicles (EVs) management, to tackle grid congestion challenges. Ultimately, this chapter aims to provide policymakers, energy planners, researchers, and practitioners with valuable insights into the potential of ILECs and integrated flexibility as essential tools in the quest for a sustainable, efficient, and resilient energy system. By embracing the principles of decentralization, community engagement, and advanced technologies, ILECs can pave the way for a future where congestion management becomes an integrated and harmonized aspect of our energy infrastructure, facilitating a smooth transition toward a cleaner and more sustainable energy future.

7.2 Congestion Management in Distribution Systems

7.2.1 The Distribution System and The System Operator

The electricity network can typically be defined into two interconnected parts: the transmission system and the distribution system. The transmission system is responsible for the long-distance, high-voltage transport of electricity from large power plants to local distribution systems. It includes high-voltage power lines, transformers, and other equipment that typically operate at voltage levels exceeding 110 kV. In contrast, the distribution system facilitates the delivery of electricity from the transmission system to the end-user, such as homes, businesses, and industrial facilities. It includes medium-voltage and low-voltage power lines, transformers, and other equipment that operate at voltage levels of 10–50 kV (medium voltage) or less than 1 kV (low voltage). Figure 7.1 provides a schematic overview of the electricity grid in terms of these two systems.

Beyond disparities in voltage levels, several other noteworthy distinctions characterize the transmission system and the distribution system. Table 7.1 presents a few of these key differences. It shows that the distribution systems contain more network nodes, but the topology is typically less complex due to their radial

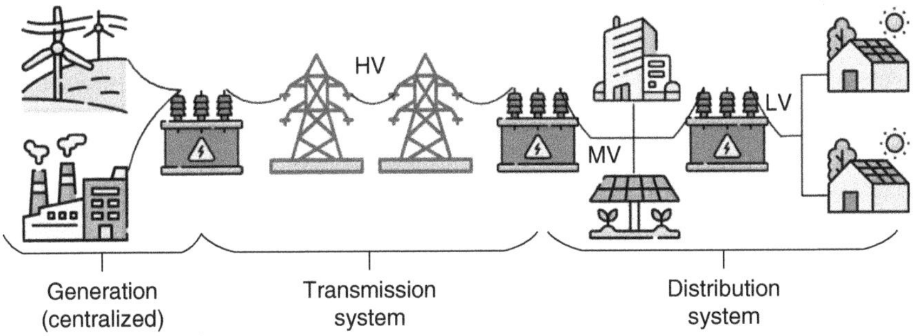

Figure 7.1 Schematic overview of the electrical power system.

Table 7.1 Key characteristics of the transmission and distribution system.

	Transmission system	**Distribution system**
Number of nodes	10–1000	100–10 000
Asset properties	Mainly inductive	Mainly resistive
Topology	Meshed	Radial
Loading aggregation level	Highly aggregated	Nonaggregated
Loading stochasticity	Low-stochastic	High-stochastic
Measurement availability	Redundant	Limited, smart meters
Nature of data	Nonpersonal	Partly personal

structure. Furthermore, the loads in a distribution system emanate from individual small-scale end-users, featuring load profiles that heavily depend on the human behavior of individuals. In contrast, in transmission systems, these loads are aggregated, and individual decision-making is typically smoothened out, resulting in lower stochasticity. Lastly, due to the large number and size of distribution networks, measurement devices are scarcer, and measurements (e.g. from smart meters) involve highly personal data. Regulations about data use and privacy therefore often prohibit the use of this data for grid operations.

In the latter part of the twentieth century, the electricity grids were typically operated by vertically integrated utilities that were responsible for the entire electricity supply chain, from power generation to distribution and supply to end-users. This model was criticized for being inefficient and leading to high electricity prices, as the utilities had little incentive to reduce costs or improve efficiency. This prompted calls for the introduction of competition in the electricity sector, and the establishment of separate entities to operate the transmission and/or distribution systems. The legal basis for such entities, being the transmission system operators (TSOs) and distribution system operators (DSOs), was established in the European Union (EU) in 1996 [1]. In the United States, the operation of both grids became the responsibility of independent system operators (ISOs) [2] or regional transmission organizations (RTOs) in the years 1996 and 1999, respectively [3].

In Europe, the Third Energy Package defines the DSO and its responsibilities as follows: “a natural or legal person responsible for operating, maintaining and developing the distribution system in a safe, secure and efficient manner, and, where appropriate, for developing and operating the relevant parts of an intelligent network” [4].

The package also sets out specific obligations for DSOs, including:

- Ensuring the security, quality, and continuity of electricity supply in their distribution areas.
- Contributing to the achievement of national and EU energy and climate policy objectives, such as reducing greenhouse gas emissions and increasing the share of renewable energy in the electricity mix.
- Providing nondiscriminatory access to their distribution networks to all market participants, including new entrants and renewable energy producers.
- Coordinating with TSOs to ensure the efficient and secure operation of the overall electricity system.

In summary, the primary objective of DSOs is to guarantee the reliability, sustainability, availability, and affordability of the distribution system. This task presents a formidable challenge as it often involves navigating complex trade-offs without the availability of straightforward solutions. Moreover, the evolving global landscape exerts considerable pressure on DSOs, necessitating swift adaptation to the dynamic shifts in distribution grid usage and operation.

7.2.2 Developments in the Distribution Grid

The distribution grid in numerous countries worldwide is subject to the influence of various significant societal trends: the energy transition, the reduced costs of energy technologies, and the digitization of modern life.

The most important one is probably the energy transition. The transition from fossil fuels to renewable energy sources is primarily motivated by combatting climate change, mitigating resource depletion, and achieving energy independence [5]. Recent reports from the Intergovernmental Panel on Climate Change have demonstrated that global carbon emissions should be reduced to prevent damages from climate change to both nature and humankind [6]. As a result, 190 countries signed the Paris Agreement during the United Nations Climate Change Conference in December 2015, in order to reduce carbon emissions and keep the global average temperature increase below 2.0° with respect to the year 1990. The result has been a combination of supportive policies and subsidies for renewable technologies. These measures, and other factors, have resulted in a second trend, being the great reduction in the cost of energy technologies over the past few years. This trend is illustrated by the reduction in costs for lithium-ion batteries and photovoltaic (PV) models in the past five years in Figure 7.2.

A third global trend that can be identified is the digitization of modern life. New advances in communication technology, computing power, and data storage are changing the way our critical infrastructure is observed, operated, and protected.

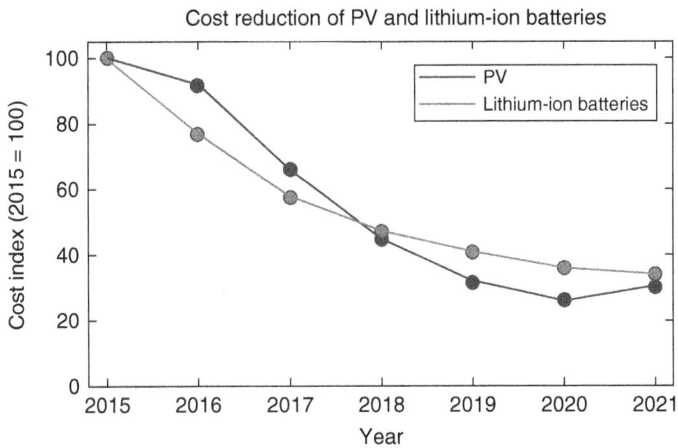

Figure 7.2 Cost index of PV and lithium-ion batteries. Source: Adapted from Ref. [7].

These three ongoing societal trends impact the distribution grid and its operation in several ways.

7.2.2.1 Growth of Distributed Renewable Energy Resources

Due to a combination of policies offering subsidies and reduced costs, DSOs in many countries have seen a rapid growth of PV and wind turbines in their distribution grids [8]. This is a paradigm shift from a system in which electricity was produced in a few large power plants to a more decentralized model with renewable energy resources distributed over the grid. Now every household can be a net producer of electricity, whereas in the past only electricity was consumed in the distribution system. As a result, power flows traditionally going top-down from the LV–MV transformer to the households can now be reversed at moments of high solar production [9]. Not only can this result in component overloading during peak hours, but voltage levels can change as a result of, e.g. PV in the distribution grid as well. Where traditionally consumption reduces the voltage level along feeders, production increases the voltage level, possibly resulting in over-voltage issues in the grid.

7.2.2.2 Electrification of Demand

As a result of the energy transition, more and more fossil-based energy demand is now being replaced by electrical demand. This transformation is driven by heightened recognition of the adverse impacts of climate change and air pollution. Consequently, EVs are gradually displacing their traditional internal combustion engine counterparts, and conventional gas-based heating and cooking systems are being substituted by electricity-based alternatives, such as heat pumps (HPs) and induction cooktops. These trends are expected to continue as several countries and institutions have introduced new laws preventing the sale of CO_2-intensive technologies. An example is the EU, inhibiting the sale of new CO_2-emitting cars from 2035 in its "Fit for 2055" deal [10]. Another example is the Netherlands, where from 2026 old boilers that need to be replaced should be replaced by more sustainable alternatives, like hybrid HPs, full-electric HPs, or a connection to a heat network if possible [11]. Despite the increasing efficiency of these types of appliances, they are highly electricity intensive and will cause a strong growth in the electricity (peak) demand.

7.2.2.3 Improved Metering

Grid metering is vital for the flexible and intelligent operation of electricity grids. Not only metering is important for billing customers, but the resulting data can provide insights into the operation in the grid or can even be used for real-time control. Due to the size of distribution grids, metering devices have typically been very scarce compared to the transmission grid. However, over the past few years, smart meters have been introduced in households in many countries around the world. Figure 7.3 shows the number of smart meters installed in the United States from 2007 to 2020. Also, more and more DSOs in Europe are metering LV–MV transformers and substations.

7.2.2.4 Higher Controllability of Assets

In recent decades, substantial research efforts have been dedicated to the development of control algorithms for assets within the built environment and industrial sectors. However, a significant limitation existed wherein assets such as PVs, EVs, and HPs lacked user or external controller control capabilities. Instead, the control functions were embedded within the firmware provided by the asset manufacturers. Progress was made with the emergence of "smart appliances," enabling consumers to utilize smartphone applications for certain functionalities such as "smart charging" of their EVs according to personal preferences. Notably, smart charging is primarily feasible for EV owners with private charging stations, while challenges persist for public charging stations due to the involvement of multiple

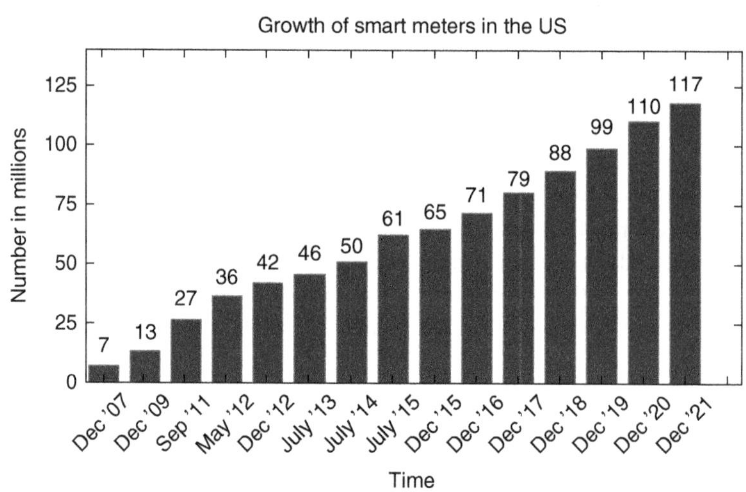

Figure 7.3 Number of electric smart meters installed in the United States from 2007 to 2020. Source: Adapted from Ref. [12].

stakeholders and the lack of a globally standardized communication protocol between EVs and charging infrastructure. Promisingly, efforts in the United States have centered around establishing the open charge point protocol (OCPP) and open charge point interface (OCPI) as prevailing standards, whereas the EU has adopted a distinct approach focusing on the ISO 15118 standard, also known as the "combined charging system" (CCS) protocol.

Furthermore, the market now offers home energy management system (HEMS), which enable monitoring, control, and optimization of energy flows within households. Section 7.3.3 provides an example of an algorithm for HEMS optimizing both costs and CO_2 emissions. However, the absence of a universally adopted protocol for interconnecting various asset types within HEMS necessitates that all assets within a household originate from the same manufacturer.

7.2.3 Congestion in Distribution Grids

The increase in distributed renewable energy resources and the ongoing electrification of demand results in larger, more volatile, and possibly reversed power flows in the distribution grid. This trend introduces not only challenges for the TSO to balance supply and demand in the grid at all times but also new challenges for DSOs to operate the distribution grid safely within its operational limits. There are various operational limits a DSO should consider (power quality, voltage regulation, reactive power management, fault detection and isolation, etc.) but this chapter focuses specifically on the emerging problem of congestion in distribution grids worldwide.

According to the Grid Code of the United Kingdom, congestion in the distribution grid can be defined as

> "A situation on a Transmission System or Distribution System where there is insufficient available Transfer Capacity to meet the desired transfer of electricity. Congestion occurs when the sum of the requested transactions in energy or available Transfer Capacity exceeds the capacity of the elements of the Transmission System or Distribution System involved. (Source: National Grid Electricity Transmission, Grid Code, Section 3, Issue 16, April 2019, p. 49)"

If the capacity of elements like cables and transformers is exceeded, thermal overloading takes place, affecting the element's insulation lifetime [13]. In transformers, the principal factor in determining the expected life of oil-immersed transformers is the hottest-spot temperature, which is the highest temperature of the winding at the operating condition, according to the IEEE standard C57.91-2011 [14]. Although temporary overloading of transformers and cables is generally not a problem as long as thermal limits are not violated, regular overloading will cause lifetime degradation and is therefore costly. If the exceedance of the total network capacity becomes too high, circuit breakers might trip to protect the network and as such cause an outage.

It is important to recognize that the distribution network design process is based on long-term planning, considering that most components have lifetimes ranging

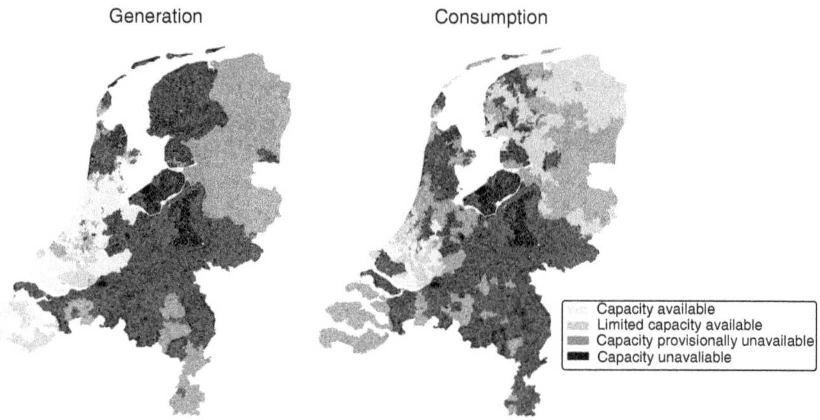

Figure 7.4 A map of the Netherlands showing regions in which there is limited availability of new grid connections $\geq 3 \times 80$ A for producers (left) and consumers (grid) in the distribution grid. The second to lightest color indicates limited capacity and the color darker denotes temporarily no capacity available, depending on the results of the congestion management investigation. The darkest areas implies no capacity is available and that there is also no possibility for congestion management. Source: Adapted from Ref. [16].

from 20 to 40 years. To illustrate, Eurelectric and Deloitte estimated that in 2020, 65–70% of the LV power lines in the EU were older than 20 years [15]. In long-term grid planning studies, the anticipated impact of demand projections on cable and transformer loading is assessed to determine appropriate component sizing. Due to assumptions and uncertainties in calculations, a significant over-dimensioning factor is typically incorporated. However, the aforementioned developments have revealed that this significant over-dimensioning has proven inadequate in several European countries.

In their yearly monitoring report, the Federal Network Agency, the German regulatory office for electricity, gas, telecommunications, post and railway markets, publishes the yearly amount of energy curtailed for congestion management purposes. In their latest report of 2022, they shared that between 2020 and 2021 the cost of curtailment for congestion management increased from €1.4bn to €2.3bn. It was estimated that about 63% of these curtailments were in the distribution system. In the Netherlands, congestion issues are now so pressing that in large parts of the country both consumers and producers of electricity cannot be granted access to the distribution grid anymore. Figure 7.4 presents the map shared by the National Association of DSOs, which shows where there is limited availability of new grid connections larger than 3×80 A for consumers and producers. The darkest areas indicate where no new connections can be granted (in September 2023). This is a big issue, as new solar parks and businesses cannot be connected, thereby limiting both sustainable and economic development.

7.2.4 Traditional Mitigation Strategies for Congestion

Traditionally, there have been several approaches for DSOs to mitigate congestion in their distribution grid. In this subsection, we will discuss the most important ones:

grid reinforcements, grid reconfiguration, using reserve capacity, and rejecting grid access. Each of them has its own downsides as congestion is rapidly emerging, and new solutions are thus required.

7.2.4.1 Grid Reinforcements

The most common approach to tackling congestion in distribution grids is replacing (future) overloaded components with ones having a larger capacity, which can thus better accommodate the present and the anticipated load. As discussed earlier, the longevity of most distribution grid components is planned for a minimum of 20 years, although component reinforcement may be executed earlier if congestion issues arise. However, it is important to note that this approach can be financially burdensome for DSOs since overloaded components are replaced before their complete depreciation. Moreover, grid reinforcements necessitate significant capital investments, availability of trained technical personnel, considerations for limited urban space, and adherence to possibly long spatial procedures. Consequently, the comprehensive reinforcement of all components projected to experience overload in the coming decades poses an immense challenge for DSOs.

Using the Netherlands as an illustrative example, a study conducted by the Association of Energy Network Operators reveals that by 2050, approximately 11 490–15 000 MV–LV stations will require expansion, and the construction of 7500–12 000 new stations will be necessary. Furthermore, calculations indicate that the current total cable length of 260 400 km in distribution systems will need to be augmented by approximately 23.6–31.8% to address anticipated demand [17].

7.2.4.2 Grid Reconfiguration

Grid reconfiguration encompasses the dynamic modification of the topology or configuration of an electrical grid to optimize its performance and effectively manage power flows. This process entails adjusting the interconnections between grid components, such as lines, transformers, and switches, with the aim of achieving specific objectives. In the domain of distribution networks, grid reconfiguration involves manipulating the routing of power flows within the network through actions like line switching or load transfer between phases. This approach addresses operational challenges, enhances system reliability, and facilitates congestion management. Figure 7.5 provides the visual representation of this concept, demonstrating how altering the placement of switches can significantly impact the distribution grid's topology, inspired by the work of Allan et al. [18].

While grid reconfiguration serves as a valuable tool for DSOs, its application in existing grids can pose challenges. Many grids lack remotely controlled switches or do not possess the necessary topology to accommodate their installation. Moreover, implementing grid reconfiguration is a complex undertaking in practical scenarios. It entails data collection to inform decision-making and necessitates the establishment of a robust communication infrastructure for effective switch control.

7.2.4.3 Using Reserve Capacity

A similar approach is to use the backup capacity that is in the MV grid in case faults and short-circuits occur. Most MV grids consist of extra components and switches,

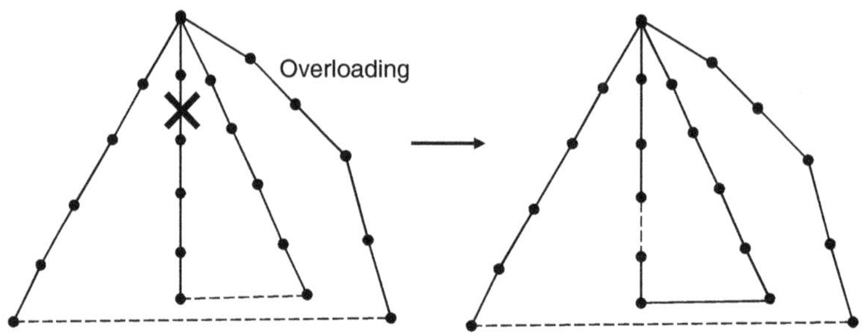

Figure 7.5 Schematic example of grid reconfiguration as a measure for cable overloading. The dashed lines represent cables that are not connected due to the opening of a switch.

such that if a fault occurs or if there is maintenance, the customers can still be provided with electricity. As these events are relatively rare, there is some redundant capacity outside of the main corridors in the grid that can be used more effectively for congestion management.

7.2.4.4 Rejecting Grid Access

One last, albeit undesirable, strategy employed to alleviate congestion is the refusal of new customers seeking access to the electricity grid. This approach directly contradicts the fundamental principle that DSOs are obligated to provide nondiscriminatory access to their distribution networks for all market participants, as discussed in Section 7.2.1. Therefore, resorting to this measure should always be avoided. However, in many countries, it can be employed in cases where a potential new connection poses a risk to the stability of electricity supply for existing connected parties.

In recent years, DSOs have recognized that traditional methods employed for congestion management in the distribution grid are insufficient to address the anticipated congestion challenges in the coming decades. As a result, the notion of leveraging integrated flexibility within distribution grids is now being actively explored as a contemporary solution. The subsequent section delves into this concept of integrated flexibility, exploring how ILECs can offer this flexibility to deliver congestion management services to grid operators.

7.3 Integrated Flexibility in ILECs

This section introduces the concept of integrated flexibility in ILECs. The latter can serve several purposes within the ILEC but also outside the entity, e.g. by providing services to a grid operator for congestion management. Section 7.3.1 provides a general description of this concept and describes its multidimensional applicability within ILECs. The chapter then presents two case studies to show how integrated flexibility can be coordinated within an ILEC and how the flexibility can be extracted to serve multiple purposes. In detail, Section 7.3.2 presents a case study in which the flexibility of a fleet of EVs is utilized. The EVs in the fleet have both grid-to-vehicle

(G2V) and vehicle-to-grid (V2G) capabilities and are charged and discharged such that monetary gains from exploiting flexibility in the wholesale market are optimized. Section 7.3.3 provides a case study where a smart HEMS within an ILEC unlocks integrated flexibility in an optimal way by considering import/export price signals from the grid, while minimizing CO_2 emissions. These import/export prices can include both prices for energy and incentives for congestion management.

7.3.1 The Concept of Integrated Flexibility

Flexibility is a key concept in multicarrier energy systems (MCESs), as it enables the system to dynamically adapt to changes in energy supply and demand, optimize resource utilization, and enhance the integration of renewable energy sources. This section aims to elaborate on the concept of integrated flexibility within MCESs and highlights its importance in enabling a sustainable and resilient energy future and a stable utility operation [19, 20].

Integrated flexibility refers to the system's ability to adjust the generation, conversion, and consumption of different energy carriers in response to varying conditions. Figure 7.6 shows integrated flexibility characterization and highlights its key aspects (inspired by [20]). It encompasses the temporal, spatial, and operational dimensions of energy systems, enabling efficient and reliable energy flow management.[21]

Scope fosters both spatial and temporal attributes. Spatial flexibility relates to the distribution and allocation of energy resources across different geographical locations within an MCES. Therefore, spatial flexibility studies encompass various levels, including system-level, district-level, building-level, and sector-level

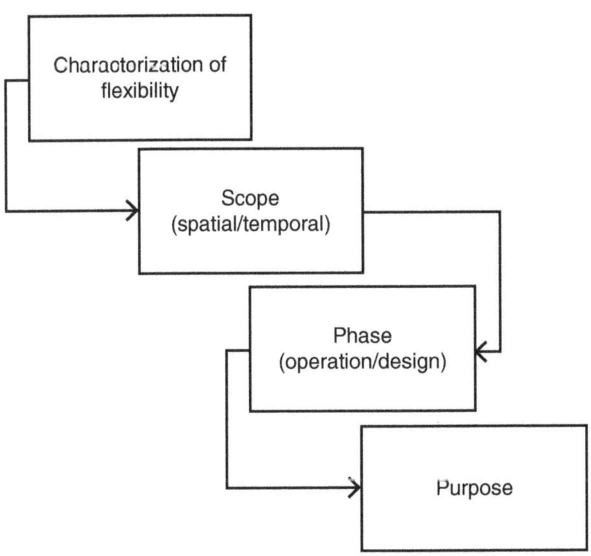

Figure 7.6 Attributes characterizing flexibility.

investigations [22]. Temporal flexibility refers to the system's ability to accommodate variations in energy supply and demand over time. This involves managing fluctuations in renewable energy generation, demand patterns, and storage dynamics coming from all carriers. By optimizing the timing of energy generation and consumption and the interplay among carriers, MCESs can balance supply and demand, reduce peak loads, and increase overall system efficiency.

Based on previously identified spatial levels, the phase (design or operation of MCES) and the purpose are tackled subsequently. When it comes to the operation phase, the purpose can be energy management, control strategies, and potential evaluation for optimal flexibility unlocking and extraction. Of course, different methods, e.g. heuristic, optimal, and predictive can be used for each purpose to serve operation efficiently. When it comes to the design phase, the purpose can include parameter selection for appropriate modeling and system design optimization.

Phase and purpose are briefly discussed below per level.

At the system level, the operation phase may include studies focusing on evaluating flexibility potential, system control strategies, and energy management for optimal extraction. The evaluation of flexibility potential involves heuristic or optimization-based approaches, where simulations or objective functions are used to estimate the benefits of operational strategies. District-level studies [23] mainly examine the energy flexibility potential of communities or coordinated buildings, assessing operational strategies and centralized control under the operation phase. Similarly, at the building level, flexibility studies can reside in both the design [24] and operational phases [25]. Parametric design modeling is commonly employed to identify ideal systems and building characteristics that can achieve flexibility. Control strategy comparisons and energy management evaluations are also conducted using heuristic or optimal methods.

Furthermore, sector-level studies, as, for instance, in [23], analyze the broader impacts of energy flexibility, informing policymaking, building codes, and standards. These studies explore the technical potential and benefits of flexibility measures on grid metrics, occupant comfort, load shifting, and energy efficiency.

What is to be highlighted is that in most of the spatial scopes identified above, the operation phase is predominantly or solely tackled by the literature in comparison to the design phase. This is expected as the integrated flexibility concept is primarily connected with the operation phase of a system.

The most pertinent approach to unlocking and utilizing integrated flexibility for grid congestion purposes is the **integrated demand response (IDR) programs** [26, 27]. The integration of multiple energy carriers allows for a more efficient utilization of resources and enables cross-sector optimization as will be shown in Sections 7.3.2 and 7.3.2. For example, during periods of high electricity demand, demand response programs can not only incentivize consumers to reduce their electricity consumption but also encourage them to shift their heating or cooling loads to off-peak hours. So, based on different price signals, the consumers can arbitrage their consumption among carriers and respond efficiently in demand response programs while having profit. This flexible demand management across various energy carriers helps balance supply and demand, alleviating grid congestion and reducing the need for additional infrastructure investments. Storage systems also play a vital role in IDR successful application. For example, excess energy generated during periods of low demand or high renewable energy production can be efficiently stored (electrical, thermal, chemical storage) and can be intelligently redirected and utilized for heating, cooling, or charging EVs, thus maximizing the utilization of clean energy resources. This flexible energy balancing approach helps alleviate congestion by reducing strain on the grid and optimizing resource utilization.

The subsections below analyze two significant cases where integrated flexibility harvesting has succeeded in ILECs through (i) a fleet of EVs (Section 7.3.2) and (ii) residential buildings (Section 7.3.3):

(1) The increasing adoption of EVs presents an opportunity for the local communities and the citizens to contribute to flexibility in MCESs. Through V2G technology, EVs can not only draw electricity from the grid but also supply excess energy back to the grid when required. The section presents a simulation study in which flexibility is utilized to maximize profits in the wholesale market. The available capacity of the grid is not taken into account in this use case. The results show that a significant profit increase can be achieved for the ILEC operator if the flexibility is procured.
(2) Integrated flexibility coming from assets of different carriers in residential buildings has a huge potential in the ILEC bottom-up model. In specific, by leveraging the flexibility of residential loads, storage, and DERs, the local MCES can manage peak demands, reduce strain on the grid, and optimize energy generation and consumption. We present a case study where a large variety of flexible assets in a building are modeled, by showing how these could be optimized together by a home-energy management system having multiple objective functions. The first objective is minimizing cost, which could consist of both the energy costs and simple price incentives for congestion management. The second objective is CO_2 reduction.

7.3.2 Optimal Flexibility Management of Electric Vehicles in ILEC Context

The penetration of EVs worldwide is growing fast, as the percentage of EVs will reach the 60% of vehicles sold globally by 2030, according to the International Energy Agency (IEA) [28]. By the end of 2021, over 16.5 million EVs were on the road. The global stock of EVs is expected to grow to around 350 million units by 2030. However, future growth will depend on efforts to diversify battery production and supply of key minerals to reduce the risk of supply shortages and price hikes. Electrification of the mobility sector brings many benefits, such as reducing greenhouse gas emissions as well as reducing dependence on fossil fuels [29]. It also plays an important role in facilitating the transition to more sustainable energy systems through smart management of EV charging and discharging processes [30]. In fact, mass adoption of EVs may lead to unexpected overloads for the distribution network. One of the

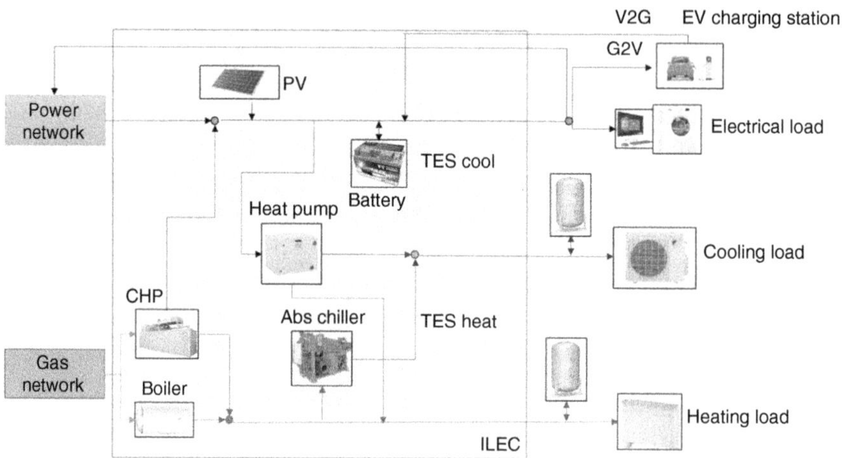

Figure 7.7 Overview of an ILEC interacting with EVs in both G2V and V2G modes. Source: Adapted from Ref. [32].

major challenges remains to foresee the time of EV charging process to guarantee a continuous supply of energy and install enough power capacity within each service area of the distribution grid [31]. Therefore, beyond the benefits, it is crucial to consider the global impact of this massive EV adoption on existing network infrastructure in terms of grid capacity and increasing demand of electricity due to the increasing growth of electricity consumption from EVs.

If properly managed in charging/discharging processes, EVs can become a flexibility resource for the energy system, thereby being transformed from passive loads to distributed storage, and this is even more true when the EVs are equipped with both G2V and V2G technology [32, 33]. In the context of ILECs and thus in the presence of multiple energy carriers, optimal management of EVs allows unlocking the flexibility potential coming from the interplay of electricity and mobility with the other energy carriers. In the following, a use case aiming at the optimal flexibility management of EVs in ILEC context will be discussed. An illustration of the use case is shown in Figure 7.7, where EVs can interact in both G2V and V2G modes with the ILEC, which provides electricity, heating, and cooling to a building cluster.

The EVs are supposed to be divided into clusters, each of them characterized by a specific number of EVs and specific values for (i) battery capacity, (ii) arrival and departure times at/from the charging stations, (iii) the state of charge (SOC) at the arrival time, and (iv) the SOC desired at the departure time. The main assumption is that the ILEC operator is responsible for management of EVs through the charging stations and can sell into the wholesale market the flexibility collected from distributed technologies and from EVs when operating in V2G mode. Therefore, the EV model can be expressed as follows [34]:

$$e_{v,t}^{\mathrm{EV}} = e_{v,t-1}^{\mathrm{EV}} + \eta_v^{\mathrm{EV,ch}} p_{v,t-1}^{\mathrm{EV,ch}} \Delta t - p_{v,t-1}^{\mathrm{EV,dis}} \Delta t, \quad \forall v \in \mathcal{V}, t \in \left(t_a, t_d\right] \tag{7.1}$$

$$e_{v,t}^{\mathrm{EV}} = 0, \quad \forall v \in \mathcal{V}, t \in (t_d, t_a) \tag{7.2}$$

$$e_{v,t}^{\mathrm{EV}} = e^{\mathrm{EV,ini}}, \quad \forall v \in \mathcal{V}, t = t_0, t_n \tag{7.3}$$

$$e_{v,t}^{\mathrm{EV}} = e_v^{\mathrm{EV,arr}}, \quad v \in \mathcal{V} \text{ if } t = t_a \tag{7.4}$$

$$e_{v,t}^{\mathrm{EV}} \geq e_v^{\mathrm{EV,dep}}, \quad v \in \mathcal{V} \text{ if } t = t_d \tag{7.5}$$

$$\underline{e_v}^{\mathrm{EV}} \leq e_{v,t}^{\mathrm{EV}} \leq \overline{e_v}^{\mathrm{EV}}, \quad \forall v \in \mathcal{V}, t \in \mathcal{T} \tag{7.6}$$

$$p_{v,t}^{\mathrm{EV,ch}} \leq r_v^{\mathrm{EV,ch}} u_{v,t}^{\mathrm{EV}}, \quad \forall v \in \mathcal{V}, t \in [t_a, t_d] \tag{7.7}$$

$$p_{v,t}^{\mathrm{EV,ch}} = 0, \quad \forall v \in \mathcal{V}, t \in (t_d, t_a) \tag{7.8}$$

$$0 \leq p_{v,t}^{\mathrm{EV,dis}} \leq r_v^{\mathrm{EV,dis}} \left(1 - u_{v,t}^{\mathrm{EV}}\right), \quad \forall v \in \mathcal{V}, t \in [t_a, t_d] \tag{7.9}$$

$$p_{v,t}^{\mathrm{EV,dis}} = 0, \quad \forall v \in \mathcal{V}, t \in (t_d, t_a) \tag{7.10}$$

$$p_{v,t}^{\mathrm{EV,u}} + p_{v,t}^{\mathrm{EV,s}} = \eta_v^{\mathrm{EV,dis}} p_{v,t}^{\mathrm{EV,dis}}, \quad \forall v \in \mathcal{V}, t \in \mathscr{T} \tag{7.11}$$

where the symbols are explained in the nomenclature list at the end of the chapter.

The objective function of the optimization problem formulated through mixed-integer linear programming (MILP) approach is to maximize the ILEC operator's profit and has several terms such as the revenue for selling the electricity provided by distribution technologies into the wholesale market, the revenue for selling flexibility collected from EVs in V2G mode into the wholesale market, and the cost for the electricity and natural gas bought from external networks to satisfy the ILEC's energy needs. The optimization is carried out on an hourly basis for a typical winter day of January. The technical data of EVs in the case study are in Table 7.2. Moreover, reference is made to the Italian context, with an ILEC constituted of office buildings in the city of Turin. The hourly prices of electricity and gas markets are obtained from the hourly data of the Italian electricity and gas market in January 2020 [33].

Optimization results show that the ILEC operator profit derived from selling EVs flexibility into the wholesale market in the winter day under study is equal to €194.95, whereas the bidding strategies related to flexibility collected from EVs are

Table 7.2 Technical data of EVs in the case study.

Cluster no.	Number of EVs	Battery capacity (kWh)	Parking duration		SOC	
			Arrival	Departure	Initial	Desired
1	30	50	8 : 00	18 : 00	0.2	0.8
2	20	50	7 : 00	17 : 00	0.2	0.8
3	10	80	9 : 00	19 : 00	0.2	0.8

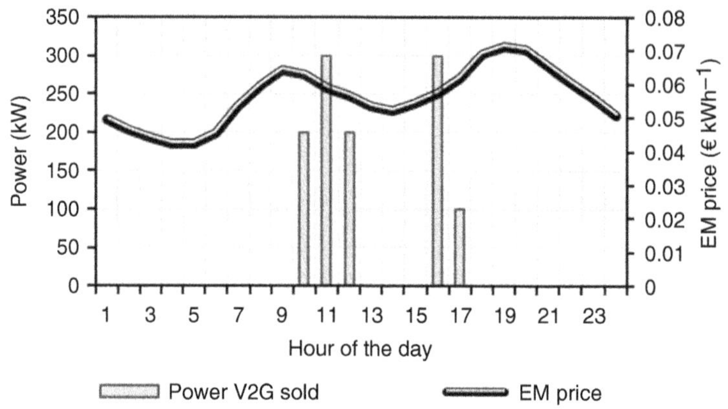

Figure 7.8 Bidding strategies related to flexibility collected from EVs into wholesale market.

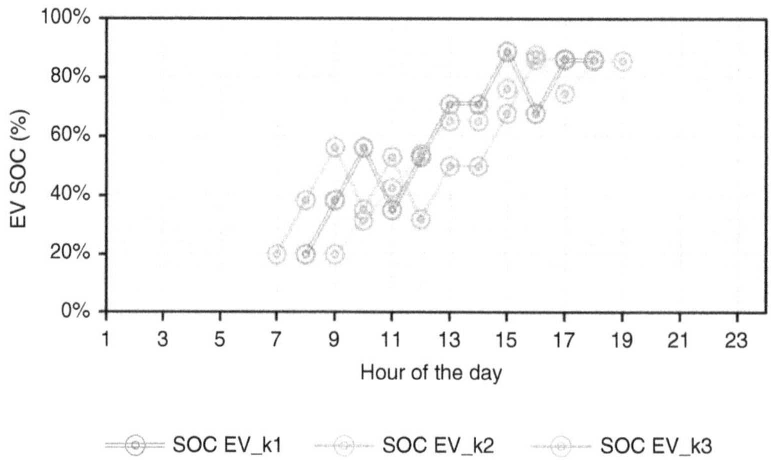

Figure 7.9 Optimized strategies for EVs management during the parking period at charging station.

shown in Figure 7.8. It can be observed that power discharged by EVs in V2G mode is all sold into the market in correspondence with high electricity market prices, allowing to maximize the revenue of the ILEC operator's profit. Figure 7.9 shows the optimized strategies for EVs management during parking at charging stations, where it can be observed that the charging and discharging strategies change for the EVs in different clusters, but all EV owners are satisfied in terms of desired SOC at departure time from the charging station, which is at least equal to 80%.

Simulation results obtained in the absence of EVs show that the ILEC's operator profit decreases by about 13% for the day under consideration, since there is no revenue for selling flexibility from EVs into the market and because the EVs cannot be used for self-consumption in the ILEC when operating in V2G mode, meaning that in the absence of interaction with e-mobility sector, external resources need to be used with additional costs.

7.3.3 Optimal Home (Building) Energy Management System

The emergence of ILECs has opened up opportunities for households to engage in two-way energy transactions with electricity and district heating providers. By selling excess energy generated from renewable energy sources like PVs and solar thermal (ST) systems, households can reduce costs while network operators can incentivize flexibility to ensure a safe operation of the network, e.g. through the congestion management [27, 35]. To achieve these objectives, an effective HEMS is required to schedule domestic appliances in real time. The design of the HEMS depends on the technology of the appliances, which can involve multiple carriers such as electricity, natural gas, district heating, and cooling [35].

The integration of different carriers can be facilitated by technologies like HPs, which interconnect power and heating/cooling networks, as well as by hybrid energy storage systems like battery energy storage system (BESS) and thermal storage system (TSS). This interplay, along with the participation of EVs in V2G or G2V modes, enables households to achieve higher flexibility in response to dynamic energy demand and price fluctuations, with the goal of congestion management from the network side, ultimately leading to varying levels of self-sufficiency from the household side [36]. The schematic diagram of such a smart building or home is outlined in Figure 7.10.

An effective approach to designing HEMS is the adoption of MILP, which models household appliance characteristics to achieve optimal scheduling throughout the day to minimize the cost. The cost is here taken to be a difference in the imported and exported amount of energy times some price. This price can include both energy and grid components for congestion management. In this section, we will assume that the incentives for congestion management come in the form of a

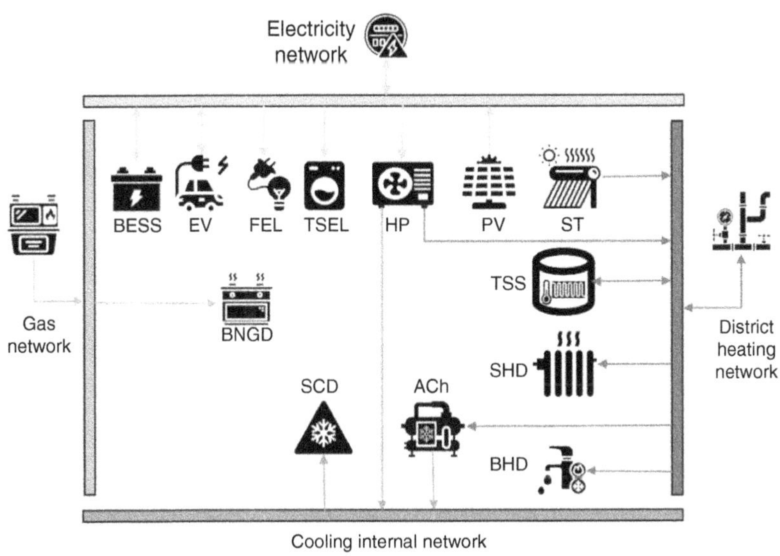

Figure 7.10 Technology schematic of multi-carrier home.

simple time-dependent energy price, but Section 7.4 will discuss other measures for congestion management in more detail. Apart from the cost, we consider the CO_2 emissions as a second objective in the HEMS' MILP model by formulating it as a multiobjective optimization problem (MOOP) [34, 36]. The mathematical formulation of the HEMS technologies is provided in the following sections.

7.3.3.1 Mathematical Model

(a) **PV and ST:** The energy production by the PV and ST is dependent on the area of the modules, their efficiency, and the solar irradiance during the day. Therefore, the following equations can be developed for them:

$$p_t^{\mathrm{PV}} = A^{\mathrm{PV}} I_t \eta^{\mathrm{PV}}, \quad \forall t \in \mathcal{T} \tag{7.12}$$

$$p_t^{\mathrm{PV}} = p_t^{\mathrm{PV,u}} + p_t^{\mathrm{PV,s}}, \quad \forall t \in \mathcal{T} \tag{7.13}$$

$$h_t^{\mathrm{ST}} = A^{\mathrm{ST}} I_t \eta^{\mathrm{ST}}, \quad \forall t \in \mathcal{T} \tag{7.14}$$

$$h_t^{\mathrm{ST}} = h_t^{\mathrm{ST,u}} + h_t^{\mathrm{ST,s}}, \quad \forall t \in \mathcal{T} \tag{7.15}$$

(b) **BESS:** One of the key technologies for the HEMS to provide flexibility is BESS which its energy state at each time step is dependent on the power charged or discharged in the previous time step. The important thing to note here is that they cannot be charged and discharged at the same time and this is expressed in the following constraints [34]:

$$e_t^{\mathrm{BESS}} = e_{t-1}^{\mathrm{BESS}} + \eta^{\mathrm{BESS,ch}} p_{t-1}^{\mathrm{BESS,ch}} \Delta t - p_{t-1}^{\mathrm{BESS,dis}} \Delta t, \quad \forall t \in \mathcal{T} - \{t_0\} \tag{7.16}$$

$$e_t^{\mathrm{BESS}} = e^{\mathrm{BESS,ini}}, \quad \text{if } t = t_0 \tag{7.17}$$

$$\underline{e^{\mathrm{BESS}}} \le e_t^{\mathrm{BESS}} \le \overline{e^{\mathrm{BESS}}}, \quad \forall t \in \mathcal{T} \tag{7.18}$$

$$0 \le p_t^{\mathrm{BESS,ch}} \le r^{\mathrm{BESS,ch}} u_t^{\mathrm{BESS}}, \quad \forall t \in \mathcal{T} \tag{7.19}$$

$$0 \le p_t^{\mathrm{BESS,dis}} \le r^{\mathrm{BESS,dis}} \left(1 - u_t^{\mathrm{BESS}}\right), \quad \forall t \in \mathcal{T} \tag{7.20}$$

$$p_t^{\mathrm{BESS,u}} + p_t^{\mathrm{BESS,s}} = \eta^{\mathrm{BESS,dis}} p_t^{\mathrm{BESS,dis}}, \quad \forall t \in \mathcal{T} \tag{7.21}$$

(c) **EV:** Genuinely, the EV is similar to the BESS in a sense that it can be charged and discharged. However, the main difference is that at some hours of the day it is not available. Therefore, the mathematical model for a single EV can be derived from the model expressed in Section 7.3.2.

(d) **FEL:** One of the key performance indexes for the HEMS is curtailing the loads for flexibility provision. It is evident that there is a capacity for power in the curtailment or extension of the FEL. Also, the flexibility is scaled with respect to the scheduled base load. To this end, the FEL can be modeled as follows [34]:

$$e_t^{\mathrm{FEL}} = e_{t-1}^{\mathrm{FEL}} + \Delta t \left(p_{t-1}^{\mathrm{FEL}*} - p_{t-1}^{\mathrm{FEL}}\right), \quad \forall t \in \mathcal{T} - \{t_0\} \tag{7.22}$$

$$-\overline{e^{\text{FEL}}} \leq e_t^{\text{FEL}} \leq \overline{e^{\text{FEL}}}, \quad \forall t \in \mathcal{T} \tag{7.23}$$

$$\kappa^- p_t^{\text{FEL}*} \leq p_t^{\text{FEL}} \leq \kappa^+ p_t^{\text{FEL}*}, \quad \forall t \in \mathcal{T} \tag{7.24}$$

(e) **Time-shiftable electric load (TSEL):** There is another type of load that cannot be curtailed or extended, but it can be shifted in time to provide flexibility for the building and upstream grid. Such a load can be modeled as [36]

$$p_{d,t}^{\text{TSEL}} = \sum_{p \in \mathcal{P}} u_{d,p,t}^{\text{ph}} p_{d,p}^{\text{ph}}, \quad \forall d \in \mathcal{D}, \text{t} \in \mathcal{T} \tag{7.25}$$

$$\sum_{p \in \mathcal{P}} u_{d,p,t}^{\text{ph}} \leq 1, \quad \forall d \in \mathcal{D}, t \in \mathcal{T} \tag{7.26}$$

$$y_{d,p,t}^{\text{ph}} = z_{d,p,\left(t+T_{d,p}^{\text{dur}}\right)}^{\text{ph}}, \quad \forall d \in \mathcal{D}, p \in \mathcal{P}, t \in \mathcal{T} \tag{7.27}$$

$$y_{d,p,t}^{\text{ph}} - z_{d,p,t}^{\text{ph}} = u_{d,p,t}^{\text{ph}} - u_{d,p,t-1}^{\text{ph}}, \quad \forall d \in \mathcal{D}, p \in \mathcal{P}, t \in \mathcal{T} - \{t_0\} \tag{7.28}$$

$$z_{d,p,t}^{\text{ph}} = y_{d,p+1,t}^{\text{ph}}, \quad \forall d \in \mathcal{D}, p \in \mathcal{P} - \{p_n\}, t \in \mathcal{T} \tag{7.29}$$

$$\sum_{t \in \mathcal{T}} y_{d,p,t}^{\text{ph}} = N_m, \quad \forall d \in \mathcal{D}, p \in \mathcal{P} \tag{7.30}$$

(f) **HP:** One of the most efficient energy conversion technologies on the market is reversible HP. Their COP is usually higher than one which makes them a suitable alternative for conventional technologies such as air conditioners or natural gas boilers. The corresponding mathematical modeling for this technology that can operate either in cooling mode or in heating mode is expressed in the following equations [37]:

$$h_t^{\text{HP}} = \text{COP}^{\text{HP,h}} p_t^{\text{HP,h}}, \quad \forall t \in \mathcal{T} \tag{7.31}$$

$$0 \leq p_t^{\text{HP,h}} \leq \overline{p^{\text{HP,h}}} u_t^{\text{HP}}, \quad \forall t \in \mathcal{T} \tag{7.32}$$

$$h_t^{\text{HP}} = h_t^{\text{HP,u}} + h_t^{\text{HP,s}}, \quad \forall t \in \mathcal{T} \tag{7.33}$$

$$c_t^{\text{HP}} = \text{COP}^{\text{HP,c}} p_t^{\text{HP,c}}, \quad \forall t \in \mathcal{T} \tag{7.34}$$

$$0 \leq p_t^{\text{HP,c}} \leq \overline{p^{\text{HP,c}}} \left(1 - u_t^{\text{HP}}\right), \quad \forall t \in \mathcal{T} \tag{7.35}$$

(g) **TSS:** The TSS is also similar to the BESS and it can be modeled as follows [38]:

$$e_t^{\text{TSS}} = e_{t-1}^{\text{TSS}} + \eta^{\text{TSS,ch}} h_{t-1}^{\text{TSS,ch}} \Delta t - h_{t-1}^{\text{TSS,dis}} \Delta t, \quad \forall t \in \mathcal{T} - \{t_0\} \tag{7.36}$$

$$e_t^{\text{TSS}} = e^{\text{TSS,ini}}, \quad \forall t = t_0, t_n \tag{7.37}$$

$$0 \leq h_t^{\text{TSS,ch}} \leq r^{\text{TSS,ch}} u_t^{\text{TSS}}, \quad \forall t \in \mathcal{T} \tag{7.38}$$

$$0 \leq h_t^{\text{TSS,dis}} \leq r^{\text{TSS,dis}} \left(1 - u_t^{\text{TSS}}\right), \quad \forall t \in \mathcal{T} \tag{7.39}$$

$$\underline{e^{\mathrm{TSS}}} \leq e_t^{\mathrm{TSS}} \leq \overline{e^{\mathrm{TSS}}}, \quad \forall t \in \mathcal{T} \tag{7.40}$$

$$h_t^{\mathrm{TSS,u}} + h_t^{\mathrm{TSS,s}} = \eta^{\mathrm{TSS,dis}} h_t^{\mathrm{TSS,dis}}, \quad \forall t \in \mathcal{T} \tag{7.41}$$

(h) **Space heating/cooling demand (SHD/SCD):** The space cooling/heating system of the building can be developed by adopting the following thermodynamic equation [36]:

$$T_t^{\mathrm{in}} = \left(1 - \frac{\Delta t}{R^{\mathrm{in}} C^{\mathrm{in}}}\right) T_{t-1}^{\mathrm{in}} + \left(\frac{T_{t-1}^{\mathrm{out}}}{R^{\mathrm{in}} C^{\mathrm{in}}} + \frac{h_{t-1}^{\mathrm{shd}} / \left(-c_{t-1}^{\mathrm{SCD}}\right)}{C^{\mathrm{in}}}\right) \Delta t, \quad \forall t \in \mathcal{T} - \{t_0\} \tag{7.42}$$

$$T_t^{\mathrm{in}} = T_{\mathrm{ini}}^{\mathrm{in}}, \quad \forall t_0 \tag{7.43}$$

$$\underline{T^{\mathrm{in,set}}} \leq T_t^{\mathrm{in}} \leq \overline{T^{\mathrm{in,set}}}, \quad \forall t \in \mathcal{T} \tag{7.44}$$

(i) **Absorption chiller (ACh):** In addition to the reversible HP, the absorption chillers can run on the heating energy and produce cooling energy [39]

$$c_t^{\mathrm{ACh}} = \mathrm{COP}^{\mathrm{ACh}} h_t^{\mathrm{ACh}}, \quad \forall t \in \mathcal{T} \tag{7.45}$$

$$0 \leq h_t^{\mathrm{ACh}} \leq \overline{h^{\mathrm{ACh}}}, \quad \forall t \in \mathcal{T} \tag{7.46}$$

(j) **Energy balance equations:** The energy balance for the electricity, heating, and cooling is expressed in the following. It is important to note that the HEMS cannot import and export energy (electricity and heating) at the same time.

$$\begin{aligned} p_t^{\mathrm{BESS,ch}} &+ p_t^{\mathrm{EV,ch}} + p_t^{\mathrm{FEL}} + \sum_{d \in D} p_{d,t}^{\mathrm{TSEL}} + p_t^{\mathrm{HP,h}} + p_t^{\mathrm{HP,c}} - p_t^{\mathrm{PV,u}} \\ &- p_t^{\mathrm{BESS,u}} - p_t^{\mathrm{EV,u}} = p_t^{\mathrm{imp}}, \quad \forall t \in \mathcal{T} \end{aligned} \tag{7.47}$$

$$p_t^{\mathrm{PV,s}} + p_t^{\mathrm{BESS,s}} + p_t^{\mathrm{EV,s}} = p_t^{\mathrm{exp}}, \quad \forall t \in \mathcal{T} \tag{7.48}$$

$$0 \leq p_t^{\mathrm{imp}} \leq \overline{p_t^{\mathrm{imp}}} u_t^p, \quad \forall t \in \mathcal{T} \tag{7.49}$$

$$0 \leq p_t^{\mathrm{exp}} \leq \overline{p_t^{\mathrm{exp}}} \left(1 - u_t^p\right), \quad \forall t \in \mathcal{T} \tag{7.50}$$

$$h_t^{\mathrm{TSS,ch}} + h_t^{\mathrm{ACh}} + h_t^{\mathrm{shd}} - h_t^{\mathrm{HP,u}} - h_t^{\mathrm{TSS,u}} - h_t^{\mathrm{ST,u}} = h_t^{\mathrm{imp}}, \quad \forall t \in \mathcal{T} \tag{7.51}$$

$$h_t^{\mathrm{HP,s}} + h_t^{\mathrm{TSS,s}} + h_t^{\mathrm{ST,s}} = h_t^{\mathrm{exp}}, \quad \forall t \in \mathcal{T} \tag{7.52}$$

$$0 \leq h_t^{\mathrm{imp}} \leq \overline{h_t^{\mathrm{imp}}} u_t^{\mathrm{h}}, \quad \forall t \in \mathcal{T} \tag{7.53}$$

$$0 \leq h_t^{\mathrm{exp}} \leq \overline{h_t^{\mathrm{exp}}} \left(1 - u_t^{\mathrm{h}}\right), \quad \forall t \in \mathcal{T} \tag{7.54}$$

$$c_t^{\mathrm{HP}} + c_t^{\mathrm{ACh}} = c_t^{\mathrm{SCD}}, \quad \forall t \in \mathcal{T} \tag{7.55}$$

(k) **Objective functions:** The primary objective for the HEMS is to minimize the daily cost of energy and account for the minimization of the environmental impact by means of CO_2 reduction.

$$\min \text{cost} = \sum_{t \in \mathcal{T}} \left\{ \lambda_t^{p,\text{imp}} p_t^{\text{imp}} - \lambda_t^{p,\text{exp}} p_t^{\text{exp}} + \lambda_t^{h,\text{imp}} h_t^{\text{imp}} - \lambda_t^{h,\text{exp}} h_t^{\text{exp}} \right\} \Delta t \tag{7.56}$$

$$\min CO_2 = \sum_{t \in \mathcal{T}} \xi \left\{ p_t^{\text{imp}} + h_t^{\text{imp}} \right\} \Delta t \tag{7.57}$$

The given MOOP for the HEMS model can be solved by adopting augmented epsilon constraint method (AUGMECON) and the Technique for Order of Preference by Similarity to the Ideal Solution (TOPSIS) as multiattribute decision-making (MADM) methodology can be used to find the final solution from the Pareto frontier [40, 41].

7.3.3.2 Case Study and Results

In this section, the presented model of HEMS showcases its optimization results within a household for a single day. Additionally, Figure 7.11a displays the energy prices for the representative day. The technological data utilized in the model is obtained from Refs. [35, 40]. Furthermore, in the Netherlands, specific values are assigned to $\xi = 0.15$ ($kgCO_2$ kWh^{-1}) and $LHV_g = 8.792$ ($kWh\,m^{-3}$), as indicated in [35]. The model is developed using Python, and the Pyaugmecon package, in conjunction with Gurobi [42], is employed to solve the MOOP for 15 grid points, resulting in 15 distinct problems.

The AUGMECON-based Pareto optimal solution, displayed in Figure 7.11c, illustrates an inverse relationship between CO_2 emissions and cost. Transitioning from higher carbon emission solutions to lower ones incurs higher costs for the representative day. By applying equal weights to the CO_2 emissions and cost objective functions using MADM's TOPSIS method, the trade-off solution is identified at point E in the figure.

Figure 7.11d presents the FEL's behavior relative to the energy price and solar irradiance profile shown in Figure 7.11b. The analysis reveals that as the energy price decreases in the afternoon, the FEL significantly increases power consumption to optimize low-cost electricity and abundant PV energy generation. However, as evening approaches and energy prices rise, PV production ceases, leading the FEL to reduce power consumption. This dynamic response allows for enhanced flexibility in both the HEMS and power grids simultaneously.

The domestic energy profiles, namely electricity and heating, are illustrated in Figure 7.11e,f, respectively. During the early hours of the day, the BESS discharges to supply power to the FEL and the HP. The heat generated from the HP is utilized as a SHD to warm up the house during colder periods. The surplus heat produced by the HP is exported to the district heating network, while lower-cost electricity is imported between 4 and 8 a.m. Additionally, during this time frame, the BESS is discharged to its minimum level, and a significant portion of the imported electricity is utilized to charge the EV that departs at 8:30 a.m.

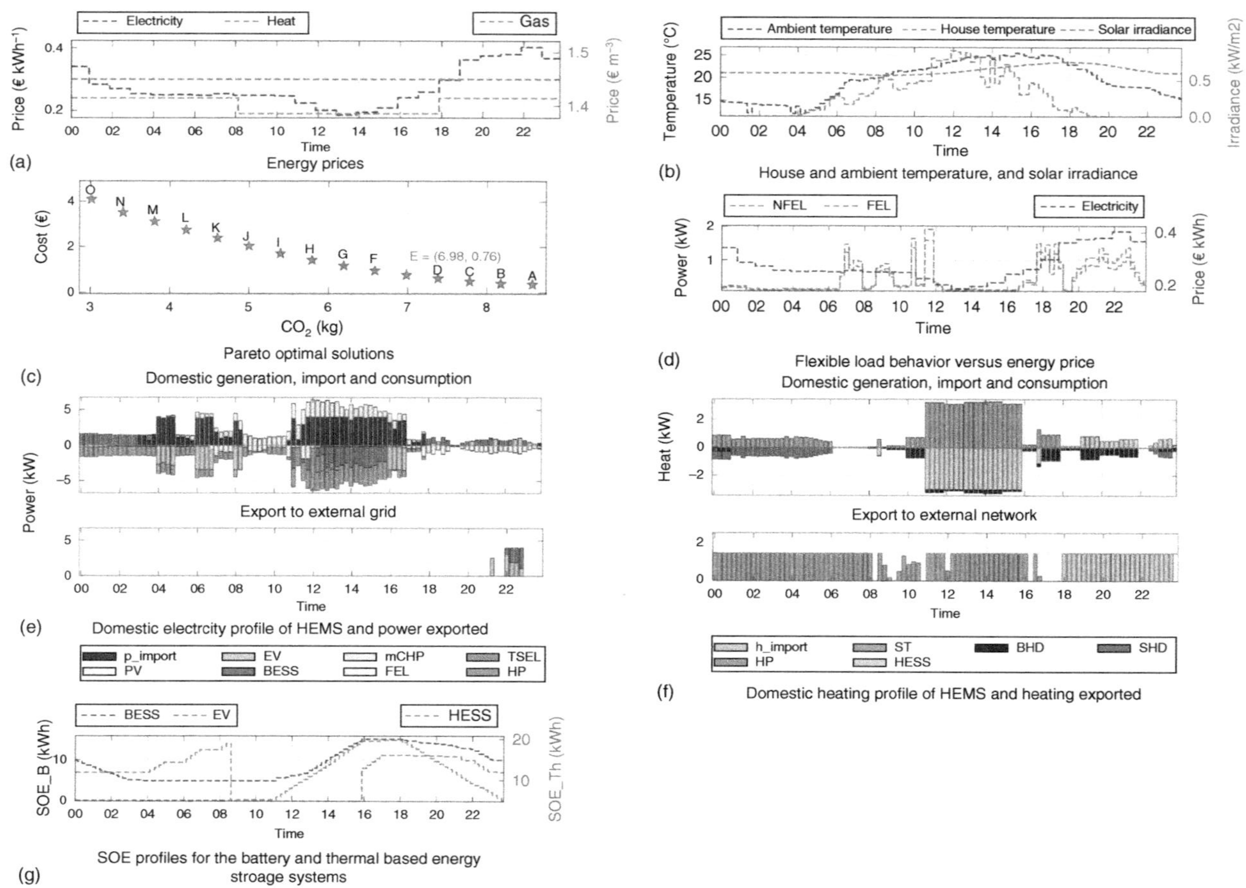

Figure 7.11 Results of the case study.

In the late morning, the PV starts contributing to domestic energy production, reaching its peak at 12 p.m. As the cost of electricity decreases in the early afternoon, a higher amount of electrical energy is imported and stored in the BESS, along with the energy generated by the PV. Simultaneously, the heat produced by the HP is stored in the HESS. Additionally, due to the favorable energy prices, the TSEL operates at noon. Progressing further in time, the EV arrives at 4 p.m., and considering the anticipated higher energy prices, it is immediately charged. Ultimately, in the evening, the HEMS supplies power to the FEL while attempting to offset costs by exporting excess stored energy to the heating and electricity networks.

In conclusion, the implemented HEMS offers enhanced flexibility for the building and energy network while considering the environmental impacts. By efficiently managing diverse energy sources and loads, the system optimizes resource utilization, cost savings, and energy efficiency. Besides, the network imposes dynamic prices to ensure the safe operation by means of possible congestion in the network and/or voltage violations. Leveraging the BESS, EV, HP, and PV, the HEMS balances electricity demand, charges the EV strategically, and utilizes surplus heat for heating. It also stores excess energy and exports it during peak periods, bolstering network stability. Overall, the HEMS provides optimal resource management, cost savings, reduced CO_2 emission, and increased flexibility for both the building and the network.

7.4 Instruments for Flexibility Activation for Congestion Management

The electricity system and the energy system, in general, are interconnected between different countries in most parts of the world. This is due not only to physical connections between the systems but also to coupled energy markets. Therefore, regulations regarding renewable energy targets, climate plans, and energy markets are often not just a national affair but also one that is affected by international politics and agreements. The instruments for flexibility activation for congestion management can therefore differ by region. This section will give a brief overview of developments in Europe and give some examples of instruments that are in development and already implemented in practice for flexibility activation for congestion management.

7.4.1 Developments in Europe

Within the EU, the different member states must adhere to the so-called Clean energy for all Europeans package (CEP), which was published in 2019. This collection of measures, regulations, and directives aims to accelerate the transition toward renewable energy sources and energy efficiency to achieve the EU's commitments for the Paris Agreement and decarbonize the EU's energy system according to the objectives of the European Green Deal. EU member states can adapt the directives and regulations of the CEP into their national regulations [43].

Part of the CEP focuses on the establishment of a modern design for the European electricity markets in order to accommodate a more flexible, market-based integration of a greater share of renewables. The directive on common rules for the internal market for electricity states that EU member states shall make it possible for DSOs to procure flexibility services for congestion management. It is further specified that congestion problems in the distribution networks shall be addressed with transparent, nondiscriminatory market-based solutions [44, 45]. Practical examples of market-based congestion management solutions are given in Section 7.4.3.

Under certain conditions, when market-based solutions are not sufficient, the system operator (SO) could resort to nonmarket-based redispatching of demand, generation, or storage. This last resort can only be used by SOs when market-based solutions are (i) not sufficient to resolve all congestion (even if all market-based resources are already used); (ii) when there is a lack of competition due to limited participants; or (iii) when the congestion is so predictable that market-based solutions could worsen the congestion problems by gaming [44].

Apart from market-based instruments, it is indicated that tariff instruments could also be used to incentivize efficient consumer and producer behavior. The network tariff is charged to network users for the service of using the network and having the electricity transported from the location of generation to the connection of the user. The main objective of network tariffs is to recoup the costs of the transmission and distribution of electricity [44, 45]. Although congestion problems could, in the short term, increase the costs of the electricity network and its operation, the main objective of the network tariff is not the mitigation or management of congestion. Therefore, the possibilities of using network tariff instruments for congestion management are limited. This, however, does not mean that they are ineffective in the prevention and mitigation of congestion problems. In principle, the network tariff sends price signals to the network users, which are able to respond to these price signals by adjusting their electricity demand or generation. The behavior of the network users can thus be incentivized through tariff instruments. It should be noted that this effect is limited due to the relation to cost reflectivity, which means that this incentive cannot grow indefinitely when congestion problems arise [46].

7.4.2 Examples of Market-Based Instruments in Practice

7.4.2.1 GOPACS

An operational example of congestion management in the form of market-based redispatch is GOPACS. GOPACS is a platform that can be used by grid operators in the Netherlands to reduce congestion in both the distribution and transmission grids. GOPACS is now connected to multiple energy trading platforms, including ETPA and EPEX SPOT Localflex, enabling market parties to contribute to congestion management through several trading platforms [47].

After the closing of the day-ahead (DA) market, the Dutch network operators determine if, where, and when congestion problems are expected in their networks. To resolve these expected problems, the GOPACS platform is used to request market parties to temporarily reduce their consumption or generation of electricity. Using

Table 7.3 Adjusted example of a GOPACS market message based on a market message from 10 February 2023.

Request for buy orders in *congestion location*
Request for sell orders in *all*

Date	10-02-2023							
Hours	08				09			
Quarters	00	15	30	45	00	15	30	45
Required in MW	101.8	101.8	101.8	101.8	−0.043	−0.043	−0.043	−0.043

Extra information/message:

Type: TRANSPORT PROBLEM, Grid operator: TSO, Compliance: Voluntary, *State: Closed*

Source: Adapted from Ref. [48].

a market message, as shown in Table 7.3, which contains information about the location of the expected congestion, the times, and the amount of requested flexibility is published on the GOPACS platform around 14 : 00 hours. Since this process is applied after the closing of the DA market, the redispatch actions will cause unbalance in the portfolios of the involved market parties. Therefore, every requested action within a congestion area is combined with an opposite order from a market party outside of the congestion areas, while checking that these actions do not cause congestion problems elsewhere in the system. After the publication of the buy and sell requests, the connected market parties have until 16 : 00 hours to place bids for the different redispatch actions, after which the market is closed, and potentially suitable bids are matched based on the merit order principle. If the different orders are matched, the SOs will pay the spread between the two orders, thereby solving the expected congestion problems.

7.4.2.2 Piclo Flex

Another example of a market-based flexibility solution is Piclo Flex. Piclo Flex is an autonomous marketplace that facilitates the procurement and operation of local flexibility services, connecting SOs and flexibility service providers (FSPs). On the platform, FSPs with flexible assets, such as batteries, fleets of EV, and demand response, can qualify to participate in the flexibility market. Within this qualification, apart from company qualifications, the flexible assets are evaluated on criteria including capacity and location. A SO can initiate a competition on the platform for the procurement of flexibility. This competition includes various specifications such as the structure of the competition, locational boundaries, the designated service windows (e.g. weekday mornings from 9 : 00 to 11 : 00 hours during the period from 1 April to 30 September), the power type, service requirements, and direction as well as the required voltage level for the asset connections. When qualified, the FSPs are able to submit bids into a competitive auction for both long-term and short-term flexibility services initiated by the SOs. These bids must include information about the capacity, expressed in MW, and the maximum run time specifying the duration the assets can deliver the submitted capacity. Depending on the structure of the competition, additional details like availability prices, utilization prices, service fees, and fixed prices can be specified in the bids. The SO will review the bids submitted by the qualified FSPs and based on considerations including price make decisions to accept or reject bids and enter into Flexibility Services Agreements with the accepted FSPs. The platform also facilitates the operation and the settlement of the flexibility services; it allows the SOs to monitor the availability of the assets until the time of dispatch and facilitates the exchange of dispatch instructions and scheduled messages. Additionally, the SOs are notified of any changes and potential impacts on the contracts, and using measurements, the performance of the FSP services can be evaluated and settled [49].

Since the launch of the flexibility marketplace in the United Kingdom in 2018, the platform and its application have grown over the years, introducing more flexibility auctions in the United Kingdom, as well as Lithuania, Italy, Portugal, and the United States (state of New York), as well as the introduction of Piclo Exchange, which enables participants to manage their contractual obligations using secondary trading in the capacity market.

7.4.3 Examples of Tariff Instruments in Practice

7.4.3.1 Capacity Tariff

To ensure that the electricity network will not be overloaded, a new capacity tariff component has been added to the network tariffs in Belgium. This capacity tariff aims to incentivize prosumers to avoid high peaks and spread out their load. The base rate for the distribution network is thus composed of two main components: the amount of active energy consumed, expressed in kWh, and the consumed average monthly peak, expressed in kW. This average monthly peak is determined based on the rolling average of the 12 most recent monthly peaks, which is the maximum consumed quarter-hour power of each billing month. Based on network sizing and simultaneity in low-voltage network consumption, a minimum value of 2.50 kW is maintained [50].

7.4.3.2 Bandwidth Tariff

Within the Netherlands, among other different potential new network tariff options, the Dutch DSOs are currently investigating the so-called bandwidth (BW) network tariff. This BW network tariff works with a capacity subscription model, in which consumers are subscribed to a certain predetermined capacity BW. Apart from the different BW capacity limits and associated costs, various forms of this network tariff could be implemented, including a static and a dynamic version. For the static BW network tariff, consumers can use their electricity connection and consume or feed in electricity without limitations or additional costs within the predefined capacity BW. However, whenever more electricity is used or produced, resulting in a peak that exceeds the predetermined BW capacity limit, additional charges will be applied for the amount of energy used above the BW capacity limit, as indicated in Figure 7.12. For the dynamic implementation, optional components that are time- and/or location-dependent are added to the network tariff, where network users are

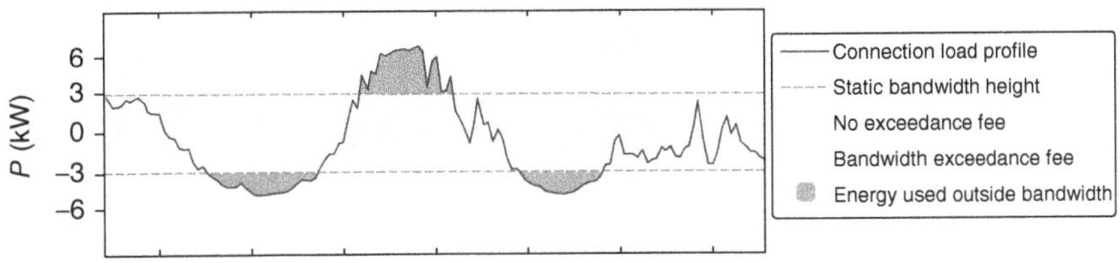

Figure 7.12 An illustration of the bandwidth tariff model.

informed about the specific conditions of the BW for all moments during a day well in advance. Here again, different options are possible, where the BW is, for example, only operational when network or congestion problems are expected, as is implemented in the simulation case study in Section 7.4.4, or the BW capacity limit is lowered or increased depending on the loading of the network. Additionally, this BW approach could also be implemented for aggregated connections or ILECS, where the BW is set for the collective, allowing for peak use of individual users within the ILEC, as long as this does not exceed the BW capacity limit of the aggregated connections. In the following section, a simulation case study will be provided comparing a static and dynamic BW network tariff, to illustrate the possible effects and merits of the implementation of the different options.

7.4.4 Simulation Case Study

In this case study, we consider a DSO that contracts an ILEC behind a MV–LV transformer with either a static or a dynamic BW tariff [51]. This implies that connections part of the ILEC pay some flat grid tariff if their consumption stays within some power BW of $[-P^{BW}, P^{BW}]$ for some power level $P^{BW} \geq 0$. If some amount of energy is consumed/produced outside of this range, some extra amount of π_t^{BW} euros per volume of energy is charged. This price is indexed by some time index $t = 0, 1, \ldots$ as for the dynamic BW tariff, the penalty will be varied between $\pi_t^{BW} = 0$ if no congestion is expected at t the day before delivery, and some amount π^{BW} that is defined in the contract. For the static BW model, it holds that $\pi_t^{BW} = \pi^{BW}$ for all $t = 0, 1, \ldots$. Let us assume that our time resolution Δt is 15 minutes, as this is a typical resolution for congestion management applications.

An overview of the agents and their interactions in the simulation is presented in Figure 7.13. The physical system consists of the LV network, households being part of the ILEC and having a connection to the network, and the local weather. The households all contain an inflexible base load, and possibly a selection of flexible PVs, HPs, and BESSs. Each of the connections will also have an HEMS that will optimize the dispatch of the connected flexible assets based on the BW tariff information from a DSO, DA prices from the DA market, and weather forecasts.

For the dynamic BW tariff, the DSO's knowledge about the future state of the network needs to be modeled. We assume that the ILEC sends a DA load prognosis to the DSO once a day at 12 p.m., right after the DA market closes. Based on these prognoses, the DSO performs a load flow calculation to determine possible cable/transformer overloading the next day. In our simulation, this is done using the

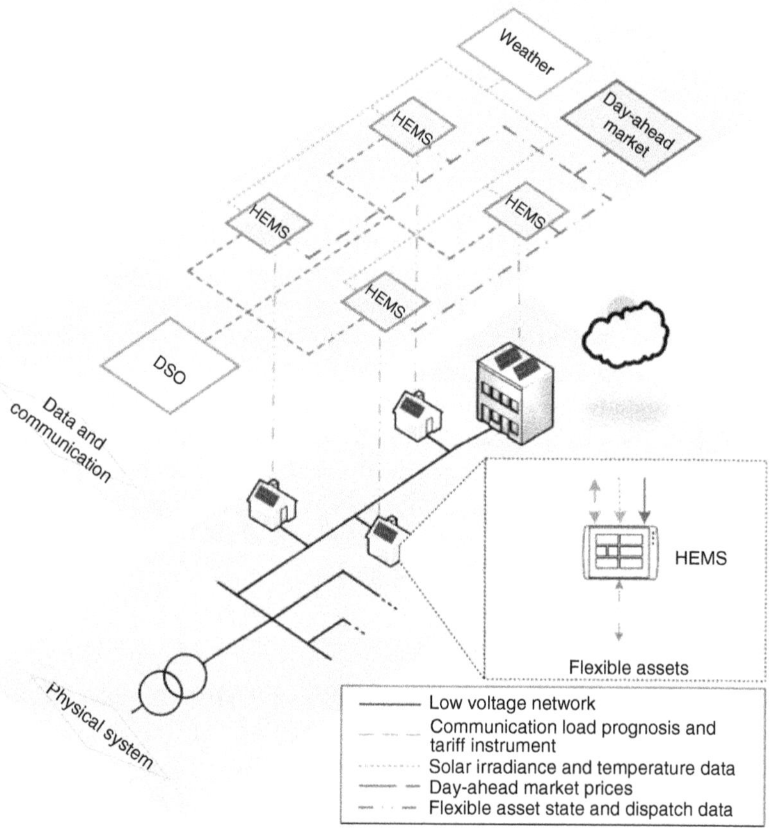

Figure 7.13 Overview of the agents and interactions in the simulation case. Source: Ref. [51] / Eindhoven University of Technology.

openDSS software [52]. Whenever the prognosis load flow calculation from the DSO indicates that overloading problems are expected at certain time steps, the nonzero BW penalty is communicated to the ILEC and to the HEMSs afterward at those time steps. For convenience, we consider the ILEC to have perfect knowledge of the future DA prices and weather conditions. We furthermore assume that the HEMSs use a perfect model for the asset behavior. The horizon of the rolling optimization is chosen to be 12 hours. Only for the prognosis of the DA schedule of the ILEC, the horizon is 36 hours. Since we use similar asset models as in Section 7.3.3, the resulting control problem is very similar to the problem presented there. The major difference is the way the BW tariff enters the objective function as an extra penalty term Π_t^{BW} for every t. The term Π_t^{BW} is then characterized by the following equations:

$$\Pi_t^{\mathrm{BW}} \geq \pi_t^{\mathrm{BW}} \left(- \left[E_t^{\mathrm{buy}} - E_t^{\mathrm{sell}} \right] - P^{\mathrm{BW}} \Delta t \right) \tag{7.58}$$

$$\Pi_t^{\mathrm{BW}} \geq 0 \tag{7.59}$$

$$\Pi_t^{\mathrm{BW}} \geq \pi_t^{\mathrm{BW}} \left(\left[E_t^{\mathrm{buy}} - E_t^{\mathrm{sell}} \right] - P^{\mathrm{BW}} \Delta t \right) \tag{7.60}$$

where E_t^{buy} and E_t^{sell} denote the net energy bought and sold during that PTU. This way, the penalty punishes the exceedance of P^{BW} in either production or consumption. In the case the BW is not exceeded, the minimization of the penalty reduces the penalty to 0.

We will consider a Dutch low-voltage network with 37 houses, of which 22 houses have a PV installation, half of those have a battery, and 18 of the houses have a full-electric HP that is used for space heating. For the dynamic BW tariff, the results are going if P^{BW} is chosen to be 3.0 kW.

Figure 7.14 presents the result of the simulation. It can be seen that for both the dynamic and static BW, when a higher price is charged for the amount of power above the BW limit, the dispatch of power during congestion moments is minimized below the BW limit. For lower prices, the dispatch will be lowered, but not fully minimized depending on the asset state and market prices. Furthermore, it can be seen that a dynamic BW network tariff, as applied in this study, is not always able to ensure that congestion problems are resolved, even if the price for exceeding the BW is high enough to incentivize the lowering of the dispatch power. Due to a rebound effect in the dispatch of flexible assets and the influence of market prices, it can also

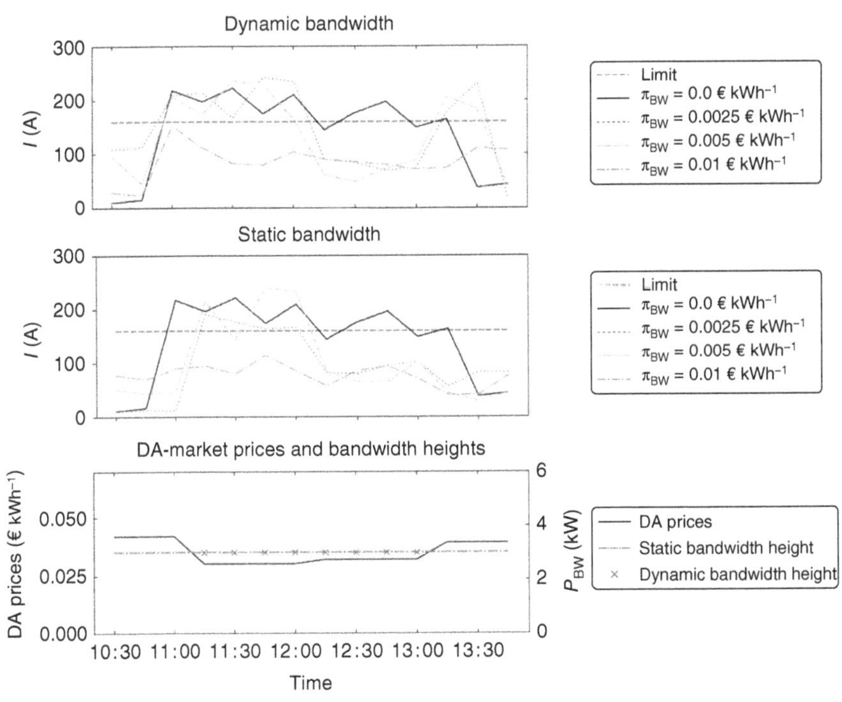

Figure 7.14 The effect of different network tariff costs for the dynamic and static bandwidth on the overloading of one of the main cables of the study network. The bandwidth height when active and corresponding DA prices for the 9th of September 2020 are also presented.

cause a shift in the moment of congestion to moments where the dynamic BW limit is not active. This is not the case for the static BW. It can therefore be stated that although the static BW model is less directly responsive to what is happening in the network at given moments, since it is also present when no congestion problems are expected, it provides more stability and certainty from a DSO point of view compared to the dynamic BW model, which is communicated to network users well in advance.

7.5 Challenges and Outlook

This chapter so far has demonstrated that in theory integrated flexibility from ILECs can be used effectively to mitigate congestion in distribution grids and that with the new Clean Energy Package there is a first legal basis to apply this measure in practice in Europe. However, there are still significant challenges to overcome before applying this solution on a large scale. There are several studies that present existing barriers for activating (integrated) flexibility in the distribution grid in general and for congestion management in particular [53–58]. These barriers are usually country specific, but general themes can be identified. This section presents these barriers in the categories of market barriers, regulatory barriers, technical barriers, and barriers related to customers.

7.5.1 Market Barriers

ILECs currently face some market barriers that hinder their ability to provide flexibility for congestion management. First, the monetary value attributed to flexibility is currently low, limiting the incentives for participation in service provision. The establishment of the necessary infrastructure and acquisition of requisite knowledge entail upfront costs that must be recouped over time. This financial limitation undermines the attractiveness of engaging in providing flexibility services, unless value stacking of multiple revenue streams can be tapped into. However, the process of value stacking is not optimally facilitated for ILECs, particularly in terms of accessing, e.g. intraday markets. The complexity and lack of streamlined procedures in participating in these markets present obstacles that impede ILECs' efficient engagement in value-stacking strategies.

7.5.2 Regulatory Barriers

While the potential of integrated flexibility is widely recognized, DSOs often face limitations imposed by regulatory frameworks. Many DSOs operate within heavily regulated environments, which restrict their ability to fully capitalize on the benefits of integrated flexibility. In Europe, for instance, countries are currently in the process of transposing the European directive into their national legislation and grid codes. As a result, legal constraints prevent many DSOs from introducing new tariff structures or contract types aimed at mitigating congestion. However, it is expected that governments will make significant strides in the near future to address these barriers and provide DSOs with a greater regulatory flexibility.

Privacy regulations related to metering data also present regulatory barriers in some countries. These regulations impose requirements for flexibility activation and settlement, which can complicate the integration of flexibility services. Additionally, existing regulations, such as netting schemes, can undermine the financial incentives for activating integrated flexibility, thus limiting the development of such services in certain countries.

Overcoming these regulatory barriers requires a proactive approach from policymakers and regulatory authorities. Governments need to facilitate the introduction of flexible regulatory frameworks that empower DSOs to leverage integrated flexibility for congestion management effectively. Addressing privacy concerns and streamlining regulations surrounding data access and sharing is also crucial. By removing these barriers, governments can foster an environment conducive to the widespread deployment and utilization of integrated flexibility for congestion management.

7.5.3 Technological Barriers

In the realm of integrated flexibility for congestion management, several technological barriers pose challenges that need to be addressed to enable its effective implementation. One such barrier is the lack of standardized communication protocols. The interoperability of various devices and systems within integrated energy communities requires well-defined protocols to ensure seamless communication and data exchange. Additionally, the controllability of distributed energy resources becomes crucial for effective congestion management. The ability to monitor and control them in real time is essential to optimize their participation and flexibility in alleviating grid congestion. Another technological barrier lies in developing accurate baseline methods that establish a reference point for measuring the flexibility and deviations in energy consumption or production. Ensuring the cybersecurity of integrated flexibility systems is paramount, as the increasing digitalization and connectivity expose the energy grid to potential cyber threats and attacks. Lastly, the widespread deployment of smart meters plays a vital role in collecting accurate and timely data for real-time monitoring, control, and billing purposes. Overcoming these technological barriers requires advancements in communication protocols, robust controllability mechanisms, reliable baseline methods, stringent cybersecurity measures, and the integration of smart meters into the energy infrastructure.

7.5.4 Customer-Related Barriers

In the realm of integrated flexibility for congestion management, several customer-related barriers must be addressed to encourage the adoption and utilization of flexible assets. One significant barrier is the initial investment cost of acquiring and installing flexible assets for customers. The upfront expenses associated with implementing technologies like energy storage systems or demand response capabilities can be a deterrent for customers, especially for those with limited financial resources. Additionally, customers' willingness to offer flexibility becomes crucial. Some customers may be unaware of the potential benefits or may have concerns about the impact on their daily routines or comfort. Perceived inconvenience due to flexibility activation can be a barrier, as customers may fear disruptions or inconveniences associated with adjusting their energy consumption patterns. Moreover, customers may be hesitant to allow an external party to control their assets, concerned about privacy, data security, or loss of control over their energy usage. Lastly, customers might have difficulty identifying which flexibility services align best with their interests and needs, leading to uncertainty about the potential benefits and suitability of participation. In the context of ILECs, the ILEC operator could play a significant role in reducing these barriers.

Despite these current challenges, it is clear that integrated flexibility will play an important role in congestion management in the distribution grids of the future. The European directives, national legislations, and grid codes provide grid operators with increasing possibilities to use integrated flexibility for congestion management. Furthermore, electricity markets are now opening up to smaller parties, which allow new business cases for monetizing flexibility in distribution systems. It is therefore expected that new services and software for flexibility portfolio management will develop. This can support ILECs to use the integrated flexibility within the community not only for self-sufficiency but also for supporting grid operators in keeping the electricity grid reliable, sustainable, available, and affordable.

Nomenclature

Abbreviations

Ach	absorption chiller
AUGMECON	augmented epsilon constraint method
BESS	battery energy storage system
BW	bandwidth
CEP	Clean Energy Package
DSO	distribution system operator
EU	European Union
EVs	electric vehicles
FEL	flexible electric load
FSP	flexibility service provider
G2V	grid-to-vehicle
HEMS	home energy management system
HP	heat pump
HV	high voltage
IDR	integrated demand response
IEA	International Energy Agency
ILEC	integrated local energy community

ISO independent system operator
LV low voltage
MADM multiattribute decision-making
MCES multicarrier energy system
MILP mixed-integer linear program
MOOP multiobjective optimization problem
MV medium voltage
PV photovoltaics
RTO regional transmission organization
SCD space cooling demand
SHD space heating demand
SO system operator
SOC state of charge
ST solar thermal
TOPSIS Technique for Order of Preference by Similarity to the Ideal Solution
TSEL time-shiftable electric load
TSO transmission system operator
TSS thermal storage system
V2G vehicle-to-grid

Sets

$v \in \mathcal{V}$ set of electric vehicles
$d \in \mathcal{D}$ set of TSEL devices
$p \in \mathcal{P}$ set of operation phases for TSEL
$t \in \mathcal{T}$ set of time periods
$x \in \mathcal{X}$ set of technologies used in the model

Parameters

$\overline{h_t^{\mathrm{exp}}}/\overline{h_t^{\mathrm{imp}}}$ maximum importable/exportable heat (kW)
$\underline{e^{\mathrm{BESS}}}/\overline{e^{\mathrm{BESS}}}$ minimum/maximum energy capacity of BESS (kWh)
$\underline{e_v{}^{\mathrm{EV}}}/\overline{e_v{}^{\mathrm{EV}}}$ minimum/maximum energy capacity of EV (kWh)
$\overline{e^{\mathrm{FEL}}}$ energy capacity of FEL (kWh)
$\underline{e^{\mathrm{TSS}}}/\overline{e^{\mathrm{TSS}}}$ minimum/maximum energy capacity of TSS (kWh)
$\overline{p^{\mathrm{HP},h}}/\overline{p^{\mathrm{HP},c}}$ HP maximum heating/cooling generation capacity (kW)
$\overline{p_t^{\mathrm{imp}}}/\overline{p_t^{\mathrm{exp}}}$ maximum importable/exportable power (kW)
h_t^{ST} ST heating production (kW)
$\eta^{\mathrm{BESS,\,ch}}/\eta^{\mathrm{BESS,\,dis}}$ BESS efficiency
$\eta_v{}^{\mathrm{EV,\,ch}}/\eta_v{}^{\mathrm{EV,\,dis}}$ charging/discharging efficiency of EV
η^{PV} efficiency of PV

η^{ST}	efficiency of ST
$\eta^{\mathrm{TSS,\,dis}}/\eta^{\mathrm{TSS,\,ch}}$	TSS charging/discharging efficiency
κ^{-}/κ^{+}	minimum/maximum curtailing ratio of FEL with respect to base load
$\lambda_t^{h,\mathrm{imp}}/\lambda_t^{h,\mathrm{exp}}$	heating import/export price (€ kWh^{-1})
$\lambda_t^{p,\mathrm{imp}}/\lambda_t^{p,\mathrm{exp}}$	electricity import/export price (€ kWh^{-1})
Π_t^{BW}	bandwidth cost (€)
A^{PV}	area of PV system (m^2)
I_t	solar irradiance (kW m^{-2})
N_m	number of times that TESL can operate in a day
P^{BW}	bandwidth height (kW)
$R^{\mathrm{in}}C^{\mathrm{in}}$	equivalent thermal resistance capacitance of the house (h)
T_t^{out}	ambient temperature (°C)
$e^{\mathrm{BESS,\,ini}}$	initial energy state of BESS (kWh)
$e_v^{\mathrm{EV,\,arr}}$	energy state of EV in arrival (kWh)
$e_v^{\mathrm{EV,\,dep}}$	energy state of EV in departure (kWh)
$e_v^{\mathrm{EV,\,ini}}$	initial energy state of EV (kWh)
$e^{\mathrm{TSS,\,ini}}$	initial capacity of TSS (kWh)
$p_{d,p}^{\mathrm{ph}}$	phase power of TSEL (kW)
$p_t^{\mathrm{FEL}*}$	base load of FEL (kW)
p_t^{PV}	PV power production (kW)
$r^{\mathrm{BESS,\,ch}}/r^{\mathrm{BESS,\,dis}}$	charging/discharging rate of BESS (kW)
$r_v^{\mathrm{EV,\,ch}}/r_v^{\mathrm{EV,\,dis}}$	charging/discharging rate of EV (kW)
$r^{\mathrm{TSS,\,ch}}/r^{\mathrm{TSS,\,dis}}$	TSS charging/discharging capacity (kW)
π_t^{BW}	bandwidth price (€ kWh^{-1})
Δt	time-step length
ξ	carbon intensity (kgCO$_2$ kWh^{-1})
$\mathrm{COP}^{\mathrm{ACh}}$	ACh coefficient of performance
$\mathrm{COP}^{\mathrm{HP},\,h}/\mathrm{COP}^{\mathrm{HP},\,c}$	HP coefficient of performance

Variables

h_t^{ACh}	ACh heating consumption (kW)
$h_t^{\mathrm{HP,s}}/h_t^{\mathrm{HP,u}}$	HP heating sold/used (kW)
$h_t^{\mathrm{ST,s}}/h_t^{\mathrm{ST,u}}$	ST heating portion sold/used (kW)

$h_t^{\mathrm{TSS,ch}}/h_t^{\mathrm{TSS,dis}}$	TSS charging/discharging power (kW)
$h_t^{\mathrm{TSS,s}}h_t^{\mathrm{TSS,u}}$	TSS heating portion sold/used [kW]
$h_t^{\mathrm{imp}}/h_t^{\mathrm{exp}}$	heating import/export (kW)
E_t^{buy}	energy bought (kWh)
E_t^{sell}	energy sold (kWh)
T_t^{in}	indoor temperature (°C)
c_t^{ACh}	ACh cooling production (kW)
c_t^{HP}	HP cooling production (kW)
c_t^{SCD}	SCD cooling power (kW)
e_t^{BESS}	BESS stored energy (kWh)
$e_{v,t}^{\mathrm{EV}}$	EV stored energy (kWh)
e_t^{FEL}	FEL stored energy (kWh)
e_t^{TSS}	TSS stored energy (kWh)
$p_{d,t}^{\mathrm{TSEL}}$	TSEL power (kW)
$p_t^{\mathrm{BESS,ch}}/p_t^{\mathrm{BESS,dis}}$	BESS charging/discharging power (kW)
$p_t^{\mathrm{BESS,s}}/p_t^{\mathrm{BESS,u}}$	BESS power sold/used (kW)
$p_{v,t}^{\mathrm{EV,ch}}/p_{v,t}^{\mathrm{EV,dis}}$	EV charging/discharging power (kW)
$p_{v,t}^{\mathrm{EV,s}}/p_{v,t}^{\mathrm{EV,u}}$	EV power sold/used (kW)
p_t^{FEL}	FEL power (kW)
$p_t^{\mathrm{HP,c}}/p_t^{\mathrm{HP,h}}$	HP power used for cooling/heating production
$p_t^{\mathrm{PV,s}}/p_t^{\mathrm{PV,u}}$	PV power sold/used (kW)
$p_t^{\mathrm{imp}}/p_t^{\mathrm{exp}}$	power import/export (kW)
$u_{d,p,t}^{\mathrm{ph}}/y_{d,p,t}^{\mathrm{ph}}/z_{d,p,t}^{\mathrm{ph}}$	TSEL operational/start/end phase binary state
u_t^h/u_t^p	binary for heating and electricity carriers
$u_{v,t}^{\mathrm{EV}}$	EV binary
u_t^{HP}	HP binary
u_t^{TSS}	TSS binary

References

1 European Parliament (1997). Directive 96/92/EC of the European Parliament and of the Council of 19 December 1996 concerning common rules for the internal market in electricity. Official Journal L 027, Art. no. 31996L0092. https://eur-lex.europa.eu/LexUriServ/LexUriServ.do?uri=CELEX%3A31996L0092%3AEN%3AHTML (accessed 04 April 2024).

2 Federal Energy Regulatory Commission (1996). Order No. 888. https://www.ferc.gov/industries-data/electric/industry-activities/open-access-transmission-tariff-oatt-reform/history-oatt-reform/order-no-888 (accessed 04 April 2023).

3 Regional transmission organizations (1999). Order no. 2000, RTO final rule part 1 of 4. https://www.ferc.gov/sites/default/files/2020-06/RM99-2-00K_1.pdf (accessed 04 April 2023).

4 European Parliament (2009). Directive 2009/72/EC of the European Parliament and of the Council of 13 July 2009 concerning common rules for the internal market in electricity and repealing Directive 2003/54/EC. Eur-lex Europa. https://eur-lex.europa.eu/legal-content/EN/ALL/?uri=celex%3A32009L0072 (accessed 04 April 2023).

5 Solomon, B.D. and Krishna, K.J. (2011). The coming sustainable energy transition: history, strategies, and outlook. *Energy Policy* 39 (11): 7422–7431. https://doi.org/10.1016/j.enpol.2011.09.009.

6 Masson-Delmotte, V., Zhai, P., Pörtner, H.O. et al. (2022). *Global Warming of 1.5 C: IPCC Special Report on Impacts of Global Warming of 1.5 C Above Pre-industrial Levels in Context of Strengthening Response to Climate Change, Sustainable Development, and Efforts to Eradicate Poverty*. Cambridge University Press.

7 IEA (2022), Critical minerals threaten a decades-long trend of cost declines for clean energy technologies – analysis – IEA. https://www.iea.org/commentaries/critical-minerals-threaten-a-decades-long-trend-of-cost-declines-for-clean-energy-technologies (accessed 04 April 2023).

8 Lacal-Arántegui, R. and Jäger-Waldau, A. (2018). Photovoltaics and wind status in the European Union after the Paris Agreement. *Renewable & Sustainable Energy Reviews* 81: 2460–2471. https://doi.org/10.1016/j.rser.2017.06.052.

9 Asare-Bediako, B., Kling, W.W., and Ribeiro, P.F. (2014). Future residential load profiles: scenario-based analysis of high penetration of heavy loads and distributed generation. *Energy and Buildings* 75: 228–238. https://doi.org/10.1016/j.enbuild.2014.02.025.

10 Press Corner European Commission (2022). Zero emission vehicles: first 'Fit for 55' deal will end the sale of new CO_2 emitting cars in Europe by 2035. European Commission. https://ec.europa.eu/commission/presscorner/detail/en/ip_22_6462 (accessed 04 April 2023).

11 Ministerie van Binnenlandse Zaken en Koninkrijksrelaties (2022). Hybride warmtepomp de nieuwe standaard vanaf 2026. Nieuwsbericht | Rijksoverheid.nl. https://www.rijksoverheid.nl/actueel/nieuws/2022/05/17/hybride-warmtepomp-de-nieuwe-standaard-vanaf-2026 (accessed 04 April 2023).

12 Statista (2023). U.S. cumulative smart meter installations 2022 | Statista. https://www.statista.com/statistics/676472/number-of-smart-meter-installations-in-the-united-states/#:~:text=A%20total%20of%20124%20million,meters%20installed%20in%20the%20U.S. (accessed 04 April 2023).

13 Isha, M.T. and Wang, Z. (2008). Transformer hotspot temperature calculation using IEEE loading guide. *2008 International Conference on Condition Monitoring* *and Diagnosis*. IEEE. 21-24 April, Beijing, China, https://doi.org/10.1109/CMD.2008.4580455

14 IEEE Standards Association (2012). C57.91-2011 – IEEE guide for loading mineral-oil-immersed transformers and step-voltage regulators. IEEE Standard | IEEE Xplore. https://ieeexplore.ieee.org/document/6166928 (accessed 04 April 2023).

15 Monitor Deloitte, E. DSO, and Eurelectric (2021). Connecting the dots: distribution grid investment to power the energy transition. https://www2.deloitte.com/content/dam/Deloitte/ch/Documents/energy-resources/deloitte-ch-en-eurelectric-connecting-the-dots-study.pdf (accessed 04 April 2023).

16 Netbeheer Nederland (2021). Capaciteitskaart invoeding elektriciteitsnet. https://capaciteitskaart.netbeheernederland.nl/ (accessed 04 April 2023).

17 Netbeheer Nederland (2023). Het energiesysteem van de toekomst: de II3050-scenario's. https://www.netbeheernederland.nl/publicatie/ii3050-eindrapport (accessed 04 April 2023).

18 Allan, R.N., Billinton, R., Sjarief, I. et al. A reliability test system for educational purposes-basic distribution system data and results. *IEEE Transactions on Power Systems* 6 (2): 813–820. https://doi.org/10.1109/59.76730.

19 Morch, A., Di Somma, M., Papadimitriou, C. et al. (2022). Technologies enabling evolution of integrated local energy communities. *2022 IEEE International Smart Cities Conference (ISC2) IEEE*, 26-29 September, Pafos, Cyprus, https://doi.org/10.1109/isc255366.2022.9922568.

20 Papadimitriou, C., Di Somma, M., Charalambous, C. et al. (2023). A comprehensive review of the design and operation optimization of energy hubs and their interaction with the markets and external networks. *Energies* 16 (10): 4018. https://doi.org/10.3390/en16104018.

21 Li, H., Wang, Z., Hong, T., and Piette, M.A. (2021). Energy flexibility of residential buildings: a systematic review of characterization and quantification methods and applications. *Advances in Applied Energy* 3: 100054. https://doi.org/10.1016/j.adapen.2021.100054.

22 Finck, C., Li, R., and Zeiler, W.W. (2020). Optimal control of demand flexibility under real-time pricing for heating systems in buildings: a real-life demonstration. *Applied Energy* 263: 114671. https://doi.org/10.1016/j.apenergy.2020.114671.

23 Luc, K.M., Li, R., Xu, L. et al. (2020). Energy flexibility potential of a small district connected to a district heating system. *Energy and Buildings* 225: 110074. https://doi.org/10.1016/j.enbuild.2020.110074.

24 Ren, H., Sun, Y., Albdoor, A.K. et al. (2021). Improving energy flexibility of a net-zero energy house using a solar-assisted air conditioning system with thermal energy storage and demand-side management. *Applied Energy* 285: 116433. https://doi.org/10.1016/j.apenergy.2021.116433.

25 Ramos, J.S., Mc, P., Moreno, M. et al. (2019). Potential of energy flexible buildings: evaluation of DSM strategies using building thermal mass. *Energy and Buildings* 203: 109442. https://doi.org/10.1016/j.enbuild.2019.109442.

26 Brahman, F., Honarmand, M., and Jadid, S. (2015). Optimal electrical and thermal energy management of a residential energy hub, integrating demand response and energy storage system. *Energy and Buildings* 90: 65–75. https://doi.org/10.1016/j.enbuild.2014.12.039.

27 Kazemi, M., Papadimitriou, C., Paterakis, N.G. et al. (2023). Optimal design of multi-carrier and multi-objective home energy management system. *2023 International Conference on Smart Energy Systems and Technologies (SEST)*, 4-6 September, Mugla, Turkey. pp. 1–6. https://doi.org/10.1109/SEST57387.2023.10257384

28 IEA. By 2030 EVs represent more than 60% of vehicles sold globally, and require an adequate surge in chargers installed in buildings – analysis – IEA. IEA. https://www.iea.org/reports/by-2030-evs-represent-more-than-60-of-vehicles-sold-globally-and-require-an-adequate-surge-in-chargers-installed-in-buildings (accessed 04 April 2023).

29 Di Somma, M. and Graditi, G. (2022). Challenges and opportunities of the energy transition and the added value of energy systems integration. *Technologies for Integrated Energy Systems and Networks* 1–14. https://doi.org/10.1002/9783527833634.ch1.

30 Piazza, G., Bracco, S., Delfino, F. et al. (2023). Impact of electric mobility on the design of renewable energy collective self-consumers. *Sustainable Energy, Grids and Networks* 33: 100963. https://doi.org/10.1016/j.segan.2022.100963.

31 Brenna, M., Foiadelli, F., Zaninelli, D. et al. (2021). Chapter 10: The integration of electric vehicles in smart distribution grids with other distributed resources. In: *Distributed Energy Resources in Local Integrated Energy Systems* (ed. G. Graditi and M. Di Somma), 315–345. Elsevier https://doi.org/10.1016/b978-0-12-823899-8.00006-6.

32 Di Somma, M., Ciabattoni, L., Comodi, G., and Graditi, G. (2020). Managing plug-in electric vehicles in eco-environmental operation optimization of local multi-energy systems. *Sustainable Energy, Grids and Networks* 23: 100376. https://doi.org/10.1016/j.segan.2020.100376.

33 Varshosaz, F., Moazzami, M., Fani, B., and Siano, P. (2019). Day-ahead capacity estimation and power management of a charging station based on queuing theory. *IEEE Transactions on Industrial Informatics* 15 (10): 5561–5574. https://doi.org/10.1109/tii.2019.2906650.

34 Cusenza, M.A., Guarino, F., Longo, S. et al. (2020). Environmental assessment of 2030 electricity generation scenarios in Sicily: an integrated approach. *Renewable Energy* 160: 1148–1159. https://doi.org/10.1016/j.renene.2020.07.090.

35 Khorasany, M., Najafi-Ghalelou, A., Razzaghi, R., and Mohammadi-Ivatloo, B. (2021). Transactive energy framework for optimal energy management of multi-carrier energy hubs under local electrical, thermal, and cooling market constraints. *International Journal of Electrical Power & Energy Systems* 129: 106803. https://doi.org/10.1016/j.ijepes.2021.106803.

36 Paterakis, N.G., Erdinc, O., Bakirtzis, A.G., and Catalao, J.P.S. (2015). Optimal household appliances scheduling under day-ahead pricing and load-shaping demand response strategies. *IEEE Transactions on Industrial Informatics* 11 (6): 1509–1519. https://doi.org/10.1109/tii.2015.2438534.

37 Yousefi, M., Hajizadeh, A., Soltani, M.N., and Hredzak, B. (2020). Predictive home energy management system with photovoltaic array, heat pump, and plug-in electric vehicle. *IEEE Transactions on Industrial Informatics* 17 (1): 430–440.

38 Kazemdehdashti, A., Mohammadi, M., Seifi, A.R., and Rastegar, M. (2020). Stochastic energy management in multi-carrier residential energy systems. *Energy* 202: 117790. https://doi.org/10.1016/j.energy.2020.117790.

39 Yuan, X., Liu, Y., and Bucknall, R. (2020). A novel design of a solid oxide fuel cell-based combined cooling, heat and power residential system in the UK. *IEEE Transactions on Industry Applications* 57 (1): 805–813. https://doi.org/10.1109/TIA.2020.3034073.

40 Mavrotas, G. (2009). Effective implementation of the ε-constraint method in multi-objective mathematical programming problems. *Applied Mathematics and Computation* 213 (2): 455–465. https://doi.org/10.1016/j.amc.2009.03.037.

41 Paterakis, N.G., Mazza, A., Santos, S.F. et al. (2015). Multi-objective reconfiguration of radial distribution systems using reliability indices. *IEEE Transactions on Power Systems* 31 (2): 1048–1062. https://doi.org/10.1109/TPWRS.2015.2425801.

42 pyaugmecon (2023). PyPI. https://pypi.org/project/pyaugmecon/ (accessed 04 April 2023).

43 European Commission (2019). Clean energy for all Europeans package. Publications Office. https://data.europa.eu/doi/10.2833/9937 (accessed 04 April 2023).

44 European Parliament and Council of the European Union (2019). Regulation (EU) 2019/943 of the European Parliament and of the Council – of 5 June 2019 – on the internal market for electricity. http://data.europa.eu/eli/reg/2019/943/oj (accessed 04 April 2023).

45 European Parliament and Council of the European Union (2012). Directive (EU) 2019/944 of the European parliament and of the council – of 5 June 2019 – on common rules for the internal market for electricity and amending directive 2012/27/EU. http://data.europa.eu/eli/dir/2019/944/oj (accessed 04 April 2023).

46 The European Union Agency for the Cooperation of Energy Regulators (ACER) (2021). Report on distribution tariff methodologies in Europe. https://www.acer.europa.eu/Official_documents/Acts_of_the_Agency/Publication/ACER%20Report%20on%20D-Tariff%20Methodologies.pdf (accessed 04 April 2023).

47 GOPACS (2023). About GOPACS – GOPACS. GOPACS. https://en.gopacs.eu/about-gopacs/ (accessed 04 April 2023).

48 GOPACS (2023). Public announcements – GOPACS. GOPACS. https://en.gopacs.eu/public-announcements/ (accessed 04 April 2023).

49 Piclo (2018) Piclo Flex. https://support.picloflex.com/ (accessed 04 April 2023).

50 VREG (2022). Wat zijn de nieuwe nettarieven en hoe worden ze berekend? https://www.vreg.be/nl/wat-zijn-de-nieuwe-nettarieven-en-hoe-worden-ze-berekend (accessed 04 April 2023).

51 van der Holst, B., Verhoeven, G., Matthijssen, E. et al. (2023). A dynamic bandwidth tariff assessment in a Dutch distribution network using a novel scalable distributed simulation framework. *The 27th International Conference and Exhibition on Electricity Distribution (CIRED).* https://doi.org/10.1049/icp.2023.0722

52 EPRI (2021), DSS extensions: multi-platform OpenDSS extensions. https://dss-extensions.org/ (accessed 04 April 2023).

53 Good, N., Ellis, K.A., and Mancarella, P. (2017). Review and classification of barriers and enablers of demand response in the smart grid. *Renewable & Sustainable Energy Reviews* 72: 57–72. https://doi.org/10.1016/j.rser.2017.01.043.

54 Yang, C.-J. (2017). Opportunities and barriers to demand response in China. ideas.repec.org, https://ideas.repec.org/a/eee/recore/v121y2017icp51-55.html (accessed 04 April 2023).

55 Vallés, M., Reneses, J., Cossent, R., and Frías, P. (2016). Regulatory and market barriers to the realization of demand response in electricity distribution networks: a European perspective. *Electric Power Systems Research* 140: 689–698. https://doi.org/10.1016/j.epsr.2016.04.026.

56 Weck, M.H.J., Van Hooff, J., and Van Sark, W.G.J.H.M. (2016). Review of barriers to the introduction of residential demand response: a case study in the Netherlands. *International Journal of Energy Research* 41 (6): 790–816. https://doi.org/10.1002/er.3683.

57 Minniti, S., Haque, A.N.M.M., Nguyen, P.H., and Pemen, A.J.M. (2018). Local markets for flexibility trading: key stages and enablers. *Energies* 11 (11): 3074. https://doi.org/10.3390/en11113074.

58 Venegas, F.G., Petit, M., and Perez, Y. (2021). Active integration of electric vehicles into distribution grids: barriers and frameworks for flexibility services. *Renewable and Sustainable Energy Reviews* 145: 111060. https://ideas.repec.org/a/eee/rensus/v145y2021ics1364032121003488.html.

8

Peer-to-Peer Energy Trading Approaches: Maximizing the Active Participation of the Prosumers in the Multi-carrier Energy Communities

Andrés F. Cortés-Borray, Amaia González-Garrido, Ander Z. Gómez, Joseba J. Huarte, and Nerea R. Carames

TECNALIA, Basque Research and Technology Alliance (BRTA), Astondo Bidea, Building 700, Derio E-48160, Spain

8.1 Introduction

The traditional wholesale electricity market has long been dominated by large-scale utilities and energy retailers; the latter determines the price at which they supply electricity to end-users. Within this scheme, different network pricing schemes can be offered to the consumers who purchase electricity through fixed or time-of-use (ToU) tariffs. However, the energy generation and consumption landscape is undergoing a significant transformation. The increasing deployment of distributed energy resources (DERs), such as rooftop solar photovoltaic (PV), small wind turbines, electric vehicles (EVs), energy storage systems (ESSs), and even thermal technologies like heat pumps (HPs) or combined heating and power (CHP) units have paved the way for a new paradigm shift in the energy sector, by placing active end-users at heart. Looking at possibilities for a consumer-centric market has led to the rise of new energy trading schemes among peers, known as peer-to-peer (P2P). Here, end-users are aware of and are in control of establishing the conditions of transactions and handling the exchange of goods and services, which would yield significant benefits for themselves and the power system [1].

Consumers can benefit by reducing electricity costs, shaving peak load, or maximizing on-site renewable usage while participating in P2P energy trading [2]. Besides, this type of scheme can help power network planning and operation by reducing technical losses and peak demand, lowering investment and operational costs, minimizing reserve requirements, and enhancing energy supply and quality [3, 4]. However, it also poses relevant challenges in a multi-agent system encompassing several users [1, 5], e.g., encouraging end-users to cooperate among themselves, implementing a suitable decision-making process with a minimum conflict of interests for all users, or dealing with a number of actors in the network requesting services with different objectives in mind. Hence, this highlights the importance of designing P2P energy trading procedures supported by policies that encourage the active participation of end-users and their DERs.

Integrated Local Energy Communities: From Concepts and Enabling Conditions to Optimal Planning and Operation, First Edition. Edited by Marialaura Di Somma, Christina Papadimitriou, Giorgio Graditi, and Koen Kok.

Different approaches can be employed for P2P energy trading based on the power network configuration and end-user characteristics, but the primary objective remains maximizing energy efficiency, cost management, and reduction of supply expenses.

The remainder of this chapter is organized as follows: Section 8.2 presents an overview of the regulatory framework, previous initiatives, and future trends in the P2P approach. It also introduces the P2P energy trading concept along with its associated challenges and actors in light of the power system. A comparison of different P2P energy trading architectures (market design and implementation) is explained in Section 8.3. A detailed literature review is presented in Section 8.4, covering P2P trading architectures, upstream markets and ancillary services, and network concerns. An outline of the deployed P2P architecture in the H2020 EU project eNeuron [6] is presented in Section 8.4.2. Finally, conclusions summarizing the most relevant aspects of this chapter are drawn in Section 8.4.3.

8.2 Background and P2P Concept

8.2.1 Background, Alternatives, Initiatives, and Trends

The electricity sector has evolved in the past decades due to the massive deployment of local generation, controllable loads, storage systems, and digitalization as the backbone of these technologies. Consequently, concepts such as microgrids (MGs) and virtual power plants (VPPs) have emerged to manage diverse DERs in a centralized manner.

In this sense, an MG can be defined as a group of interconnected loads and DERs within clearly defined electrical boundaries that acts as a single controllable entity with respect to the external electrical network [7], operating in either a grid-connected or island mode by following the guidelines of IEC TS 62898-2:2018. Modern MGs provide effective solutions to establish a robust internal operation, including a broader flexibility exchange among their end-users [8]. Besides MGs, a VPP [9] consists of a cluster of distributed generation or consumption facilities remotely monitored and managed by an advanced control system to achieve more efficient operation. Unlike MGs, the VPP concept may cover a large geographic area and, therefore, could be a bridge to optimize the dispatch in the wholesale market or other services via aggregation.

In view of the European ambition to transform the electrical system into a more consumer-centred and the increasing deployment of residential solar PVs, according to Directive 2019/944 [10], an active consumer is defined as a final customer who self-consumes or stores electricity generated within its premises. Hence, this fact leads to a broader scope through collective self-consumption, which is a term that could refer to any collective form of joint energy production and consumption [11], usually restricted to a limited geographical location.

Recently, there has been an ongoing harmonization of the use of the term "community" in the European Union (EU) internal energy market (formerly named citizen energy community). This makes reference to a legal entity, based on the open and voluntary participation of its members, with primary environmental and social purposes, which may engage assets related to generation, distribution, supply, consumption, and storage. Therefore, this fact aligns the energy communities to pave the way for new market actors and more digitalized solutions, for instance, the emergence of local energy markets (LEMs) to procure balancing services from local distributed resources and demand-response aggregators by means of a (market) platform where system operators can inform their specific local needs [12].

The Council of European Energy Regulators [13] describes P2P trading as "a marketplace for producers and consumers to buy and sell electricity to each other directly." De Almeida et al. [11] include the use of an electric platform or marketplace to differentiate from collective self-consumption or other approaches. The Working Group of the European Commission further shows the importance of P2P electricity trading in the "Digitalizing the energy system – EU action plan" [14] to empower consumers, but at the same time, they declare the need to explore multiple business models and identify the legal and technical prerequisites.

Consequently, further research is needed to propose innovative business models, design an efficient regulatory and policy framework, and clarify the role of involved actors [15]. Indeed, governmental support (e.g., financial aid and proper regulations) is mandatory to facilitate the barriers to deploying new pilots [16]. Other legal challenges regarding property law, contractual arrangement, responsibilities, and data law are presented and discussed in De Almeida et al. [11] and Wu et al. [12]. Additionally, third-party market "aggregator" shall play a key role in the successful P2P deployment [8].

In this sense, P2P energy trading appears to be one of the viable solutions in which consumers and prosumers can buy or sell energy from/to other peers [15], having the possibility to address their own particular preferences beyond the traditional wholesale markets [17]. Despite the not-standardized framework of P2P, there are few practical implementations shown in [3, 15, 17, 18]. Various pilot projects and research are being carried out at different levels worldwide in this regard. Most research projects are specially oriented to develop and validate P2P market trading platforms like *Enerchain* for a decentralized blockchain-based P2P trading platform, *P2P3M* for a P2P trading platform in South Korea and the United Kingdom, *Trans-Active Grid* for an automatic electricity trading platform in the United States, and *the Energy Collective* for local P2P markets in Denmark. Additionally, Bjarghov et al. [17] state that *Piclo*, *Sonnen Community*, and *Vandebron* are examples of projects or "start-ups" that approach peer energy sharing, develop new business models from a regulatory perspective, and utilize hybrid market structures to match electricity supply and demand: *Piclo* as an independent marketplace for flexibility services in the United Kingdom, *Sonnen Community* for sharing energy among peers and creating a VPP to participate in the wholesale markets in Germany, and *Vandebron* as a marketplace which connects directly households and sustainable renewable producers in the Netherlands. In addition, the *Brooklyn Microgrid* project allows a local marketplace for peers (consumers and prosumers) to trade energy locally without a centralized coordinator. Concretely, prosumers can transact energy autonomously in near-real time on the *Exergy* platform.

On the other hand, the P2P framework is starting to be extended not only to trade electricity but also to other carriers like heat, cooling, gas, and hydrogen. For instance, Shi et al. [19] and Wang et al. [20] proposed a P2P market mechanism for a multi-energy system to trade electricity and heat among peers. In the former, each market is cleared separately, and in the latter, both markets are cleared through a marginal cost-based pricing mechanism. Gao et al. [21] introduced a two-stage distributed robust optimization schema based on a P2P trading of electricity, heat, and gas between MGs. Therefore, these types of works give an insight into what trend is leading the research regarding P2P energy trading to pave the way to a decarbonized energy system.

8.2.2 P2P Energy Trading Concept and Challenges

According to the International Renewable Energy Agency [22], P2P electricity trading is a business model that utilizes an interconnected platform as an online marketplace that enables direct trading of electricity between consumers and producers without any intermediaries aiming to increase renewable generation deployment and provides ancillary services to the power network. Similar to an open market economy, suppliers aim to secure the highest possible price while considering their costs and profits by also seeking the lowest price that meets their needs. Here, trades occur when supply and demand offers are aligned, i.e., represented by sell and buy bids.

In contrast to traditional power supply, where energy suppliers typically purchase electricity from generation utilities on behalf of final consumers, P2P trading allows prosumers (those who both produce and consume electricity) or self-consumers to sell excess electricity back to the network at a buy-back rate [23]. However, the consumer tariffs for electricity supply are generally higher than the buy-back rates prosumers receive, and these tariffs do not account for the additional benefits of renewable generation for the power system. This aspect highlights how relevant it is that prosumers have the autonomy to make individual choices regarding their energy-sharing preferences by determining the quantity of energy they want to share, setting the price at which they are willing to share it, and selecting the peers and timing for sharing [4]. This level of independence and self-control is a crucial aspect of P2P sharing. For example, while a prosumer with a rooftop PV system can charge their EV using the generated electricity, their neighbor would rely on electricity from a centralized power plant located elsewhere. If the neighbor charged their EV using PV power generated within the neighborhood, the prosumer with the PV installation would receive compensation for the electricity injected into the distribution network. Nonetheless, note that this compensation fails to consider the costs derived from reducing technical losses and congestion provided by local generation. Although the allocation of electricity costs to managing the network infrastructure varies from country to country to supply power to the customers' premises, a P2P model could potentially reduce these costs.

In terms of deployment, the P2P scheme can be implemented among neighboring households in a local community or on a larger scale involving multiple communities, as long as different key components in a virtual, physical, and miscellaneous layer are considered. According to Tushar et al. [1, 4], the virtual layer should contain (i) an *information system* element that provides security for both the user's data and distribution network, market monitoring, and equal participation among users; (ii) a *market operation* element that settles the energy allocation, payment rules, and bidding formats to display an efficient trading process close to real-time; (iii) a *pricing mechanism* for enabling users to reap more benefits by adequate settling prices for their energy; and (iv) an *energy management system* that meets the required energy supply to the user taking into account the employed bidding mechanism.

On the other hand, the physical layer must involve (i) the *distribution network connection* to guarantee the energy balance between generation and consumption; (ii) the *metering infrastructure* with the capability of deciding when to participate in the P2P platform based on the available information regarding demand, generation, prices, and network status; and (iii) the *communication infrastructure* with proper latency, throughput, reliability, and security features to exchange information among users.

The miscellaneous layer should contain (i) a representative number of *market participants* willing to take part in the trading platform and (ii) a suitable *energy policy regulation* that defines transparent rules regarding the market design, distribution of fees and taxes, and the role, rights, and duties of the actors involved in the P2P market in relation to the current energy market.

However, all these aspects bring associated a series of technical, economic, and social challenges to face [1, 3, 18, 24], such as:

- coordination among users for balancing local generation and demand,
- addressing the conflicts of interest among several stakeholders,
- promoting and engaging user participation in low-cost energy trading,
- security of transactions,
- short-term forecasting of energy resources and prices,
- consider the depreciation of renewable systems and the infrastructures associated with them, and distribution system energy losses within the internal trading prices,
- development of an adequate price mechanism,
- the impact on the distribution network operation, mainly in terms of voltage limits, loading levels, and reliability indexes,
- a reliable and secure communication infrastructure, and
- provision of ancillary services via P2P.

In a P2P energy trading context, several actors are involved, and according to Domènech Monfort et al. [25], these can be categorized into four groups: *Manager*, *Administrator*, *Supervisor*, and *Broker*. The first term stands for the responsible stakeholder for controlling the energy trading activities within a specific group of peers, ensuring the equilibrium of a local energy community. The second term gathers the responsible party for coordinating the electricity trading activities within a P2P market without control action on the participants in the P2P market.

The third term picks the responsible actor for supervising the P2P market, which includes authorizing transactions and ensuring compliance with the physical constraints of the distribution network. The last term stands for the responsible party for facilitating communication between peers involved in the energy trading process or acting as a negotiator between end-users and a manager or administrator.

In a nutshell, P2P energy trading represents a transformative shift in the electricity market, empowering prosumers, promoting renewable energy adoption, and fostering a more sustainable and resilient energy system.

8.3 P2P Methods and Logical Architecture

In this section, two fundamental aspects of the business layer of P2P markets will be described in detail: *market design* and *market implementation* mechanisms. Market design defines the structure of the market and the energy trading method among prosumers [26]. In contrast, market implementation mechanisms include different techniques for computing peer decision-making, transaction price determination, and market simulation processes. Both P2P market features are depicted in Figure 8.1.

8.3.1 P2P Market Design

Three P2P market designs have been mainly reported in the literature (see Figure 8.2) according to the decision-making process on the energy trading process

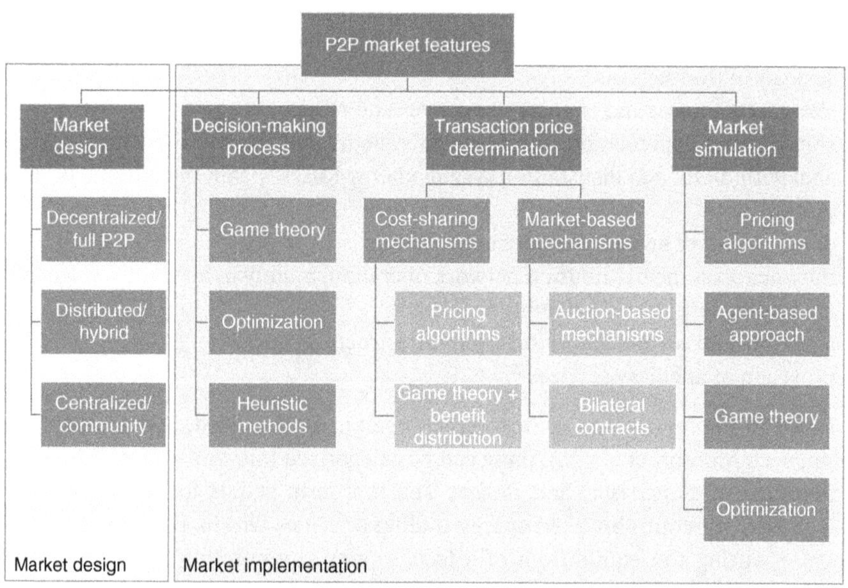

Figure 8.1 Classification of P2P market designs and implementation techniques.

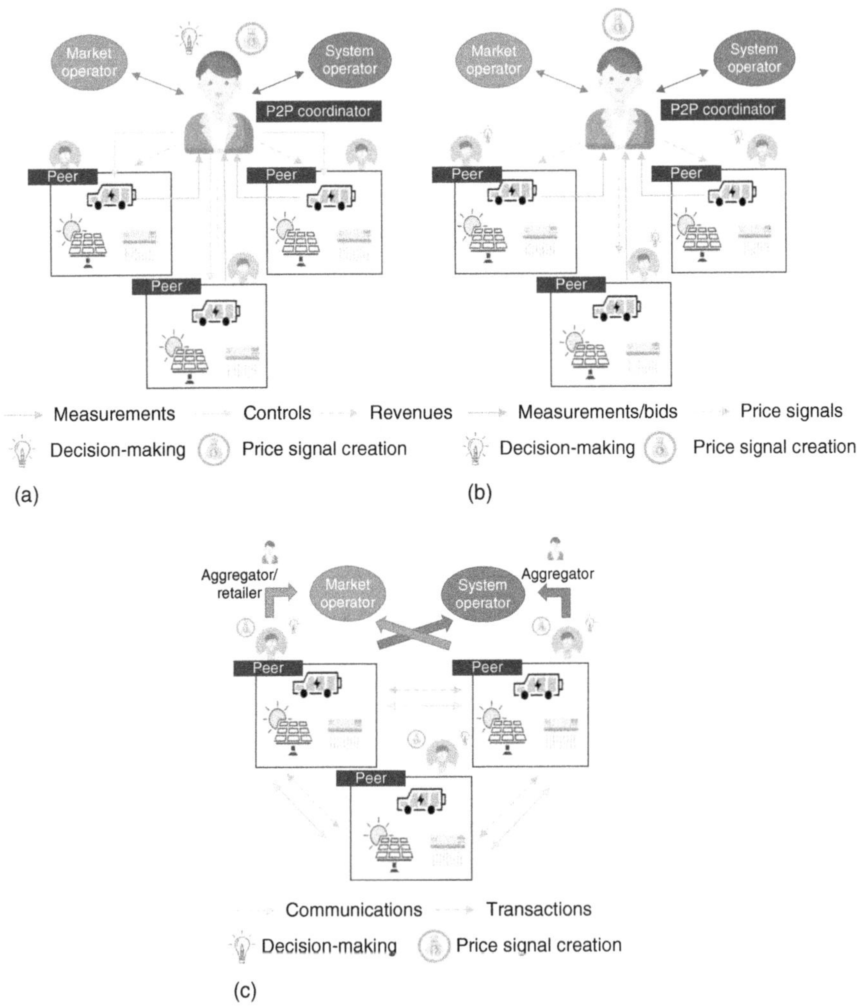

Figure 8.2 P2P market designs: (a) centralized, (b) distributed, and (c) decentralized.

and the communications and control architecture [15, 17, 18]: community-based or centralized markets, full P2P or decentralized markets, and hybrid or distributed markets.

8.3.1.1 Community-Based or Centralized Markets

In community-based market design, a central entity named the community manager intervenes. Its role is to set the prices and manage the energy trading and market settlement processes within the P2P market, as well as to interface with the rest of the system and each peer. In these markets, the decision-making process is usually executed using centralized optimization techniques that aim to maximize either individual or collective benefits. As an advantage, these markets allow a coordinated response that can ease grid operation. Data management, data aggregation, and fair energy sharing between peers remain the key challenges for the community manager [18].

8.3.1.2 Full P2P or Decentralized Markets

In full P2P markets, peers negotiate prices and energy volumes with each other, directly trading energy without centralized supervision. The decision-making process aims to minimize individual electricity costs, which is carried out by each peer, usually applying distributed optimization techniques. The key advantage of this design is the freedom of choice for both supplier selection and bid definition, as well as the high data privacy levels due to bilateral energy trading. Nevertheless, technical challenges like higher deployment of information and communications technologies (ICTs) infrastructure, complexity for consensus between peers, or lack of control/supervision of grid operators remain [18].

8.3.1.3 Hybrid or Distributed Markets

Hybrid markets are a combination of the two aforementioned market designs. The coordinator does not have control over individual peers and the energy traded but influences the decision-making process using price signals computed using pricing algorithms (see Section 8.3.2.2). Hereinafter, peers carry out the scheduling process using optimization according to the received price signal [27]. This design presents the advantage of combining the advantages of the two previous market designs but implies an additional complexity for the community managers when coordinating internal and external energy trading.

8.3.2 P2P Market Implementation

An effective market implementation is reached when demand and supply are matched within the LEM by determining energy prices and volumes at each time step, i.e., when market clearing occurs [28]. However, the scientific community has no consensus on the most efficient market mechanism [16]. This is a complex process that depends not only on the selected market design but also on individual peer preferences, potential electrical and thermal network limitations, available flexibility of existing DERs, or the selected price creation mechanism. In the P2P market implementation process, three main aspects can be distinguished: peer decision-making, trading price, and market simulation.

8.3.2.1 Peer Decision-Making Process

The individual decision-making process takes place where each peer decides the energy and/or price bids to be communicated, which will result in the scheduled energy profiles being met [29]. In distributed and decentralized markets, this process is executed by peers, following price signals sent by the coordinator and defining both energy and price bids, respectively. Nonetheless, in centralized markets, it is the coordinator in charge of scheduling and controlling assets. According to the reviewed literature, some commonly used approaches are optimization techniques, heuristic methods, and game theoretic approaches.

Discrete optimization techniques like mixed-integer linear programming (MILP) and mixed-integer nonlinear programming (MINLP), continuous optimization approaches like quadratic programming (QP) or linear programming (LP), and/or distributed optimization approaches like alternating direction method of multipliers (ADMM) or Lagrangian relaxation are suitable to define peer energy schedules and calculate the prices to be sent [27]. MILP problems are hard-to-solve linear problems that include integer or binary variables used to model some constraints (e.g., yes/no and on/off). MINLP includes nonlinear functions that make problem-solving more complex. ADMM approaches are based on solving convex optimization problems by decomposing them into smaller sub-problems [30]. In a similar way, the Lagrangian multiplier method is used when constrained optimization problems are formulated.

Other alternatives for peer decision-making are *heuristic methods*, which can be defined as a procedure for solving a mathematical problem by an intuitive approach, in which the structure of the problem can be interpreted and exploited intelligently to obtain a reasonable solution but without guaranteeing an optimal solution. In the literature, these can be found divided into simple heuristics or meta-heuristics [27]. While simple heuristics serve to find a near or optimal solution, meta-heuristic approaches try to go beyond a local optimum. Meta-heuristic approaches are usually inspired by biological behavior, known as bio-inspired, such as genetic algorithms, evolutionary algorithms, or particle swarm optimization.

Game-theoretic approaches are commonly reported in the literature for both decision-making processes and market simulation purposes since they represent how the actions of individual agents influence others in a competitive situation [31]. Based on their interaction, two main categories are distinguished: cooperative and noncooperative game theory. In cooperative game theory, players act as a unit, and their actions aim to maximize the collective benefit. Nash bargaining and coalition games are commonly used for convergence purposes. Conversely, in noncooperative games, participants try to maximize their benefit, leading to a collective result that is not always optimal for the whole, which is named Nash equilibrium.

8.3.2.2 Transaction Price Determination Process

Individual decisions are usually based on economic signals that can be generated using different mechanisms, which can be further divided into market-based and cost-sharing mechanisms [20]. Depending on the number of iterations needed to obtain the internal price, one-shot and iterative implementation approaches have been identified.

Market-based mechanisms usually rely on iterative bidding. Among these approaches, auction-based mechanisms and bilateral contracts stand out. Unlike bilateral contracting, auction mechanisms do not involve direct negotiation between peers. A commonly reported auction-based mechanism in P2P markets is the continuous double auction (CDA) mechanism, where aggregated supply and demand curves are built from individual bids, and their intersection determines the energy traded and the market-clearing price, i.e., marginal price. This price definition mechanism, which is used in European day-ahead markets, is known as *pay-as-clear* or uniform double auction. Nonetheless, alternative mechanisms like *pay-as-bid* or discriminatory double auctions are widely utilized [32]. Conversely, in reverse auction (RA) mechanisms, buyers send a request for energy, and sellers bid for an amount and price they are willing to pay. As a result, the cheapest bid is the winner [33]. Bilateral contracting is used in decentralized P2P markets, and it is based on the clearing of selling and buying bids when these are compatible, meaning that each transaction is priced differently and, therefore, a consensus-based price exists. Consensus is a key challenge in this approach due to individual peer preference matching complexities.

Pricing algorithms are widely used cost-sharing mechanisms on distributed P2P markets, which are based on algorithms that compute transaction prices considering different criteria like participant opportunity costs and/or individual retail prices or energy offer/demand availability at each period. Prices can be used for market settlement resulting from either energy bids or an optimized central schedule of the community, hence influencing the peer decision-making process [34].

Hereafter, several pricing algorithms used in the literature will be described and formulated briefly. Among these pricing algorithms, bill sharing (BS), mid-market rate (MMR), intermediate market rate (IMR), or supply–demand ratio (SDR) will be detailed.

BS is a price-creation mechanism based on a proportional share of the total energy bill according to the energy generation or consumption of each market period [35]. As a result of it, common LEM buying (pb_t) and selling (ps_t) prices will be obtained. This method does not fully encourage energy selling since sellers do not receive an additional payment for their energy with respect to their feed-in tariff (FiT) prices.

These buying and selling prices might vary depending on the individual supply and FiT prices, respectively. Equations (8.1) and (8.2) represent local market selling and buying prices when different time-varying supply and the existence of FiT.

$$\mathrm{ps}_t = \min_n \left(\mathrm{PS}_{n,t}\right) \cdot \frac{\max\left(0, \sum_{n,t}\left(G_{n,t} - D_{n,t}\right)\right)}{\sum_{n,t} G_{n,t}} \tag{8.1}$$

$$\mathrm{pb}_t = \max_n \left(\mathrm{PB}_{n,t}\right) \cdot \frac{\max\left(0, \sum_{n,t}\left(D_{n,t} - G_{n,t}\right)\right)}{\sum_{n,t} D_{n,t}} \tag{8.2}$$

where $\mathrm{PS}_{n,t}$ represents individual retailer selling or FiT price, $\mathrm{PB}_{n,t}$ depicts individual retailer buying or supply tariff prices, and $D_{n,t}$ and $G_{n,t}$ describe peer energy demand and generation, respectively.

MMR is a price creation mechanism based on the calculation of the P2P market average price of participants' supply and FiTs. Therefore, by applying this mechanism, buyers and sellers benefit in an equal manner since they buy at a cheaper price and sell at a higher price, respectively. When all members have the same time-varying supply and FiT, the resulting price (p_t) is calculated according to Eq. (8.3).

$$p_t = \frac{\mathrm{PS}_t + \mathrm{PB}_t}{2} \tag{8.3}$$

IMR is a variant of the MMR mechanism that seeks to approximate LEM's price closer to supply or FiT prices in order to benefit either sellers or buyers. To obtain the resulting price, minimum buying, and maximum selling prices are combined in a convex linear equation using an α coefficient that ranges between 0 and 1. This mechanism is formulated as presented in Eq. (8.4).

$$p_t = \alpha \cdot \min_n \left(\mathrm{PB}_{n,t}\right) + (1-\alpha) \cdot \max_n \left(\mathrm{PS}_{n,t}\right) \tag{8.4}$$

SDR is a price calculation mechanism based on the ratio between the total surpluses and deficits within the LEM at each time step (sdr_t) [36], as expressed in Eq. (8.5).

$$\mathrm{sdr}_t = \frac{\sum_{n,t} \max\left(0, \sum_{n,t} \left(G_{n,t} - D_{n,t}\right)\right)}{\sum_{n,t} \max\left(0, \sum_{n,t} \left(D_{n,t} - G_{n,t}\right)\right)} \tag{8.5}$$

According to Eq. (8.5), four SDR scenarios can happen: (i) surplus energy is sold to the grid ($\mathrm{sdr}_t > 1$), (ii) all surplus energy will be sold within the LEM, and no energy will be exchanged with the grid ($\mathrm{sdr}_t = 1$), (iii) part of the energy needs to be bought from the grid ($0 < \mathrm{sdr}_t < 1$), and (iv) there is no surplus energy, and all has to be purchased from the grid ($\mathrm{sdr}_t = 0$).

The sdr_t is inversely related to the resulting LEM price, as shown in Eq. (8.6). Therefore, the higher the sdr_t, the lower the LEM due to higher excess energy available and, on the contrary, the lower the sdr_t, the higher the LEM price.

$$p_t = \frac{\mathrm{PS}_t \cdot \mathrm{PB}_t}{\left(\mathrm{PB}_t - \mathrm{PS}_t\right) \cdot \mathrm{sdr}_t + \mathrm{PS}_t} \tag{8.6}$$

In order to incentivize prosumers to sell within the P2P market, Long et al. [37] proposed to introduce a compensating factor λ bounded between 0 and $\mathrm{PB}_t - \mathrm{PS}_t$. As a result, the internal P2P market price can be calculated as in Eq. (8.7):

$$p_t = \begin{cases} \frac{(\mathrm{PS}_t+\lambda)\cdot \mathrm{PB}_t}{(\mathrm{PB}_t-\mathrm{PS}_t-\lambda)\cdot \mathrm{sdr}_t+\mathrm{PS}_t+\lambda}, & 0 < \mathrm{sdr}_t < 1 \\ \mathrm{PS}_t + \frac{\lambda}{\mathrm{sdr}_t}, & \mathrm{sdr}_t \geq 1 \end{cases} \tag{8.7}$$

8.3.2.3 Market Simulation and Clearing

During market clearing, the corresponding bids are matched, and energy exchanges and economic transactions are completed. Different techniques like optimization, pricing algorithms, agent-based approaches, and game theory are used, and they depend widely on the market design. These techniques have already been detailed in Section 8.3.2.1.

In fully decentralized P2P markets, approaches like pricing algorithms cannot be used for market simulation since peers negotiate and trade bilaterally. Due to that fact, peers' behavior and preferences need to be represented in decision-making and clearing processes [15]. For this, agent-based and game-theoretic approaches are utilized. The first scope is based on iterative algorithms that try to achieve equilibrium by adapting bids according to some learning strategies that aim to optimize individual objectives. In game theory, a simultaneous optimization problem of individual peers' benefit is solved. Centralized optimization techniques are also used, but at the expense of being less accurate since they do not represent peers' behavior [34].

Depending on the selected technique for market simulation, single-level or multilevel implementation models can be distinguished. In single-level approaches, decision-making, pricing and/or clearing processes are simulated in a joint manner, whereas in the multilevel implementation, these processes are simulated separately. For instance, agent-based approaches adopted by Zhou et al. [33] and Cintuglu et al. [38] computed individual decision-making, bidding, and market clearing in a joint manner. Another common approach is combining game theory for decision-making and ADMM optimization techniques for market clearing [39, 40].

8.4 Literature Review

8.4.1 Market Design and P2P Trading

In recent years, there has been an increasing focus on research related to P2P energy sharing. This has led to defining different models that encompass a range of approaches, including direct energy trading between peers, energy-sharing communities, and blockchain-based platforms. Zhou et al. [41] explored these P2P energy-sharing models by discussing the underlying mechanisms, benefits, and limitations of each model, considering factors such as energy pricing, market mechanisms, system scalability, and security. Additionally, in Tushar et al. [42], different game-theoretic approaches proposed in the literature for P2P energy trading were discussed, including cooperative and noncooperative game models.

In this sense, different authors have expanded the application of these approaches to multiple scenarios based on economic and/or technical objectives with various actors and users involved to assess their benefits and drawbacks. For instance, Guerrero et al. [43], aiming to minimize the collective cost of supply, formulated a MILP problem for optimal household scheduling and market clearing, considering some market prices determined using the CDA method. It internalizes grid usage costs estimated by solving the optimal power flow (OPF) problem for the distribution system operator (DSO). Similarly, Espadinha et al. [44] presented a two-stage iterative optimization problem that minimizes total community costs where, first, energy exchanges and, second, new trading prices are calculated. Energy trading is executed using a bilateral contract scheme, whose price limits were retailer selling and buying prices. In [29], a three-level P2P energy, ancillary service, and residual balancing market participation framework are presented. First, individual peer scheduling is executed using a MILP approach, and energy transactions are simulated using the CDA method. Once P2P trading is executed, ancillary and balancing reserves are estimated and provided by solving a MILP problem for total cost minimization. Hu et al. [45] proposed a MINLP optimization problem that determines the capacity allocation of each nano grid (NG) to the different markets. In the bilateral P2P market, the bid matching process is executed by the NG coordinator to mitigate individual imbalances, allowing energy trading among NG, whereas in balancing and ancillary service markets, a balancing responsible party is in charge of ensuring the balance of the NG as a whole. In these, three auction mechanisms are assessed: pay-as-bid, marginal price, and pay as the opportunity cost. In [46], MILP problems are solved to calculate each energy hub's (EH's) schedules, aiming to minimize daily operation costs. For market settlement, the CDA method is proposed where each hub sends its bids to the coordinator, and this determines the hubs participating, the energy to be traded, and the price of each transaction. In [36], a distributed iterative optimization approach is used to execute a centralized P2P market where a virtual entity coordinates energy exchanges between peers. A dynamic price is utilized based on the SDR pricing method, which seeks to maximize individual peer benefits. In [47], a decentralized P2P market is presented where EVs are scheduled using LP techniques. Energy trading and price definition are carried out by using an iterative QP approach. Yang et al. [48] present a three-stage multi-energy sharing strategy for gas and electricity EHs. In the first stage, LP optimization is used for EH scheduling. In the second stage, bilateral negotiations are solved using a noncooperative approach (Raiffa–Kalai–Smorodinsky bargaining solution). In the last stage, market clearing is reached using LP optimization approaches.

As most of the works mentioned above are based on the usage of centralized optimization models due to their straight implementation, some authors employ more complex approaches like distributed optimization techniques to tackle the privacy issue in the P2P scheme. Zhang et al. [49] utilized the ADMM approach to compute individual peer scheduling and market equilibrium in a coordinated manner for P2P energy and ancillary service DSO markets. Resulting market prices are iteratively computed until consensus is reached, which considers grid usage and peer scheduling. In [50], the coordinator utilizes the Lagrange multiplier method to solve a constrained optimization problem that aims to create the optimal trading schedule to maximize social welfare considering potential imbalance costs on individual revenues. An internal trading price is set to guarantee mutual profits by using the MMR pricing method. In [51], a two-stage optimization problem is solved using distributed algorithms and cooperative game theory. In the first stage, peer participation and optimal schedules are defined using the ADMM approach. In the second stage, the optimal trading prices are estimated, and market clearing is reached using the Nash bargaining theory. Similarly, Lu et al. [52] focused on market simulation, proposing a two-stage P2P market clearing model solved using ADMM as a distributed optimization technique. The model aims to maximize energy exchange and collective welfare while reducing deviations. In [53], an electricity-carbon P2P trading scheme is proposed, which is solved using a bi-level optimization model. At the upper level, the ADMM approach is used to minimize individual transaction costs, while at the lower level, electricity trading and carbon emission flow are computed based on upper-level results and distribution network costs.

Other P2P approaches are implemented for multi-energy carriers. In [19], an energy scheduling scheme is presented for electricity-hydrogen-heat clusters using the ADMM approach for the prosumer decision-making process and market clearing optimization. A marginal cost-based pricing mechanism is proposed where such marginal costs are defined by the power grid and CHP unit costs for buying and selling, respectively. Similarly, Gao et al. [21] suggest a two-stage

ADMM-based distributed robust optimization strategy for electricity, heat, and gas trading between MGs where Nash bargaining theory is applied to model bilateral negotiations. In [54], a bi-level multi-energy P2P trading framework is proposed based on a modified Nash bargaining theory and using a distributed optimization approach (based on ADMM) for energy scheduling in the lower level and applying network reconfigurations for satisfying operational constraints and minimizing losses, obtaining the resulting payments.

Alternatively, other works focus their effort on analyzing the P2P scheme through the multi-agent concept by implementing game-theoretic models, as these kinds of approaches can be combined with centralized or distributed optimization techniques. In [2], a rule-based decision-making model was proposed in order to determine individual prosumer schedules and overall market clearing according to received P2P market price signals, surplus, and storage availability. Trading among peers is executed as long as the P2P market price is between FiT and retailer buying price, allowing peers to negotiate bilaterally on their own. In [38], a centralized architecture is proposed where peer decision-making and market-clearing processes are executed using a multi-agent-based approach and two heuristic techniques for convergence facilitation. As to prices, the BS, MMR, and SDR algorithms are assessed and compared. Similarly, in [33], a multi-agent-based competitive game theoretic approach is considered for DER scheduling in a grid-connected MG. Market prices and clearing processes are coordinated by an aggregator following an RA mechanism. An agent-based approach based on a decentralized partially observed Markov decision process for modeling peers' behavior stochasticity is followed by Qiu et al. [55]. The authors prove this approach for heat, cool, and electricity trading using a CDA for market price determination method in a decentralized manner. A multi-leader-multi-follower game is formulated by [56] for modeling the risk behavior and flexibility options in a multi-energy community-based P2P market, considering communities as leaders and market operators and grid operators as followers.

In a similar way, a coalition game-based centralized market is presented in [20], where electricity and heat coalitions are formed. These two markets are cleared separately, and the prices are created by adjusting a guiding price computed with the MMR method according to the electrical or thermal balance of the coalition. In [57], a coalition game theory is proposed to establish a centralized P2P trading market where the coordinator schedules each peer using MILP. An improved MMR is implemented to set an optimal price according to such schedules and the principle that cost is equal to revenue. In [39], a two-stage P2P trading scheme is proposed. In the first stage, total energy cost is minimized by optimizing energy sharing within buildings. In the second stage, clearing prices are computed according to mutual energy sharing in a noncooperative game by using a fast ADMM-based approach. In [58], a noncooperative game theoretic approach based on the Stackelberg game model was applied for heat and electricity P2P trading. The decision-making process and competition are represented as a noncooperative Nash game that aims to maximize individual benefit. A distributed optimization is applied to solve the convex quadratic problem that leads to market clearing.

Based on the literature review, the main findings on market design and implementation are summarized in Table 8.1 and described as follows:

- Half of the reviewed papers proposed a decentralized or full P2P market approach. In these markets, distributed optimization techniques like ADMM are often combined with game theoretic approaches to model peers' behavior and market clearing.
- Bilateral contracts are the most utilized approach for transactions due to their suitability for the decentralized market design.
- The decision-making process and, mainly, peer scheduling and bidding are developed with centralized optimization techniques like MILP or NLP in distributed and centralized markets.
- Distributed market approaches are widely used (8/20 papers) mainly due to their balance between peer decision-making independence and market clearing facilitation. Among revised papers, pricing mechanisms like MMR or SDR are employed for price creation, whereas optimization techniques stand out for decision-making and market simulation.
- Several papers proposed distributed market designs for their suitability in multi-market cases where P2P energy trading and ancillary service provision are combined.

8.4.2 Interaction with the Wholesale and Multi-energy Markets

Different studies and current initiatives presented in Section 8.4.1 have provided evidence of the positive impact of P2P sharing on prosumers. Despite these findings, the widespread adoption of P2P sharing models in current electricity markets remains limited. The P2P market has close interaction with the traditional wholesale electricity markets. Furthermore, P2P energy trading will co-exist in the near future with conventional wholesale-retail markets and services upward the peers [60] or with other multi-energy markets. Therefore, it is highly important to explore the relationships and influence on these existing energy markets and the interaction between all involved actors (market, retailers, electrical network, P2P coordinator, and peers).

Regarding the influence of peer transactions in the wholesale markets, the peers are usually assumed to act as "price takers" without the strength to influence these prices (limited P2P adoption). In this case, the wholesale market is used as a "residual balancer" [60], where the residual buying and selling needs of peers are satisfied. Most authors consider selling and purchase prices mainly set by traditional retailers. In this line, Yang et al. [48] computed P2P market prices, taking as reference average utility and natural gas wholesale prices for buying and feeder prices for selling, respectively, after multi-bilateral negotiations. Other authors like Gao et al. [21] and Zou et al. [54] considered DSO buying-selling prices as limit reference values and network usage fees related to congestion issues for P2P price calculation.

The roles of the power system operator, electricity supplier/retailer, and P2P coordinator agent are conceived as separate actors. Usually, the P2P coordinator agent

Table 8.1 Literature review of P2P energy architecture.

References	P2P market design	Decision-making method	Pricing method	Market simulation method
[47]	Decentralized	Optimization (QP)	Bilateral contracts	Optimization (QP)
[2]	Decentralized	Simple heuristics	Bilateral contracts	Simple heuristics
[33]	Centralized	Agent-based approach	Auction-based (RA)	Agent-based approach
[39]	Decentralized	Non-cooperative game/Optimization (ADMM)	Bilateral contracts	Noncooperative game/Optimization (ADMM)
[44]	Decentralized	Optimization (LP)	Bilateral contracts	Optimization (LP)
[21]	Decentralized	Noncooperative game/Optimization (ADMM)	Bilateral contracts	Noncooperative game/Optimization (ADMM)
[56]	Distributed	Noncooperative game/Optimization (MILP)	Bilateral contracts	Noncooperative game/Optimization (MILP)
[43]	Decentralized	Optimization (MILP)	Auction-based (CDA)	Optimization (MILP)
[59]	Decentralized	Optimization (ADMM)	Bilateral contracts	Optimization (ADMM)
[50]	Distributed	Optimization (Lagrangian method)	Pricing algorithm (MMR)	Optimization (Lagrangian method)
[45]	Centralized	Optimization (NLP)	Auction-based (CDA)	Optimization (NLP)
[57]	Distributed	Cooperative game/Optimization (MILP)	Pricing algorithm (MMR)	Cooperative game/Optimization (MILP)
[51]	Decentralized	Cooperative game/Optimization (ADMM)	Cooperative game + benefit distribution	Cooperative game/Optimization (ADMM)
[46]	Distributed	Optimization (MILP)	Auction-based (CDA)	Optimization (MILP)
[36]	Distributed	Optimization (NLP)	Pricing algorithm (SDR)	Optimization (NLP)
[52]	Decentralized	Optimization (ADMM)	Bilateral contracts	Optimization (ADMM)
[55]	Decentralized	Agent-based approach	Auction-based (CDA)	Agent-based approach
[19]	Decentralized	Optimization (ADMM)	Pricing algorithm (marginal price)	Optimization (ADMM)

Table 8.1 (Continued)

References	P2P market design	Decision-making method	Pricing method	Market simulation method
[53]	Distributed	Optimization (ADMM)	Bilateral contracts	Optimization (ADMM)
[58]	Decentralized	Noncooperative game/Optimization (QP)	Bilateral contracts	Non-cooperative game/Optimization (QP)
[20]	Distributed	Cooperative game/Optimization (NLP)	Pricing algorithm (MMR)	Cooperative game/Optimization (NLP)
[48]	Decentralized	Noncooperative game/Optimization (LP)	Bilateral contracts	Noncooperative game/Optimization (LP)
[49]	Decentralized	Optimization (ADMM)	Bilateral contracts	Optimization (ADMM)
[38]	Distributed	Agent-based approach	Pricing algorithm	Agent-based approach
[29]	Distributed	Optimization (MILP)	Auction-based (CDA)	Optimization (MILP)
[54]	Decentralized	Cooperative game/Optimization (ADMM)	Bilateral contracts	Cooperative game/Optimization (ADMM)

is limited to the role of a local market operator rather than a power system operator [38]. That is, the P2P coordinator agent is responsible for managing the P2P energy transactions but also trading with the retailer to purchase or sell electricity to balance the energy deficit or surplus.

On the other hand, Zhou et al. [38] affirmed that the retailer agent may act as a passive agent who provides pre-announced electricity prices to the P2P coordinator. With this assumption, the internal P2P prices have to be bounded between market and retail prices. For example, Azim et al. [2] assumed the off-peak and peak electricity retail prices for unmet consumers' electricity and flat FiT rate for solar surplus. A similar P2P market price bundling between ToU and FiT tariffs has been proposed by Qiu et al. [55] and Shi et al. [19], where fixed internal prices are first set based on ToU prices, and then the CDA mechanism is employed to compute the final prices. Khorasamy et al. [46] proposed a local P2P energy trading among multi-carrier energy systems (electricity, thermal, and cooling), which trade their energy surplus/deficit to other EHs at prices bounded by buying and selling prices of "district markets" (upstream markets). The authors considered market-indexed hourly selling prices and flat daily prices for the purchase of surplus electricity (as the minimum selling price of each day). Similarly, Jiang et al. [51] considered ToU selling prices and flat daily prices to purchase electricity. Wang et al. [58] evaluated the impact of the forecasted spot market, ToU, and peak-valley pricing schemes as input reference prices to the power-heating contract transaction mechanism. The same approach regarding retail prices is modeled in [61, 62].

In the case of massive P2P deployment, the hourly demand curve of the whole system will be modified to a certain extent, impacting a higher uncertainty on the retail energy volume and revenues [60], which fosters a rethink of the business model and role. However, it is important to remember that retailers have been implementing demand-response measures, such as direct load control, dynamic pricing mechanisms, or ToU tariffs, without the existence of a P2P market.

On the other hand, some authors propose new P2P actors (related to the operation, market, or retail sector). Nonetheless, there is no consensus about all the relationships, rights, and duties among all involved P2P actors. For example, in [52], a deviation punishment settlement mechanism of P2P trading is settled, similar to a bilateral contract, to penalize a certain degree of load deviation. In a higher stage of the P2P market, this deviation penalty tends to be reduced coordinately. In [63], the energy sharing coordinator (ESC) acts as the operator of the community P2P sharing mechanisms and clears the unmatched electricity supply or demand through the power grid. Similarly, Wang et al. [20] introduced the electricity utility company (EUC), a new role that combines retailers and wholesalers in the future energy system who trades electricity with prosumers on the distribution level, in this case, referring to day-ahead prices and ToU retail tariffs. Wang et al. [20] also suggested that in the case of a profit-driven market or network operator, it is necessary to introduce extra costs to represent market operation costs, network tariffs, taxes, and other derived costs. Liu et al. [36] proposed that the energy service provider (ESP) could charge a service fee from the peers, fair to energy sharing and directly proportional to the amount of shared PV energy, in order to cover expenditures, sharing activities, and technical losses. The P2P internal prices are again bounded between the purchase electricity prices and FiT rates. Similarly, Huang et al. [57] evaluate that the main monetary benefit of MG P2P market operators comes from registration (entry) fees for users and services charges (i.e., generation, operation, and maintenance services). MG users are able to engage in P2P trading among peers while purchasing electricity from the electrical grid at TOU price and selling surplus energy at a FiT scheme. Nguyen et al. [64] also consider a scenario with profit-driven actors in which the P2P trader margin is applied as a percentage to the P2P internal price and local use of network service, similar to retailers paid for the network charges in the traditional bid.

Wang et al. [65] presented the *crowdsourcee* peers who are able to give control of their resources to the utility or system operator for day-ahead planning or real-time adjustments. In return, the operator provides socio-economic incentives or a discount on the electricity bill. The authors relied on time-varying electricity prices for users based on distributed locational marginal price (DLMP). Cui et al. [39] evaluated the trading costs contracted to the retail energy provider (REP), being indexed to the market prices and additional taxes on carbon emissions. Melendez et al. [66] declared that peers are required to buy needed electricity from the distribution network (paying the market price) and to trade P2P electricity using plain or swing option contracts (based on strike prices, quantity of energy, or degree of flexibility) among loads and aggregation of EVs. In this case, the loads' entities pay a strike price and "option value" fee to the aggregation of EVs.

On the other hand, Heo et al. [50] raised concern about existing market barriers to small-scale customers, such as households, who are not accustomed to the active role of being a player in the electricity market. Under this paradigm, the authors proposed that the P2P operator decides the trading schedules of peers and their marginal/trading prices. The peers' bill is influenced by electricity tariffs (several types of time-based tariffs, net billing [NB] mechanism or purchasing agreement [PA] for surplus energy), distribution network usage charges (NUCs), and P2P platform service charges (PSCs). Zhang et al. [49] also determined the DLMP and grid usage prices (GUPs) established by the ancillary market for each P2P transaction, acting as incentive signals for the P2P market to support the grid operation in terms of loss reduction, voltage support, and congestion management. Similarly, Ghaemi et al. [56] computed DLMPs after the market clearing, which include DSO network congestion and losses based on OPF.

Espadinha et al. [44] conducted a sensitivity analysis of the wholesale market prices to evaluate how the spread and variability affect the P2P trades when buying and selling electricity from/to the grid. The P2P trading prices are also influenced by different network charges and electricity tariffs for each type of peer (industrial, urban/rural municipalities). Guerrero et al. [43] considered the range of P2P trading prices between ToU and FiT rates, and peers will pay a charge for using the physical network by assigning congestion charges. Similarly, Ruan et al. [67] included network usage fees based on the amount of electricity trading and the electrical distance between prosumers during the electricity trading process, whose unit price can be adjusted with security constraints while considering buying and selling prices from the grid.

As can be concluded, all consulted literature assumed that peers are acting as "price takers" without influencing the wholesale market prices and retailer business model, as seen in Table 8.2. With a higher deployment, the economic and technical impacts can no longer be ignored [60]. To conclude, the following issues need further clarification in future research:

- Peers should be considered like "price makers" in the wholesale market and retail business.
- Retailers should consider the impact of P2P energy trading on their businesses.
- Complex interactions between retailers and P2P coordinators shall be investigated.
- Retail imbalances directly produced by P2P energy trading should be suitably evaluated.
- The P2P coordinator needs to design an attractive business model, including service fees, P2P margin, or other charges, in order to cover expenditures and market-sharing activities.
- Network service usage should be adequately allocated among peer transactions.

Table 8.2 Literature review of P2P energy trading interaction with the wholesale market, retailers, and network operator.

References	Retail selling price	Retail purchase prices	P2P fees/ charges	Network charges
[2]	ToU	FiT	—	—
[39]	Market-indexed	Market-indexed	Carbon emissions	—
[44]	Retail tariffs	Retail tariffs	—	Network charges
[21]	ToU (3 periods)	ToU (3 periods)	—	Network usage fee
[56]	Market/ToU	—	—	Network charges (congestion and losses)
[43]	ToU	FiT	—	Congestion charges
[50]	ToU	NB, PA	PSC	NUC
[57]	ToU	FiT	Entry fee, PSC	—
[51]	ToU (3 periods)	Daily flat	—	—
[46]	Market-indexed	Daily flat	—	—
[36]	ToU	FiT	Service fee	—
[52]	Day-ahead price	—	Deviation penalty	—
[66]	Market/Flat price	—	Option value fee	—
[64]	Retail tariffs	FiT	P2P margin	Network service use
[55]	ToU	FiT	—	—
[67]	Market-indexed	Market-indexed	—	Network usage fee
[19]	ToU	FiT	—	—
[65]	Market/DLMP	Market-indexed	—	Incentives/Discount
[58]	Market/ToU	—	—	—
[20]	Market/ToU	—	—	—
[48]	Market/Flat price	Market/Flat price	—	—
[49]	Market/DLMP	Market-indexed	—	Network usage prices
[63]	Retail tariffs	FiT	—	—
[54]	Market/ToU	—	—	Network usage fee

8.4.3 Contribution to Ancillary Services

Generally, P2P trading has mainly involved energy transactions among peers, i.e., in a community, but there are only a few studies that shine a light on ancillary services in a P2P market concept.

Most of these works are focused on compensating for peers' energy imbalances due to uncertainty. For instance, Guo et al. [59] stated that none of the existing literature works considers clearing energy and reserve via a P2P market mechanism. The authors proposed a novel P2P joint energy and reserve market in which each agent can negotiate with neighboring agents to determine the quantities and prices of traded energy and reserve. These reserve needs are allocated in order to compensate for the renewable uncertainty of prosumers but not to support the upstream power grid. Hu et al. [45] considered simultaneous energy and balancing services trading among different NGs inside an islanded MG based on P2P trading. The balancing service comprises power imbalances due to inaccurate forecasting and real-time frequency regulation service provision. Several capacity remuneration mechanisms are considered to clear both ancillary services, including marginal and pay-as-bid pricing, but these are not based on P2P mechanisms. Similarly, Lu et al. [52] proposed that agents (in an alliance as sellers and buyers) purchase reserve capacity from mobile energy storage suppliers to hedge the risk of deviation penalty in the P2P market commitments and, thus, to increase the bilateral welfare of prosumers. Finally, Wang et al. [65] outlined the design of *crowdsourcee* resources that provide near real-time grid management to balance demand shortage or surplus at the feeder notified by the operator. They proposed a design of monetary incentives that reward *crowdsourcees* in a P2P scheme.

Other works conceived ancillary services using a centralized approach after the P2P trading. Zhang et al. [49] proposed an ancillary market cleared in a centralized way by the DSO to remove any existing violation of grid constraints during negotiations. A generic product is oriented to procure load flexibility for real-time grid stability (congestion, voltage, loss, peak shaving). Similarly, Zhou et al. [29] proposed ancillary service provisions not associated with specific purposes or services, with different time resolutions to adapt to P2P energy trading. The power utility procures flexibility for foreseen grid problems, which is average among peers.

Highlighting the work of Marques et al. [68], a full P2P scheme in which system operators inform and differentiate the needed products and peers have the chance to re-negotiate the bilateral transactions made was proposed. They solved congestion issues at the DSO level, based on the Euclidean distance between peers; voltage constraints, based on the sensitivity of the offered flexibility; and frequency regulation, for imbalances between production and consumption. A summary of the above-mentioned review is presented in Table 8.3.

8.4.4 Consideration of Multi-energy Network Constraints

Although, in most cases, P2P trading is envisaged as virtual energy trading, in a multi-energy system, this energy exchange will vary consumption and generation patterns, which ultimately may affect the energy flow in power, heat, cooling, or gas networks, provoking unexpected events on these distribution grids (i.e., voltage profiles and higher losses or grid congestion). Therefore, not just the electric power networks need to be ready between peers so that the agreed-upon electricity can be delivered [60], but also the networks of other energy vectors. Table 8.4 summarizes the most interesting research papers addressing this topic.

Table 8.3 Literature review of ancillary services provision in the framework of P2P energy trading.

Reference	Ancillary service	Target actor	Description
[59]	Reserve	Among peers	Reserve needs to compensate for renewable uncertainty
[45]	Balancing	Among nano grids	Power imbalances due to inaccurate forecasting and real-time frequency regulation service provision
[52]	Reserve	Peers and mobile energy storage	Agents purchase reserve capacity from mobile energy storage suppliers to reduce P2P deviation
[65]	Balancing	Among peers	Balance demand shortage or surplus at the feeder
[49]	Ancillary	DSO	Centralized market to solve multiple DSO needs
[29]	Ancillary	Power utility	Procure multiple needs average among peers
[68]	Congestion Balancing	DSO TSO	Full P2P approach for congestion and voltage issues Frequency regulation oriented to energy imbalances

Table 8.4 Literature review of P2P energy trading, including power network constraints.

References	Network	Power losses	Voltage	Power flow	Heat flow
[2]	IEEE 8500-node test network	✓	✓	—	—
[56]	Netherlands MV network, DHN	✓	✓	✓	✓
[53]	4-Bus test power network, DHN	✓	✓	✓	✓
[43]	UK low-voltage network	✓	✓	✓	—
[59]	IEEE 9-bus test network	—	✓	✓	—
[50]	IEEE 18-bus test network	—	—	✓	—
[68]	IEEE 37-bus test network	—	✓	✓	—
[67]	148 and 297-bus test network	✓	✓	✓	—
[53]	IEEE 33-bus test network	—	✓	✓	—
[65]	SCE 56-bus test feeder	—	✓	✓	—
[49]	144-bus network	✓	✓	✓	—
[54]	IEEE 33-bus test network/DHN	✓	✓	✓	✓
[19]	5-Bus power and heat networks	—	✓	✓	✓

For example, Azim et al. [2] investigated the impact of P2P trading on power losses and nodal voltages on the IEEE 8500-node test feeder, where network losses increase at peak demand periods under the P2P scheme. They also suggested that the network operators may be required to limit the number of prosumers to reduce the impact on extreme voltage values. Guerrero et al. [43] presented a sensitivity analysis to assess the impact of P2P on a low-voltage network in the United Kingdom and to ensure that network constraints were not violated during the energy exchanges. As all transactions had to be validated during the bidding process, they proposed to internalize extra costs associated with the network sensitivity coefficients (voltage, loss, and power flow) in the P2P exchange. Similarly, Marques et al. [68] proposed a tariff design to handle and solve network issues at the DSO level: voltage issues (sensitivity factor between the peers and the overloaded node) and congestion issues (Euclidean distance between peers), while power flow validation is performed on the IEEE 37-node network to ensure the feasibility of P2P transactions. Ghaemi and Anvari-Moghaddam [56] analyzed potential network constraints on a representative medium-voltage network in the Netherlands as well as thermal energy flow adequacy on a district heating network (DHN).

On the other hand, Ruan et al. [67] proposed to adjust network usage fees based on the amount of electricity trading and the electrical distance between peers derived from the network power loss and security constraints. The P2P electricity trading is evaluated in a 148-bus and 297-bus grid, where a small penalty is added to the trading amount in case of infeasibility of OPF. Zhang et al. [49] presented a novel P2P market design in which the distribution locational marginal prices are determined by solving the OPF problem. Consequently, GUPs for each P2P transaction are calculated, acting as incentive signals to support the grid operation in terms of loss reduction, voltage support, and congestion management. Shi et al. [19] presented a marginal cost-based pricing mechanism for P2P transactions in an electricity-heat market employing a second-order cone programming relaxation for the OPF and a linear approximation for heat energy flow.

Other authors have also considered a network model to validate the P2P energy exchange: Guo et al. [59] applied a linearized DC OPF on the IEEE 9-bus test system. Heo et al. [50] verified the performance of the proposed P2P trading scheme on the IEEE 18-bus network. G. Sun et al. [53] aimed to minimize the operating costs of the IEEE 33-bus distribution network, including electricity-carbon P2P transactions. Wang et al. [65] validated blockchain-based P2P energy trading. Zhang et al. [62] were oriented to the power transmission grid. Zou et al. [54] minimized line power losses in the IEEE 33-bus distribution network, also considering thermal energy (heat) flows in a DHN. C. Sun et al. [18] proposed a decentralized P2P framework for a multi-energy system based on the dynamic update of power and heat loss factors to seek market equilibrium using a four-node electric and heat network.

As can be concluded, considering network constraints (losses, voltage thresholds, and line thermal capacity flows) to validate the P2P energy exchange is essential as well as evaluating the impact of P2P transactions on technical losses, undesired voltage, or grid congestion.

8.5 P2P Approach in the eNeuron Project

8.5.1 Overview of the eNeuron Concept

The H2020 project eNeuron (greEN Energy hUbs for local integRated energy cOmmunities optimization) is aimed at developing innovative tools for the optimal design and operation of integrated local energy communities (ILECs) based on multi-carrier energy systems [69]. To achieve this objective, a conceptual architecture of the ILEC that includes two levels of operation: (i) the EH and (ii) the micro-Energy Hub (mEH) has been defined (see Figure 8.3).

The EH represents the overall *energy community*. It is the main architectural and operation solution for managing the energy flows among the multi-carrier energy systems within the ILEC. The EH also acts as an intermediary between the ILEC and the rest of the system, such as electricity suppliers or wholesale markets, when satisfying the energy needs of the ILEC or the DSO when participating in the provision of congestion management services.

The mEH represents the *prosumer* that is located at the lower architectural level and comprises different technologies related to multiple energy vectors (generation, conversion, and storage). mEHs can cooperate to satisfy the energy needs of the entire community represented by the EH by exchanging energy among them. These are equipped with the required energy management systems that optimize locally the operation of their multi-carrier energy systems to fulfill their own objectives and the overall goals of the community (e.g., minimize energy cost, maximize self-consumption, or minimize CO_2 emissions).

Based on these concepts, the eNeuron project has developed an integrated toolbox that includes the required tools and functionalities to support the two mentioned levels of operation, the EH and mEH, as depicted in Figure 8.3.

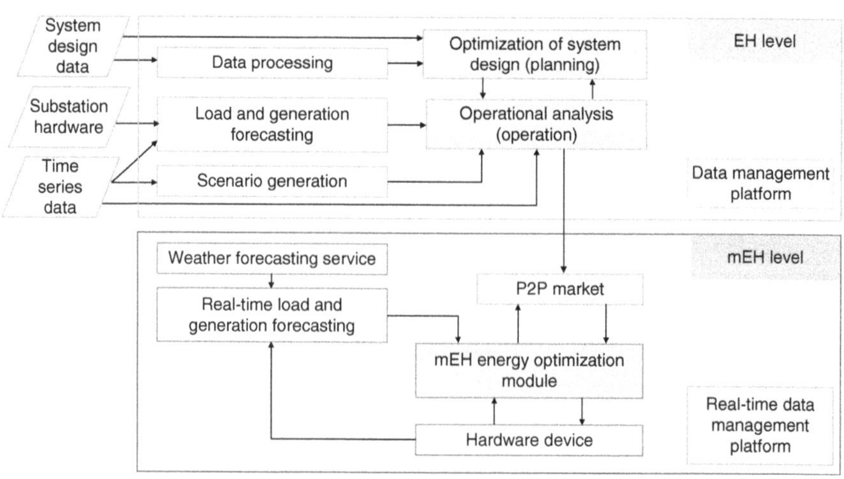

Figure 8.3 eNeuron toolbox conceptual scheme.

At the *EH level*, the tools developed contribute to two different processes: (i) the *planning phase* that is aimed at selecting the optimal combination and sizing of multi-carrier energy system technologies within the LEC and their operation strategies based on economic and sustainability objectives and (ii) the *operational phase* that, based on a multi-objective optimization approach that takes into account uncertainties on renewable generation, market prices, and loads, determines the day-ahead optimal operation dispatch of the EH [70].

At the *mEH level*, the toolbox includes a *P2P market* platform to manage P2P energy trading in a local, intra-day market. This platform aims to operate the mEHs (peers) within the EH (ILEC) to fulfill the global objective of matching the day-ahead optimal dispatch defined at the EH level as closely as possible. The P2P market platform works interactively with the energy management systems of the mEHs (*energy optimization module* based on Kazemi's work [71]) to decide the real-time scheduling of each mEH according to the P2P internal prices broadcasted by the P2P market.

The day-ahead operation profile of the EH is defined on a daily basis by the operational analysis tool at the EH level, taking into account the optimized participation of the ILEC in the day-ahead electricity market. Consequently, the ultimate goal of the P2P market platform is to set a market mechanism that facilitates the management of the energy resources at the mEH level to minimize the deviation of the aggregated power profile of the mEHs in real-time from the objective power profile. As a result, it minimizes the imbalance of costs for the ILEC.

In the majority of the solutions found in the literature [72], the objective of the developed P2P market is to maximize self-consumption within the ILEC, that is, minimize the amount of energy that the ILEC buys/sells to/from the external energy system (e.g., supplier/retailer). The developed approach in the eNeuron project, as the combination of a day-ahead optimization with intra-day P2P energy sharing, enables the participation of an ILEC in different electricity markets and in the provision of ancillary services (e.g., congestion management), which represents a gap in the current literature as mentioned in Section 8.4.3.

8.5.2 eNeuron P2P Market Platform

8.5.2.1 Architecture

The eNeuron P2P market platform is based on a distributed P2P architecture, as depicted in Figure 8.4. Within this architecture, no communication exists among peers. They directly communicate with a P2P coordinator who manages a local energy trading market to enable energy transactions among them. In this approach, the P2P coordinator has no direct control over the energy resources within the ILEC. It just decides the optimal price signal (*Local P2P pricing module*) to be sent to all peers to incentivize them to modify their power consumption/generation schedules to reach the overall objective, i.e., minimize the deviations from the day-ahead energy schedule for the community, which is dictated by the EH level. The decision-making process is located at the mEH level (*energy optimization module*), which decides the optimal real-time scheduling of each mEH, based on a multi-objective optimization approach according to the received price signal.

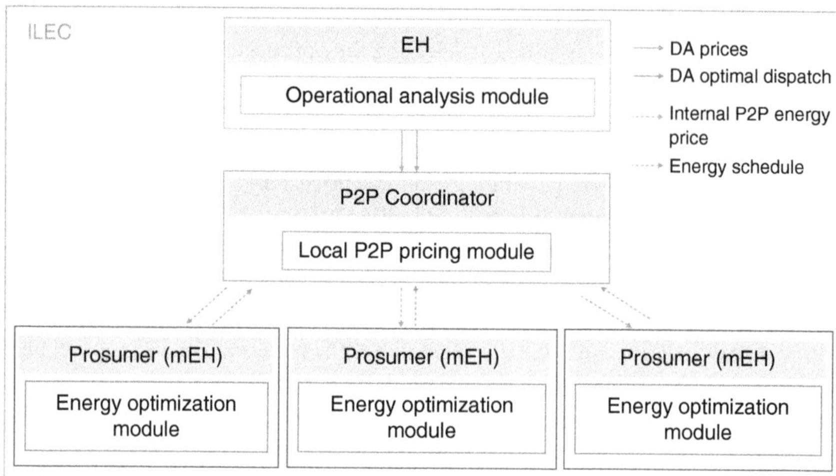

Figure 8.4 eNeuron P2P market architecture.

8.5.2.2 Algorithm Definition

The P2P market implements an iterative mechanism that seeks to reduce the error in each iteration between the aggregated optimal energy profiles for all mEHs in real-time and the aggregated day-ahead energy profile of the EH. This process is repeated until the objective is reached.

This P2P market is run with a rolling time horizon of 24 hours, with every control cycle of 15 minutes of the current day. However, these values are fully parameterizable and can be adapted to the requirements of the considered ILEC.

The main steps of the P2P market are as follows:

1. The P2P coordinator sends the day-ahead market price profile to the mEHs in the ILEC.
2. Each mEH optimizes its energy resources (batteries, EVs, HPs, CHP, etc.) according to it and sends back to the P2P coordinator its optimal net energy profile, calculated as the difference between the aggregated consumption and generation.
3. The P2P coordinator aggregates all the mEH energy profiles and checks if the aggregation of the real-time operation profiles of the mEHs matches the aggregated objective profile for the EH or whether there have not been significant changes in the aggregated real-time power profile since the last iteration. If any of the mentioned checks are true, the iterative process ends; otherwise, the process continues to step 4.
4. The P2P coordinator computes a new internal energy price profile to incentivize mEHs to change their energy profiles so that the real-time EH's energy profile is better aligned with the day-ahead EH profile.
5. The new internal energy price profile is sent to the mEHs, and the process is repeated from step 2.

The internal energy price profile computed by the P2P coordinator in each iteration is calculated proportionally to the deviation of the aggregated mEH power profiles in real-time from the day-ahead power profile for the EH. For this purpose, the day-ahead market price profile is taken as a reference, and this is modified in each time step proportionally to the calculated deviation. If there is an upward deviation, the price is increased to incentivize power generation increases or power consumption decreases. If there is a downward deviation, the price decreases to encourage power generation reduction or increase in power consumption. If there is no deviation at a certain time step, the considered internal P2P price would be the reference one, that is, the day-ahead price value corresponding to that time step.

As mEHs are allowed to modify their power profiles according to the dynamic price signal in each iteration, the variation in the resultant aggregated power profile of the EH with respect to the objective power profile can be significant from one iteration to another, causing heavy oscillations in the internal energy price. These could sometimes lead to difficulties in finding convergence of the solution, making it necessary to implement specific convergence techniques at the mEH optimization level, such as the step length control [38]. This technique is based on limiting the flexibility between iterations by constraining the power variation of each mEH from one iteration to another or by limiting the number of mEH allowed to react in each iteration.

8.6 Conclusion

This chapter highlights the relevance of the P2P energy trading concept, mechanisms, and market features that empower consumers to actively participate in such transaction models. P2P trading facilitates direct transactions between energy producers and consumers, unlocks new economic opportunities, enhances consumer choice, and contributes to the global commitment to a cleaner and sustainable energy future. However, addressing regulatory, technical, and market challenges will be crucial to realizing the full potential of P2P mechanisms and driving the energy transition toward a decentralized and democratized energy landscape. The main challenges include integrating new DERs into the multi-energy system networks, understanding the impact on these networks in terms of technical constraints, providing ancillary services for the electric grid, establishing a reliable and secure information management infrastructure and platform, and strategically selecting the appropriate energy market type for efficient energy exchange. In order to overcome these hurdles and validate the advantages of the P2P market approach, various pilot projects and research initiatives have been conducted worldwide at different levels. These endeavors aim to demonstrate the technical and economic viability of P2P energy trading while paving the way for a more resilient, sustainable, and inclusive energy landscape.

Acknowledgment

The analysis and definition of the P2P framework presented in the chapter have been obtained in the eNeuron project. The eNeuron project has received funding from the

European Union's Horizon 2020 Research and Innovation Programme under Grant Agreement No. 957779.

List of Abbreviations

ADMM	alternating direction method of multipliers
BS	bill sharing
CDA	continuous double auction
CHP	combined heating and power
DER	distributed energy resource
DHN	district heating network
DLMP	distributed locational marginal price
DSO	distribution system operator
EH	energy hub
ESC	energy sharing coordinator
ESP	energy service provider
ESS	energy storage system
EU	European Union
EUC	electricity utility company
EV	electric vehicle
FiT	feed-in tariff
GUPs	grid usage prices
HP	heat pump
ICT	information and communication technology
ILECs	integrated local energy communities
IMR	intermediate market rate
LEM	local energy market
LP	linear programming
mEH	micro-Energy Hub
MG	microgrid
MILP	mixed-integer linear programming
MINLP	mixed-integer nonlinear programming
MMR	mid-market rate
NB	net billing
NG	nano grid
NUCs	network usage charges
OPF	optimal power flow
P2P	peer-to-peer
PA	purchasing agreement
PSCs	P2P platform service charges
PV	photovoltaic
QP	quadratic programming
RA	reverse auction
REP	retail energy provider

SDR supply–demand ratio
ToU time-of-use
VPP virtual power plant

References

1 Tushar, W., Saha, T.K., Yuen, C. et al. (2020). Peer-to-peer trading in electricity networks: an overview. *IEEE Transactions on Smart Grid* 11 (4): 3185–3200. https://doi.org/10.1109/TSG.2020.2969657.

2 Azim, M.I., Tushar, W., and Saha, T.K. (2020). Investigating the impact of P2P trading on power losses in grid-connected networks with prosumers. *Applied Energy* 263: https://doi.org/10.1016/j.apenergy.2020.114687.

3 Bukar, A.L., Hamza, M.F., Ayub, S. et al. (2023). Peer-to-peer electricity trading: a systematic review on current developments and perspectives. *Renewable Energy Focus* 44: 317–333. https://doi.org/10.1016/J.REF.2023.01.008.

4 Tushar, W., Yuen, C., Saha, T.K. et al. (2021). Peer-to-peer energy systems for connected communities: a review of recent advances and emerging challenges. *Applied Energy* 282: 116131. https://doi.org/10.1016/J.APENERGY.2020.116131.

5 Ye, Y., Tang, Y., Wang, H. et al. (2021). A scalable privacy-preserving multi-agent deep reinforcement learning approach for large-scale peer-to-peer transactive energy trading. *IEEE Transactions on Smart Grid* 12 (6): 5185–5200. https://doi.org/10.1109/TSG.2021.3103917.

6 European Commission (2020). greEN Energy hUbs for local integRated energy cOmmunities optimizatioN. https://cordis.europa.eu/project/id/957779 (accessed 05 September 2023).

7 Hirsch, A., Parag, Y., and Guerrero, J. (2018). Microgrids: a review of technologies, key drivers, and outstanding issues. *Renewable and Sustainable Energy Reviews* 90: 402–411. https://doi.org/10.1016/j.rser.2018.03.040.

8 Wu, Y., Wu, Y., Cimen, H. et al. (2022b). Towards collective energy community: potential roles of microgrid and blockchain to go beyond P2P energy trading. *Applied Energy* 314: https://doi.org/10.1016/j.apenergy.2022.119003.

9 Naval, N. and Yusta, J.M. (2021). Virtual power plant models and electricity markets – a review. *Renewable and Sustainable Energy Reviews* 149: https://doi.org/10.1016/j.rser.2021.111393.

10 European Parliament and Council of the EU (2019). Directive (EU) 2019/944 on common rules for the internal market for electricity and amending directive 2012/27/EU. *Official Journal of the European Union* 1–75.

11 De Almeida, L., Cappelli, V., Klausmann, N., and Van Soest, H. (2021). Peer-to-peer trading and energy community in the electricity market analysing the literature on law and regulation and looking ahead to future challenges.

12 Wu, Y., Wu, Y., Cimen, H. et al. (2022a). P2P energy trading: blockchain-enabled P2P energy society with multi-scale flexibility services. *Energy Reports* 8: 3614–3628. https://doi.org/10.1016/j.egyr.2022.02.074.

13 CEER (2019). Consultation on dynamic regulation to enable digitalisation of the energy system: conclusions paper.

14 European Commission (2022). Commission staff working document. Digitalising the energy system – EU action plan. https://doi.org/10.2760/358256 (accessed 05 September 2023).

15 Suthar, S., Cherukuri, S.H.C., and Pindoriya, N.M. (2023). Peer-to-peer energy trading in smart grid: Frameworks, implementation methodologies, and demonstration projects. *Electric Power Systems Research* 214: https://doi.org/10.1016/j.epsr.2022.108907.

16 Seyedhossein, S.S. and Moeini-Aghtaie, M. (2022). Risk management framework of peer-to-peer electricity markets. *Energy* 261: https://doi.org/10.1016/j.energy.2022.125264.

17 Bjarghov, S., Loschenbrand, M., Ibn Saif, A.U.N. et al. (2021). Developments and challenges in local electricity markets: a comprehensive review. *IEEE Access* 9: 58910–58943. https://doi.org/10.1109/ACCESS.2021.3071830.

18 Sousa, T., Soares, T., Pinson, P. et al. (2019). Peer-to-peer and community-based markets: a comprehensive review. *Renewable and Sustainable Energy Reviews* 104: 367–378. https://doi.org/10.1016/J.RSER.2019.01.036.

19 Shi, M., Wang, H., Xie, P. et al. (2023). Distributed energy scheduling for integrated energy system clusters with peer-to-peer energy transaction. *IEEE Transactions on Smart Grid* 14 (1): 142–156. https://doi.org/10.1109/TSG.2022.3197435.

20 Wang, N., Liu, Z., Heijnen, P., and Warnier, M. (2022). A peer-to-peer market mechanism incorporating multi-energy coupling and cooperative behaviors. *Applied Energy* 311: 118572. https://doi.org/10.1016/J.APENERGY.2022.118572.

21 Gao, J., Shao, Z., Chen, F. et al. (2023). Distributed robust operation strategy of multi-microgrid based on peer-to-peer multi-energy trading. *IET Energy Systems Integration* https://doi.org/10.1049/ESI2.12107.

22 International Renewable Energy Agency (2020). Peer-to-peer electricity trading – innovation landscape brief. www.irena.org

23 Liu, Y., Wu, L., and Li, J. (2019). Peer-to-peer (P2P) electricity trading in distribution systems of the future. *The Electricity Journal* 32 (4): 2–6. https://doi.org/10.1016/J.TEJ.2019.03.002.

24 Ableitner, L., Tiefenbeck, V., Meeuw, A. et al. (2020). User behavior in a real-world peer-to-peer electricity market. *Applied Energy* 270: 115061. https://doi.org/10.1016/J.APENERGY.2020.115061.

25 Domènech Monfort, M., De Jesús, C., Wanapinit, N., and Hartmann, N. (2022). A review of peer-to-peer energy trading with standard terminology proposal and a techno-economic characterisation matrix. *Energies* 15 (23): https://doi.org/10.3390/en15239070.

26 Parag, Y. and Sovacool, B.K. (2016). Electricity market design for the prosumer era. *Nature Energy* 1 (4): 1–6. https://doi.org/10.1038/nenergy.2016.32.

27 Papadimitriou, C., Di Somma, M., Charalambous, C. et al. (2023). A comprehensive review of the design and operation optimization of energy hubs and their interaction with the markets and external networks. *Energies* 16 (10): 4018. https://doi.org/10.3390/en16104018.

28 Mengelkamp, E., Gärttner, J., Rock, K. et al. (2018). Designing microgrid energy markets: a case study: the Brooklyn microgrid. *Applied Energy* 210: 870–880. https://doi.org/10.1016/j.apenergy.2017.06.054.

29 Zhou, Y., Wu, J., Song, G., and Long, C. (2020). Framework design and optimal bidding strategy for ancillary service provision from a peer-to-peer energy trading community. *Applied Energy* 278: 115671. https://doi.org/10.1016/J.APENERGY.2020.115671.

30 Liu, Q., Shen, X., and Gu, Y. (2019). Linearized ADMM for nonconvex nonsmooth optimization with convergence analysis. *IEEE Access* 7: 76131–76144. https://doi.org/10.1109/ACCESS.2019.2914461.

31 Pilz, M. and Al-Fagih, L. (2019). Recent advances in local energy trading in the smart grid based on game-theoretic approaches. *IEEE Transactions on Smart Grid* 10 (2): 1363–1371. https://doi.org/10.1109/TSG.2017.2764275.

32 Lin, J., Pipattanasomporn, M., and Rahman, S. (2019). Comparative analysis of auction mechanisms and bidding strategies for P2P solar transactive energy markets. *Applied Energy* 255: 113687. https://doi.org/10.1016/j.apenergy.2019.113687.

33 Cintuglu, M.H., Martin, H., and Mohammed, O.A. (2015). Real-time implementation of multiagent-based game theory reverse auction model for microgrid market operation. *IEEE Transactions on Smart Grid* 6 (2): 1064–1072. https://doi.org/10.1109/TSG.2014.2387215.

34 Mello, J., de Lorenzo, C., Campos, F.A., and Villar, J. (2023). Pricing and simulating energy transactions in energy communities. *Energies* 16 (4): 1949. https://doi.org/10.3390/en16041949.

35 Long, C., Wu, J., Zhang, C. et al. (2017). Peer-to-peer energy trading in a community microgrid. *2017 IEEE Power & Energy Society General Meeting*, Chicago, IL, USA (16-20 July 2017), pp. 1–5. https://doi.org/10.1109/PESGM.2017.8274546 (accessed 05 September 2023).

36 Liu, N., Yu, X., Wang, C. et al. (2017). Energy-sharing model with price-based demand response for microgrids of peer-to-peer prosumers. *IEEE Transactions on Power Systems* 32 (5): 3569–3583. https://doi.org/10.1109/TPWRS.2017.2649558.

37 Long, C., Wu, J., Zhou, Y., and Jenkins, N. (2018). Peer-to-peer energy sharing through a two-stage aggregated battery control in a community microgrid. *Applied Energy* 226: 261–276. https://doi.org/10.1016/j.apenergy.2018.05.097.

38 Zhou, Y., Wu, J., and Long, C. (2018). Evaluation of peer-to-peer energy sharing mechanisms based on a multiagent simulation framework. *Applied Energy* 222: 993–1022. https://doi.org/10.1016/J.APENERGY.2018.02.089.

39 Cui, S., Wang, Y.W., and Xiao, J.W. (2019). Peer-to-peer energy sharing among smart energy buildings by distributed transaction. *IEEE Transactions on Smart Grid* 10 (6): 6491–6501. https://doi.org/10.1109/TSG.2019.2906059.

40 Sun, G., Zhang, Z., Zhou, Y. et al. (2023). Bi-level model for integrated energy service providers in joint electricity and carbon P2P market. *Journal of Cleaner Production* 393: 136303. https://doi.org/10.1016/J.JCLEPRO.2023.136303.

41 Zhou, Y., Wu, J., Long, C. et al. (2017). Performance evaluation of peer-to-peer energy sharing models. *Energy Procedia* 143: 817–822. https://doi.org/10.1016/j.egypro.2017.12.768.

42 Tushar, W., Yuen, C., Mohsenian-Rad, H. et al. (2018). Transforming energy networks via peer-to-peer energy trading: the potential of game-theoretic approaches. *IEEE Signal Processing Magazine* 35 (4): 90–111. https://doi.org/10.1109/MSP.2018.2818327.

43 Guerrero, J., Chapman, A., and Verbic, G. (2018). Decentralized P2P energy trading under network constraints in a low-voltage network. http://arxiv.org/abs/1809.06976 (accessed 05 September 2023).

44 Espadinha, J., Baptista, P., and Neves, D. (2023). Assessing P2P energy markets contribution for 2050 decarbonization goals. *Sustainable Cities and Society* 92: https://doi.org/10.1016/j.scs.2023.104495.

45 Hu, Q., Zhu, Z., Bu, S. et al. (2021). A multi-market nanogrid P2P energy and ancillary service trading paradigm: mechanisms and implementations. *Applied Energy* 293: https://doi.org/10.1016/j.apenergy.2021.116938.

46 Khorasany, M., Najafi-Ghalelou, A., Razzaghi, R., and Mohammadi-Ivatloo, B. (2021). Transactive energy framework for optimal energy management of multi-carrier energy hubs under local electrical, thermal, and cooling market constraints. *International Journal of Electrical Power & Energy Systems* 129: https://doi.org/10.1016/j.ijepes.2021.106803.

47 Alvaro-Hermana, R., Fraile-Ardanuy, J., Zufiria, P.J. et al. (2016). Peer to peer energy trading with electric vehicles. *IEEE Intelligent Transportation Systems Magazine* 8 (3): 33–44. https://doi.org/10.1109/MITS.2016.2573178.

48 Yang, J., Xu, W., Ma, K., and Li, C. (2023). A three-stage multi-energy trading strategy based on P2P trading mode. *IEEE Transactions on Sustainable Energy* 14 (1): 233–241. https://doi.org/10.1109/TSTE.2022.3208369.

49 Zhang, K., Troitzsch, S., Hanif, S., and Hamacher, T. (2020). Coordinated market design for peer-to-peer energy trade and ancillary services in distribution grids. *IEEE Transactions on Smart Grid* 11: 2929–2941. https://www.researchgate.net/publication/338501038.

50 Heo, K., Kong, J., Oh, S., and Jung, J. (2021). Development of operator-oriented peer-to-peer energy trading model for integration into the existing distribution system. *International Journal of Electrical Power & Energy Systems* 125: https://doi.org/10.1016/j.ijepes.2020.106488.

51 Jiang, A., Yuan, H., and Li, D. (2021). A two-stage optimization approach on the decisions for prosumers and consumers within a community in the peer-to-peer energy sharing trading. *International Journal of Electrical Power & Energy Systems* 125: https://doi.org/10.1016/j.ijepes.2020.106527.

52 Lu, Z., Wang, J., Yong, W. et al. (2021). Coordinated P2P electricity trading model with aggregated alliance and reserve purchasing for hedging the risk of deviation penalty. *Energy Reports* 7: 426–435. https://doi.org/10.1016/j.egyr.2021.01.050.

53 Sun, C., Liu, Y., Li, Y. et al. (2023). Network-aware P2P multi-energy trading in decentralized electric-heat systems. *Applied Energy* 345: 121298. https://doi.org/10.1016/J.APENERGY.2023.121298.

54 Zou, Y., Xu, Y., Feng, X., and Nguyen, H.D. (2023). Peer-to-peer transactive energy trading of a reconfigurable multi-energy network. *IEEE Transactions on Smart Grid* 14 (3): 2236–2249. https://doi.org/10.1109/TSG.2022.3223378.

55 Qiu, D., Wang, J., Dong, Z. et al. (2023). Mean-field multi-agent reinforcement learning for peer-to-peer multi-energy trading. *IEEE Transactions on Power Systems* 38 (5): 4853–4866. https://doi.org/10.1109/TPWRS.2022.3217922.

56 Ghaemi, S. and Anvari-Moghaddam, A. (2023). Local energy communities with strategic behavior of multi-energy players for peer-to-peer trading: a techno-economic assessment. *Sustainable Energy, Grids and Networks* 34: 101059. https://doi.org/10.1016/J.SEGAN.2023.101059.

57 Huang, H., Nie, S., Lin, J. et al. (2020). Optimization of peer-to-peer power trading in a microgrid with distributed PV and battery energy storage systems. *Sustainability* 12 (923): https://doi.org/10.3390/su12030923.

58 Wang, L., Gu, W., Wu, Z. et al. (2020). Non-cooperative game-based multilateral contract transactions in power-heating integrated systems. *Applied Energy* 268: https://doi.org/10.1016/j.apenergy.2020.114930.

59 Guo, Z., Pinson, P., Chen, S. et al. (2021). Chance-constrained peer-to-peer joint energy and reserve market considering renewable generation uncertainty. *IEEE Transactions on Smart Grid* 12 (1): 798–809. https://doi.org/10.1109/TSG.2020.3019603.

60 Zhou, Y., Wu, J., Long, C., and Ming, W. (2020). State-of-the-art analysis and perspectives for peer-to-peer energy trading. *Engineering* 6 (7): 739–753. https://doi.org/10.1016/j.eng.2020.06.002.

61 Wang, Z., Yu, X., Mu, Y. et al. (2021). Peer-to-peer energy trading strategy for energy balance service provider (EBSP) considering market elasticity in community microgrid. *Applied Energy* 303: https://doi.org/10.1016/j.apenergy.2021.117596.

62 Zhang, W.Y., Chen, Y., Wang, Y., and Xu, Y. (2023). Equilibrium analysis of a peer-to-peer energy trading market with shared energy storage in a power transmission grid. *Energy* 274: https://doi.org/10.1016/j.energy.2023.127362.

63 Zhao, F., Li, Z., Wang, D., and Ma, T. (2023). Peer-to-peer energy sharing with demand-side management for fair revenue distribution and stable grid interaction in the photovoltaic community. *Journal of Cleaner Production* 383: https://doi.org/10.1016/j.jclepro.2022.135271.

64 Nguyen, S., Peng, W., Sokolowski, P. et al. (2018). Optimizing rooftop photovoltaic distributed generation with battery storage for peer-to-peer energy trading. *Applied Energy* 228: 2567–2580. https://doi.org/10.1016/j.apenergy.2018.07.042.

65 Wang, S., Taha, A.F., Wang, J. et al. (2019). Energy crowdsourcing and peer-to-peer energy trading in blockchain-enabled smart grids. *IEEE Transactions on Systems, Man, and Cybernetics: Systems* 49 (8): 1612–1623. https://doi.org/10.1109/TSMC.2019.2916565.

66 Melendez, K.A., Das, T.K., and Kwon, C. (2020). A Nash-bargaining model for trading of electricity between aggregations of peers. *International Journal of Electrical Power & Energy Systems* 123: https://doi.org/10.1016/j.ijepes.2020.106185.

67 Ruan, H., Gao, H., Qiu, H. et al. (2023). Distributed operation optimization of active distribution network with P2P electricity trading in blockchain environment. *Applied Energy* 331: https://doi.org/10.1016/j.apenergy.2022.120405.

68 Marques, J., Soares, T., and Morais, H. (2023). P2P flexibility markets models to support the coordination between the transmission system operators and distribution system operators. *Sustainable Energy, Grids and Networks* 34: https://doi.org/10.1016/j.segan.2023.101055.

69 Di Somma, M., Papadimitriou, C., Morch, A. et al. (2023). An innovative toolbox for the optimal design and operation of integrated local energy communities. *27th International Conference on Electricity Distribution (CIRED 2023)*, Rome, Italy (12–15 June 2023). pp. 676–680. https://doi.org/10.1049/ICP.2023.0460 (accessed 05 September 2023).

70 Askeland, M., Morch, A., Papadimitriou, C. et al. (2023). Workflow-based architecture for optimal planning of integrated local multi-energy systems. *2023 International Conference on Smart Energy Systems and Technologies, SEST 2023*, Mugla, Turkiye (04–06 September 2023). https://doi.org/10.1109/SEST57387.2023.10257503 (accessed 05 September 2023).

71 Kazemi, M., Papadimitriou, C., Paterakis, N.G. et al. (2023). Optimal design of multi-carrier and multi-objective home energy management system. *2023 International Conference on Smart Energy Systems and Technologies (SEST)*, Mugla, Turkiye (04–06 September 2023). pp. 1–6. https://doi.org/10.1109/SEST57387.2023.10257384 (accessed 05 September 2023).

72 Papadimitriou, C., Charalampous, C., Bracho, J. et al. (2022). Report on the energy hub concept and the multi objective programming approach of an energy hub. https://eneuron.eu/resources/ (accessed 05 September 2023).

9

Integration of Multiple Energy Communities: Transaction Prices, Reactive Power Control, and Ancillary Services

Alberto Borghetti[1], Tohid Harighi[1], Carlo Alberto Nucci[1], Giorgio Graditi[2], Marialaura Di Somma[2], and Martina Caliano[2]

[1] *University of Bologna, Department of Electrical, Electronic and Information Engineering, Viale Risorgimento, 2, 40136, Bologna, Italy*
[2] *Italian National Agency for New Technologies, Energy and Sustainable Economic Development, Lungotevere Thaon di Revel, 76, 00196, Rome, Italy*

9.1 Introduction

Energy communities are established to increase the local balance between production and consumption, allowing direct transactions between users who can function as consumers or producers through distributed generation (DG) [1]. They are expected to play a nonnegligible role in achieving the energy transition mandated by the Conference of Parties and the European Union (EU) Green Deal (e.g. [2–4]). Communities are also expected to facilitate the implementation of sector coupling solutions through electromobility, by providing cheaper solutions for electric vehicle (EV) charging and the exploitation of the flexibility services that can be offered. Community participants may include parking lots equipped with several EV-charging stations. These parking lots may be operated through optimization models/tools to exploit the storage capability of the vehicle batteries, especially if the charging stations are bidirectional. The current regulatory framework, including EU 2019/944 Electricity Market Directive and 2018/2001/EU Revised Renewable Energy Directive, allows for the presence of multiple communities in a single distribution network. This framework grants users the freedom to decide whether to join a community and independently choose their energy provider [5]. The presence of different providers with specific tariffs and DG power production costs affects transaction prices among community participants.

The emerging regulations do encourage the participation of end-users, whether as either individuals or aggregated collectives, in both the energy market and ancillary services markets. For instance, the Italian Regulatory Authority for Energy, Networks, and Environment (ARERA) has issued resolutions promoting the provision of local ancillary services by microgrids and energy communities, supporting the operation of distribution networks. These services involve qualified generation and storage units, reactive power compensation devices, and demand

Integrated Local Energy Communities: From Concepts and Enabling Conditions to Optimal Planning and Operation, First Edition. Edited by Marialaura Di Somma, Christina Papadimitriou, Giorgio Graditi, and Koen Kok.

response techniques. In some cases, energy communities may also act as distribution system operators (DSOs), providing services to connected users and the transmission system operator (TSO).

The chapter is structured as follows. Section 9.2 describes the proposed optimization model that considers the presence of multiple energy communities in the same distribution network with different energy providers. Section 9.3 focuses on calculating the limits for providing reactive power flexibility services. Section 9.4 reviews the support of energy communities for electro-mobility. Section 9.5 concludes the chapter and summarizes the key findings.

9.2 Multiple Energy Communities

This section proposes an optimization model that analyzes the scenario of multiple energy communities and diverse providers. It is assumed that members of each community can establish direct energy transactions among them, while energy sharing between different communities is not allowed. The model calculates fair transaction prices among community participants using shadow prices associated with relevant balance constraints. The model examines the presence of various numbers of communities with different characteristics (user proximity, size of generation, and storage units), considering the Institute of Electrical and Electronics Engineer (IEEE) 123 bus test feeder. The results compare day-ahead scheduling with and without communities and show the influence of two different energy providers on transaction prices within each community.

9.2.1 Optimization Model

The optimization model used to analyze the presence of multiple communities connected to the same power distribution network is basically an optimal power flow procedure for the next day, which includes direct transactions among the participants of each community. For this purpose, variable P_{LEC} is included that represents the effect of these direct exchanges, while variable P_{grid} denotes the exchange of each prosumer with its external energy provider. Obviously, as the meter of each prosumer is unique, P_{LEC} and P_{grid} at each prosumer should share the same sign, i.e. reselling is not possible. Each prosumer may choose a different external energy provider characterized by different price profiles during the day, which are considered known parameters in the optimization.

The power flow model utilized in this study adopts a convex relaxation approach, as described in previous works [6, 7], based on the DistFlow method [8]. The model assumes a three-phase network that is both radial and balanced. Each branch is represented by a T model, with series impedance denoted as $0.5(r_i + \text{j}x_i)$ and admittance of the central shunt branch as $\text{j}b_i$. For each branch i, the square of the current rms values at the two terminals (referred to as "in" and "out") is represented as u_{in} and u_{out}.

The model itself is deterministic but can be incorporated into a scenario-based stochastic approach, which can handle uncertainties, as explained in previous works [9].

As mentioned, in line with the methodology presented in Ref. [10], the interactions between participant i and any other member of the same local energy community (LEC) k at time t are expressed through variables $P_{\text{LEC }k,i,t}$, while the exchanges with the external energy provider are described using variables $P_{\text{grid }i,t}$.

The objective of the day-ahead scheduling is to minimize OF described in Eq. (9.1), which involves the cost/revenues of energy transactions between participants in the same community and their energy providers. The time horizon is divided into 96 periods of 15 minutes. Assuming the buying and selling tariffs (π_{buy} and π_{sell}, respectively) predefined by the energy provider, Eq. (9.2) represents the cost or revenues of the exchanges with the energy provider (C_{grid}), where $P_{\text{grid }i,t}$ is positive for energy bought and negative for energy sold.

$$\text{OF} = \sum_{i \in \Omega} \sum_{t \in T} \begin{pmatrix} C_{\text{grid }i,t} + \mu_{\text{loss }i,t} \ell_{i,t} + \\ \mu_{\text{BES }i,t} \ell_{\text{BES }i,t} + \mu_{\text{LEC}} \hat{P}_{\text{LEC }k,i,t} \end{pmatrix} \Delta t \tag{9.1}$$

$$\begin{gathered} C_{\text{grid }i,t} \geq \pi_{\text{buy }i,t} P_{\text{grid }i,t} \quad \text{and} \quad C_{\text{grid }i,t} \geq \pi_{\text{sell},i,t} P_{\text{grid }i,t} \\ \pi_{\text{buy }i,t} \geq \pi_{\text{sell }i,t} \end{gathered} \tag{9.2}$$

In Eq. (9.1), coefficients μ_{loss}, μ_{BES}, and μ_{LEC} represent the penalization factors for branch power loss, battery energy storage (BES) charge and discharge losses, and exchanges within the community, respectively. These values are chosen to be small enough so that the corresponding contribution to the objective function (OF) is negligible compared to C_{grid}. The OF does not include generation costs, assuming all local generation is provided by photovoltaic (PV) units.

The Joule power loss in each branch is calculated using Eq. (9.3). The power losses associated with BES discharge and charge are represented by Eq. (9.4), where $\eta_{\text{discharge}}$ and η_{charge} are efficiency factors less than one. The exchanged power P_{BES} is constrained by the maximum power limit of the battery system.

$$\ell_{i,t} = 0.5 r_i u_{\text{in }i,t} + 0.5 r_i u_{\text{out }i,t} \tag{9.3}$$

$$\ell_{\text{BES }i,t} \geq \begin{cases} (1 - \eta_{\text{dicharge }i}) P_{\text{BES }i,t} \\ (1 - 1/\eta_{\text{charge }i}) P_{\text{BES }i,t} \end{cases} \tag{9.4}$$

Nonnegative variable $\hat{P}_{\text{LEC }k,i,t}$ is defined by

$$\hat{P}_{\text{LEC }k,i,t} \geq \begin{cases} P_{\text{LEC }k,i,t} \\ -P_{\text{LEC }k,i,t} \end{cases} \text{if } i \in \Omega_k,\ 0 \ \text{otherwise} \tag{9.5}$$

where Ω_k represents the set of members belonging to community k.

The DistFlow method establishes relationships between the square root of rms voltage values at branch i terminals, at the central node mp of the T model, and the active/reactive power flows entering and leaving. These relationships are expressed in Eqs. (9.6) and (9.7). If a utility capacitor bank is present, it injects reactive power (Q_{cap}), while the user absorbs reactive power (Q_{user}).

$$\begin{aligned} v_{mp\ i,t} &= v_{in\ i,t} - r_i P_{in\ i,t} - x_i Q_{in\ i,t} + 0.25\left(r_i^2 + x_i^2\right) u_{in\ i,t} \\ v_{in\ i,t} - v_{out\ i,t} &= 2r_i P_{in\ i,t} + 2x_i Q_{in\ i,t} - 0.75\left(r_i^2 + x_i^2\right) u_{in\ i,t} \\ &\quad - 0.25\left(r_i^2 + x_i^2\right) u_{out\ i,t} + x_i b_i v_{mp\ i,t} \end{aligned} \tag{9.6}$$

$$\begin{aligned} P_{in\ i,t} &= P'_{out\ i,t} + 0.5\ r_i\ u_{in\ i,t} + 0.5\ r_i\ u_{out\ i,t} \\ Q_{in\ i,t} &= Q'_{out\ i,t} - b_i v_{mp\ i,t} + 0.5\ x_i\ u_{in\ i,t} + 0.5\ x_i\ u_{out\ i,t} \\ P'_{out\ i,t} &= P_{out\ i,t} + P_{user\ i,t} \\ Q'_{out\ i,t} &= Q_{out\ i,t} + Q_{user\ i,t} - Q_{cap\ i,t} \end{aligned} \tag{9.7}$$

The network flows resulting from direct transactions among members of community k are represented by $P_{LEC\ k}$, which is not null only if prosumer i belongs to community k.

$$P_{LEC\ k,i,t} = P_{LEC\_in\ k,i,t} - P_{LEC\_out\ k,i,t} \tag{9.8}$$

Variables u_{in} and u_{out} are constrained to be lower than the square of the maximum branch current limit $\left(I_{max}^2\right)$; v_{in} and v_{out} are bounded by the square of the minimum and maximum bus voltage limits (V_{min}^2, V_{max}^2, respectively).

To incorporate the DistFlow model into the quadratic constraint problem, the apparent power equalities are relaxed as

$$\begin{aligned} P_{in\ i,t}^2 + Q_{in\ i,t}^2 &\le v_{in\ i,t} u_{in\ i,t} \\ P_{mp\ i,t}^2 + Q'^2_{mp\ i,t} &\le v_{mp\ i,t} u_{in\ i,t} \\ P_{mp\ i,t}^2 + Q_{mp\ i,t}^2 &\le v_{mp\ i,t} u_{out\ i,t} \\ P'^2_{out\ i,t} + Q'^2_{out\ i,t} &\le v_{out\ i,t} u_{out\ i,t} \end{aligned} \tag{9.9}$$

$$\begin{aligned} P_{mp\ i,t} &= P'_{out\ i,t} + 0.5 r_i u_{out\ i,t} \\ Q_{mp\ i,t} &= Q'_{out\ i,t} + 0.5 x_i u_{out\ i,t} \\ Q'_{mp\ i,t} &= Q_{mp\ i,t} - b_i v_{mp\ i,t} \end{aligned} \tag{9.10}$$

In a feasible solution, all equations in Eq. (9.9) should be satisfied as equalities, as well as one of the equations in Eq. (9.4). The solution procedure includes an automatic check, and if the mismatch exceeds a predefined tolerance, the optimization process is repeated with two additional nonnegative penalization terms in Eq. (9.1), that become zero when a feasible solution is obtained, as described in Ref. [10].

The square rms values of voltage and total power flows at the sending end should be equal to the corresponding values at the receiving end of the connected branch. When considering multiple branches terminating and originating from the same bus, these relationships can be generalized as follows:

$$v_{out\ i\in\Omega_j^r,t} = v_{in\ i\in\Omega_j^s,t} = v_{j,t} \tag{9.11}$$

$$\sum_{i\in\Omega_j^r} P_{\text{out }i,t} = \sum_{i\in\Omega_j^s} P_{\text{in }i,t} \tag{9.12}$$

$$\sum_{i\in\Omega_j^r} P_{\text{LEC\_out }k,i,t} = \sum_{i\in\Omega_j^s} P_{\text{LEC\_in}\,k,i,t} \tag{9.13}$$

In these equations, the sets of branches connected to bus j as the sending and receiving end are represented by Ω_j^s and Ω_j^r, respectively.

The squared voltage at the connection point to the transmission network (slack bus 0) is assumed to be known and constant throughout the day. Direct transactions among community members do not result in any power flow exchange with the transmission network:

$$\sum_{i\in\Omega_0} P_{\text{LEC\_in }i,t} = 0 \tag{9.14}$$

The net power for each user can be calculated using the following equation:

$$\begin{aligned} P_{\text{user }i,t} &= P_{\text{L }i,t} - P_{\text{G }i,t} - P_{\text{BES }i,t} + \ell_{\text{BES }i,t} \\ Q_{\text{user }i,t} &= Q_{\text{L }i,t} - Q_{\text{G }i,t} - Q_{\text{BES }i,t} - Q_{\text{C }i,t} \end{aligned} \tag{9.15}$$

The active power consumption (P_{user}) is the active power requested by the local load (P_{L}), minus the active power provided by the generating unity (P_{G}), and the active power at the output of the BES system (P_{BES}), considering the power loss ℓ_{BES}. Similarly, the reactive power consumption (Q_{user}) is the sum of the reactive power requested by the local load (Q_{L}), minus the reactive power provided by the generating unity (Q_{G}), the reactive power at the output of the BES system (Q_{BES}), and the reactive power provided by a capacitor bank inside the user facility (Q_{C}) if it is present.

The BES unit follows the given model described by the following equations:

$$\begin{aligned} &E_{i,t} = E_{i,t-1} - P_{\text{BES }i,t}\,\Delta t \quad \text{for} \quad 1 < t < 96 \\ &E_{i,1} = E_{\text{max }i} - P_{\text{BES }i,t}\,\Delta t \quad \text{and} \quad E_{i,96} = E_{\text{max }i} \end{aligned} \tag{9.16}$$

The energy content is constrained by the minimum energy limit (E_{min}) and the maximum energy limit (E_{max}). In numerical tests, the initial and final energy content values are assumed to be equal.

The load is modeled using the linearized ZIP model, as shown in Eq. (9.17):

$$\begin{aligned} P_{L\ i,t} &= P_{Z\ i,t}\, v_{\text{out }i,t} + (P_{I\ i,t} + \Delta P_{I\ i,t}) + P_{P\ i,t} \\ Q_{L\ i,t} &= Q_{Z\ i,t}\, v_{\text{out }i,t} + (Q_{I\ i,t} + \Delta Q_{I\ i,t}) + Q_{P\ i,t} \end{aligned} \tag{9.17}$$

$$\begin{aligned} &2(\Delta P_{I\ i,t}\, P_{I\ i,t} + \Delta Q_{I\ i,t}\, Q_{I\ i,t}) = (v_{\text{out }i,t} - 1)\left(P_{I\ i,t}^2 + Q_{I\ i,t}^2\right) \\ &Q_{I\ i,t}\, \Delta Q_{I\ i,t} - P_{I\ i,t}\, \Delta P_{I\ i,t} = 0 \end{aligned} \tag{9.18}$$

The active power consumption (P_{L}) and reactive power consumption (Q_{L}) are determined by the known active and reactive power consumptions at the rated voltage of the constant impedance (P_Z, Q_Z), constant current (P_I, Q_I), and constant power (P_P, Q_P) components. The terms ΔP_I and ΔQ_I represent the linearized voltage dependence of the constant current component consumption.

The total net consumption (positive) or production (negative) of user i should balance the sum of the power exchanges with the energy provider and with other members of the same community k:

$$P_{\text{user } i,t} = P_{\text{grid } i,t} + P_{\text{LEC } k,i,t} \tag{9.19}$$

Both $P_{\text{grid}\, i,t}$ and $P_{\text{LEC}\, k,i,t}$ share the same sign to avoid that a participant can resell the energy bought from the grid to other community members and vice versa. $P_{\text{LEC}\, k,i,t} = 0$ if i is not part of Ω_k.

The model represents a convex quadratically constrained programming (QCP) problem. Binary variables were introduced to prevent reselling. The procedure was implemented in the Advanced Integrated Multidimensional Modeling Software (AIMMS) Developer modeling environment [11]. The Gurobi 9.5 QCP solver was used to obtain the results. The computational time for the considered cases is approximately 100 seconds when executed on a computer equipped with an Intel-i7 processor and 32 GB of RAM, running 64-bit Windows 10 operating system. The computational time increases with the number and sizes of BES units.

9.2.2 Case Study and Analysis of the Results

For the analysis, the 123-bus IEEE feeder shown in Figure 9.1a has been adopted as the case study. The medium voltage (MV) side of the substation has a rated voltage of 4.16 kV, while bus 119 has a rated voltage of 480 V. All lines and loads are assumed to be balanced, with parameters obtained by averaging values from Ref. [12]. PV units and battery units are added, and variable capacitor banks are connected to certain buses. The rated power of PV inverters is increased by 10%. The maximum power of the capacitor banks is set to 20% below the average values specified in Ref. [12].

The test cases involve a deterministic day-ahead optimization process. The load profiles used in all the test cases are obtained using the Centre for Renewable Energy Systems Technology (CREST) tool [13] with varying numbers of dwellings. The data file contains profiles of the buy and sell prices (π_{buy} and π_{sell}) for two different power providers, as well as the profile of the ratio between power output and panel surface, assumed to be the same for all PV units.

In the base case, two providers (f_1 and f_2) with different price profiles are considered. Users are randomly assigned to one of the two providers, as illustrated in Figure 9.1b.

Similarly, the users are randomly divided into three communities, and some users at specific nodes are not assigned to any community (referred to as group noEC). Each node has an equal probability of being assigned to community EC1, EC2, EC3, or not being part of any community. The association of users and communities is depicted in Figure 9.2a. Figure 9.2b displays an alternate composition of the three communities by grouping nearby users together. This grouping achieves similar values for daily demand, PV generation, and installed BES in the communities. The "noEC" group remains unchanged from the base case. The results of the two cases

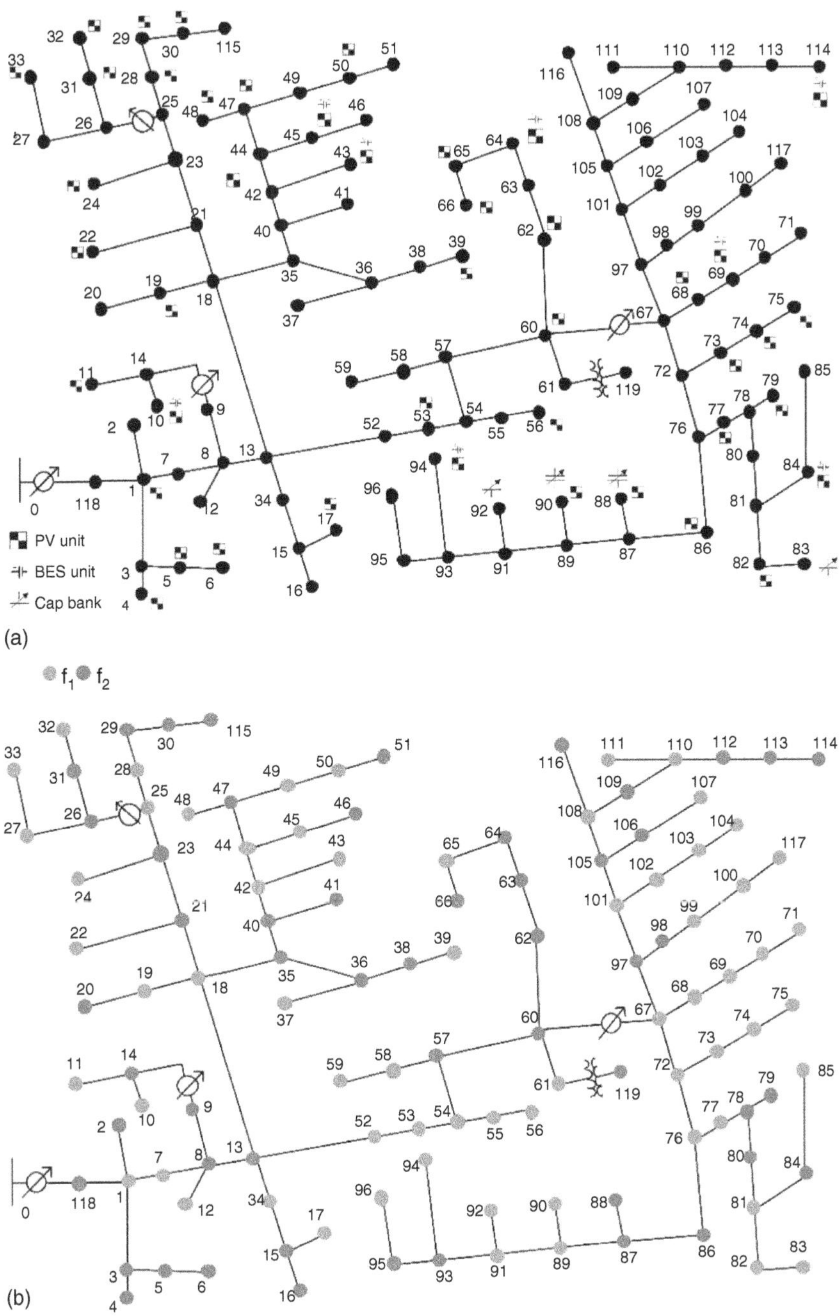

Figure 9.1 (a) Network layout for test cases. (b) Users randomly assigned to different providers f_1 and f_2.

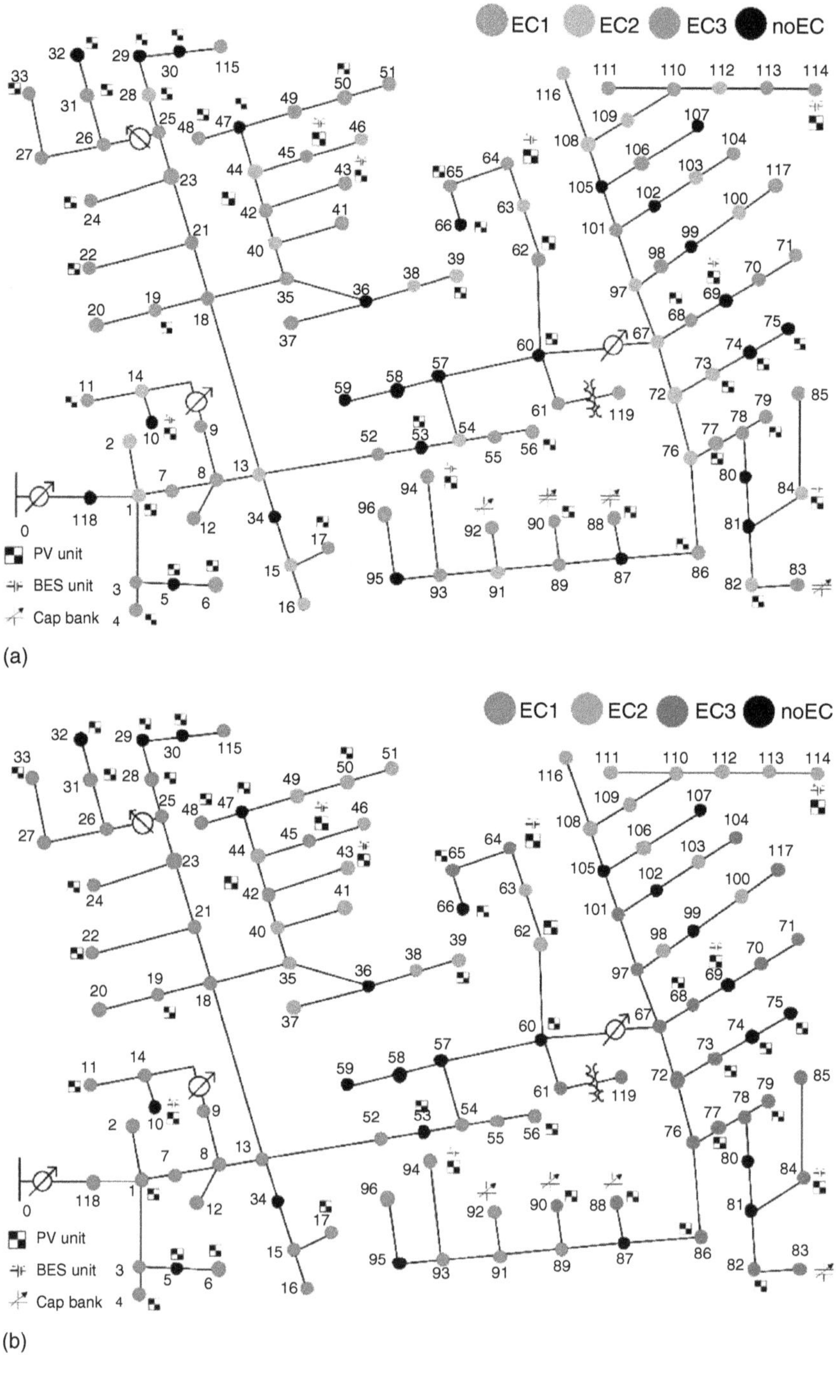

Figure 9.2 (a) Users randomly assigned to different communities, with certain nodes excluded from any community. (b) Communities formed by neighboring prosumers.

are very similar. The variations are attributed to the specific characteristics of users within each community, rather than to their proximity.

The calculation is performed for varying numbers of communities, ranging from 1 to 18. Users are randomly assigned to specific communities or excluded from communities, with a uniform probability. Figure 9.3a illustrates the sum of procurement costs for all communities, including providers f_1 and f_2 as well as the procurement costs for users not belonging to any community. Figure 9.3b displays the average costs per user in communities where exchanges are allowed, communities where exchanges are not allowed, and the group of users not assigned to any community (noEC).

The results indicate reduced procurement costs for each community when compared to scenarios disallowing direct transactions. Assuming two providers with distinct pricing profiles, the algorithm appropriately assigns internal production to consumers with higher tariffs in cases of overall power importation. Conversely, it allocates internal generation from producers with lower selling tariffs in instances of global power exportation.

Transaction prices within a community are determined by using shadow prices related to balancing constraints (9.19). As the number of communities increases, procurement costs rise due to a decrease in available transaction partners for each participant.

The method can also assess the impact of additional installations of production and storage units, and the computational effort required remains reasonably limited.

9.3 Provision of Reactive Power Compensation Services

This section specifically focuses on the provision of reactive power compensation services by a LEC comprising prosumers connected to the same MV distribution network. The literature in this field addresses reactive power control for individually managed final users, prosumers, and communities. These studies often consider compliant and noncompliant zones of operation for active and reactive power, including penalties for noncompliant reactive power absorption. The control of distributed generators, particularly PV systems, is also explored, along with the relationship between reactive power control, active power curtailment, and transformer control [15–18].

Reactive power flexibility is a crucial auxiliary service for the operation of distribution networks, as it helps improve voltage profiles, potentially eliminating or delaying the need for costly voltage control devices like static var compensators and voltage regulators. Coordinating variations in user reactive power injections or absorptions with the control of transformers equipped with on-load tap changers (OLTCs) is essential [19]. Optimization approaches have been developed to manage reactive power in grids with renewable energy sources (RESs), considering the P–Q inverter capability curves and compliant regions defined in different countries [20–23]. Preliminary calculations of maximum deviations from reference values are important for DSO and TSO decisions regarding the provision of ancillary services

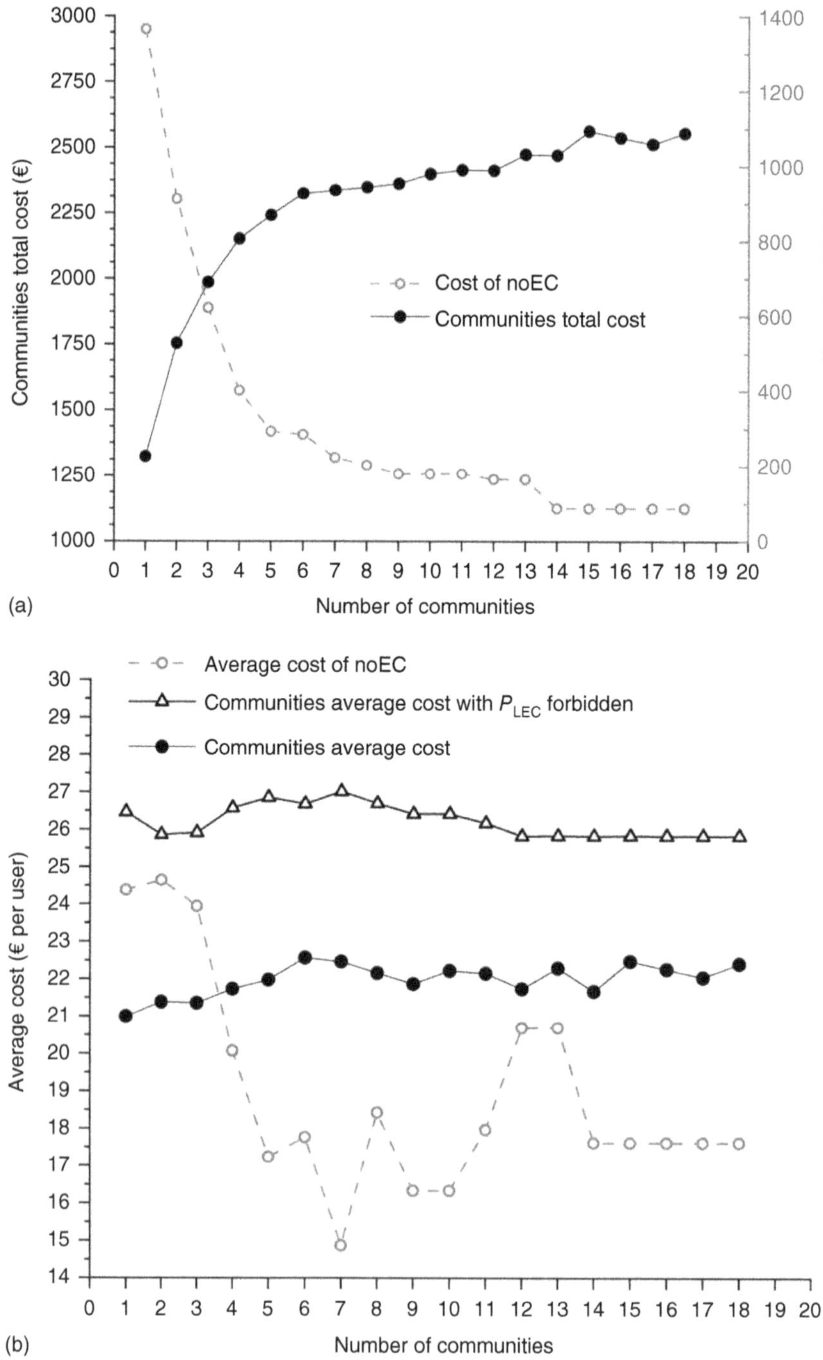

Figure 9.3 (a) Variation of total procurement costs for communities and the noEC group with the number of communities. (b) Average costs per user in the communities, with P_{LEC} forbidden in the same communities, and in the noEC group, for varying numbers of communities. Source: From Harighi et al. [14]/IEEE.

[24, 25]. In the context of energy communities equipped with energy storage systems and aggregated distributed energy resources (DERs), Refs. [26, 27] focus on active load flexibilities, storage capacity sharing, and voltage ancillary services, employing phasor measurement units for control coordination.

9.3.1 Optimization Model

The adopted optimization model has been presented in Section 9.2.1. A deterministic day-ahead scheduling problem is considered, which involves optimizing the allocation of energy resources and control mechanisms for the upcoming day with a 15-minute resolution. This optimization is based on the availability of accurate forecasts of load and PV production.

The permission of direct reactive power exchanges among the participants in the community may be effective, i.e. a participant that absorbs too much reactive power with respect to active one (as it operates at a low power factor), can reduce the penalty by the help of the reactive power injections of other participants. The model is conceived so that these internal reactive power exchanges among the community participants are balanced to avoid excessive reactive power exchanges with the grid, both positive and negative.

For this purpose, in the OF (9.1), C_{grid} is replaced by $C_{grid} + \mu_{PF}Q_{PF}$, where μ_{PF} is the penalization of the noncompliance of the minimum power factor Q_{PF} that is obtained from the model of the power factor limits described in Figure 9.4.

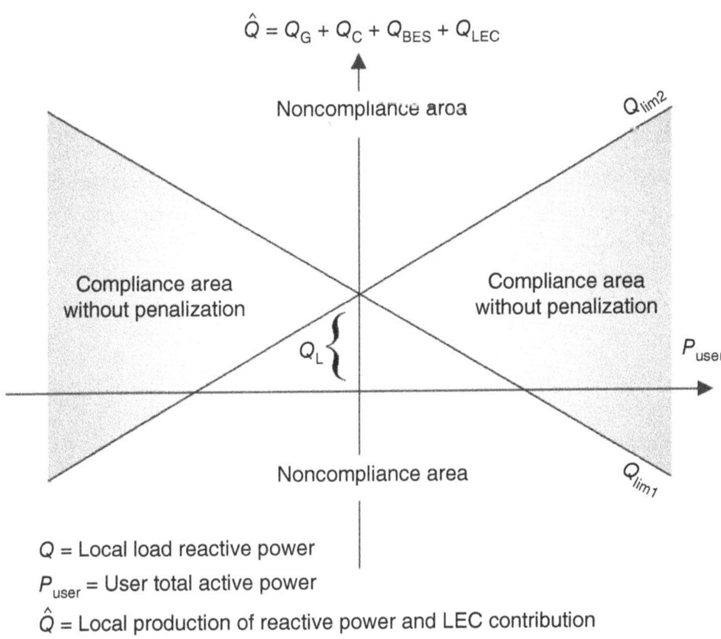

Figure 9.4 Operating region that complies with the minimum power factor.

The compliance area, illustrated in Figure 9.4, represents the region where the sum of reactive power decision variables (of generator Q_G, battery storage unit Q_{BES}, capacitor bank Q_C) and reactive power exchanges among users (Q_{LEC}) satisfy the requirements. Any operating point outside this area incurs a penalty proportional to the degree of noncompliance Q_{PF}. Certain regulatory frameworks allow for a fixed reactive power exchange without penalties for cases of low active power consumption or production, as discussed in Refs. [19, 23] concerning Swiss and Belgian regulations. To maintain simplicity, the specificities mentioned, such as customer penalizations for bus voltage violations, are not incorporated into the model presented here. Reactive power outputs Q_G and Q_{BES} are subject to constraints based on the minimum power factor of the local generator and energy storage system. Additionally, Q_C is either fixed or limited by the maximum capacity of the switchable capacitor bank (discrete switching is not represented in the model for simplicity).

The variable representing noncompliance (Q_{PF}) is assumed to be nonnegative, as there are no rewards associated with operating within the compliance area. Due to the nonconvex nature of the compliance area depicted in Figure 9.4, the representation of Q_{PF} includes a condition related to the sign of Q_{PF}. This condition is handled by introducing binary variables:

$$Q_{PF\ i,t} \geq \begin{cases} Q_{\lim 1\ i,t} - \widehat{Q}_{i,t} \text{ if } P_{user\ i,t} \geq 0 \text{ or } -Q_{\lim 1\ i,t} - \widehat{Q}_{i,t} \text{ if } P_{user\ i,t} < 0 \\ -Q_{\lim 2\ i,t} + \widehat{Q}_{i,t} \text{ if } P_{user\ i,t} \geq 0 \text{ or } Q_{\lim 2\ i,t} - \widehat{Q}_{i,t} \text{ if } P_{user\ i,t} < 0 \end{cases} \tag{9.20}$$

where

$$\begin{aligned} Q_{\lim 1\ i,t} &= Q_{L\ i,t} - \tan[\arccos(pf_{\min})]P_{user\ i,t} \\ Q_{\lim 2\ i,t} &= Q_{L\ i,t} + \tan[\arccos(pf_{\min})]P_{user\ i,t} \end{aligned} \tag{9.21}$$

The following equilibrium is enforced at the beginning and end of each branch, relevant to reactive power direct transactions Q_{LEC} among the participants:

$$\sum_{i \in \Omega_j^r} Q_{LEC\_out\ i,t} = \sum_{i \in \Omega_j^s} Q_{LEC\_in\ i,t} \tag{9.22}$$

$$\sum_{i \in \Omega_0} Q_{LEC\_in\ i,t} = 0 \tag{9.23}$$

The model includes the representation of transformers equipped with OLTCs. The constraints are formulated below:

$$t_{\min}^2 v'_{out\ i,t} \leq v_{out\ i,t} \leq t_{\max}^2 v'_{out\ i,t} \tag{9.24}$$

In the formulation, the upper and lower bounds of the transformer ration t_{OLTC} are denoted by t_{max} and t_{min}, respectively. The current formulation does not explicitly account for discrete steps.

Regarding the operation of the OLTC, two approaches are considered: (i) the optimization process jointly optimizes the OLTC ratio alongside other decision variables to minimize the OF and (ii) the OLTC tap is selected to regulate the voltage at the secondary side based on two linearized constraints:

$$\begin{aligned} v_{\text{out } i,t} &= V_{\text{ri}}^2 + 2\ V_{\text{ri}}\ \Delta V_{i,t} \\ \Delta V_{i,t} &= s_{\text{R}} P'_{\text{out } i,t} + s_{\text{X}} Q'_{\text{out } i,t} \end{aligned} \tag{9.25}$$

In the first constraint, the term involving ΔV^2 is neglected, while in the second constraint, v_{out} is assumed to be close to 1 pu. In Eq. (9.25), V_{r} is the rated voltage at the secondary side of the transformer, s_{R} and s_{X} are positive parameters that represent the regulator compensation settings. To consider $t_{\max}$ and $t_{\min}$, three binary variables are included in Eq. (9.24), each corresponding to $v_{\text{out } i,t} = t_{\max}^2 v'_{\text{out } i,t}$, $v_{\text{out } i,t} = t_{\min}^2 v'_{\text{out } i,t}$, and $v_{\text{out} i,t}$ between $t_{\min}^2 v'_{\text{out } i,t}$ and $t_{\max}^2 v'_{\text{out } i,t}$, whose sum must be one.

As mentioned, the model represents the possibility that each participant can directly exchange reactive power with the other members of the community. As in Eq. (9.19) for the active power, the total net reactive power consumption (positive) or production (negative) of user i should balance the sum of the power exchanges with the energy provider and with other members of the community. The energy provider does not absorb reactive power, i.e. $Q_{\text{grid}\,i,\,t} \geq 0$.

$$\begin{aligned} Q_{\text{user } i,t} &= Q_{\text{grid } i,t} + Q_{\text{LEC } i,t} && \text{if } Q_{\text{LEC } i,t} \geq 0 \\ Q_{\text{user } i,t} &= Q_{\text{LEC } i,t} && \text{if } Q_{\text{LEC } i,t} < 0 \end{aligned} \tag{9.26}$$

The $Q_{\text{LEC}\,i,\,t}$ flows in the network are represented by

$$Q_{\text{LEC } i,t} = Q_{\text{LEC\_in } i,t} - Q_{\text{LEC\_out } i,t} \tag{9.27}$$

9.3.2 Calculation of the Maximum Up and Down Flexibilities

The implemented procedure comprises the following three steps.

1. **Reference optimization:** This step involves calculating the optimal scheduling of both active and reactive power resources for the entire community by using the model described in Section 9.3.2. The objective is to minimize the community's energy procurement costs by taking into account the tariffs for active power exchanges with the external provider. Additionally, penalties are considered for participants operating with a power factor below a predefined limit.
2. **Q_{down} optimization:** In this step, the focus is on determining the maximum decrease in reactive power absorption or maximum increase in reactive power injection at the terminals of each community participant. The goal is to minimize energy procurement costs while considering the revenues generated from providing a reactive power flexibility service. This service involves reducing reactive power compared to the reference profile Q_{ref} calculated in the reference optimization step, based on requests from the DSO. The revenues are determined by a predefined price π_{down} (in € kvarh^{-1}).
   The OF becomes

$$OF_{Q_{\text{down}}} = \sum_{i\in\Omega}\sum_{t\in T}(C_{\text{grid } i,t} - \pi_{\text{down}} Q_{\text{down } i,t} + \mu_{\text{loss } i,t}\ell_{i,t} + \mu_{\text{BES } i,t}\ell_{\text{BES } i,t})\ \Delta t \tag{9.28}$$

where

$$Q_{\text{down }i,t} = Q_{\text{ref }i,t} - Q_{\text{user }i,t} \tag{9.29}$$

3. $\mathbf{Q_{up}}$ **optimization:** This step involves calculating the maximum increase in reactive power absorption or maximum decrease in reactive power injection at the terminals of each community participant. Similar to the Q_{down} optimization, the objective is to minimize energy procurement costs, taking into account the revenues generated from increasing reactive power compared to the reference profile Q_{ref}.
   The OF becomes

$$\text{OF}_{Q_{\text{up}}} = \sum_{i \in \Omega} \sum_{t \in T} (C_{\text{grid }i,t} - \pi_{\text{up}} Q_{\text{up }i,t} + \mu_{\text{loss }i,t} \ell_{i,t} + \mu_{\text{BES }i,t} \ell_{\text{BES }i,t})\, \Delta t \tag{9.30}$$

where

$$Q_{\text{up }i,t} = Q_{\text{user }i,t} - Q_{\text{ref }i,t} \tag{9.31}$$

The Q_{down} and Q_{up} optimizations are considered independent of each other due to the absence of constraints associated with reactive power compensation decisions made at different times.

In the Q_{down} and Q_{up} optimizations, the voltage control and reactive power compensation mechanisms of the distribution network, such as OLTCs and variable capacitor banks, can be utilized with different objectives.

One possibility is to operate these control means in order to maximize the rewards. This scenario can be applicable in the case of an energy community that also serves as an operator for the specific section of the distribution system. Here, the voltage control and reactive power compensation means are optimized to maximize the financial benefits or rewards obtained.

Alternatively, the voltage control and reactive power compensation means can be operated to maintain the voltage profile near the rated value. This is commonly observed in the typical situation where the energy community and the DSO are separate entities. In this case, the control mechanisms are employed to ensure that the voltage profile remains close to the desired or rated value, improving the stability and reliability of the distribution network.

As for the model used in Section 9.2, the models for the calculation of the reactive power flexibility limits are also represented by mixed-integer QCP problems and have been implemented in AIMMS using Gurobi solver.

9.3.3 Case Study and Analysis of the Results

As in Section 9.2, all the described optimization models have been implemented in AIMMS Developer modeling environment, using Gurobi solvers. Here, two test cases are considered:

- the 14-bus network, adapted from Ref. [28], in which three feeders are connected to the same substation;
- the 123-bus test system already described in Section 9.2.2.

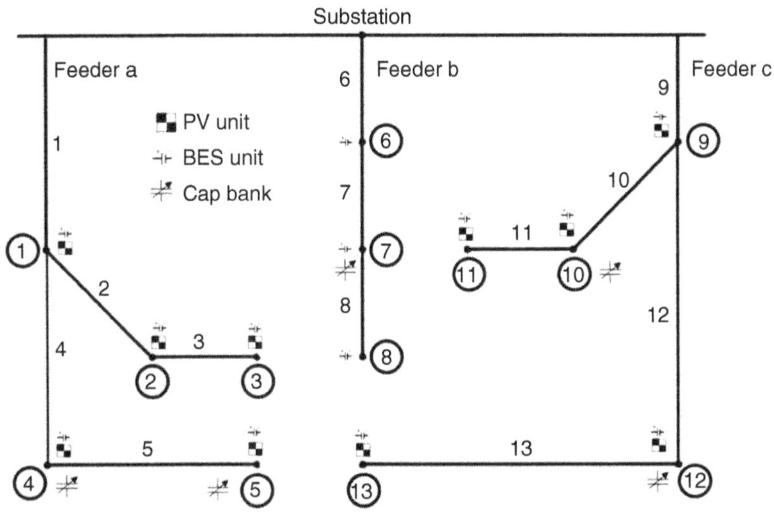

Figure 9.5 Layout of the network of the first test case.

Each participant in the energy community, referred to as a prosumer, is equipped with a PV system, a load, and a BES unit. These prosumers collectively form the energy community.

For noncompliance penalties, the tariff μ_{PF} is set at 5 € kvarh^{-1}. The flexibility reward tariffs for reactive power adjustments, denoted as π_{down} and π_{up}, are both 3 € kvarh^{-1}, in accordance with Ref. [23].

To meet the requirements, the minimum allowable power factor is assumed to be 0.9. The bus voltages must remain within the interval of 0.9–1.1 pu.

Figure 9.5 shows the scheme of the first test case. The BES units within the system have the capability to either inject or absorb reactive power, as determined by the optimization procedure. The reactive power limits for these units range from −48.43% to +48.43% of the rated value, assuming a power factor of 0.9. On the other hand, all PV units operate at a power factor of unity.

In terms of energy requirements, the forecasted total energy demand for the day is 349.4 MWh. The PV energy generation is 52.4 MWh, accounting for 15% of the load. Furthermore, the total installed storage capacity amounts to 5.5 MWh, equivalent to 10.5% of the daily PV generation.

The calculations are conducted for four scenarios, each distinguished by the operation of the capacitor banks and whether exchanges involving reactive power Q_{LEC} are permitted:

- **Scenario 0:** all capacitor banks owned by the utility, without community;
- **Scenario 1:** all capacitor banks owned by prosumers, without reactive power exchanges;
- **Scenario 2:** all capacitor banks owned by prosumers, and reactive power exchanges permitted;
- **Scenario 3:** all capacitor banks owned by the utility, and reactive power exchanges permitted;

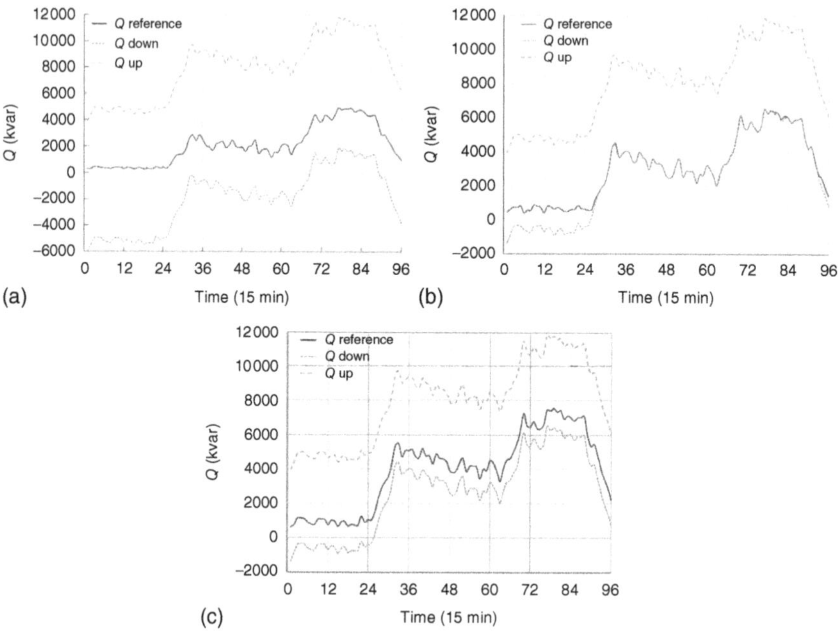

Figure 9.6 Profiles of the cumulative value of the community reactive power calculated by the reference, Q_{up}, and Q_{down} procedures for scenarios: (a) 2 (like 1 not shown), (b) 3, and (c) 4 (like 0 not shown). First test case.

- **Scenario 4:** all capacitor banks owned by the utility, without reactive power exchanges.

Figure 9.6 illustrates the comparison of the profiles of the sum of Q_{user}, obtained from the reference, Q_{down}, and Q_{up} procedures across the different scenarios. The figure shows the margins available in each period, which can be utilized by the community to provide the reactive power flexibility service.

Optimizing the utilization of capacitor banks by the prosumers (scenarios 1 and 2) proves advantageous primarily for the provision of Q_{down} reserves, while the penalties in the reference case remain unaffected. This is shown by the comparison of results between scenarios 1 and 4.

For all scenarios, the costs associated with energy procurement from the external provider are similar.

When the variable capacitor banks are owned by the utility and Q_{LEC} exchanges are permitted, the participants' capacity to provide reactive power is fully utilized in the reference optimization to minimize noncompliance penalties. As a result, there is a minimal margin available for offering Q_{down} services for most of the day, as shown in Figure 9.6b.

Figure 9.7 refers to the second test case (123-bust test network). It compares the profiles of the sum of the Q_{user} values calculated by the reference, Q_{down}, and Q_{up} procedures. The OLTCs and capacitors operated by the community and Q_{LEC} transactions are allowed. The comparison of the results of this scenario with those of

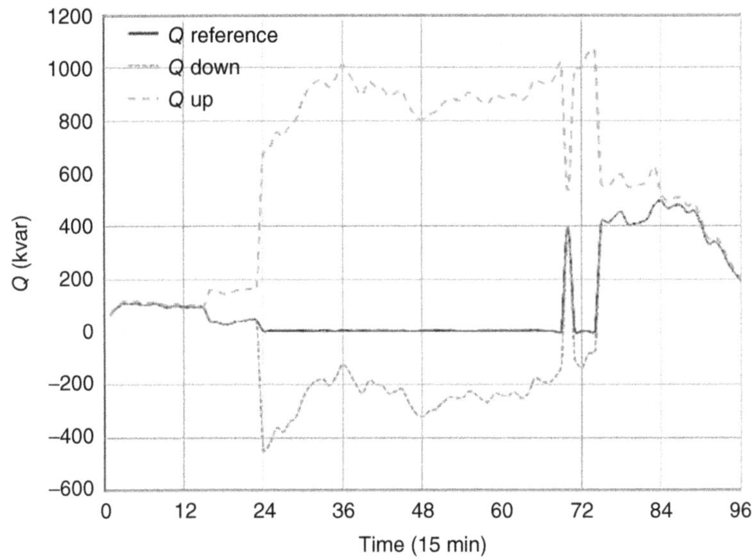

Figure 9.7 Profiles of the cumulative value of the reactive powers of the community participants calculated by the reference, Q_{up} and Q_{down} procedures for scenarios, for the second test case.

the other scenarios (OLTCs and capacitors operated by the utility, with and without active and reactive power exchanges) shows that, in this case, considering the capacitor banks as owned by the prosumers does not provide significant advantages for the community due to the limited number and size of the banks. The operation as a community reduces, however, the energy procurement costs. The possibility to exchange reactive power among the community participants significantly reduces the noncompliance penalties. The values of the energy procurement cost, associated with the active power exchanges with the external provider, are almost the same independently of the different reactive power optimization models.

The results have been obtained using the same computer as in Section 9.2.2. Computational times were limited to a few seconds for the first test case and tens of seconds for the second.

9.4 Electromobility Integration

In the context of community-based energy framework, EV-charging stations play a dual role. They act as loads during the charging phase in the grid-to-vehicle (G2V) mode and, when equipped with vehicle-to-grid (V2G) technology, function as distributed energy storage to supply power to other community loads [29, 30]. Extensive efforts in the literature have been dedicated to address the integration and management of EV charging in local energy systems. These results highlight the potential of EV-charging stations to provide demand response services, including V2G [31], and to complement RESs to enhance energy efficiency. A review of

EV-charging technologies and their effect on smart grid integration was presented in Ref. [32], along with a detailed exploration scheduling and optimal management approaches in Ref. [33].

The integration of EV charging in power systems has been addressed by using different optimization techniques and control strategies, especially in the context of renewable energy integration and participation in electricity markets [34]. Challenges and opportunities related to EV integration through V2G charging infrastructures have also been discussed, emphasizing their potential role in providing ancillary services and frequency regulation in smart distribution systems [35]. Moreover, EVs impact on the optimal energy management of renewable microgrids, which may include PV units, fuel cells, wind turbine, microturbines, and batteries, has been a subject of research [36]. Results from these studies show that optimized EV-charging strategies can significantly reduce the total operating cost of the microgrids. Another proposed framework addresses the impact of EV-charging process in energy hubs, analyzing their effects on system performance and customer preferences regarding charging patterns [37].

The impact of electric mobility on the design of renewable energy collective self-consumers was addressed in [38]. In detail, the integration of renewable sources and storage systems with charging infrastructures for EVs was analyzed through stochastic optimization modeling, assessing different scenarios with different levels of electromobility penetration. Results showed that the presence of EVs allows increasing the optimal installed size of the RES plants significantly, from 15% up to 25%, highlighting the potential for EVs to foster the RES penetration in the context of local energy systems as communities.

In this context, an approach to day-ahead scheduling within an energy community has been presented in Ref. [39]. The approach harmonizes the operations of microgrids and EV-charging stations, by using a two-stage optimization process. In the first stage, a distributed scheme based on the alternating direction method of multipliers (ADMM) is employed to schedule the activities of each community participant. This process offers economic benefits even for participants with relatively low or without energy-selling capacity. All EV batteries connected to the charging stations within a microgrid are aggregated into a single storage unit during this stage. The second stage focuses on developing detailed scheduling for each charging station, aligning with the energy transactions defined in the first stage. This stage also leverages V2G services to optimize energy utilization. The results of the considered case studies indicate that EV-based microgrids efficiently utilize available energy resources within the community. Furthermore, the potential for revenue generation from EV discharging encourages the adoption of V2G services.

The model has been further developed to address the issue of managing fluctuations in PV generation within each EV-based microgrid, as described in Ref. [40]. A set of specific chance constraints is used to determine the optimal up and down reserve levels of the stored energy in the EV batteries and dedicated BES units, ensuring the community's ability to cover random PV generation variations. The study has evaluated scenarios where reserves are provided solely by BES units or jointly by BES units and EV batteries. Both cases ensure positive and negative reserve amounts throughout the day, but the scheduling and costs differ. When both BES units and EV batteries provide reserves, scheduling becomes more efficient, leading to reduced daily costs for EV-based microgrids.

9.5 Conclusion and Key Learnings

The chapter focuses on two distinctive procedures for day-ahead scheduling within the context of energy communities that allow direct exchanges among their participants. Direct transactions of active power substantially reduce energy procurement costs, leading to a more efficient scheduling of available energy resources. This is especially beneficial when purchase tariffs significantly exceed selling rates.

The first procedure addresses scenarios where multiple energy communities operate within the grid, granting users the freedom to decide their community affiliations and energy providers based on their preferences. Direct transactions are confined within each community. Results obtained through simulations on the IEEE 123-bus test feeder demonstrate the significant cost reductions that each community can achieve by internalizing their transactions and optimizing the allocation of internal production. However, as the number of communities increases, there is an associated rise in procurement costs, mainly due to the reduced number of participants available for transactions in each independent community.

The second procedure specifically addresses the issue of low power factor operation, which is becoming increasingly important with the proliferation of PV units that reduce active power consumption during daylight hours. By enabling both active and reactive power transactions among the users, the procedure also supports the provision of reactive power flexibility services by the community. Calculations for the maximum and minimum reactive power deviations, coupled with predetermined tariffs, replace noncompliance penalties. The results obtained for various test cases consistently indicate that scheduling reactive power compensation resources has a limited impact on energy procurement costs, making the procedures economically advantageous.

These computational models exhibit reasonable time efficiency, rendering them suitable for integration into stochastic procedures that account for uncertainties in PV production and load consumption profiles.

The final part of the chapter is devoted to the description of the support that energy communities can provide to electromobility and of the integration of EV-charging scheduling into their efficient operation.

Acknowledgments

The authors thank Adriano Calzolari for useful discussions and comments. Thanks also go to Filippo Conti for his valuable help in performing the calculations. This chapter summarizes the results obtained in research activities funded by the Italian Ministry for +University and Research through grant J35F21003140006 (PhD program 2014IT16M2OP005, industrial partner: Inrete Distribuzione

Energia S.p.A.), project PRIN-2017K4JZEE "Planning and flexible operation of micro-grids with generation, storage and demand control," followed by project PRIN-20228E3T44 "Smart and/or sustainable mobility". These activities have been carried out in collaboration with Fabio Napolitano, Fabio Tossani (co-authors of the papers that present some of the obtained results).

List of Abbreviations

ADMM	alternating direction method of multipliers
AIMMS	Advanced Integrated Multidimensional Modeling Software
ARERA	Italian Regulatory Authority for Energy, Networks, and Environment
BES	battery energy storage
CREST	Centre for Renewable Energy Systems Technology
DERs	distributed energy resources
DG	distributed generation
DSO	distribution system operator
EU	European Union
EV	electric vehicle
G2V	grid-to-vehicle
IEEE	Institute of Electrical and Electronics Engineers
LEC	local energy community
MV	medium voltage
OF	objective function
OLTC	on-load tap changer
PV	photovoltaic
QCP	quadratically constrained programming
TSO	transmission system operator
V2G	vehicle-to-grid

References

1 European Committee of the Regions, Commission for the Environment, Climate Change and Energy (2018). *Models of Local Energy Ownership and the Role of Local Energy Communities in Energy Transition in Europe*. European Committee of the Regions. https://data.europa.eu/doi/10.2863/603673.

2 Cappellaro, F., D'Agosta, G., De Sabbata, P. et al. (2022). Implementing energy transition and SDGs targets throughout energy community schemes. *Journal of Urban Ecology* 8: 1–9.

3 Boulanger, S.O.M., Massari, M., Longo, D. et al. (2021). Designing collaborative energy communities: a European overview. *Energies* 14: 8226–8226.

4 Cunha, F.B.F., Carani, C., Nucci, C.A. et al. (2021). Transitioning to a low carbon society through energy communities: lessons learned from Brazil and Italy. *Energy Research & Social Science* 75: 101994: 1–19.

5 Inês, C., Guilherme, P.L., Esther, M.G. et al. (2019). Regulatory challenges and opportunities for collective renewable energy prosumers in the EU. *Energy Policy* 138 (December): 2020. https://doi.org/10.1016/j.enpol.2019.111212.

6 Jabr, R.A. (2006). Radial distribution load flow using conic programming. *IEEE Transactions on Power Apparatus and Systems* 21 (3): 1458–1459. https://doi.org/10.1109/TPWRS.2006.879234.

7 Low, S.H. (2014). Convex relaxation of optimal power flow – part I: formulations and equivalence. *IEEE Transactions on Control of Network Systems* 1 (1): 15–27. https://doi.org/10.1109/TCNS.2014.2309732.

8 Baran, M.E. and Wu, F.F. (1989). Network reconfiguration in distribution systems for loss reduction and load balancing. *IEEE Transactions on Power Delivery* 4 (2): 1401–1407. https://doi.org/10.1109/61.25627.

9 Orozco, C., Borghetti, A., De Schutter, B. et al. (2022). Intra-day scheduling of a local energy community coordinated with day-ahead multistage decisions. *Sustainable Energy, Grids and Networks* 29: 100573. https://doi.org/10.1016/j.segan.2021.100573.

10 Harighi, T., Borghetti, A., Napolitano, F., and Tossani, F. (2023). Provision of reactive power services by energy communities in MV distribution networks. *Sustainable Energy, Grids and Networks* 34: 101038.

11 Roelofs, M. and Bisschop, J. (2011). *AIMMS User's Guide*. Paragon Decision Technology B.V.

12 S. Carneiro, et al. (1992). IEEE PES Distribution System Analysis Subcommittee. IEEE 123 Node Test Feeder. https://cmte.ieee.org/pes-testfeeders/resources/ (accessed 23 March 2024).

13 McKenna, E., Thomson, M., and Barton, J. (2015). CREST demand model. Loughborough University. Dataset. https://doi.org/10.17028/rd.lboro.2001129.v8 (accessed 23 March 2024).

14 Harighi, T., Borghetti, A., Napolitano, F., and Tossani, F. (2023). Optimization model for the analysis of multiple energy communities in the same distribution network with different providers. *2023 IEEE Belgrade PowerTech*, June 25-29, Belgrade, Serbia. pp. 1–6. https://doi.org/10.1109/PowerTech55446.2023.10202985

15 Demirok, E., González, P.C., Frederiksen, K.H.B. et al. (2011). Local reactive power control methods for overvoltage prevention of distributed solar inverters in low-voltage grids. *IEEE Journal of Photovoltaics* 1 (2): 174–182. https://doi.org/10.1109/JPHOTOV.2011.2174821.

16 Kabiri, R., Holmes, D.G., McGrath, B.P., and Meegahapola, L.G. (2015). LV grid voltage regulation using transformer electronic tap changing, with PV inverter reactive power injection. *IEEE Journal of Emerging and Selected Topics in Power Electronics* 3 (4): 1182–1192. https://doi.org/10.1109/JESTPE.2015.2443839.

17 Su, X., Masoum, M.A.S., and Wolfs, P.J. (2014). Optimal PV inverter reactive power control and real power curtailment to improve performance of unbalanced four-wire LV distribution networks. *IEEE Transactions on Sustainable Energy* 5 (3): 967–977. https://doi.org/10.1109/TSTE.2014.2313862.

18 Safayet, A., Fajri, P., and Husain, I. (2017). Reactive power management for overvoltage prevention at high PV penetration in a low-voltage distribution

system. *IEEE Transactions on Industry Applications* 53 (6): 5786–5794. https://doi.org/10.1109/TIA.2017.2741925.

19 Valverde, G., Shchetinin, D., and Hug-Glanzmann, G. (2019). Coordination of distributed reactive power sources for voltage support of transmission networks. *IEEE Transactions on Sustainable Energy* 10 (3): 1544–1553. https://doi.org/10.1109/TSTE.2019.2892671.

20 Kekatos, V., Wang, G., Conejo, A.J., and Giannakis, G.B. (2015). Stochastic reactive power management in microgrids with renewables. *IEEE Transactions on Power Apparatus and Systems* 30 (6): 3386–3395. https://doi.org/10.1109/TPWRS.2014.2369452.

21 Contreras, D.A. and Rudion, K. (2018). Improved assessment of the flexibility range of distribution grids using linear optimization. *2018 Power Systems Computation Conference (PSCC)* June 11-15, Dublin, Ireland, IEEE. pp. 1–7. https://doi.org/10.23919/PSCC.2018.8442858

22 Karagiannopoulos, S., Mylonas, C., Aristidou, P., and Hug, G. (2021). Active distribution grids providing voltage support: the Swiss case. *IEEE Transactions on Smart Grid* 12 (1): 268–278. https://doi.org/10.1109/TSG.2020.3010884.

23 Avramidis, I.I., Capitanescu, F., Evangelopoulos, V.A. et al. (2022). In pursuit of new real-time ancillary services providers: hidden opportunities in low voltage networks and sustainable buildings. *IEEE Transactions on Smart Grid* 13 (1): 429–442. https://doi.org/10.1109/TSG.2021.3112925.

24 Capitanescu, F. (2018). TSO–DSO interaction: active distribution network power chart for TSO ancillary services provision. *Electric Power Systems Research* 163: 226–230. https://doi.org/10.1016/j.epsr.2018.06.009.

25 Silva, J., Sumaili, J., Bessa, R.J. et al. (2018). Estimating the active and reactive power flexibility area at the TSO-DSO interface. *IEEE Transactions on Power Apparatus and Systems* 33 (5): 4741–4750. https://doi.org/10.1109/TPWRS.2018.2805765.

26 Nagpal, H., Avramidis, I.I., Capitanescu, F., and Madureira, A.G. (2022). Local energy communities in service of sustainability and grid flexibility provision: hierarchical management of shared energy storage. *IEEE Transactions on Sustainable Energy* 1–13. https://doi.org/10.1109/TSTE.2022.3157193.

27 Rousis, A.O., Tzelepis, D., Pipelzadeh, Y. et al. (2021). Provision of voltage ancillary services through enhanced TSO–DSO interaction and aggregated distributed energy resources. *IEEE Transactions on Sustainable Energy* 12 (2): 897–908. https://doi.org/10.1109/TSTE.2020.3024278.

28 Civanlar, S., Grainger, J.J., Yin, H., and Lee, S.S.H. (1988). Distribution feeder reconfiguration for loss reduction. *IEEE Transactions on Power Delivery* 3 (3): 1217–1223. https://doi.org/10.1109/61.193906.

29 Shafie-Khah, M., Vahid-Ghavidel, M., Di Somma, M. et al. (2020). Management of renewable-based multi-energy microgrids in the presence of electric vehicles. *IET Renewable Power Generation* 14 (3): 417–426: https://doi.org/10.1049/iet-rpg.2019.0124.

30 Di Somma, M., Ciabattoni, L., Comodi, G., and Graditi, G. (2020). Managing plug-in electric vehicles in eco-environmental operation optimization of local multi-energy systems. *Sustainable Energy, Grids and Networks* 23: 100376. https://doi.org/10.1016/j.segan.2020.100376.

31 Kempton, W. and Tomić, J. (2005). Vehicle-to-grid power implementation: from stabilizing the grid to supporting large-scale renewable energy. *Journal of Power Sources* 144 (1): 280–294. https://doi.org/10.1016/j.jpowsour.2004.12.022.

32 Yong, J.Y., Ramachandaramurthy, V.K., Tan, K.M., and Mithulananthan, N. (2015). A review on the state-of-the-art technologies of electric vehicle, its impacts and prospects. *Renewable and Sustainable Energy Reviews* 49: 365–385. https://doi.org/10.1016/j.rser.2015.04.130.

33 Yang, Z., Li, K., and Foley, A. (2015). Computational scheduling methods for integrating plug-in electric vehicles with power systems: a review. *Renewable and Sustainable Energy Reviews* 51: 396–416. https://doi.org/10.1016/j.rser.2015.06.007.

34 Hota, A.R., Juvvanapudi, M., and Bajpai, P. (2014). Issues and solution approaches in PHEV integration to the smart grid. *Renewable and Sustainable Energy Reviews* 30: 217–229. https://doi.org/10.1016/j.rser.2013.10.008.

35 Brenna, M., Foiadelli, F., Zaninelli, D. et al. (2021). Chapter 10: The integration of electric vehicles in smart distribution grids with other distributed resources. In: *Distributed Energy Resources in Local Integrated Energy Systems: Optimal Operation and Planning*, 315–345. https://doi.org/10.1016/B978-0-12-823899-8.00006-6.

36 Kavousi-Fard, A., Abunasri, A., Zare, A., and Hoseinzadeh, R. (2014). Impact of plug-in hybrid electric vehicles charging demand on the optimal energy management of renewable micro-grids. *Energy* 78: 904–915. https://doi.org/10.1016/j.energy.2014.10.088.

37 Moeini-Aghtaie, M., Abbaspour, A., Fotuhi-Firuzabad, M., and Dehghanian, P. (2015). Optimized probabilistic PHEVs demand management in the context of energy hubs. *IEEE Transactions on Power Delivery* 30 (2): 6926867: 996–1006. https://doi.org/10.1109/TPWRD.2014.2348918.

38 Piazza, G., Bracco, S., Delfino, F. et al. (2023). Impact of electric mobility on the design of renewable energy collective self-consumers. *Sustainable Energy, Grids and Networks* 33: 100963. https://doi.org/10.1016/j.segan.2022.100963.

39 Marasciuolo, F., Orozco, C., Dicorato, M. et al. (2021). Two-stage scheduling of electrical vehicle charging station clusters in a community of DC microgrids. *2021 IEEE International Conference on Environment and Electrical Engineering and 2021 IEEE Industrial and Commercial Power Systems Europe (EEEIC/I&CPS Europe)* September 7–10, Bari, Italy, IEEE. https://doi.org/10.1109/EEEIC/ICPSEurope51590.2021.9584584

40 Marasciuolo, F., Orozco, C., Dicorato, M. et al. (2023). Chance-constrained calculation of the reserve service provided by EV charging station clusters in energy communities. *IEEE Transactions on Industry Applications* 59 (4): 4700–4709. https://doi.org/10.1109/TIA.2023.3264965.

10

Validation of Energy Hub Solutions Through Simulation and Testing in a Lab Environment and Real World

Gabriele Comodi[1], Mosè Rossi[1], Alessandro Romagnoli[2], Alessio Tafone[3], and Andreas Tuerk[4]

[1] *Università Politecnica delle Marche, Faculty of Engineering, Department of Industrial Engineering and Mathematical Sciences (DIISM), Via Brecce Bianche 12, 60131, Ancona, Italy*
[2] *Nanyang Technological University, School of Mechanical and Aerospace Engineering, 50 Nanyang Ave, Singapore, 639798, Singapore*
[3] *Technical University of Munich at the Campus for Research Excellence and Technological Enterprise (TUM-CREATE), 1 Create Way, Singapore, 138602, Singapore*
[4] *Institute for Climate, Energy Systems and Society, Waagner-Birostrase 100, 8020, Gaz, Austria*

10.1 Introduction

In recent years, there has been a strong acceleration toward global decarbonization. The Agenda for Sustainable Development, approved by the UNN in 2015, sets 17 objectives, many of which are linked, directly or indirectly, to the world of energy [1]: Clean Water and Sanitation (SDG6); Affordable and Clean Energy (SDG7); Sustainable Cities and Communities (SDG11); Climate Action (SDG13); Industry, Innovation, and Infrastructure (SDG9). In Europe, the "Clean Energy for all Europeans" package adopted in 2019 set a comprehensive energy policy framework based on the following pillars [2]: energy performance in buildings, renewable energy, energy efficiency, governance regulation, electricity market design, and non legislative initiatives including measures to define and better monitor energy poverty.

In this context, some important challenges emerge, among many [3]: (i) the decarbonization of the transport, heating, and hard-to-abate sectors; (ii) the ever-increasing use of renewable sources characterized by intrinsic limitations in technology such as availability, uncertainty, and unpredictability. As regards the first challenge, a commonly accepted path is the so-called sector coupling, which includes two main strategies: end-user and cross-vector sector coupling. The former consists of the electrification of final consumption within the technical/economic limits of both the technologies and the electricity grid; the latter consists of the integration of different energy networks and in the use of more energy vectors or carriers. As regards the second challenge, the main solutions proposed mainly concern self-consumption or local consumption of renewable sources, the use of energy storage and, in general, greater flexibility in final uses. In any case, all

Integrated Local Energy Communities: From Concepts and Enabling Conditions to Optimal Planning and Operation, First Edition. Edited by Marialaura Di Somma, Christina Papadimitriou, Giorgio Graditi, and Koen Kok.

solutions have in common the increasingly active role of end-users/prosumers. This is why the European Union (EU) directive on the single energy market defines a consumer-centric electricity market, giving the final consumer or prosumer the same dignity as energy producers or distributors. Precisely as a result of this central role entrusted to consumers, new market players are emerging such as aggregators or virtual power plants (VPPs), which have the aim of grouping together the potential flexibility on the energy demand side (controllable loads, electric vehicles (EVs), batteries, etc.) and on the distributed generation side (small-sized plants that are managed in a coordinated manner so as to appear to the grid as a single larger power plant). These emerging business models need a new architecture. This chapter presents the energy hub (EH) as an architectural and operational solution for coupling multiple energy carriers at different scales. Those presented in this chapter are the results of the research activities carried out by the authors in several National and European projects.

The paper is organized as follows: after the introduction, the architecture of the EH is presented, also introducing typologies of users/prosumers and technologies; subsequently, the results of applications of EH architectures in European projects both in living labs and in real cases are presented; the last part of the chapter will focus on presenting the EH architecture as the most suitable for managing energy communities.

10.2 Energy Hub and Micro Energy Hub Architecture

This section presents the EH architecture as proposed by eNeuron H2020 project [4]. EH is an architectural and operational solution for coupling multiple energy carriers at different scales to identify the potential benefits achievable for the energy community and its stakeholders. An EH is considered an abstract unit where multiple energy carriers can be converted, conditioned, stored, and consumed. It represents an interface between the community and external energy infrastructures and/or loads. EHs exchange energy at their interfaces, e.g. electricity and natural gas networks, for achieving an optimal balance within the EHs themselves from both technical and economical points of view. Within the hub, energy is converted and conditioned by technologies such as storage, combined heat and power (CHP), transformers, power-electronic devices, compressors, and heat exchangers. The EH concept is not limited to a certain size of the system: the concept enables the integration of an arbitrary number of energy carriers and products providing significant flexibility in system modeling architecture and operation. Due to this high flexibility, various real facilities can be modeled as EHs. In the eNeuron project, the integrated local energy communities (ILECs) will form the EH at a higher level while micro energy hub (mEH) concept is offered to capture the different levels of application. The EH definition is thus seen as a conceptual model for controlling and managing multienergy carrier systems across multiple scales, in order to optimize their architecture and operation. Inherent flexibility potentials of EHs allow introducing the concept of mEH representing the prosumer (industrial, commercial, or residential)

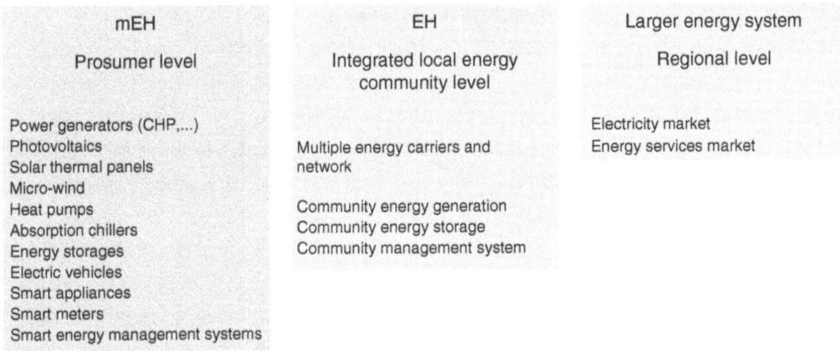

Figure 10.1 mEH and EH concept.

within the community. Under this assumption, each mEH will represent an integrated energy system consisting of multienergy generation, conversion, and storage technologies to satisfy its own energy needs. In the ILEC, mEHs cooperate by sharing all energy carriers, with the aim to satisfy the energy needs of the entire community represented by the EH. Community generation and community storage systems are also involved through dedicated community energy management systems. The EH promotes local balancing as well as strategic exchanges with electrical external grids through coordination of exchange. In this way, the EH will always have interactions with larger systems, sustaining access to the largest pool of external resources possible, while leaving open the possibility of local resource optimization and provision of services to other hubs or the bulk system. The system as a whole will allow synergies between different sectors such as electricity, heat, cooling, and transport (electric and hydrogen) as well as between different technologies. The interactions among mEHs within the ILEC and with the larger system under eNeuron concept are shown in Figure 10.1. mEHs can exchange energy locally. The connection to the larger system will allow covering the residual demand and selling excess or procuring shortfalls of energy as well as providing system services to the grid operators.

10.2.1 "mEH" and "EH"

This section describes the most common mEH and EH configurations in real contexts together with related technologies, energy vectors, and energy networks involved.

10.2.1.1 mEH

mEH can be seen as single entities featured by single PCC (point of common coupling) with energy networks. mEH can be passive or active consumers or even prosumers. Sometimes the difference between EH and mEH is not well defined and, depending on the case, an entity can be classified as EH or mEH. This is the case, for example, of condominiums which, due to size, could be considered mEH but which can actually be considered EH of apartments (which in this case would be mEH). In this chapter, condominiums are described as mEH.

10.2.1.1.1 Apartment/Detached House/Condominiums

In recent years, there has been a transformation of final users from passive consumers to active consumers, if not even prosumers, with the possibility of self-producing electricity thanks to distributed generation technologies, such as photovoltaic (PV) rooftop panels. At the residential level, the active users can participate in the demand/response programs that monitor their energy production and consumption [5]. The demand/response program objectives address the reduction of peak electricity demand and injection from and into the national grid, thus increasing the self-consumption of the renewable sources. The optimal management of the large appliances, like washing machine, dishwasher, and dryer, can be implemented in order to demonstrate the load shifting and peak shaving potentials of the selected users [6–8]. The interaction through the dedicated app could allow the users to acquire greater awareness of their consumption and interact with other players (such as the local distribution system operators [DSOs], aggregators, ILECs, or other peers), thus empowering final consumers/prosumers for participation in flexibility programs [9]. The apartment mEH introduces several benefits including electric consumption monitoring in real time, organizing electrical billing, identifying energy efficiency measures, and determining the cost/benefit ratio of measures or corrective actions aimed at reducing energy consumption.

Apartments in the same building (condominiums) are often served by centralized heating systems, which enable the opportunity of installing technologies not suitable for single apartments and detached houses, due to size limit (e.g. CHP or combined cooling, heat, and power [CCHP]), or to be connected to a district heating (DH) and cooling system [10].

10.2.1.1.2 Office Buildings

Office buildings are large tertiary sector buildings. They differ from condominiums mainly for two reasons: (i) the consumption profiles of the building are much more standard (working day/weekend; summer day/winter day) being typical of the tertiary sector and (ii) the management and maintenance of energy systems are often entrusted to energy service companies interested in energy efficiency strategies [11]. Due to their size, they are able to host distributed energy resource (DER) such as PV plants (in rooftops or in parking shelters) or CHP systems. Thanks to the installation of building energy management system (BEMS), demand-side management and demand-response strategies are usually deployed for optimal scheduling of controllable loads and power generation units [12].

10.2.1.1.3 Factories

The industrial sector is the highest energy consumer in most of the countries worldwide. The opportunity of integrating renewables (especially rooftops PV), CCHP units, energy storages, and EV-charging station has increased the flexibility that can be provided by factories. The demand-side management program, driven by energy management systems/tools, boosted the efficiency improvement of industries in the recent years mainly because the loads involved in this sector are more predictable than the ones related to the other sectors. Generally speaking, the use of a proper energy management system (EMS) is anyway required for performing energy efficiency and flexibility improvements in the industrial sector. Factories are ideal mEHs in an EH context, for the size of consumption (often equivalent to hundreds or thousands of residential users), the capacity of self-consumption of DER, and for investment capacity [13, 14].

10.2.1.1.4 Campuses

Campuses can be considered mEHs due to the presence of different energy vectors like electric, natural gas, thermal/cooling energy, and water (actually a non energy vector) [15]. In particular, consumption patterns can be clustered in recognized load profiles (such as winter/summer, working/weekend days, and lectures/holiday period), which can be easily translated in standardized energy management/flexibility actions. As for the factories, are ideal mEHs, for the size of consumption and capacity of self-production of DER and for the presence of several energy vectors [16]. Investment capacity can be improved through third-party financing by involving energy service companies (ESCo's) [17, 18].

10.2.1.1.5 mEH Technologies

Table 10.1 reports the main energy conversion and storage technologies, which are usually installed in mEH. Some of them are competing technologies (heat pumps versus natural gas boilers); in other cases, they can be combined (e.g. hybrid heat pumps consist of a heat pump coupled with a natural gas boiler). The choice of the suitable set of technologies is usually the result of a techno-economic analysis, which takes into account the price of commodities, investment and operational costs, and the presence of incentives. In recent years, the energy market is experiencing a growth in business models linked to the increase in flexibility of final consumption, which is pushing toward an electrification of consumption (heat pumps and EVs) and toward the use of energy storage.

Table 10.1 mEH technologies.

mEH	Technologies involved	Energy carriers involved	Energy networks which could be involved
Apartments, detached houses, condominiums, office buildings, factories, campuses	Heat pumps Hybrid heat pumps Natural gas boilers Electric boilers Batteries PV EV-charging stations (wall box) Centralized heating/cooling Micro-CHP CHP/CCHP Steam boilers	Electricity Hot water Chilled water Natural gas	Electricity Natural gas District heating/cooling

10.2.1.2 EH

EHs are entities that group together different mEHs to coordinate them as a single entity capable of participating in the electricity market and/or that of energy services. This subchapter presents three types of EH.

10.2.1.2.1 Energy Communities

Energy communities are heterogeneous groups of consumers, producers, and prosumers who come together for a common purpose in the energy field. The objectives, in fact, can be different, often linked to the exploitation of economies of scale, which are difficult for individual consumers to reach, such as (i) buy energy or certified renewable energy at a more convenient price and (ii) buy renewable plants at lower prices. As better explained in the second part of the chapter, energy communities can provide a framework for apartments, detached houses, condominiums, office buildings, factories, and campuses to optimize energy consumption patterns, provide flexibility to the energy network, and increase the local exploitation of renewable energy. The main feature of communities is that they often (not always) are geographically constrained.

10.2.1.2.2 Aggregators

Aggregators are market subjects that group single entities together in order to reach a size that allows them to participate in the energy market. An example, already introduced in the previous paragraph, is the aggregator of consumer groups who negotiate, on their behalf, a better energy price. In recent times, other types of aggregators such as VPPs or flexibility aggregators have entered the energy market.

VPPs are a set of small- and medium-sized DERs connected to low voltage (LV)/medium voltage (MV) network bundled and managed as a single biggest size plant in order to participate in the power or service markets.

Flexibility aggregators are flexibility service providers that aggregate and coordinate the flexibility-provided resources such as programmable electric loads, batteries, CHP plants, and EVs.

10.2.1.2.3 Multisite Consumers

EH architecture can also be applied to those entities in the industrial or tertiary sector characterized by having multiple production or administrative sites (factories, branches, offices, warehouses, and distribution and sales points) distributed throughout and within the same market area. Multisite consumers are aggregators characterized by grouping entities that refer to the same company. Examples of multisale consumers are sales points, distribution points, offices, and warehouses of large-scale distribution (supermarkets and hypermarkets) or large industries with different production sites and office buildings.

10.2.1.2.4 EH Technologies

Table 10.2 reports the main energy conversion and storage technologies, which are usually involved in EH. EH technologies differ from mEH technologies in their size

Table 10.2 EH technologies.

EHs	Technologies involved	Energy carriers involved	Energy networks which could be involved
Energy communities, aggregators, multisite consumers	Heat pumps Hybrid heat pumps Natural gas boilers Electric boilers Batteries PV EV-charging stations Centralized heating/cooling Micro-CHP	Electricity Hot water Chilled water Natural gas	Electricity Natural gas District heating/cooling

and functional location within the EH. In fact, these technologies are installed in the networks that connect mEHs and can be considered mEHs themselves. For example, a CHP system or a heat pump in a DH network or batteries installed on the LV/MV network or a network of charging stations on the distribution network are technologies which by size and functionality belong to the EH but which the EH manages as mEH.

10.3 EH and mEH Validation Through Simulation and Testing in Lab Environment

10.3.1 EH of Università Politecnica delle Marche

Università Politecnica delle Marche (UnivPM) is located in the Marche Region, Central Italy. It has several campuses spread in the region, four in the city of Ancona. UnivPM can be considered an EH with four mEHs in different sites over the city of Ancona (Italy). UnivPM accounts for a total of around 17 000 people among students and staff. These sites consist mostly of schools and offices, and among these the site of Monte Dago is a multienergy microgrid (mEH).

EH: UnivPM University

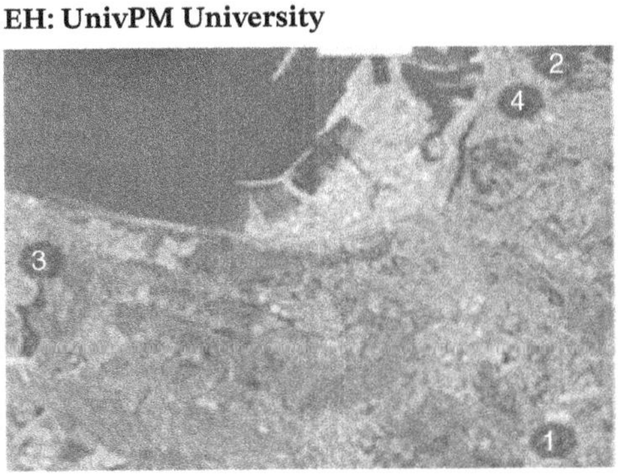

UnivPM mEHs:

1. Montedago multienergy microgrid (three faculties: Engineering, Life Sciences and Agriculture);
2. Faculty of Economics;
3. Faculty of Medical Sciences;
4. UnivPM Rectorate (headquarter).

mEHs 2, 3, and 4 are almost passive users with no DER and little room for flexibility. mEH 1, instead, is a multienergy microgrid with several technologies that enables opportunities of flexibility. The potential for flexibility and the development of an EH and mEH platform are the objectives of the European eNeuron project of which the UnivPM University is one of the demonstration sites [4].

10.3.1.1 Technologies, Energy Vectors, and Networks

The main energy service demand of the five mEH of UnivPM is electricity, heating, and cooling. In mEH 2, 3, and 4, this demand is met by electricity and natural gas withdrawn by the local distribution networks. Also, mEH 1 is served by natural gas and electricity distribution network but it consists of several buildings connected by internal electricity, district heating/cooling network served by a 575 kW_e CHP system fed by natural gas, whose electricity production is totally self-consumed within the mEH; thermal energy is produced by CHP and by natural gas boilers and it is distributed through a DH network; some buildings are equipped with absorption chillers which use DH thermal energy to locally produce cooling energy. mEH 1 is also equipped with (i) a hydrogen lab with 23 kW electrolyser, 1 kW_e fuel cell and with metal hydrates hydrogen storage; (ii) a battery energy storage lab where testing flexibility strategies with electric storage; (iii) EV-charging stations. Table 10.3 reports those technologies, involved in the UnivPM EH, and their interaction with the related energy (or non energy) networks. Figure 10.2 reports a scheme showing the interactions among energy networks, energy conversion, and storage technologies involved in mEH 1. Finally, Figure 10.3 shows the site map and the dashboard of the monitoring system of mEH 1. The EMS of UnivPM provides an overview of the aggregated energy production and consumption of all the UnivPM EH as a unique (potential/virtual) user in the energy market; at the same time, it allows to monitor and manage the mEH (in particular, mEH) by setting controls and parameters of single flexibility resources such as CHP systems or batteries.

Table 10.3 Technologies involved in UnivPM EH.

	Electricity	Natural gas	Heating	Cooling	Hydrogen	Mobility	Water
Electricity	—	CHP	CHP Heat pumps	Heat pumps	Electrolyzer Fuel cell	EV-charging station	Circulation pumps Electrolyzer
Natural gas	'—	—	CHP	—	—	—	—
Heating	—	—	—	Absorption chillers	Fuel cell	—	—
Cooling	—	—	—	—	—	—	—
Hydrogen	—	—	—	—	—	—	Electrolyzer
Mobility	—	—	—	—	—	—	—
Water	—	—	—	—	—	—-	—

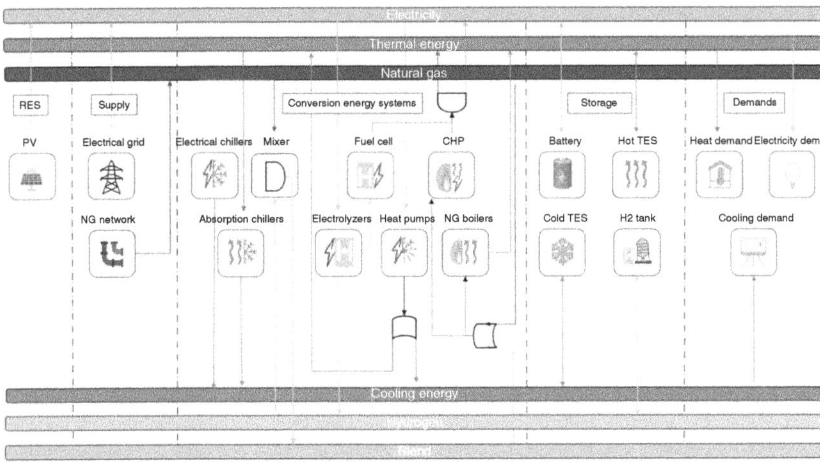

Figure 10.2 Technologies and networks involved in UnivPM mEH 1. Source: Ref. [18]/ Elsevier/CC BY 4.0.

10.3.2 EH for Aggregators: EV-Charging Station

Generally speaking, EH and mEH architecture is suitable for whatever type of aggregators. This section presents the application of the architecture to an aggregator of EV-charging station in which each charging station is managed as a mEH and a set of EV-charging stations are managed as a unique virtual battery providing flexibility in the energy market. Those presented are part of the results of Comesto project [19], an Italian national-funded project addressing energy storage within energy communities. In particular, once defined the number of EV-charging station, the consumption pattern of each of them (mEH) was simulated (Figure 10.4); then, all the information was grouped together in order to be seen (by the electric market) and managed as a unique virtual battery [20, 21].

Figure 10.4 shows the simulations of three charging stations (mEH) differing for nominal charging power such as 3 kW (a), 4.5 kW (b), and 6 kW (c), battery state of charge (SOC) of vehicle connected, and charging profile. Figure 10.4 shows also the parameter named minimum state of charge (MINSOC). In a context of flexibility, MINSOC is the minimum SOC requested by the final users when they disconnect their car: the difference between SOC and MINSOC can be considered as a reserve of flexibility.

Figure 10.5 reports the aggregated profiles of the EH that is the whole group of EV-charging stations. Figure 10.5a reports the SOC and the MINSOC of all the EV connected to the charging stations, while Figure 10.5b reports the flexibility potential of the virtual battery hub. In particular, FLEX_DOWN is the flexibility that can be achieved by charging all the EV connected, while FLEX_UP is the flexibility that can be achieved by discharging all the EV connected in the hypothesis that vehicle-to-grid (V2G) operation is allowed. Finally, Figure 10.5c reports the consumption profile of the EH.

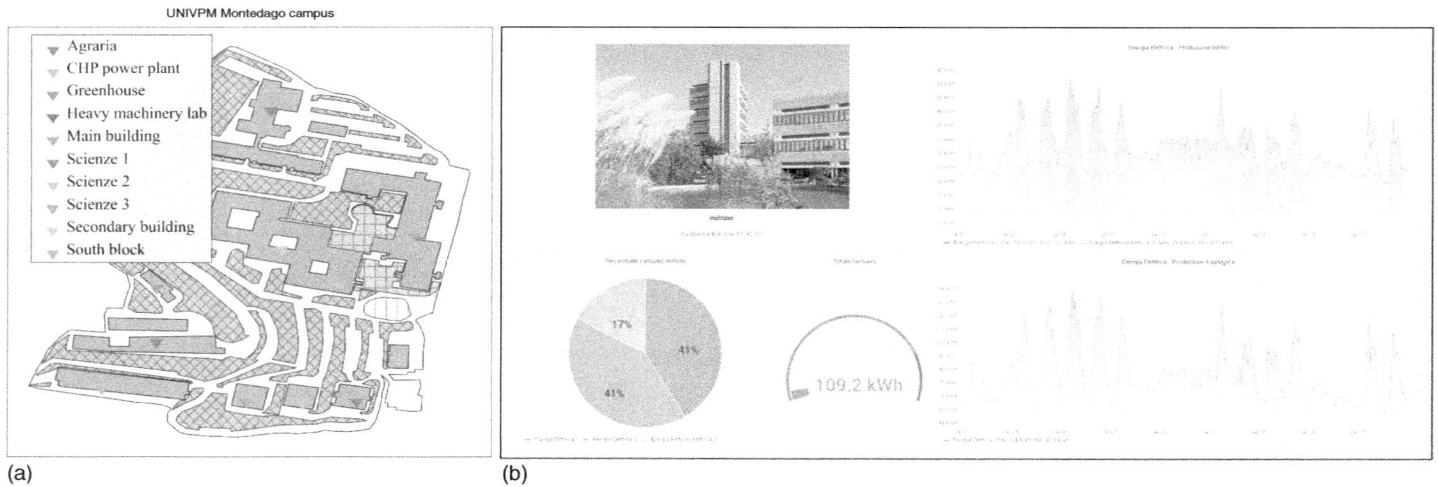

Figure 10.3 UnivPM Montedago mEH: site map (a) and user dashboard (b). Source: Marche Polytechnic University.

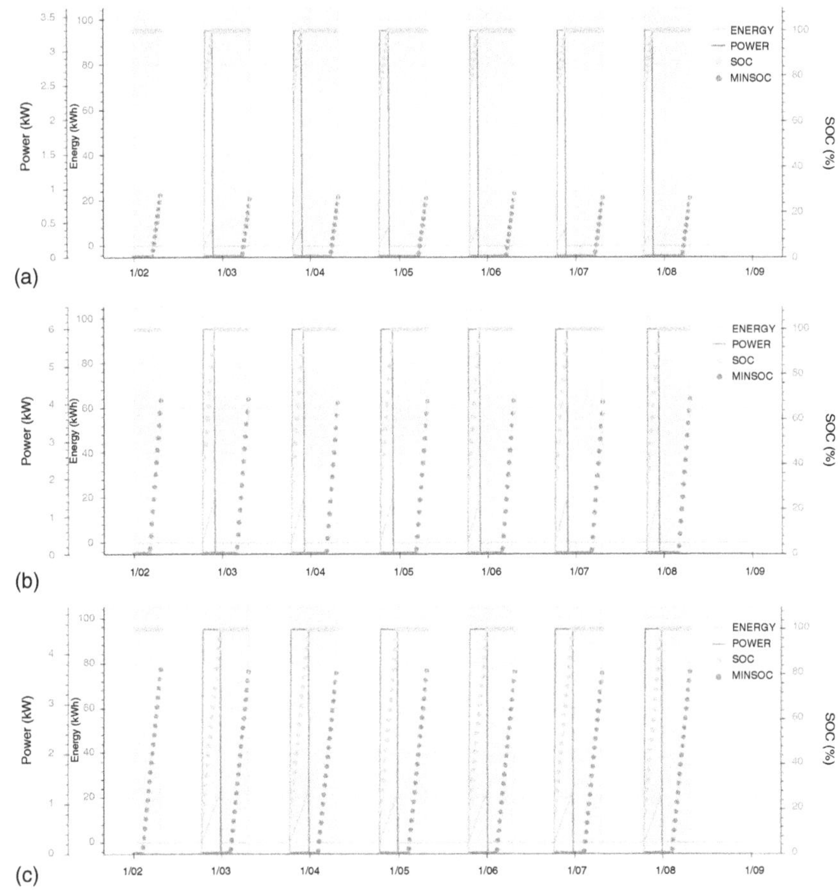

Figure 10.4 Results of simulation of three EV-charging stations.

10.3.3 SMES Project in Singapore

Singapore is an urban city in tropics and its climate is characterized by high and uniform temperature (25–31 °C), high humidity (80–90%), and abundant rainfall. The utility grid is dominantly driven by imported natural gas accounting for approximately 95% of the fuel mix [22]. Since Singapore's first liquefied natural gas (LNG) terminal was inaugurated in 2014, interest aroused in the free cold released by LNG regasification process as a business opportunity for cooling applications through district cooling grids. Due to the high and uniform temperature and humidity throughout the year, the cooling load is significantly quite high and constant and shows an increasing trend due to an enhanced urban heat island effect [23].

Within its energy strategy toward 2050 [24], Singapore aims to halve emissions from its peak to 33Mton$_{CO_2eq}$ by 2050, potentially achieving net-zero emissions with the adoption of different strategies to decarbonize the grid while guaranteeing energy security and affordability. Directly responding to this ambitious target, two novel EH projects have been developed and will be described as follows.

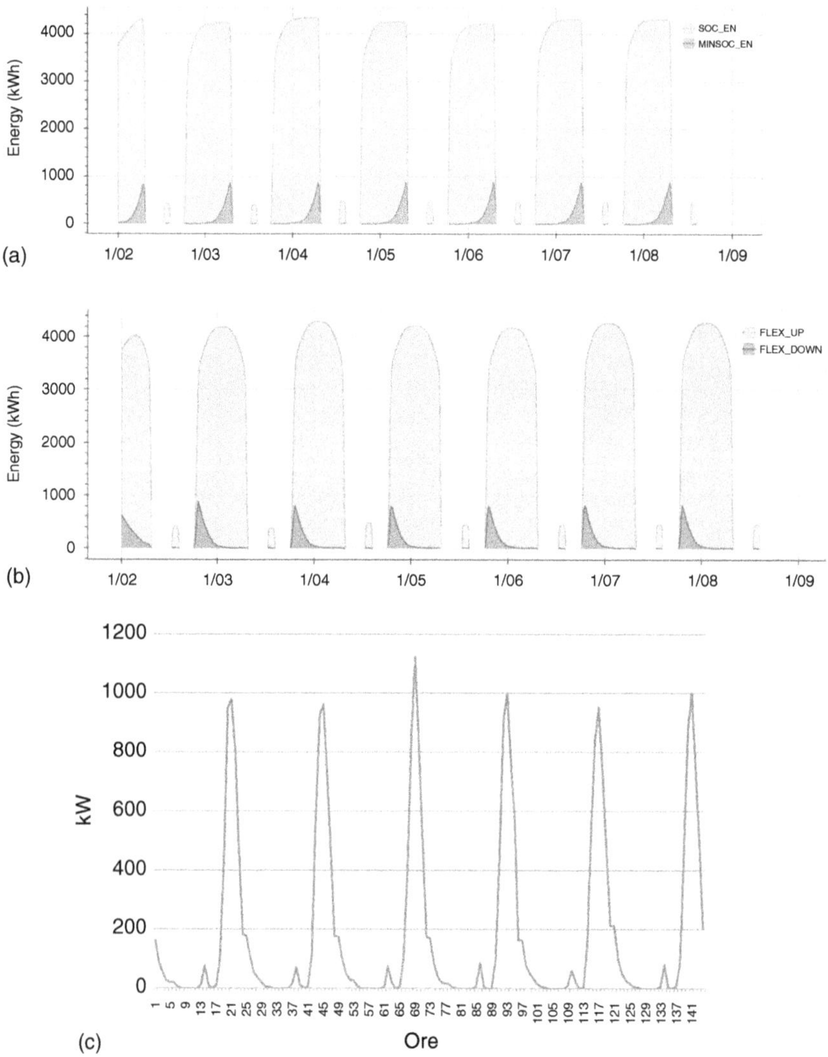

Figure 10.5 SOC and flexibility potential of the virtual battery hub.

10.3.3.1 The Cryo-polygeneration Demonstrator (CPD) for Nanyang Technological University

A novel district-scale demonstrator has been currently proposed and developed in Singapore [25], aiming to significantly contribute achieving the goal imposed by the country's national agenda on sustainable development [26]. Based on the concept of distributed cryo-polygeneration system (Figure 10.6), namely an integrated rapidly deployable and highly energy-efficient solution that utilizes cold energy from LNG and waste heat from power generation, the project might help in meeting the growing energy needs of urbanization and industrialization, especially in tropical and subtropical areas. Indeed, the recovery of premium-quality physical exergy in LNG at cryogenic temperature has rightly attracted considerable interest in the last decades and is becoming crucial in view of energy efficiency improvement and, above all,

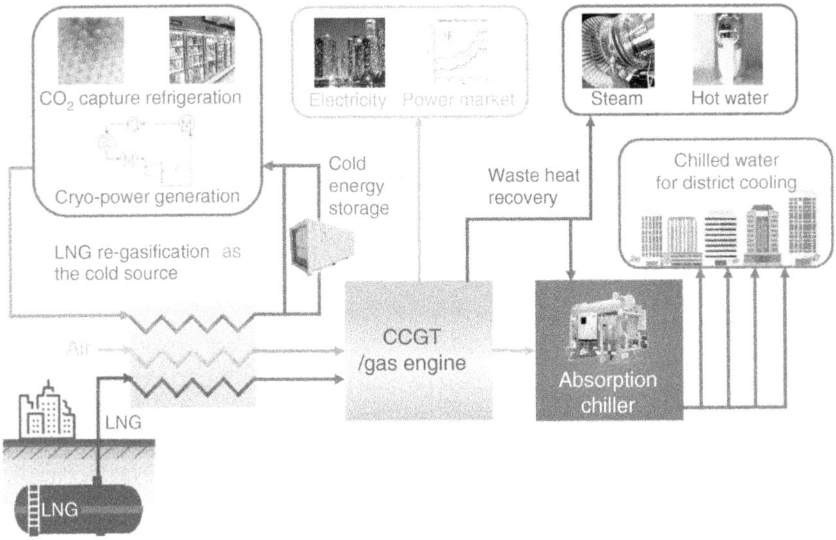

Figure 10.6 Superstructure of the cryo-polygeneration system. Source: Refrigeration image: N-Lange.de/Wikimedia Commons/CC BY SA 4.0. Electricity image: AstroChel – Science and Space/YouTube. Steam image: SIEMENS/Wikimedia Commons/CC BY SA 3.0. Hot water image: Processing Instruction nikkytok/Adobe Stock Photos. Cold energy storage image: CRS Mobile Cold Storage. Absorption chiller image: Trane.

environmental impact reduction. Indeed, if wasted during the regasification in open rack vaporizers, as it often occurs at the majority of LNG import terminals [27], the cold energy will be detrimentally dispersed to the environment. Conversely, it might be recovered for a variety of different processes and applications [28].

Making full use of the LNG regasification cold energy by implementing dedicated cold-to-cold applications, the district-scale demonstration will be developed at Nanyang Technological University (NTU) in Singapore, becoming the first showcase of systems integration that can be exported overseas as an urban solution. The CPD will therefore be a novel one-stop solution that encompasses all technologies, concepts by which power generation, cold energy harnessing, cold export, cryogenic power generation, and city gas generation that can be jointly generated from one solution based on the customer's needs. In order to maximize the benefits from the cryo-polygeneration system while still providing the same quality of energy services produced by conventional energy system, the optimal design of the plant layout was carried out [29] and it required the techno-economic analysis of single components in order to maximize the efficiency of the available energy resources. As above mentioned, the main case study for which the cryo-polygeneration was designed, and successively electrically and thermally connected, comprises a buildings' complex within NTU campus in Singapore (Figure 10.7). Belonging to the high-tension small category, the buildings are all connected to a district cooling network served by dedicated electrical chillers. The yearly electricity and cooling consumptions are available as measured data with half-hourly resolution due to smart meters installed in the buildings' complex. Figure 10.8 shows the representative weekly electrical (without air conditioning consumption) and cooling demand, a short

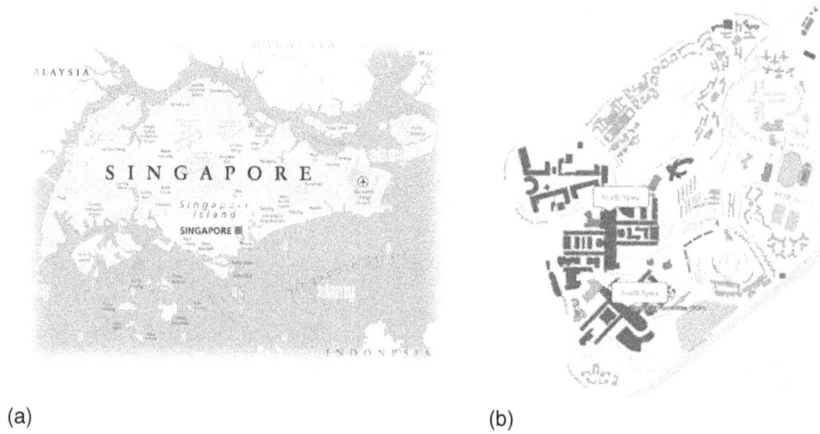

Figure 10.7 Singapore geo-localization (a) and Nanyang Technological University's map (b). Source: Adapted from Ref. [30].

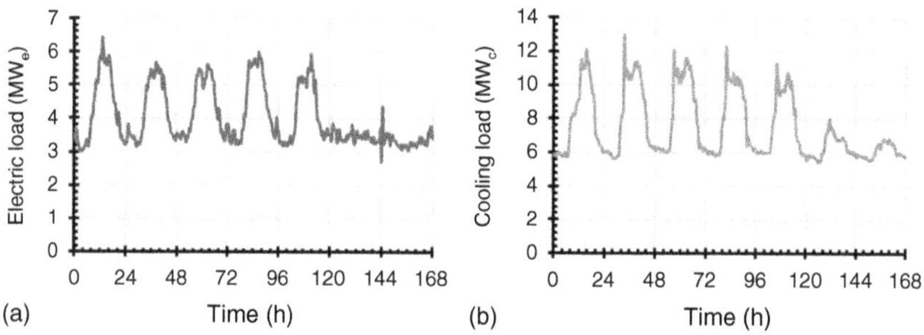

Figure 10.8 Weekly electricity w/o AC consumption (a) and cooling (b) demands for the buildings located in NTU campus.

portion of the yearly demands considered in this study. The annual electricity demand is 36.39 GWh_e with a peak of the power demand equal to 8.36 MW_e. The annual cooling demand is 65.77 GWh_c with a peak of the cooling power demand equal to 13.98 MW_c.

The optimal design and sizing of the cryo-polygeneration components was carried out through an in-house developed optimization tool based on a three-level configuration: the simulation level (TRNSYS [31]), the "genetic algorithm" (ga) level [32] and the interface for the communication between those two levels (Matlab [33]). In particular, TRNSYS is dynamic simulation software whose solver calls the subroutines present in the input file and tries to solve the equations for each simulation time step. "Ga" is a single-objective optimization algorithm that minimizes the objective function (namely the total annualized cost) subject to a set of constraints. The MATLAB code developed by the authors acts as an interface between TRNSYS simulator and the "ga" algorithm and streamlines the optimization process. The single-objective genetic algorithm in-built in MATLAB, i.e. the ga function, is based on the very popular Non-dominated Sorting Genetic

Table 10.4 Optimal technologies capacity from design optimization.

	Case scenarios	
Components	**BC**	**3**
Gas turbine (kW_e)	—	5 463
PV (kW_p)	—	2 500
Battery (kWh_e)	—	0
Absorption chiller (kW_c)	—	12 974
Electric chiller (kW_c)	20 471	8 583
TES (kWh_c)	—	806
Total annual cost (M$)	6.97	4.21
ΔCO_2 (%)	—	−19

Algorithm – II [32] and is frequently utilized in the literature [34] for the accuracy of the optimization function.

The techno-economic and environmental performance of the cryo-polygen scenario has been comparatively assessed with a baseline case scenario, namely a business as usual configuration where all the demands of the buildings are supplied by the electricity in a grid connected environment (Table 10.4).

The results confirm the workability of the framework that effectively determines the optimal design of the system. The application of the model indicates that the implementation of the cryo-polygeneration system has significant energetic, economic, and environmental benefits over the baseline case scenario. Indeed, the results promote the integration of both cold thermal energy storage and PV system into the cryo-polygeneration system since their implementation has significantly decreased both the primary energy consumption and the total annualized cost, thus reducing the global CO_2 emissions. A 2.5 MW_p PV system supported by a natural gas fed GT of 5.2 MW_e has been selected to provide the required electricity to the electric bus. The BESS does not belong to the optimal solution space due to the relatively high capital cost and the limited PV capacity constrained by the space requirements. A LNG regasification unit of 200 kW_c, an absorption chiller and a vapor compression chiller with capacities of 9 and 2.8 MW_c, respectively, supported by a 5 MWh_c cold thermal energy storage system have been selected to fulfill the cooling load required by the buildings' complex. Compared to the baseline case scenario, this DPS configuration achieves a 38.3% total annualized cost savings with a positive ΔNPV of 45.4 MUSD over 30 years. Notably, a 19% CO_2 reduction has been achieved due to the significantly lower primary energy consumption (PES = 17.5%).

10.3.3.2 Hydrogen Hub for the Decarbonization of Singapore Industrial Port Area (IPA)

One of the major international maritime transport hubs is represented by the city-state Singapore, a country characterized by a typical tropical climate. Indeed, the Port of Singapore is currently the world's second busiest port in terms of container throughput, with ship arrival tonnage exceeding 2.8 billion gross tonnes in 2021 [35]. In the more general context of Singapore decarbonization strategies, first in 2011 with the Maritime Singapore Green Initiative program [36] and recently through the "Maritime Singapore Decarbonization blueprint"[35], Singapore has promoted the adoption of the following sustainability practices within the port and shipping sectors: (i) the use of green electricity either locally produced in port area or imported from regional grid; (ii) the use of alternative marine fuels (i.e. hydrogen and ammonia) for both cargo handling equipment and tugboats; (iii) the replacement of diesel-powered cranes with electric-automated systems; and (iv) the introduction of energy efficiency measures for buildings and warehouses.

Nevertheless, although port decarbonization is surely considered a topic of great interest both for Singapore's government and more in general in the fields of energy and environmental engineering, there is a lack of comprehensive studies that analyze and synthesize the various energy demands (mechanical, electrical, heating, and cooling) of IPAs. Moreover, the development of dedicated tools for the design and optimization of alternative energy systems for ports is often overlooked at the global system level.

Directly responding to those needs, a novel hydrogen-based EH configuration has been proposed [37] considering that the electric demand of the port warehouses, buildings, and cranes is fulfilled by (i) the PV system, (ii) the national electric grid, and (iii) the fuel cell, whereas the cooling demand by (i) the absorption chiller, thermally driven by the waste heat from hydrogen compression, (ii) the vapor compression chiller, (iii) the liquefied hydrogen (LH_2) regasification process, and (iv) the cold thermal energy storage (Figure 10.9). Concurrently, the hydrogen demands of the tugboats and yard tractors operating in port, and the co-located combined cycle gas turbine (CCGT) power plant, planning to operate with a blend of 30% hydrogen and 70% natural gas [38], are satisfied by the hydrogen (i) locally imported as ammonia (NH_3), LH_2, and liquid organic hydrogen carrier (LOHC) and (ii) locally produced through an alkaline electrolyser.

The EH is optimized through a dedicated multi-objective design and operational optimization tool [37] based on a mixed-integer linear programming (MILP) algorithm to concurrently optimize techno-economic and environmental objective functions.

The preliminary results have shown that hydrogen is a key element in decarbonization, especially when utilized by the CCGT power plant. This leads to a reduction in carbon emissions by up to 50%. Among the various hydrogen carriers, NH_3 exhibits superior performance. LOHC and locally produced hydrogen are not economically feasible; hence, NH_3 is imported to almost fully satisfy the total hydrogen demand of the EH. LH_2 is imported for the remaining hydrogen demand, and the waste cold generated during LH_2 regasification is recovered and utilized to fulfill the cooling requirements of the port's buildings. The heating recovered from the hydrogen compressors is efficiently used in an absorption chiller to meet the total cooling demand.

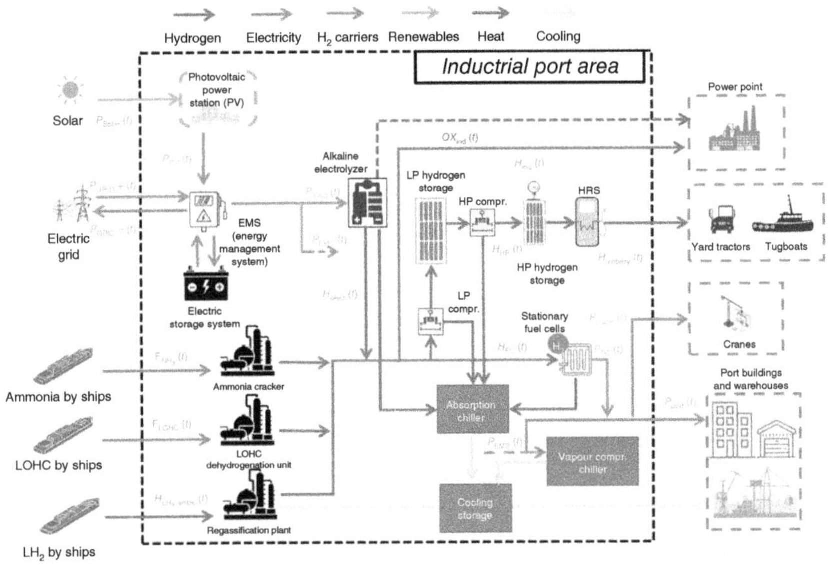

Figure 10.9 Superstructure of the EH proposed for the industrial port area. Source: Adapted from Ref. [37].

10.4 EH and mEH Validation Through Simulation and Testing in Real World

10.4.1 Italian City Scale Multi-energy Microgrid (Osimo)

The city of Osimo is a town located in the Marche Region, Central Italy. Its population is about 35 000 inhabitants. It is a town scale energy island/minigrid having only a single common point of coupling with national transmission system operator (TSO) Terna, through a primary cabin (pointer in Figure 10.10). This feature has become of particular interest in the last year, since the new Italian incentive for renewable energy communities (RECs) has set boundaries of RECs behind the primary cabin. As a consequence of that, the whole town can be considered a REC.

The urban minigrid is featured by a large amount of distributed energy systems, mainly PV plants (more than 30 MW_p of PV), meeting 29–32% of the yearly electric energy demand. The urban minigrid profile has the distinctive "duck curve" shape, even facing overgeneration for more than 800 hours per year, with an inversion of electricity flow in the primary cabin. Moreover, it is a multi-energy system with several energy and non-energy networks involved: electricity, natural gas, DH served by a cogeneration plant, urban water distribution network, and mobility. The town has a 1.2 MW_e CHP plant serving a DH network of roughly 1250 users. Most of energy and non-energy networks are managed by the same local multi-energy utility, so that it is easier to test synergies between several energy networks. An important feature of the Osimo town is that final users are equipped where possible with smart meters; in particular:

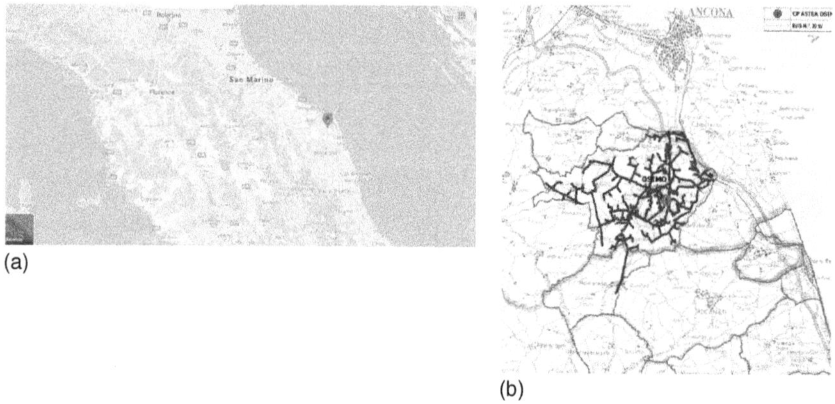

Figure 10.10 Geographical position of the Italian demo site (a) Source: Google Map; location of the municipal mini grid (b).

- all the final users of electricity have a smart meter;
- all the final users (1300) of the DH network have a smart thermal energy meter;
- several final users (1600) of the water distribution network have a smart meter.

10.4.1.1 EH Architecture

EH and mEH architecture has been and still is applied to the town of Osimo. Indeed, the architecture was developed during two H2020 EU projects [39, 40] in order to be applied to coordinate Local Energy communities, aggregators, and large consumers in order to provide flexibility services to the electric market. Today, the developed architecture is still used by ASTEA, the local utility (EH), to monitor and coordinate all its assets (mEH), namely: CHP power plant, water pumping station, batteries, and buildings. Figure 10.11 shows the architecture.

10.4.1.2 mEH: Large User (CHP)

The CHP plant serving the DH network is a flexibility resource participating in the Italian balancing market. It is managed by the local utility ASTEA and can provide 1.2 MW_e of flexibility with respect to the standard operational program. Cogeneration plant is a pivotal technology in the decarbonization of multi energy systems since it connects at least three energy networks: electricity, natural gas, and DH. Moreover, in future, it would be possible to feed CHP plants with a blend of natural gas and green hydrogen; although possible from a technical point of view, at the moment this possibility is not feasible because it is not convenient from an economic point of view. Figure 10.12 shows the mEH monitoring platform in which the EH manager can see both the location of the plant and real-time operating parameters.

10.4.1.3 mEH: Battery in the MV/LV Network

The local energy utility has installed two batteries in the DSO's LV/MV network in order to reduce network losses related to congestion management. Figure 10.13

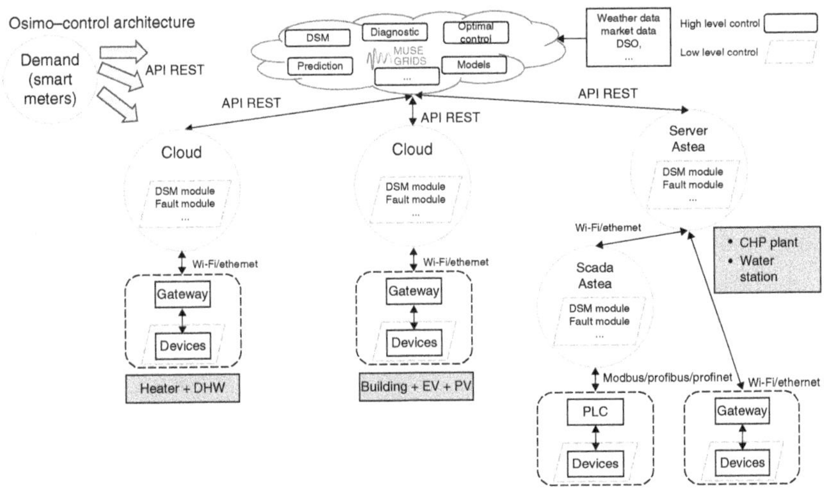

Figure 10.11 EH architecture for the city of Osimo.

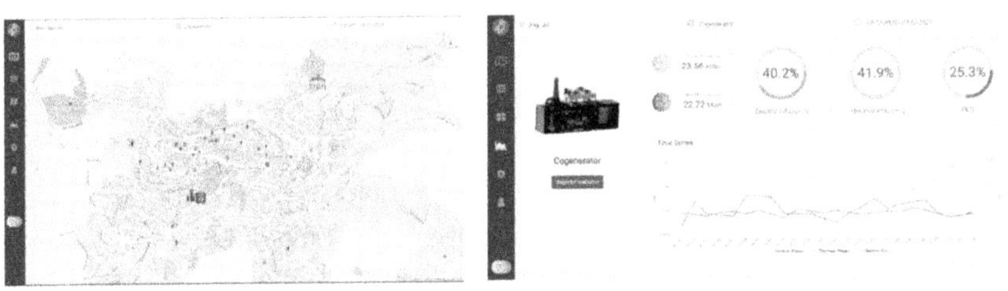

Figure 10.12 EH monitoring platform of the CHP power plant.

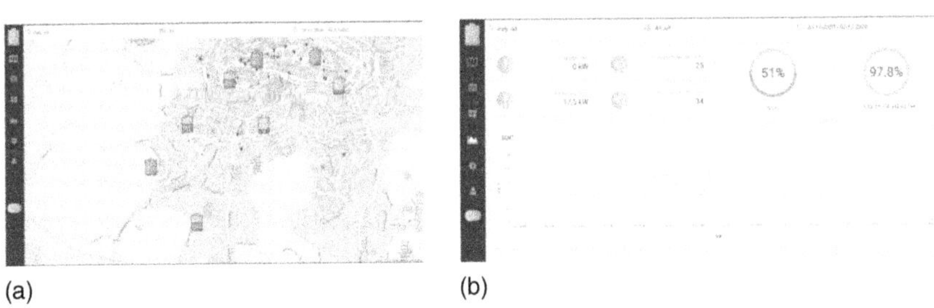

Figure 10.13 EH monitoring platform for batteries.

shows the dashboard of the EH monitoring platform: Figure 10.13a reports the position of the batteries in the DSO network while Figure 10.13b reports the data relating to one battery (mEH). It is worth noting that Figure 10.13a shows more batteries than actually installed, the excess ones were simulated for the purposes of the project.

10.4.1.4 mEH: Smart Building

Another mEH in Osimo town is the local energy utility's headquarter, which is a smart building equipped with PV plant and several flexibility resources: three

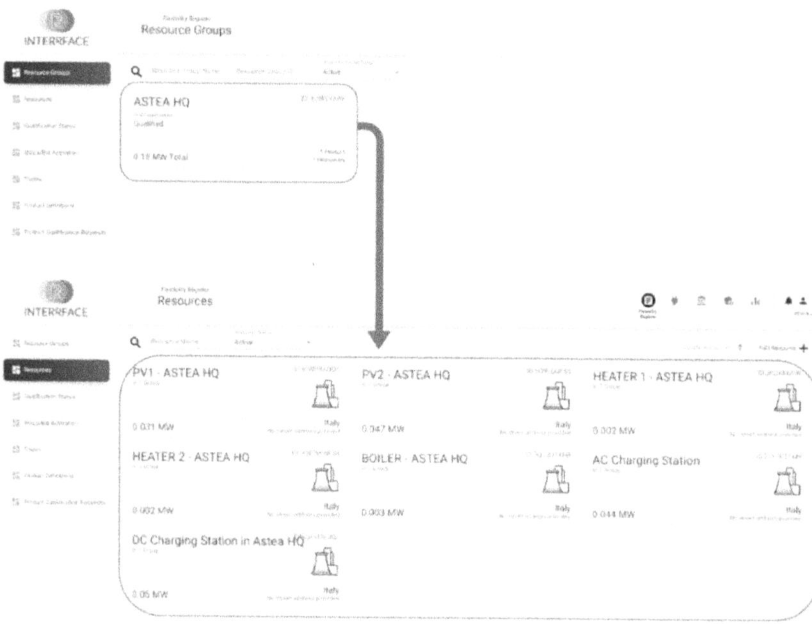

Figure 10.14 mEH monitoring platform for smart building (ASTEA HQ).

electric devices for hot water production coupled with thermal energy storage, one alternate current (AC), and one direct current (DC) EV-charging station. These flexibility resources are part of a strategy of electrification of final uses in order to fully exploit the self-consumption of electricity produced by PV plant. Figure 10.14 shows the platform developed in the INTERRFACE project for grouping together all the resources within a mEH.

10.4.1.5 mEH: Local Energy Community

In the INTERRFACE project, an REC architecture was developed. In this case, the level of EH and mEH can be translated differently. In the INTERRFACE project, the ILEC was considered a mEH providing flexibility to the EH manager in the same way as the CHP plant, batteries, or the smart building. This was mainly due to the small size of the ILEC itself (in the order of tens of kW). Generally speaking, a large ILEC is usually assumed to be an EH, all the participating final active consumers/prosumers acting as mEH. Figure 10.15 reports the mEH monitoring platform with location of flexibility resources, active consumers/prosumers (Figure 10.15a), the aggregated consumption and PV production patterns of the whole ILEC (Figure 10.15b), and the dashboard in which the ILEC manager can monitor the mEH resources.

10.4.1.6 Flexibility Results

All mEHs previously presented for the town of Osimo are in operation and provide flexibility services for the electricity market (CHP plant), for congestion management (batteries) or for increasing self-consumption of PV electricity avoiding to

Figure 10.15 mEH monitoring platform for local energy community.

inject it back to the local LV/MV network (ASTEA HQ). The only mEH not in operation is the LEC since the Italian incentive mechanism has not started yet. The overall flexibility provided by the Osimo EH can be summarized as follows:

- 1.2 MW provided by the CHP of the large user;
- 51 kW provided by a smart building;
- 105 kW provided by a battery;
- 70 kW provided by a small local REC;
- 57 kW provided by a second small local REC.

Finally, the use of batteries allowed to defer distribution infrastructure upgrades with a total avoided cost calculated by the DSO in €149 980.00.

10.5 EH and mEH: An Architecture for Renewable Energy Communities

Previous sections showed the application of EH architecture in lab environment, in simulated environment, as a concept for industrial districts, in real-world application. The EH architecture is one of the more promising concepts for flexibility providers such as aggregators, for VPPs, and for multisite consumers in the industrial and tertiary sectors. This section emphasizes EH as an enabling architecture for energy communities.

The EU envisages energy communities as a bottom-up approach for the future consumer-centric electricity market. In this context, energy communities can play a pivotal role in raising awareness and engagement and to empower final users as active consumers or prosumers [41, 42].

The EH and mEH architecture can be the tool for ILEC and for the promotion of renewable energy at the local scale [42–44].

10.5.1 An Overview of Energy Communities Rollout in Europe

The Clean Energy Package includes two types of energy communities: "renewable energy communities" (defined in the recast of the Renewable Energy Directive, 2018) and "citizen energy communities" (defined in the Electricity Market Directive, 2019), which allow citizens, public authorities, and specific types of companies to collectively organize their participation in the energy system including energy generation, self-consumption, sharing, storage, and sales of energy. RECs address all types of renewable energy and have a local character. Citizen energy communities (CECs), on the other hand, can operate over a larger area and have an emphasis on non-discriminatory access to the electricity markets, either directly or through aggregation. The Renewable Energy Directive also defines "Renewables self-consumers," enabling collective self-consumption (CSC) in the same building or multi-apartment block or even beyond; they are not bound to a specific governance structure. The EU framework leaves many details of the transposition process to the national level.

By mid of 2023, most EU member states have transposed the EU provisions on RECs and CSC, and some have transposed the concept of CECs. The concept of RECs was easier to implement as it has similarities to the well-known concept of energy cooperatives. The idea of CECs that can directly act on electricity markets posed more regulatory and technical challenges.

The transposition of RECs however is highly heterogeneous. Some member states have just transposed the text of the EU directive into national law, without defining details; others have created tailored national frameworks. Some member states have focused rather on CSC than on RECs or are creating hybrid solutions. Many countries have made exceptions from the EU provisions, e.g. Austria allows third-party ownership in RECs, which is only provided in the EU directive for CSC. Spain and France have focused on CSC rather than on RECs, but widened the scope to a district level and not to the multi-apartment level as most other member states do. Also, the way member stated subsides RECs is quite different. So far RECs rely on investment subsidies, reduced grid tariffs, reduced taxes, or a combination of these. In the context of reduced tariffs, member states follow the principle that only those grid levels should be paid for that are actually used for exchanging electricity. In Spain, the grid fee is even zero in the case of CSC [45].

RECs require proximity of their members to the renewable energy generators. Proximity can be determined using several approaches, such as administrative boundaries (e.g. municipalities or regions) or grid typologies.

Proximity based on typology of the public grid facilitates, for example, the implementation of local grid tariffs. While this is more in line with existing grid management of DSOs, REC activities may be interrupted if they, for example, cross a LV area. Also defining energy communities as electricity-related concept may prevent the inclusion of non-electricity related activities. Administrative structures (e.g. municipalities) aim at a better consideration of settlement and community structures that may not coincide with grid limitations. Another approach is defining a distance as boundary; however, REC activities may be interrupted if the distance is set too limited. Case-by case-judgements based on certain criteria, taking into account activities planned by RECs may not interrupt activities, but lead to an ex-ante uncertainty whether an REC will be approved if the criteria are unclear [45].

In a few EU countries, there is a large increase in energy communities. There are 400 registered RECs in Austria and more than 180 in France [46]. Most are residential energy communities; there are only few cases of commercial or mixed set ups. The emerging energy communities often have a strong focus on self-consumption; there is hardly any flexibility provision as the national energy and reserve markets do not provide suitable frameworks for flexibilities from decentralized assets or local flexibility provisions to DSOs [46]. So far RECs rely on subsidies, reduced grid tariffs, reduced taxes, or a combination of these.

Austria	LV, MV
Belgium Wallonia	LV, MV, and distance
Belgium Flanders	LV, MV, and activity
Hungary	LV
Slovenia	MV/HV
Italy	LV, MV
Croatia	Municipality
Lithuania	Municipality
Greece	Regions
Ireland	Communities
France	Distance
Spain	Distance (500 m)
Portugal	Distance (2 km) and activities

Source: Joanneum Research (2023).

Also the electricity allocation keys are very different among member states. Most countries have proposed static and dynamic sharing coefficients; some also have variable sharing coefficients. Static coefficients mean the allocation of energy is fixed for participants independent of their actual energy needs; variable coefficients take into account forecasted demand patterns while dynamic coefficients are based on real-time data and specific allocation formulas [47]. In Spain, the allocation mechanism is set beforehand based on variable coefficients combined with the possibility to change them every three months. In most other countries, sharing coefficients are changed less frequently. However, in some countries, the modus of dynamic sharing is de facto set by DSOs such as in Austria, where energy communities in practice have no free choice of the sharing mechanism despite the legislation does not restrict any sharing approach. DSOs across the EU are mentioning the high costs for implementing a new IT system that enable data processing and allocation and therefore take a rather restrictive approach.

There is a range of activities proposed in the literature that energy communities could implement. These include next to collective production and energy sharing also mobility communities, flexibility service to markets and the DSO, demand-side management, energy and energy efficiency services, e.g. solar-as-a-service, heat-as-a-service, or community-led renovation. Among the range of possible activities that energy communities could implement only a few are being realized at the moment, mostly sharing of PV electricity preventing viable business cases [47]. Often this limitation is given by the regulatory frameworks that consider energy communities as electricity-related concept only. Despite support mechanisms, the savings per households are in the range of only €50–€200 per year at current electricity prices, with lower values below €100 common in middle and northern Europe. Self-consumption rates are often below 40% as battery storages are still too expensive [48].

Key to a wide deployment and improved economics of Energy Communities will be a combination of technologies related to electricity, heat, cooling, and transport as well as mixed residential, commercial, and industrial setups with different load profiles increasing self-sufficiency. In particular, seasonal storage solutions should be incentivized as currently a surplus of renewable electricity is fed into the grid in summer, and a large amount of electricity is being purchased in winter, thus using the grid as a storage. This comes along with the observation made is that in several EU member states multiple small energy communities are emerging that have limited benefits for the energy system as a whole as they are not interacting among themselves or with the wider energy system, leading to a possible overinvestment of decentralized technologies.

Umbrella structures may be needed to increase the economics; services they could provide as well the benefits to the energy system reducing a fragmentation. Umbrella structures for REC or other collective energy actions could be CECs, municipal utilities, or other forms of organizations [47]. The interaction of RECs and such umbrella structures will require a change of regulative frameworks and advanced IT systems, such as the EH concept.

The current heterogeneous transposition of the EU directives can be considered as a big living lab, with some regulatory frameworks providing better conditions than others. Apart from incentivizing energy communities, a broader discussion is needed on their future role for the decarbonization of the energy system preventing a long-term subsidization of approaches that are inefficient from a system and thus societal perspective.

Allowing multiple energy carriers to be integrated into energy communities requires most member states to update regulative frameworks. It will also require new technical and IT concepts facilitating the selling of system services, or exchanging energy and flexibilities at different levels as aimed for in the EH concept, that will an important enabler to integrate energy communities into a broader energy system optimization.

10.6 Conclusions and Lessons Learned

This chapter describes the EH architecture as a valid solution for the activation of flexibility services and for local self-consumption of renewable energy by end consumers. In an energy market that increasingly sees the end user as the protagonist, whether active user or prosumer, the challenges today are many. First, it is necessary that the end user is accompanied on a path that leads him from a passive user to an active user through the stages of awareness, engagement, and finally empowerment. To do this, there must be a simple, accessible, and easily implementable architecture. The second step is that of the development of new market subjects based on new business models, capable of aggregating the offer of flexibility of end consumers as intermediaries between them and the energy services market: end consumers, however aware, will hardly be able to participate in the flexibility services market on their own, in terms of both size (individually they are too small) and competence. Flexibility service providers, such as aggregators and energy communities, are among the main entities that are emerging as intermediaries between the market and end consumers. The EH architecture is a valid solution for both end-users (mEH) and aggregator entities (EH). In fact, it allows both to accompany end-users (mEH) in their path of growing awareness and engagement and to aggregate them together with coordinated and managed strategies so as to be considered by size as a single entity in the energy market. In the first part of the chapter, the most common types of consumers/prosumers (mEH) and aggregators were presented. The second part of the chapter presented some experiences related to the EH architecture applied in lab environment, in simulation, or tested in real world. The final part of this chapter addressed the EH as a potential architecture suitable for managing local and RECs.

Acknowledgments

eNeuron project is funded by European Union's Horizon 2020 research and innovation programme under grant agreement No. 957779.

INTERRFACE project is funded by the European Union's Horizon 2020 research and innovation programme under grant agreement No. 824330.

MUSE GRIDS project is funded by the European Union's Horizon 2020 Research and Innovation Programme under Grant Agreement No. 824441.

Comesto project is funded by Italian Ministry of University and Research Pon "R&I" 2014–2020.

Compile project is funded by the European Union's Horizon 2020 research and innovation programme under grant agreement No. 824424.

Decide project is funded by the European Union's Horizon 2020 research and innovation programme under grant agreement No 894255.

List of Abbreviations

AC	alternate current
BEMSs	building energy management systems
BESSs	battery energy storage systems
CCGT	combined cycle gas turbine
CCHP	combined cooling, heat, and power
CECs	citizen energy communities
CHP	combined heat and power
CSC	collective self-consumption
DC	direct current
DERs	distributed energy resources
DSO	distribution system operator
EH	energy hub
EMS	energy management system
ESCo	energy service company
EU	European Union
EVs	electric vehicles
ILEC	integrated local energy community
LH_2	liquefied hydrogen
LNG	liquefied natural gas
LV	low voltage
mEH	micro energy hub
MILP	mixed-integer linear programming
MINSOC	minimum state of charge
MV	medium voltage
NPV	net present value
PCC	point of common coupling
PV	photovoltaics
REC	renewable energy community
SOC	state of charge
TSO	transmission system operator
V2G	vehicle-to-grid
VPP	virtual power plant

References

1 United Nations – A/RES/70/1 – General Assembly Distr. General 21 October 2015, Seventieth session Agenda items 15 and 116. Transforming our world: the 2030 agenda for sustainable development, 25 September 2015. https://sdgs.un.org/2030agenda.

2 European Commission, Directorate-General for Energy (2019). *Clean Energy for all Europeans*. Publications Office https://data.europa.eu/doi/10.2833/9937.
3 Van Nuffel, L., Dedecca, J.G., Smit, T., and Rademaekers, K. (2018). Sector coupling: how can it be enhanced in the EU to foster grid stability and decarbonise?, Policy Department for Economic, Scientific and Quality of Life Policies, Directorate-General for Internal Policies, PE 626.091, November 2018, https://www.europarl.europa.eu/RegData/etudes/STUD/2018/626091/IPOL_STU(2018)626091_EN.pdf (accessed 29 September 2023).
4 H2020 project, Green Energy hubs for local integrated energy communities optimization (eNeuron), 01/11/2020 – 31/10/2024, www.eneuron.eu (accessed 29 September 2023).
5 Tonellato, G., Heidari, A., Pereira, J. et al. (2021). Optimal design and operation of a building energy hub: a comparison of exergy-based and energy-based optimization in Swiss and Italian case studies. *Energy Conversion and Management* 242: 114316. https://doi.org/10.1016/j.enconman.2021.114316.
6 Najafi-Ghalelou, A., Zare, K., and Nojavan, S. (2019). Risk-based scheduling of smart apartment building under market price uncertainty using robust optimization approach. *Sustainable Cities and Society* 48: 101549. https://doi.org/10.1016/j.scs.2019.101549.
7 Setlhaolo, D., Sichilalu, S., and Zhang, J. (2017). Residential load management in an energy hub with heat pump water heater. *Applied Energy* 208: 551–560. https://doi.org/10.1016/j.apenergy.2017.09.099.
8 Fina, B., Auer, H., and Friedl, W. (2019). Profitability of active retrofitting of multi-apartment buildings: building-attached/integrated photovoltaics with special consideration of different heating systems. *Energy and Buildings* 190: 86–102. https://doi.org/10.1016/j.enbuild.2019.02.034.
9 Rahimzadeh, A., Victor Christiaanse, T., and Evins, R. (2021). Optimal storage systems for residential energy systems in British Columbia. *Sustainable Energy Technologies and Assessments* 45: 101108. https://doi.org/10.1016/j.seta.2021.101108.
10 Perera, A.T.D., Mauree, D., and Scartezzini, J.-L. (2017). The energy hub concept applied to a case study of mixed residential and administrative buildings in Switzerland. *Energy Procedia* 122: 181–186. https://doi.org/10.1016/j.egypro.2017.07.342.
11 Ghorab, M. (2019). Energy hubs optimization for smart energy network system to minimize economic and environmental impact at Canadian community. *Applied Thermal Engineering* 151: 214–230. https://doi.org/10.1016/j.applthermaleng.2019.01.107.
12 Roustai, M., Rayati, M., Sheikhi, A., and Ranjbar, A. (2018). A scenario-based optimization of smart energy hub operation in a stochastic environment using conditional-value-at-risk. *Sustainable Cities and Society* 39: 309–316. https://doi.org/10.1016/j.scs.2018.01.045.
13 Halmschlager, V. and Hofmann, R. (2021). Assessing the potential of combined production and energy management in industrial energy hubs – analysis of a

chipboard production plant. *Energy* 226: 120415. https://doi.org/10.1016/j.energy.2021.120415.

14 Taqvi, S., Almansoori, A., and Elkamel, A. (2021). Optimal renewable energy integration into the process industry using multi-energy hub approach with economic and environmental considerations: refinery-wide case study. *Computers and Chemical Engineering* 151: 107345. https://doi.org/10.1016/j.compchemeng.2021.107345.

15 Kourgiozou, V., Commin, A., Dowson, M. et al. (2021). Scalable pathways to net zero carbon in the UK higher education sector: a systematic review of smart energy systems in university campuses. *Renewable and Sustainable Energy Reviews* 147: 111234. https://doi.org/10.1016/j.rser.2021.111234.

16 Al-Ghussain, L., Darwish Ahmad, A., Abubaker, A.M., and Mohamed, M.A. (2021). An integrated photovoltaic/wind/biomass and hybrid energy storage systems towards 100% renewable energy microgrids in university campuses. *Sustainable Energy Technologies and Assessments* 46: 101273. https://doi.org/10.1016/j.seta.2021.101273.

17 Liu, H. (2020). Smart campus student management system based on 5G network and internet of things. *Microprocessors and Microsystems* 103428. https://doi.org/10.1016/j.micpro.2020.103428.

18 Jin, L., Rossi, M., Ciabattoni, L. et al. Environmental constrained medium-term energy planning: the case study of an Italian university campus as a multi-carrier local energy community. *Energy Conversion and Management* 278: 116701.

19 Pon 2015-2020, Community Energy Storage: Gestione Aggregata di Sistemi di Accumulo dell'Energia in Power Cloud (ComESto), 2014-2020, www.comesto.eu (accessed 29 September 2023).

20 Ciabattoni, L., Cardarelli, S., Di Somma, M. et al. (2021). A novel open-source simulator of electric vehicles in a demand-side management scenario. *Energies* 14: 1558.

21 Ciabattoni, L., Cardarelli, S., Di Somma, M. et al. (2020). A new hybrid software tool for the simulation of energy usage in a population of electric vehicles. *2020 IEEE International Conference on Environment and Electrical Engineering and 2020 IEEE Industrial and Commercial Power Systems Europe (EEEIC/I&CPS Europe)*, 9-12 June 2020, Madrid (Spain).

22 EMA: Singapore energy statistics | Energy transformation, 2023, https://www.ema.gov.sg/singapore-energy-statistics/Ch02/index2 (accessed 29 September 2023).

23 Aflaki, A., Mirnezhad, M., Ghaffarianhoseini, A. et al. (2017). Urban heat island mitigation strategies: a state-of-the-art review on Kuala Lumpur, Singapore and Hong Kong. *Cities* 62: 131–145. https://doi.org/10.1016/J.CITIES.2016.09.003.

24 Energy 2050 committee report charting the energy transition to 2050 foreword message executive summary section 1: changing energy landscape section 2: trends and possible pathways towards net-zero. 2022.

25 Cryo-polygeneration | Surbana Jurong-NTU corporate laboratory | NTU Singapore, 2023, https://www.ntu.edu.sg/sj-ntu/research-focus/research-projects/cryo-polygeneration (accessed 29 September 2023).

26 Singapore green plan 2030, 2021, https://www.greenplan.gov.sg/splash (accessed 29 September 2023).

27 Le, S., Lee, J.Y., and Chen, C.L. (2018). Waste cold energy recovery from liquefied natural gas (LNG) regasification including pressure and thermal energy. *Energy* 152: 770–787. https://doi.org/10.1016/j.energy.2018.03.076.

28 Khor, J.O., Dal Magro, F., Gundersen, T. et al. (2018). Recovery of cold energy from liquefied natural gas regasification: applications beyond power cycles. *Energy Conversion and Management* 174 (August): 336–355. https://doi.org/10.1016/j.enconman.2018.08.028.

29 Tafone, A., Raj Thangavelu, S., Morita, S., and Romagnoli, A. (2023). Design optimization of a novel cryo-polygeneration demonstrator developed in Singapore – techno-economic feasibility study for a cooling dominated tropical climate. *Applied Energy* 330: 119916. https://doi.org/10.1016/J.APENERGY.2022.119916.

30 NTU map, 2020. https://maps.ntu.edu.sg/#/ntu/d386ffa80e4e46f286d17f08/poi/details/de566357c26749f4b712b3c7 (accessed 29 September 2023).

31 Transient Simulation Tool (TRNSYS 18), 2018, https://www.trnsys.com/index.html (accessed 29 September 2023).

32 Deb, K., Pratap, A., Agarwal, S., and Meyarivan, T. (2002). A fast and elitist multiobjective genetic algorithm: NSGA-II. *IEEE Transactions on Evolutionary Computation* 6 (2): 182–197. https://doi.org/10.1109/4235.996017.

33 MATLAB – MathWorks – MATLAB & Simulink. https://www.mathworks.com/products/matlab.html (accessed 29 September 2023).

34 Rangaiah, G., Sharma, S., and Sreepathi, B.K. (2015). Multi-objective optimization for the design and operation of energy efficient chemical processes and power generation. *Current Opinion in Chemical Engineering* 10: 49–62. https://doi.org/10.1016/J.COCHE.2015.08.006.

35 Maritime Singapore decarbonisation blueprint | Maritime & Port Authority of Singapore (MPA), 2022. https://www.mpa.gov.sg/maritime-singapore/sustainability/maritime-singapore-decarbonisation-blueprint (accessed 29 September 2023).

36 Maritime Singapore green initiative | Maritime & Port Authority of Singapore (MPA), 2019, https://www.mpa.gov.sg/maritime-singapore/sustainability (accessed 29 September 2023).

37 Pivetta, D., Tafone, A., Mazzoni, S., et al. (2023). A multi-objective optimization approach for the design of alternative port energy systems in tropical region. *Proceedings of ECOS 2023 – the 36th International Conference on Efficiency, Cost, Optimization, Simulation and Environmental Impact of Energy Systems*, Las Palmas de Gran Canaria, Spain (25–30 June 2023). https://ecos2023.com/ (last access 29 September 2023).

38 Keppel to develop Singapore's first hydrogen-ready power plant, with construction undertaken by Mitsubishi Power Asia Pacific and Jurong Engineering consortium – Keppel Infrastructure, 2022, https://www.kepinfra.com/en/news-item.aspx?aid=16124&title=keppel-to-develop-singapores-first-hydrogen-ready-power-plant (accessed 29 September 2023).

39 H2020 project, Multi Utilities Smart Energy GRIDS (MUSE GRIDS), 01/11/2018 – 31/10/2022, https://muse-grids.eu/ (accessed 29 September 2023).

40 H2020 project, TSO-DSO-Consumer INTERFACE architecture to provide innovative grid services for an efficient power system (INTERRFACE), 01/01/2019 – 31/12/2022, 10.3030/824330, http://www.interrface.eu/ (accessed 29 September 2023).

41 Sæle, H., Morch, A., Buonanno, A. et al. (2022). Development of energy communities in Europe. *2022 18th International Conference on the European Energy Market (EEM)*, IEEE, 15-15/09/2022, Ljubljana, Slovenia. pp. 1–5. https://doi.org/10.1109/EEM54602.2022.9921054

42 Nouri, A., Khadem, S., Mutule, A. et al. (2022). Identification of gaps and barriers in regulations, standards, and network codes to energy citizen participation in the energy transition. *Energies* 15: 856.

43 Caliano, M., Buonanno, A., Di Somma, M. et al. (2022). Enabling conditions for the deployment of integrated local energy communities in Europe. *2022 Workshop on Blockchain for Renewables Integration (BLORIN)*, IEEE, 2-3/09/2022, Palermo, Italy. pp. 144–148. https://doi.org/10.1109/BLORIN54731.2022.10028420

44 Morch, A., Di Somma, M., Papadimitriou, C. et al. (2022). Technologies enabling evolution of integrated local energy communities. *2022 IEEE International Smart Cities Conference (ISC2)*, Pafos, Cyprus., IEEE, 26-29/09/2022, pp. 1–6. https://doi.org/10.1109/ISC255366.2022.9922568

45 Frieden, D., Tuerk, A., Antunes, A.R. et al. (2021). Are we on the right track? Collective self-consumption and energy communities in the European Union. *Sustainability* 13: 12494. https://doi.org/10.3390/su132212494.

46 Tuerk, A., Neumann, C., Matowska, M., and Rakocevic, L. (2023). Energy communities and collective actions: yearly policy brief on regulation, Report D3.5, DECIDE project, https://www.researchgate.net/publication/371327035_Energy_communities_and_collective_actions_Yearly_policy_brief_on_regulation (accessed 29 September 2023).

47 Tuerk, A., Neumann, C., Matowska, M. et al. (2023). Final report on business models, contractual conditions and recommendations, DECIDE project, 10.13140/RG.2.2.20246.73282.

48 Neumann, C. and Tuerk, A. (2023). Business models for energy communities. Report D7.1 of the COMPILE project, https://www.compile-project.eu/ (accessed 29 September 2023).

11

Energy Communities as an Alternative Way of Organizing the Energy System in Europe: Key Societal Aspects

Anna J. Wieczorek[1], Natascha van Bommel[1], Amira El-Feiaz[1], Nikki Kluskens[1], Irene Niet[1], Luc van Summeren[1], Johanna Höffken[1], Floor Alkemade[1], Laura van den Berghe[1], Claudia Meloni[2], Giorgio Graditi[2], and Marialaura Di Somma[2]

[1] *School of Innovation Sciences, Industrial Engineering and Innovation Sciences, Eindhoven University of Technology, PO Box 513, 5600MB, Eindhoven, the Netherlands*
[2] *Italian National Agency for New Technologies, Energy and Sustainable Economic Development, Lungotevere Thaon di Revel, 76, 00196, Rome, Italy*

11.1 Introduction

The European energy system is experiencing significant transformations due to the climate crisis, advancing digitalization, and the increasing demand for greater citizen engagement. Energy communities are recognized as crucial catalysts for the transition toward sustainable energy. They are mostly seen as groups of people who come together to improve their energy use and production for the benefit of their community [1]. They serve as effective platforms for promoting the acceptance of renewable sources and technologies within society as well as for fostering active citizen participation in innovative renewable energy (RE) solutions. Without energy communities, achieving a shift toward a more sustainable energy system would be unattainable [2].

While communities have so far mainly focused on collective generation, they are increasingly involved in providing, distributing, and balancing energy and flexibility. By doing so, energy communities are not only becoming a new actor in the energy system, but they also offer an alternative way of organizing it based on the need for independence from the energy suppliers, social cohesion, or support to the local economy [3]. These objectives are realized through self-consumption, community trading, or self-balancing activities. Communities can also participate in the energy market. In return for financial revenues, they can provide selective services to the system through an aggregator which would release the grid from increased congestion and problems with balancing.

However, neither the definition nor the many activities of energy communities are uniformly specified in the literature and legally institutionalized in practice. Although in Europe, the European Commission suggested broad contours of what

Integrated Local Energy Communities: From Concepts and Enabling Conditions to Optimal Planning and Operation, First Edition. Edited by Marialaura Di Somma, Christina Papadimitriou, Giorgio Graditi, and Koen Kok.

can be considered an energy community, its specific delineation is up to the Member States, while the worldwide literature does not always align with the European suggestions. Regarding activities, although critical to realizing the communities' primary objective of self-servicing and providing community benefits, direct market participation is not possible. The current energy system operates predominantly in a top-down, central way with large power units, transmission, and distribution operators. Market participation is based on high liquidity levels, professionalization, and large volumes that communities do not have and, mostly, do not want. The rationale for the operation of energy communities is more decentralized, more distributed, and bottom-up and often ideologically founded as opposed to the market rationale based on which the current energy system works. For communities, market participation implies the necessity to deal with many barriers and challenges. To make space for energy communities, changes are necessary in the organization of the current energy system.

In this chapter, we investigate the energy communities as a new form of organizing the European energy system and we analyze how the current energy system provides obstacles to the greater upscaling of the energy communities. We focus on gathering insights about societal aspects that facilitate energy communities as they make their way to the larger system and offer a novel alternative that is more democratic and just. Our research question is

Under what conditions can energy communities offer a viable alternative to the current way of organizing the energy system?

We address this question by mobilizing insights from the sustainability transitions (e.g. [4]) and science and technology literatures (e.g. [5]). Methodologically, this chapter builds on several existing studies in the field [1, 6–12] and collates the findings into a review of key societal aspects.

In this chapter, we first present a sociotechnical approach as a way of analyzing incumbent and emerging systems (Section 11.1). Using the sociotechnical approach, we first shortly sketch the evolution of the European energy system (Section 11.2). We then proceed to explain the emergence of energy communities as new actors in the energy field (Section 11.3), the increasing role of the information and communication technology ICT in facilitating the energy communities (Section 11.4), and how institutions such as legal and market structures create obstacles for upscaling of energy communities (Section 11.5). In Section 11.6, using three examples, we demonstrate how energy community can deal with various challenges. We conclude with a discussion in Section 11.7.

11.2 A Sociotechnical Approach

Technology, as important as it is, does not develop independently from society. It is enacted by actors and involves a sociocognitive process of technological sense making [5]. Bijker and Law [13] emphasize the entanglement of human and nonhuman factors in technological development such as policies, behaviors, norms, or values, brought together under the concept of "institution." The coevolution of technology

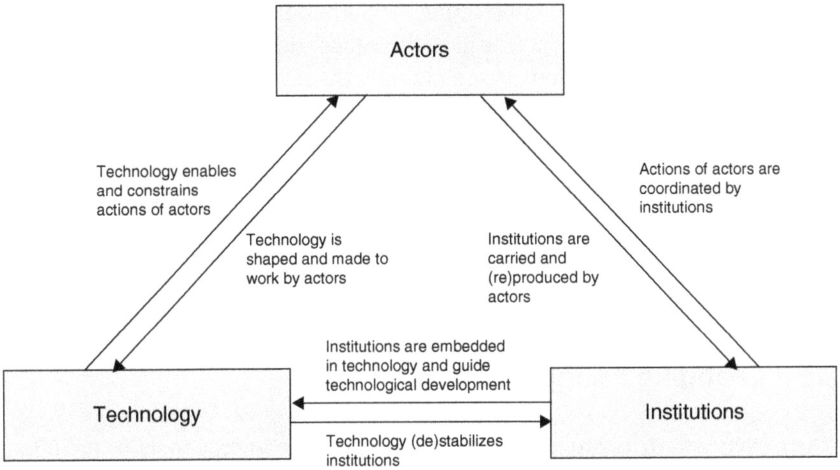

Figure 11.1 Interactions between institutions, actors, and technology. Source: Based on Van Summeren et al. [11], Rohracher [20], and Geels [21].

and society is considered to play an important role in the creation of systems through which human needs, e.g. energy, mobility, or food, are satisfied [14–18].

Such sociotechnical systems are organized and function according to a specific rationale. The rationale provides meaning to the activities of actors and helps them organize their time and space and reproduce their experiences. It also drives the selection of the technology and shapes and constrains the behavioral possibilities of actors [19]. Thereby the rationale is embedded and manifested in various institutions, which are enacted and reproduced by actors and actor networks who mobilize the technology and build infrastructure to deliver societal functions, see Figure 11.1.

Institutions refer to sets of rules, norms, values, and beliefs that regulate but do not define perceptions and activities of actors [22, 23]. According to Scott [24], institutions can be categorized into three types: regulative, normative, and cognitive. Regulative institutions specify "what is allowed" and encompass sanctions, rules, and laws. Normative institutions indicate "what is appropriate" and morally justified, and include values, norms, and expectations. Cognitive institutions describe "what is deemed thinkable" and how reality, as we experience and understand it, is constituted, and interpreted through shared cognitive categories and frames. Cognitive institutions may include symbols, belief systems, or cognitive frames and frameworks. The institutions are imprinted in material elements, technology, and infrastructures created by the actors in the given context and time.

The rationale shared by a group of actors provides a basis upon which systems operate and gives a system stability and the ability to coordinate actions of social groups by shaping what is considered allowed, legitimate, and probable [25]. Although systems are difficult to change [21], actors are not passive-rule followers or passive recipients of technologies. They can actively use rules to interpret the world, make decisions, and act [26]. They may exercise their agency by disagreeing on the basic rules, reflexively interpreting, or purposefully deviating from what is considered normal [27]. This may lead to the system overhaul, often referred to as system innovation or transition, where not only the technology and infrastructure fundamentally reorganize but also the related societal aspects such as markets, policies, user practices, and behavior [4, 28, 29].

The sociotechnical perspective and the analysis of actors, institutions, technology, and the underlying rationale prove useful to better understand the historical developments in the energy field and to discuss the type of barriers and obstacles that energy communities face in navigating their way to a larger system.

11.3 Changing Energy System

In their overview, Wieczorek et al. [12] show that the energy system, including various integrated supply sectors (electricity, heat, cooling), end-use sectors (transportation, residential, services, industry), and energy carriers (gas, electricity, heat/cold, solid, and liquid fuels) [30] has been forming over the course of more than a century. Throughout its development, different time periods have been characterized by distinct organizing principles, dominant designs of the supply systems, and a range of actors, who established regulations and set priorities. Each period's major challenges were framed in response to the realities of the preceding era.

In Europe, the inception of electricity in 1880 marked the beginning of the electricity supply industry as we know it today. The rationale of that time was "selling electricity" and it revolutionized the design of the supply system. Electricity was generated in specialized facilities known as "electricity factories" or "central (power) stations." A separate distribution network was established to transport electricity to consumers, primarily households. Utility companies were founded to facilitate the local production and retail of electricity [31].

Over time, particularly in the postwar time, there was growing criticism of the decentralized and fragmented structure of the electricity supply system. Visionary engineers and governments began to develop a nationally integrated supply system with reduced reliance on distributed generation for public electricity supply. The objective was to ensure a consistent and uninterrupted energy supply at a national level while achieving economies of scale through centralization. This approach was a response to societal demands for continuous and reliable energy flows [32]. It meant involvement of large private enterprises supplying energy to passive consumers within a regulated market competition framework.

Starting from the 1970s, the centralized and supply-focused electricity system began to transform under the influence of several developments. These included the oil crises of the 1970s, the economic recession of the 1980s, the market liberalization, deregulation, and privatization processes of the 1990s, along with the optimistic outlook on European integration. A notable outcome of these changes was the separation of the generation and retail functions from the transmission and distribution of electricity in most European countries. This decision laid the groundwork for liberalized and competitive market structures. It granted increased power not only to private corporations but also to consumers by providing them with the freedom to choose their own energy supplier [33]. As a result, the system began to accommodate both centralized and decentralized generation units but remained largely intact. Existing large-scale power units were expanded and modified to meet environmental requirements, while new smaller units were introduced. The adaptation of electricity grids was only partial [33].

Recent decades saw the energy system moving toward increased decentralization. This transformation has been driven by growing concerns about climate change and the need for decarbonization. As a result, there has been a significant rise in electricity generation from renewable sources such as photovoltaic (PV) and wind power, generated both through large-scale plants that feed into the transmission grids and through distributed renewable sources integrated into the medium- and low-voltage grids. Digitalization has played a crucial role in enabling a new organizational rationale for the energy system, emergence of new energy-consuming functions, and a greater focus on electrification. The changes have opened opportunities for more flexible and efficient energy management as well as the integration of various energy sources and technologies (e.g. energy storage or demand response) into the system [33].

The new energy system began to witness the emergence of new actors (green utilities, energy communities, aggregators, and prosumers), all taking on novel roles. With the support of government subsidies and a significant drop in PV panel prices, these new actors have gained prominence in microgeneration. Commercial aggregators have started to consolidate distributed energy resources (DERs) and offer specialized flexibility services to transmission system operators and distribution system operators (TSOs, DSOs) or balance responsible parties (BRPs) [34]. Meanwhile, individual prosumers have increasingly come together to form energy communities [35]. New ICT tools have facilitated flexible energy demand within households through various means such as heat load reduction by utilities or aggregators, and the implementation of energy management systems (EMSs) and intelligent control devices within homes. Leveraging ICTs, these communities have begun to aggregate electricity and flexibility from DER with the aim of trading it on energy and ancillary services markets. As a result, their focus has shifted from solely consuming and generating energy to actively participating in energy trading [11].

These emerging trends have been accompanied by the establishment of new laws regarding energy and renewables in Europe [2]. The laws have spurred the development of innovative markets and market mechanisms to harness demand-side flexibility. The regulatory framework, with a particular focus on the Electricity Market Directive and the Renewable Energy Directive (RED), is now being translated into national contexts. These regulations aim to create a more favorable market environment for the new actors such as energy communities and facilitate the reorganization of the energy markets around the opportunities presented by smart grids. They also reinforce existing support schemes for RE, which have led to a growing number of often small-scale and intermittent RE sources being connected to the low- and medium-voltage grids.

The ongoing transformation of the energy system is driving it toward a decentralized configuration of new actors (Section 11.3), more diverse technologies (Section 11.4), and redefined institutions (Section 11.5). This coexistence with the centralized organization of the energy system, however, presents numerous frictions and tensions. The uncertainty and intermittency associated with distributed generation from renewable sources pose operational difficulties for both the distribution and transmission grids. The rapid deployment of variable RE sources also introduces grid stability issues and disrupts the balance between supply and demand. Moreover, the increased use of ICT and the emergence of new activities like peer-to-peer trading create disruptions in traditional business models and actor roles that have characterized the energy systems in the past several decades. All these factors provide barriers for energy communities to become a serious alternative to the incumbency. We discuss them in the following sections.

11.4 Energy Communities as New Actors

Energy communities are new actors in the energy system and are increasing in numbers (see Box 11.1) but for long have not been defined legally. Embodying a new way of configuring the energy system, they face several challenges while trying to enter the market. In this section, we first discuss the notion of energy communities, their activities, and their scale of deployment; we then consider one critical condition under which they could increase their contribution to the energy transition: social acceptance.

Box 11.1 Scale of Energy Communities' Deployment in Europe

Currently, there are approximately 10 000 energy communities across Europe [58]. On aggregate, over 16 000 generation facilities are collectively managed by at least two million European citizens, with most (82%) being solar PV, with smaller percentages of wind (9%), biomass (7%), and hydropower (2%). This comes down to approximately 10 GW of renewable capacity installed; roughly 8500–11 7000 kWh produced per person involved if one factor in the efficiency losses. As such, community energy enabled renewable self-sufficiency for as many participants as are engaged in the movement. Many initiatives originated from 2010 to 2015, which coincides with the deployment of feed-in tariffs in many countries. Currently, each European country is going through its own dynamics in community energy employment; e.g. while the Dutch communities have reached a quantitative growth plateau and are focusing on quality, professionalization, and diversification, communities in Italy and Spain are growing in count.

11.4.1 Energy Communities' Characteristics, Activities, and Scale of Deployment

Literature offers an extensive set of energy community conceptualizations, ranging from community RE projects [35–38], local low-carbon energy initiatives [39, 40], (energy) sustainable communities [41, 42], energy collectives [43], local (energy) communities [44], clean energy communities [45], or grassroots energy initiatives [46–49]. Most popular is the concept of energy community [1, 50–53]. Despite the lack of uniformed definition, there is a recognition that energy communities are operating based on community-oriented rationale, mostly aiming at self-consumption and are self-organized. Suboticki et al. [54] however warn that as much as public engagement through, e.g. energy communities, is important for energy transition, it can also lead to maintenance of the energy system's status quo and the unequal distribution of burdens and benefits of the transition.

Europe is by far the furthest in defining and regulating the energy communities. The European Commission has recently published two definitions of energy communities: the citizen energy communities (CECs) in the Electricity Market Directive 2019/944 and renewable energy communities (RECs) in the RED 2018/2001 as part of the Clean Energy Package (EU 2019). We discuss the directives in more detail in Section 11.5. Based on the directives, Roberts [55] summarizes European energy communities as collective citizen initiatives in the energy system having the following characteristics. The first characteristic is a noncommercial purpose to provide environmental, social, or economic community benefits for members, shareholders, or local area – rather than financial profits. While communities in Europe are allowed to make profits and benefit from a return on investment, the revenues should be used to realize community objectives. Typically, this happens through reinvestment into local education or new renewable generation projects aiming at reduction of carbon emissions or energy poverty. The second characteristic is an open and voluntary participation. Citizens can become a member of and participate in the energy communities through nondiscriminatory and transparent processes, with a possibility to opt out at any time. The third characteristic concerns eligibility to participate, which in RECs is limited to natural persons, local authorities, and small-and-medium enterprises (SMEs), excluding large enterprises as participants. Fourth is effective control. Within CECs, this extends to individuals, local authorities, and SMEs for whom large-scale commercial ventures in the energy sector are not the primary area of economic activity. Within RECs, effective control lies with participants that are in the "proximity to the projects owned and developed by the community." Each Member State can interpret proximity in its own national legislation, which implies that the proximity may be interpreted as not only geographical nearness but also e.g. interest-based proximity. The fifth characteristic concerns democratic governance in the RECs to protect against potential abuse. It is interpreted as a minimum standard for democratic governance, which implies that no member, let alone an outside business partner, should be able to undermine the collective interests of the community through the power of voting (i.e. veto) or power of investment (i.e. threatening to divest). As the national legislation determines which options will be available within a country, the requirements of effective control can be different in each European Union (EU) country, and thereby a growing diversity of communities' governance structure, decision-making, and liabilities can be observed [56].

The most common governance structure for energy communities is an energy cooperative. It is a type of social and economic enterprise that enables citizens to collectively own and manage RE projects [56]. Cooperatives bring local actors such as residents, SMEs, or industrial partners together in an organized and legal entity where individual households or companies generate, consume, and share energy (or flexibility) and have a possibility to offer other activities such as grid balancing or congestion management. External stakeholders can invest in renewable generation in a cooperative by buying shares [35]. Since cooperatives are based on democratic governance – i.e. decisions made on a "one member – one vote" principle, the allocation of revenues from the projects is regulated by the statutes of the cooperative and related to its main purpose. Sometimes the revenues can be distributed among the members through capped dividends or specific energy benefits. For example, cooperatives can reduce the energy prices for their members through collective energy purchasing or through the supply of the energy surplus to the market. Regarding the latter, a particular form of a cooperative energy company has recently developed in the Netherlands. This entity is a hybrid, supplier-licensed energy company that combines commercial goals with a legal cooperative structure and is professionally run and organized as REC. Currently, there are two such cooperative companies active in the Netherlands and together they purchase cooperatively generated energy from – and supply back to – 208 RECs in the country. Other, less popular governance structures for energy communities include a limited partnership, community trusts and foundations, housing associations, nonprofit customer-owned enterprises, public–private partnerships, and public utility companies [57].

Regarding energy activities, European energy communities can engage in energy consumption, generation, storage, sale, both the supply to end-consumers and through power purchase agreements[1], and energy sharing including peer-to-peer trading (RECs). CECs can additionally distribute, aggregate, and provide energy efficiency services and charging services for electric vehicles.

11.4.2 Other Emerging Actors

Other emerging actors that are crucial for upscaling of energy communities in the energy system are aggregators and market intermediaries. Aggregators are

1 Power Purchase Agreements (PPA): contracts that govern the sale of electricity from a generator – such as a renewable energy community - and a licensed supplier or end consumer. Generated electricity might be exported to the electricity grid, or used directly by a consumer via a private wire connection if located close enough proximity. Regardless of whether the connection is via grid export or direct connection, PPAs apply to both import and export of electricity and set out the terms regarding the sale. Such terms typically include price, availability, duration, and the Guarantee Of Origin (GOO).

entities that facilitate integration of a diverse range of DERs such as solar panels, wind turbines, battery storage systems, electric vehicles, and demand response. By pooling these resources, aggregators offer communities services like energy balancing, grid stability support, demand-response management, and participation in wholesale electricity markets [59]. Intermediaries, on the other hand, engage in enabling and fostering the exchange of knowledge and skills through, e.g. dialogue and guidance. They connect and mediate between different actors or between different actors and artifacts such as RE technologies [60, 61]. In the Netherlands, intermediaries act as brokers between a community and other (market) actors and are represented by the so-called cooperative development agencies [62–64]: either commercial or nonprofit organizations that strive to diffuse and upscale small-scale energy community projects. In the former case, agencies focus on replicating small-scale cooperative solar roof projects on a regional, provincial, or national scale. They typically do so by following a self-developed blueprint for successful projects, applying the "streamlined" formula to cooperative development and thus shortening the timespan taken from initial idea to exploitable project. In the latter case, agencies aim to facilitate community participation in large-scale onshore wind and solar parks.

11.4.3 Community Acceptance

Community acceptance is the acceptance of renewable technologies on the local level by residents, stakeholders, and authorities [65, 66]. As such community acceptance is of key importance for upscaling of energy communities and for a successful energy transition [66]. A lack of acceptance can result in delays or even cancellation of energy transition projects and policies [67]. Community acceptance is a specific type of social acceptance, which can also be studied from a sociopolitical and market acceptance perspective [66]. Sociopolitical acceptance captures the general attitudes toward RE (technologies and policies) and is manifested by the public, key stakeholders, and policymakers [66]. Market acceptance entails the adoption of a RE technology by the market and is defined by consumers, investors, and intrafirms. What is being accepted and by who, differs across the different conceptualizations of social acceptance. While multiple actors determine social acceptance, most literature focuses on the lack of acceptance on the community level and on identifying the factors that can increase community acceptance. The interest in the community arises because at this level the concrete implementation of RE technologies and most of the potential negative impacts on others come about [68, 69].

Multiple sociopsychological, sociodemographic, and procedural nature factors that influence social acceptance in general, and community acceptance specifically, have been identified in the literature. Important sociopsychological factors are the symbolic meaning of the technology and the fit with the landscape [70]. Besides, sociodemographic factors such as age, political views, and education have been mentioned as factors influencing acceptance [71]. In addition, community acceptance is influenced by procedural factors such as a bottom-up process, participation in the process, transparency, and information provision that increase

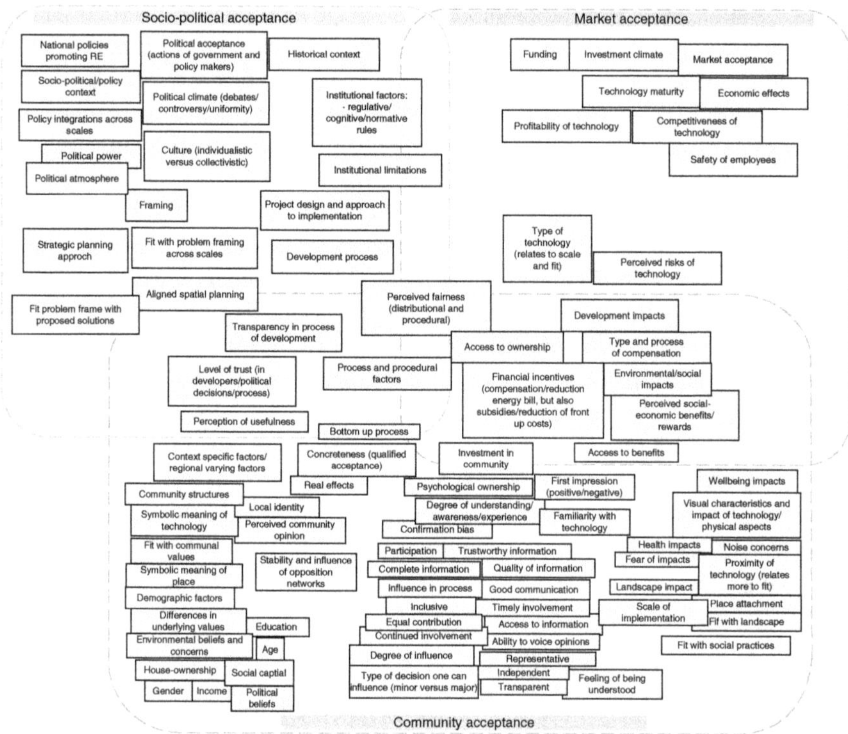

Figure 11.2 Factors hampering or stimulating acceptance-related responses of RE technologies in a renewable energy transition context [75]. This model is based on literature review of the following articles: Aas et al. [72], Bevk and Golobič [76], Bidwell [77], Devine-Wright [73, 78, 79], Huijts et al. [80], Jones and Eiser [71], Kokkinos et al. [81], Landeta-Manzano et al. [65], Larson and Krannich [74], Leer Jørgensen et al. [82], Liu et al. [83], Mjahed Hammami et al. [84], Perlaviciute et al. [70], Petrova [85], Segreto et al. [86], Soma and Haggett [87], Stenekes et al. [88], Suškevičs et al. [89], Sütterlin and Siegrist [90], Upreti and Van Der Horst [91], Whitmarsh et al. [92], Wiersma and Devine-Wright [93].

trust among the local community members toward RE technology implementation [6, 72, 73]. What is more, due process and distribution of benefits influence the perceived social–economic impacts, an important factor contributing to community acceptance of proposed RE technology implementation [74]. See Figure 11.2 for a full overview of factors contributing to community acceptance.

The notion of community in "community acceptance" differs from the notion of community in "energy communities." In the context of community acceptance, the notion of community refers to "community as a place," a group of actors bounded in a specific geographical context [94, 95]. In the context of energy communities, "community" refers to a social structure of interest consisting of networks and social relationships that is often place-based but can go beyond place-based networks [95]. A growing body of literature and EU policy considers energy communities as a tool for acceptance of the energy transition [96]. This acceptance is thought to arise from the community logic of energy communities, which is characterized by more local and democratic approaches to energy transition decision-making and implementation, crucial factors for increasing acceptance of RE technologies and behavioral change [94, 97]. The literature identifies multiple factors, in relation to energy communities and their logic, that influence acceptance of local energy solutions and implementation.

First, community engagement, and especially bottom-up engagement, where the energy community is initiated by the local community shows a higher acceptance level of local energy solutions and implementation as there is often already some form of social cohesion between its members [98, 99]. Members of a community engaging in energy community exhibit higher acceptance of renewable technologies than nonmembers, possibly because awareness also spreads through social interaction [94]. This relates to the second factor that relates to energy community logic and acceptance: trust. Local energy community often emerges in already existing place-based communities where members already know each other from earlier collaboration [99]. This increases acceptance of local energy implementation. Third, the coownership of local renewable technologies and energy production that many energy communities have resulted in increased feelings of control and self-sufficiency [100]. Coownership is also related to increased awareness of production, consumption, and environmental problems, further increasing acceptance levels of local energy transition implementation [100, 101]. Lastly, energy communities investing in local energy solutions also open possibilities for capturing benefits and reallocating benefits of energy transition implementation to the local level [100].

Energy communities thus indeed have immense potential to address several, but not all, crucial factors for acceptance of local energy solutions and implementation. The often bottom-up approaches and community logic of energy community are highly aligned with procedural and distributional justice aims, factors crucial for local acceptance [73]. However, with increased policy attention and stimulation of energy community in the energy transition, it remains crucial to pay attention to factors important for acceptance that are not automatically addressed by community logic. What is more, energy communities show indeed higher levels of acceptance from members [94], but impacts of renewable technologies can have effects beyond members of an energy community as well [8]. This addresses the need for attention to inclusivity regarding energy community, especially in relation to acceptance of local energy transition implementation.

11.5 Technology Facilitating or Hindering Energy Communities?

As discussed in Section 11.2, technology does not develop independently but in coevolution with society. Technology can be understood as "configurations that work" requiring interlinkages between artifacts, infrastructures, and social elements, such as knowledge, skills, and cultural norms [102], and resulting from interactions between different social groups [28, 103]. When considering an alternative way of organizing the energy system offered by energy communities, it is important to better understand the technological building blocks that make up the current transitioning energy system: (i) traditional technologies dominating the incumbent centralized energy system, (ii) emerging distributed or decentralized energy technologies, and (iii) digital technologies which become increasingly entangled with both [104].

In this section, we first explain the different types of technologies and then discuss the opportunities and barriers energy communities face as well as possible societal impacts in relation to the different types of technologies.

11.5.1 Three Types of Relevant Technologies

Traditional technologies tend to be developed by engineers, who are, due to their education and influence of peers, guided and coordinated by dominant cognitive, regulatory, and normative rules [102, 105, 106]. As a result, these technological configurations reinforce the rationale of centrally steered systems in both artifacts and infrastructures, making this dominant sociotechnical system more stable [21, 25]. Examples of traditional technologies are coal and gas-fired power plants, nuclear power plants, and the current electricity grid infrastructure consisting of distribution networks (low and medium voltage), a transmission network (high voltage), and larger power units connected to these central networks. Traditional technologies benefit from the centralized technologies they coshaped in the earlier societal context. The centralized energy system, however, sometimes forms a barrier for more decentralized technologies that are often facilitating energy communities' participation in the energy transition. These traditional technologies are focused on relatively central steering, while unburdening most actors in society, rendering them to passive consumers. Certain societal values, such as energy security and availability, are prioritized, while other societal values, including equality and autonomy, are less addressed [104, 107]. As a result, the centralized energy system hinders the development of alternative decentralized configurations developed by, e.g. the energy communities [1, 11, 108]. This challenge is therefore not only technical but also social.

In contradistinction to traditional energy system technologies, innovative, sustainable, and often decentralized or distributed energy technologies have become increasingly popular over the past 25 years. The collection of distributed generation (e.g. solar PV), energy storage systems (e.g. home and neighborhood batteries), and controllable loads (e.g. heat pumps and electric boilers) is often referred to as DERs [1, 109]. These technologies are developed with a logic different from the logic of centralization. They allow citizens to become prosumers and to engage in energy management by controlling DER within their home. Most technologies would fit with the logic of decentralization, as DERs are relatively underdeveloped and often connected to the distribution grid level. Furthermore, their generation is located relatively close to demand, and these technologies can be used to balance demand and supply on the local level [110]. As such, these technologies often cannot have the same benefits of economies of scale nor can they profit from centralized infrastructure in the way traditional technologies can. DERs are also more context-dependent: in some places, heat pumps are needed; in other places, batteries are more beneficial. Therefore, DER solutions are more expensive and not readily available, despite the societal context opening to decentralization in the energy system [7]. This means that, although DER technologies are becoming more prevalent, access to these solutions is limited to a few actors, often those within society with time, knowledge, skills, and financial capacities [104].

The energy system is not just transitioning from a centralized to a more decentralized system. Actors are also increasingly integrating ICTs. Digital technology is considered a configurational technology as it enables the integration of both technological and social components into new configurations [111]. By adding sensors, microprocessors, and communication devices' technologies that once operated independently can now interact and collaborate with an increasing number of "smart" technologies [112]. Initially, these digital technologies were used to collect and analyze data from the physical world, later on this expanded to include real-time intervening and the creation of new feedback loops between the physical and digital worlds [113]. Depending on the specific requirements in a different local context, these novel configurations can consist of various components and operate in specific ways, beyond predefined functionalities inscribed by engineers and designers [111, 114]. Different actors, including ICT companies, traditional energy businesses, and energy communities integrate ICT in the energy system, leading to various societal impacts. Below, we zoom in on the opportunities, barriers, and societal impacts in relation to the possible uses of digital technology.

11.5.2 Opportunities, Barriers, and Impacts of Digitalization

In the energy system, different uses of ICT can be discerned. One group of actors uses ICT to bolster their more traditional energy business and centralized technological configurations [115]. For example, ICT can be used for predictive maintenance, support electricity market decisions, or to better assess which parts of the transmission system require infrastructural investments [115, 116]. At the same time, ICT is used to deal with the challenges arising from increased decentralization and the integration of DERs. The growing number of electricity generation points has, for example, become unmanageable for system operators without ICT use [107, 117–120]. This includes, but is not limited to, the integration of a growing number of energy communities. Additionally, new actors such as energy communities are empowered by ICT to become an active participant in the energy system; without ICT to support them, many energy communities lack the resources, scale, and expertise to be an active participant [11, 121, 122]. Most often, however, ICT is implemented to create a hybrid system: fitting the newer, decentralized energy production or management technologies in the more traditional, overarching system, and infrastructure. Some communities, for example, use ICT to create a community-based virtual power plant (cVPP). Combining different DERs, the ICT-based EMS can enter energy and flexibility markets as if it is one centralized energy or flexibility provider [1, 11, 109, 120, 123]. In other words, cVPPs mimic the size and functionalities of conventional power plants that enable energy communities to blend in the centralized energy system [10, 11].

The impact of ICT integration in energy systems goes beyond the technical changes explained above. It also affects actors and institutions (see Figure 11.1). First, ICT has the potential to empower new and smaller actors, as previous consumers and other users could gain new insights and skills and gain the capacity and capabilities to become an active player in the energy system [7, 124]. In addition, ICT has the potential to strengthen collaborations among citizen initiatives, as energy communities and other prosumers can use ICT to group their DERs together and enter existing energy and flexibility markets as one virtual power plant (VPP) [1, 11, 109, 120, 123]. Without ICT-based EMS to aggregate their DERs, these actors would not meet the (scale) requirements to become active on the energy or flexibility markets [10]. In addition, the shared use of an EMS has the potential to enable also smaller and less resourceful energy communities and citizens to engage in energy management and trading, by piggybacking on technological developments of frontrunner energy communities [10]. As such, ICT has the potential to empower prosumers and energy communities to take up a larger role in the energy system. However, although these digital technologies offer advantages to end-users related to improved integration and interconnectedness of technologies, including an increase in control, convenience, and independence [125], studies also showed that these digital technologies are mostly used by a specific group of citizens, e.g. relatively young citizens who are employed, possess digital skills, and whose lifestyles and values are compatible with these new technologies leading to higher (perceived) advantages [126].

Second, the use of ICT impacts institutions. ICT use gives rise to opportunities and risks different from those embedded in incumbent institutions [127, 128]. Values, which are traditionally embedded in the energy system, are intermixed with values prioritized in the ICT sector. One such emerging value is privacy [7]. For ICT to function, a growing amount of data is necessary [129, 130]. The amount of data necessary and the sharing of that data raises concerns with numerous energy system actors: what data is needed, why is this data required, how can the data be shared, and does this still comply with existing data regulations? [131–133] Furthermore, it is not always clear based on what ICT-based EMS take decisions, for example regarding the energy or electricity market [134]. Finally, there is a risk of software and hardware lock-ins. For ICT-supported energy systems to work, it is crucial the hardware and software can communicate to (i) transfer data in a correct way and (ii) give the correct signals which follow from the data analysis. What might occur, is a software or hardware lock-in: although communities are free to switch, for example, their battery operator, this switch would result in a complete overhaul of their EMS, as their EMS can only function with this specific type of battery. These lock-ins limit the autonomy of communities and increase the chances of software and hardware monopolies [7]. The ICT companies are increasingly becoming active, competitive players in the energy system, making them reluctant to be transparent about their algorithms and functionalities [7]. The combination of increased personal data collection and lack of transparency, together with the possible lack of interoperability, can have a serious impact on the social acceptance of energy communities [135]. This also requires new legislation to ensure public values are embedded in the joined ICT and energy system [7, 107, 136].

11.6 Regulations and Markets as Key Institutional Structures

Sociotechnical systems such as energy develop over long periods of time. This evolution contributes to their stability, but it can also create path dependencies that hinder emergent alternative behaviors, new practices, and technologies that do not comply with the status quo, such as those offered by the energy communities. This path dependency is often institutionalized in the form of legal and regulatory rules and operationalized in the market through rigid market institutions, supply and demand rules, price mechanisms, competition, and user practices and routines [137].

In this section, we first delineate how current market operation challenges the energy communities' participation in the energy transition, and second, we discuss the role of the new Clean Energy Package [2].

11.6.1 Markets Reproducing Path Dependencies

Current European energy market operates based on a rationale characterized by transactions that maximize economic efficiency, profits, and shareholder value, as it finds its roots in Western capitalist ideas. The rationale of scale, professionalism, and cost-efficiency is evident in intense price competition, economies of scale driven by substantial investment costs for production assets and technological infrastructure, a higher proportion of fixed costs compared to variable costs, and the irrecoverable nature of these costs. The scale requirements can be traced across various dimensions, including production capacity, the number of community members as a proxy for potential customers, and staff size, whether voluntary or employed [138].

Van den Berghe and Wieczorek [138], Mourik et al. [108], Brummer [139], and Vallecha et al. [140] argue that this rationale along with the centralized energy regime that favors fossil fuel-based production is a source of several market barriers for energy communities. Market barriers are obstacles within the market context that impede the entry and participation of market actors in their activities [141]. In electricity markets, barriers for market entrants include declining wholesale prices, ineffective demand curves, and market power concerns ([142]). Specifically, energy communities encounter barriers such as costly grid connections [143, 144], unfavorable and uncertain support schemes [36, 143, 145, 146], the adoption of auctions for RE distribution [147–149], intense competition for resources [36], and inflexible market structures [145].

For communities to unlock their potential, there is a need to overcome the conflict between the market and the community logic. Reconciling this divide is however challenging [138]. On the one hand, energy communities have the potential to navigate the conflicting logics by employing various strategies, including collaboration, growth, diversification, and product differentiation. Collaboration can take the form of partnerships with commercial actors or established cooperatives as well as the establishment of "cooperative of cooperatives" structures. Such collaborations can assist energy communities in addressing the scale requirements associated with the market logic. Through full ownership and decision-making by all participating communities, the "cooperative of cooperatives" model allows for market participation while safeguarding the community identity. Regarding product differentiation, by focusing on local, renewable, fair, and transparent generation, product differentiation distinguishes the community's electricity from that of commercial actors. This approach allows community values to challenge the prevailing market logic, although it has primarily been observed in niche markets with limited impact on the overall market regime.

On the other hand, however, the success of the new organizational structures relies on the alignment of logics among the participating communities while growth and diversification can potentially jeopardize the community logic due to conflicts with the initial value proposition and local embeddedness. Although reaching a sufficient scale can facilitate community participation in the market, some communities express that they may never attain such a scale, and some may not even aspire to it. Energy communities therefore treat the current market regime as given and continue to operate within its boundaries, employing fit-and-conform strategies [137]. For these new actors to operate on the market, new regulatory frameworks are necessary along with business model adjustments not only by those alternatives but also by the incumbent actors.

11.6.2 Clean Energy Package – New Opportunities for European Energy Communities?

To facilitate new markets, new business models, to further expand the deployment of the RE, and to stimulate the democratization of energy system in Europe, a new legal basis of the energy and renewables law has been proposed in the Clean Energy Package [2], often referred to as CEP. We mentioned it in Section 11.3 on occasion of defining European energy communities. The essence of CEP is a set of regulations, with electricity and renewable directives (EU-2018/20017 and EU-2019/9448, respectively), which oblige Member States to provide a favorable level playing field for the new actors and to facilitate a new organization of energy markets around the opportunities offered by the smart grids. Specifically for RECs, Member States are under strict obligation to create an enabling framework by evaluating and actively removing the existing barriers; ensuring fair, proportionate, and transparent procedures; and making available tools that facilitate access to finance and information [150].

As such, the CEP is the first explicit and legal acknowledgment of energy communities and a public recognition of desired developments in Europe. It is also a necessary instrument to test whether actors have access to markets and can benefit from the latest support measures such as exemptions and new financing instruments. However, as the directives are formulated in general terms, much is open to national interpretation. The real impact of the CEP for energy communities will therefore depend on its transposition. Given the complexities described in Box 11.2, it is debatable whether the new package of European legislation will factually address the barriers RECs experience. For example, considering the focus on activity versus actor (see Box 11.2), Dutch energy cooperatives might find themselves having to meet impossible standards and will thus be hampered in undertaking the activities the CEP entitled them to in the first place. The case of the supplier license exemption (see Box 11.2) also shows that while there is an intention to address the market barriers, the exemption may not suffice to bring about substantial change for the RECs. Failing to guarantee a sustainable business case for long-term market engagement, the exemption may not be adequate to foster increased community participation in highly institutionalized markets. In that context, the efficacy of EU energy policy can be called into question because, while it asserts a central role for citizens and communities in the energy transition (community logic discussed in Section 11.5.1), it simultaneously promotes further marketization (market logic).

Box 11.2 Complexities of CEP Transposition: The Case of the Netherlands

The Dutch legislator has chosen not to adopt the REC and CEC definitions of energy communities but instead focused on describing the activities that can be performed. The "Explanatory Memorandum" explains that the right to perform activities should be guaranteed in the new Energy Act, "irrelevant" of who performs the activity [150]. There are no specific provisions for activities suggested (yet), and communities have the right to develop activities on equal footing with other market participants. This approach provides a clear legal framework but is problematic for energy communities. Confronting them with the same requirements as commercial parties leads to a disproportionately high burden for the former. For activities such as flexibility trading or collective self-balancing, meeting the "equal playground" requirements might turn out to be prohibitive.

One exception is made in the draft Energy Act: the exemption for the supplier license requirements (Article 2.19). The exemption allows energy communities that meet certain conditions to become a supplier of energy to their own members without a supplier license. It is this license requirement specifically that has driven many cooperatives to enter commercial partnerships discussed in Section 11.5.1. Looking at the conditions however, it is questionable whether this exemption is sufficient to make the activity of supply attractive to communities. One of the more problematic conditions is that a community may only supply an amount of energy that is equal to the amount of energy it produces per annum. This does not apply to communities that do not produce but only purchase electricity. It thereby does not apply to the situation in which most of the Dutch RECs are now in the combination of subsidized partnership and resale partnership, whereby cooperatively produced energy from common sources is sold to a commercial supplier or a CEC, who in turn then resells this energy back to the cooperative members via the REC itself. The community may then supply no more than what the community has collectively generated. Essentially, this means that there is a high probability that members will have to enter into another agreement with a licensed supplier for the remaining portion of energy. This can be a prohibitive activity for the community to engage in, not only because the members must enter multiple contracts but also because multiple allocation points and additional metering devices are needed.

11.7 How It Looks in Practice

In Sections 11.2–11.5, we have discussed several societal aspects that have an impact on the further development of energy communities. Using the sociotechnical lens, we observed that energy communities are an important new actor in the European energy landscape and through their novel activities and roles, they challenge the existing power relations. To become more mainstream, they highly depend on their acceptance by a broader society as a good way of organizing the energy system. For such an overhaul of the current energy system, incumbent technologies need to be phased out, DER needs greater support, and a smart use of ICT is necessary. Digital technologies can play a significant role in supporting energy communities, but they depend on who uses them and to what effect. They also create several new problems such as privacy or environmental impacts, which need to be regulated. The most profound societal aspect, however, is the rationale for operation of a new energy market where communities can play a role. Market and community logics do not go hand in hand and unless reconciled, no legal structure will be effective in facilitating the energy transition. The current CEP does not help in that regard and is highly dependent on the transposition by the Member States.

In this section, we demonstrate the practical complexity and intricacy of the interplay of all these aspects. First, we show how Spanish Energética Cooperativa (E-Coop) established itself and gained societal acceptance. Second, using the case of cVPPs in Ghent (Belgium) and Loenen (the Netherlands), we discuss the societal implications of ICTs. Using the example of local token economy (LTE) service, which is being implemented in Italy, we demonstrate how a market service can provide additional social and economic services to citizens and facilitate energy communities' participation in the market.

11.7.1 Energética Cooperativa in Spain – How to Foster Social Acceptance

The Spanish E-Coop can serve as an example of how energy cooperatives can foster social acceptance. E-Coop in short, is an energy cooperative that also functions as an intermediary organization (see Section 11.3 on actors for our take on intermediary organizations). Intermediary organizations are often applauded in the literature on energy communities, e.g. because they play an important role in "scaling up local energy activity" ([151], p. 72), their objectivity can "increase trust during the negotiation process" ([152], p. 15), and they can strengthen social equity as they "play the role of knowledge brokers to foster social learning and innovation by recombining and transferring knowledge among communities" ([45], p. 101).

The intermediating role that E-Coop has taken on manifests in several ways. First, E-Coop is dedicated to minimizing or even eliminating important barriers to start an energy community. We have two examples of such efforts, both in the city of Valladolid. In one neighborhood, E-Coop collaborated with local authorities to use public buildings for generating electricity with PV solar panels, to be used by residents of the neighborhood. To get the finances for this project, E-Coop presented the project as an investment opportunity for its members. As such, E-Coop created a chance for residents who do not have a roof, enough financial resources, or the right social network to initiate an energy community to join one. In another neighborhood, E-Coop has assisted a neighborhood association with (technical) knowledge and information to form a "solar community."

Scholarly literature often describes how initiating an energy community requires many resources: time, familiarity with energy technologies, juridical and organizational knowledge, social networks, suitable locations to install energy technologies, and financial means [8, 153–156], which creates inequality with regard to benefiting from policies focused on community initiatives and RE technologies. Through providing these resources to communities with the ambition to engage with energy, they level the playing field for energy community development in the Castilla y León region.

Second, E-Coop operates based on a certain community logic. The community logic of energy communities is often characterized by local and democratic approaches to energy (see also Section 11.4 on social acceptance), which are important factors for increasing acceptance of RE technologies. Their community logic builds upon the different community logics that their member energy communities deploy, but it also guides their member communities in forming their community logics. The community logic from which E-Coop operates is characterized by a democratic approach to energy, strong social values related to energy production and use, the idea that local ownership of electricity is more efficient and cheaper, and the logic to start with electricity saving measures before installing RE technologies. Through connecting and mediating between energy communities and these logics, they also contribute to social acceptance of RE technologies.

Third, E-Coop actively reaches out to potential new members or new energy communities. They have meetings with mayors of small rural towns, with neighborhood associations as described above, and with many other types of actors who might become, or who already are, interested in energy community development. They explain their community logic, the benefits of RE technologies, and the benefits of producing your own electricity. Through facilitating dialogue on RE, and through their outreach, they contribute to a growing social acceptance of RE technologies. During the interviews that we had with members of energy communities linked to E-Coop, we also found that the personal involvement and the perceived expertise of E-Coop employees led to trust in the cooperative, which is another ingredient for building social acceptance.

The approach of E-Coop and its intermediary role aids both energy community development and social acceptance in several ways. Through leveling the playing field, they contribute to inclusivity with regard to starting and running an energy community. This inclusivity, together with the community logic on which E-Coops' operation is based, creates the potential of E-Coop to contribute to different aspects of justice, which the literature defines as an important ingredient to creating social acceptance. Finally, through outreach activities, they spread awareness about RE technologies and their benefits, which also fosters social acceptance. Through the intermediating cooperative model of E-Coop, cooperatives can not only aid in the development of the community energy sector, but they also have the potential to steer this development in a direction that enables justice, inclusivity, and (consequently) social acceptance.

11.7.2 cVPP in Loenen Gent – How to Harness the Power of ICT

The societal implications of DER and digital technologies are explored by zooming in two energy communities that developed their own cVPP: EnerGent and Energy Cooperative Loenen (ECL). The digital heart of a cVPP is an EMS, which can aggregate and control DER, thereby enabling energy communities to collectively manage energy demand and supply within their community and participate in energy markets, in line with their community logic: community values and democratic ownership and decision-making [1].

However, interoperability issues exist between DER and EMS as a result of a lack of standards for application programming interfaces (APIs) [112, 157, 158], which hinder energy management by communities [11]. Due to a lack of standards, DER manufacturers developed their own APIs and charged high prices for gaining access. In response, EnerGent installed only one brand of batteries and solar PV inverters that can be easily controlled by their EMS. In Loenen, however, a diversity of DER brands was already installed. ECL only connected DER with the EMS that involved low costs. In addition, they listed DER types and brands that could easily be connected to their EMS, to enable members to make more informed decisions when buying a heat pump or solar PV. In the future, open standards are needed to fully unlock the full potential of "flexibility energy communities" [11].

Although the value of demand-side flexibility is increasingly acknowledged in light of growing congestion and balancing issues, there are still major constraints for (local) energy management. Second, there is a lack of financial incentives for energy management and trading on the local level, while there are high requirements and costs associated with participating in energy markets and developing an EMS, which requires cVPPs to increase in scale, beyond the size of single energy communities [9, 11, 108]. Both ECL and EnerGent are exploring the possibilities to set up an overarching cooperative aggregator that aggregates flexibility from multiple energy communities to reach the scale needed to participate in energy and emerging flexibility markets [11].

Second, these institutional barriers put pressure on energy communities to shift their focus away from activities that would generate community values (e.g. values identified as important by community members and in line with their community logic [[1, 39]]) and toward activities that help solve balancing issues on the national grid level in return for financial value [108]. However, to prepare for emerging opportunities for energy management as a result of EU directives (EU 2019) and the increasing need for demand-side flexibility in response to grid issues that result from increasing electrification of both heating and mobility and the rise of intermittent RES, both ECL and EnerGent choose a flexible EMS that allows them to engage in activities that might become possible and beneficial in the near future [11].

Third, besides the constraints of the institutional context, also the increasing complexity of the energy system hinders meaningful engagement of community members [1, 9, 10]. ECL had to put a lot of effort in organizing workshops to first inform local citizens about the challenges of the energy transition, which enabled them to collectively explore what values they wanted to create by means of their own cVPP. In contrast, EnerGent decided to limit engagement of community members to a survey, as they expected energy management would be too complex for laymen.

Finally, despite preferences for an open-source EMS, ECL bought a more mature EMS running on proprietary software because it would allow them to engage in certain activities almost immediately ([159]). It did, however, result in vendor lock-in and the need for ongoing negotiations concerning (the costs associated with) desired improvements and upgrades ([159]). EnerGent did end up with a fully open-source system, which, however, needed to be built almost from scratch. On the upside, the choice for an open-source platform did result in a sense of shared ownership among multiple energy communities, resulting in a follow-up project in which "their" EMS is being further developed [160]. As such, this open-source platform could be seen as a shared resource that could empower the whole community energy sector [10, 11]. In addition, it can be tailored to the needs and preferences of energy communities (e.g. in terms of local hardware, data storage, functionalities, and type of software; [138]).

11.7.3 Service Local Token Economy in Italy – How to Deal with Market Barriers

The service called local token economy, in short LTE, currently implemented in Italy, is inspired by the traditional time bank model, but technologically and organizationally updated, which enables the market participation of nontraditional actors such as energy communities. LTE is a service based on a marketplace where goods and services are shared by community members. Any transaction is paid by tokens that members have in their wallets and recorded on a blockchain infrastructure. Members enter the marketplace to consult and select the required goods and services and then activate the transaction. The transaction price includes a fee to cover service management costs and an incentive to be applied to goods and services with a relevant environmental and social impact. Furthermore, the community members' virtuous energy behaviors, such as flexibility and peak shaving, are rewarded with tokens. The service is controlled and managed by a community administrator who can steer the service management dashboard by means of several control parameters (fee value, membership bonus value, incentives, and demurrage) and can also view indicators on users, transactions, and wallets. Through the service, the energy community can act as a dynamic aggregate of users who can produce energy for self-consumption and/or purchase services from entities outside of the community.

The LTE model is based on the digitization of the collaborative economy, meaning that the technological levers and secure traceability of blockchain technology are leveraged to foster community development and deal with issues of energy poverty and social fragility. Such models, and in particular the reciprocity of exchanges facilitated by them, can thereby strengthen the process of sharing of broader community objectives, increase the citizens' sense of belonging, and help with inclusiveness and mobilization of goods and resources that would otherwise remain unused. By stimulating the response to the grid demands, LTE motivates development of virtuous behaviors of the users related with the reduction or reshaping of energy consumption. By activating the process of "demand and response," LTE makes the community and its users dynamic and conscious energy consumers. Energy community members become more committed to better management of local resources and their conscious use and reuse. Energy communities gain skills and capabilities to plan and optimize their energy use according to specific supply and demand needs, contributing thereby to the creation of alternative ways of energy consumption and production. A downside of such a solution is the blockchain-related energy intensity and a potential lock in of the power in a new actor – the blockchain platform owner [161].

11.8 Conclusions

Energy communities play a crucial role in facilitating the energy transition [1, 45, 106]. However, their equitable participation in these activities and their involvement in electricity markets face numerous barriers [139, 145, 162]. Apart from the barriers we identified above related with conflicting community-market logics, institutional inertia, and technological mismatches, we would like to highlight three overarching challenges related with the upscaling processes of energy communities.

First, one of the dominant assumptions is that fostering energy communities will bring about energy justice. However, this relation is not so straightforward. A review of the literature on energy communities and energy justice shows that tensions may arise unfolding *within* the internal dynamics of energy communities [8]. For example, the distribution of community energy benefits, ranging from financial and environmental to social aspects, may cause tensions between community energy and energy justice. Besides this issue of distributional justice, another tension that is well documented focuses on fairness of energy decision-making and is related to the notion of procedural justice within these communities. Procedural justice may get compromised, for example when calls for pragmatism "to get things done within a community" go at the expense of the principle of participation. Looking at energy justice and the internal dynamics of energy communities highlights the importance of closely considering actors and their differences in power and equity in energy community development.

Second, the scope of energy justice is much broader, also raising questions about justice *between* communities. Community energy initiatives are intertwined with a large variety of different institutions and actors that operate within and outside of the initiatives' borders.

Issues of procedural justice may play a role when looking at energy communities and their interactions with the institutional context. Especially uncertainties surrounding the amount of government support that community energy initiatives can expect, like a changing national emphasis on favored technologies, can negatively impact the initiatives. In addition, community energy initiatives can be experienced and interpreted as manifesting a tension between interests on local and national levels and can raise questions of distributional justice. Often, RE projects are in rural areas with the host communities carrying the burden of a RE system in their local environment (like noise disruptions, shadow flickers from windmills, or unpleasant smells) [163]. This can make communities perceive the financial and/or social benefits of community energy only as a bribe to smooth development processes and outcomes that are urgently needed to meet national policy targets. We see that the institutional context, through public policy can have a big impact on the community energy sector in a country, and uncertainties regarding regulatory support schemes and development decisions can be detrimental to community energy initiatives.

Third, when applying a broader energy justice focus it becomes clear that energy justice considerations also go *beyond* energy communities: energy community initiatives can raise justice issues that affect people beyond those stakeholders directly involved in community energy. Two examples are particularly pertinent: not all groups are equally positioned to benefit from policies focused on the development of community initiatives. Despite initiatives that work toward the inclusion of vulnerable groups or fighting energy poverty in the energy transition much of the value offered by community energy schemes is still captured by the socioeconomically well-off people. Community energy initiatives engage with RE technologies. As they stop using fossil fuels, it is rashly assumed that they contribute to a just energy system. Healy et al. [164] show that RE technologies have "embodied energy injustices," occurring transboundary: in the extraction of resources; during the processing and transport phase; at the combustion/production site; and with waste disposal. Many injustices, such as embodied energy injustices related to RE technologies, are too complex and multilayered for community energy initiatives alone to handle. However, many community energy initiatives together can potentially impact embodied energy injustices. There is potential in energy communities' combined demand and emerging position in the energy/electricity market, where they collectively contribute to shaping the contours of a just energy system.

Enabling such a collective and community-oriented development pathway in the energy system requires a systemic and coordinated approach [165]. In particular, it necessitates a clear and swift implementation of the regulations that support cooperative solutions. This is important as several more individual solutions, like heat pump adoption, are also gaining traction [166]. These individual solutions contribute to the energy transition but might aggravate existing energy injustices. Moreover, the individual pathways compete with the adoption of cooperative solutions, also in places where a cooperative solution may be more desirable from a systems perspective. A coordinated and system-oriented approach is needed to direct these different transition pathways and evaluate their justice effects.

In response to our research question about the conditions under which energy communities can become a viable alternative to the current energy system, we thus consider justice aspects in the development of energy communities and a systemic approach to energy transition as crucial to reconcile for such as a just radical change. The triple lens of within–between–beyond energy communities, along with consideration of actors – institutions and technological aspects can be used as a more holistic approach to analyzing a plethora of specific challenges energy communities face and how to address them.

List of Abbreviations

APIs	Application Programming Interfaces
BRP	Balance Responsible Party
CECs	Citizen Energy Communities
cVPP	community-based Virtual Power Plant
DERs	Decentralized Energy Resources
DSO	Distribution System Operator
ECL	Energy Cooperative Loenen
E-Coop	Energética Cooperativa
EED	Energy Efficiency Directive
EMS	Energy Management System
EU	European Union
ICT	Information and Communication Technology
LTE	local token economy
PV	photovoltaic
REC	Renewable Energy Community
RED	Renewable Energy Directive
SME	small-medium enterprise
TSO	Transmission System Operator
VPP	Virtual Power Plant

References

1 Van Summeren, L.F.M., Wieczorek, A.J., Bombaerts, G.J.T., and Verbong, G.P.J. (2020). Community energy meets smart grids: reviewing goals, structure, and roles in virtual power plants in Ireland, Belgium and the Netherlands. *Energy Research & Social Science* 63: https://doi.org/10.1016/j.erss.2019.101415.

2 EC (2019). Clean energy for all Europeans package, European Commission. https://ec.europa.eu/energy/topics/energy-strategy/clean-energy-all-europeans_en (accessed November 2023).

3 Ryghaug, M., Skjølsvold, T.M., and Heidenreich, S. (2018). Creating energy citizenship through material participation. *Social Studies of Science* 48 (2): 283–303. https://doi.org/10.1177/0306312718770286.

4 Elzen, B. and Wieczorek, A.J. (2005). Transitions towards sustainability through system innovation. *Technological Forecasting and Social Change* 72 (6): 651–661. https://doi.org/10.1016/j.techfore.2005.04.002.

5 Bijker, W.E. (1995). *Of Bicycles, Bakelites, and Bulbs: Toward a Theory of Sociotechnical Change*. Cambridge, MA: MIT Press. ISBN: 9780262023764.

6 Kluskens, N., Vasseur, V., and Benning, R. (2019). Energy justice as part of the acceptance of wind energy: an analysis of Limburg in the Netherlands. *Energies* 12 (22): https://doi.org/10.3390/en12224382.

7 Niet, I.A., Dekker, R., and van Est, R. (2022). Seeking public values of digital energy platforms. *Science, Technology & Human Values* 47 (3): 380–403. https://doi.org/10.1177/01622439211054430.

8 van Bommel, N. and Höffken, J.I. (2021). Energy justice within, between and beyond European community energy initiatives: a review. *Energy Research & Social Science* 79: 102157. https://doi.org/10.1016/J.ERSS.2021.102157.

9 van Summeren, L.F.M., Wieczorek, A.J., Verbong, G.P.J., and Bombaerts, G. (2023). Blending in, to transform the regime from within: niche hybridisation strategies of Irish energy communities. *Environmental Innovation and Societal Transitions* 48: 100761. https://doi.org/10.1016/j.eist.2023.100761.

10 Van Summeren, L.F.M., Breukers, S., and Wieczorek, A.J. (2022). Together we're smart! Flemish and Dutch energy communities' replication strategies in smart grid experiments. *Energy Research and Social Science* 89: 102643. https://doi.org/10.1016/j.erss.2022.102643.

11 Van Summeren, L.F.M., Wieczorek, A.J., and Verbong, G.P.J. (2021). The merits of becoming smart: how Flemish and Dutch energy communities mobilise digital technology to enhance their agency in the energy transition. *Energy Research & Social Science* 79: 102160.

12 Wieczorek, A.J., Rochracher, H., Bauknecht, D. et al. (2023). Decentralised energy futures. Emerging rationales of energy system organisation. *Energy Research and Social Science* https://doi.org/10.2139/ssrn.4011139.

13 Bijker, W. and Law, J. (1992). *Shaping Technology/Building Society: Studies in Sociotechnical Change*, 341. Cambridge, MA: MIT Press. ISBN: 0-262-02338.

14 Elzen, B., Geels, F.W., and Green, K. (ed.) (2004). *System Innovation and the Transition to Sustainability: Theory, Evidence and Policy*. Cheltenham: Edward Elgar.

15 Freeman, C. and Soete, L. (1987). *Technical Change and Full Employment*. Oxford: Blackwell.

16 Nelson, R. (1994). The co-evolution of technology, industrial structure, and supporting institutions. *Industrial and Corporate Change* 3 (1): 47–63.

17 Nelson, R. (1995). Recent evolutionary theorizing about economic change. *Journal of Economic Literature* 33 (1): 48–90.

18 Wieczorek, A.J. and Berkhout, F. (2009). Transitions to sustainability as societal innovations. In: *Principles of Environmental Sciences* (ed. J. Boersema and L. Reijnders), 503–512. Springer https://doi.org/10.1007/978-1-4020-9158-2_27.

19 Lounsbury, M., Geraci, H., and Waismel-Manor, R. (2002). Policy discourse, logics and practice standards: centralising the solid-waste management field.

In: *Organizations, Policy, and the Natural Environment: Institutional and Strategic Perspectives* (ed. A. Hoffman and M.J. Ventresca), 327–345. Stanford, CA: Stanford University Press.

20 Rohracher, H. (2001). Managing the technological transition to sustainable construction of buildings: a socio-technical perspective. *Technology Analysis & Strategic Management* 13: 137–150.

21 Geels, F.W. (2004). From sectoral systems of innovation to socio-technical systems. *Research Policy* 33 (6–7): 897–920. https://doi.org/10.1016/j.respol.2004.01.015.

22 Di Maggio, P. (1998). The new institutionalism: avenues of collaboration. *Journal of Institutional and Theoretical Economics* 154 (4): 696–705.

23 North, D. (1991). Institutions. *Journal of Economic Perspectives* 5 (1): 97–112.

24 Scott, W.R. (1995). *Institutions and Organizations: Ideas, Interests and Identities.* Sage.

25 Fuenfschilling, L. and Truffer, B. (2016). The interplay of institutions, actors and technologies in socio-technical systems—an analysis of transformations in the Australian urban water sector. *Technological Forecasting and Social Change* 103: 298–312.

26 Geels, F.W. and Schot, J. (2007). Typology of sociotechnical transition pathways. *Research Policy* 36: 399–417.

27 Giddens, A.T. (1984). *The Constitution of Society: Outline of the Theory of Structuration.* Cambridge: Polity Press.

28 Grin, J., Rotmans, J., and Schot, J. (2010). *Transitions to Sustainable Development: New Directions in the Study of Long Term Transformative Change.* New York: Routledge. https://www.taylorfrancis.com/books/e/9780203856598.

29 Verbong, G.P.J., Verhees, B., and Wieczorek, A.J. (2019). The role of users in sustainable innovation. In: *Handbook of Sustainable Innovation* (ed. F. Boons and A. McMeekin), 238–251. Edward Elgar.

30 IEA (2021). World energy model documentation. 2020 Version. https://iea.blob.core.windows.net/assets/bc4936dc-73f1-47c3-8064-0784ae6f85a3/WEM_Documentation_WEO2020.pdf (accessed November 2023).

31 Verbong, G., van der Vleuten, E., and Scheepers, M. (2002). Long-term electricity supply systems dynamics – a historic analysis, ECN. Eindhoven University of Technology, ECN-C-02-084.

32 Verbong, G. and Loorbach, D. (2012). Governing the energy transition. In: *Reality, Illusion or Necessity?* (ed. G. Verbong and D. Loorbach). New York: Routledge.

33 Verbong, G. and Geels, F.W. (2007). The ongoing energy transition: lessons from a socio-technical, multi-level analysis of the Dutch electricity system (1960–2004). *Energy Policy* 35 (2): 1025–1037.

34 USEF (Universal Smart Energy Framework) (2015). The independent aggregator. USEF Position Paper.

35 Walker, G. and Devine-Wright, P. (2008). Community renewable energy: what should it mean? *Energy Policy* 36: 497–500.

36 Bauwens, T. (2016). Explaining the diversity of motivations behind community renewable energy. *Energy Policy* 93: 278–290. https://doi.org/10.1016/j.enpol.2016.03.017.

37 Lowitzsch, J., Hoicka, C.E., and van Tulder, F.J. (2020). Renewable energy communities under the 2019 European clean energy package – governance model for the energy clusters of the future? *Renewable and Sustainable Energy Reviews* 122: 109489. https://doi.org/10.1016/j.rser.2019.109489.

38 Rogers, J.C., Simmons, E.A., Convery, I., and Weatherall, A. (2008). Public perceptions of opportunities for community-based renewable energy projects. *Energy Policy* 36 (11): 4217–4226. http://dx.doi.org/10.1016/j.enpol.2008.07.028.

39 Hicks, J. and Ison, N. (2018). An exploration of the boundaries of 'community' in community renewable energy projects: navigating between motivations and context. *Energy Policy* 113: 523–534. https://doi.org/10.1016/j.enpol.2017.10.031.

40 Warbroek, B., Hoppe, T., Bressers, H., and Coenen, F. (2019). Testing the social, organizational, and governance factors for success in local low carbon energy initiatives. *Energy Research & Social Science* 58: 101269. https://doi.org/10.1016/j.erss.2019.101269.

41 Rae, C. and Bradley, F. (2012). Energy autonomy in sustainable communities – a review of key issues. *Renewable and Sustainable Energy Reviews* 16 (9): 6497–6506. http://dx.doi.org/10.1016/j.rser.2012.08.002.

42 Schweizer-Ries, P. (2008). Energy sustainable communities: environmental psychological investigations. *Energy Policy* 36 (11): 4126–4135. http://dx.doi.org/10.1016/j.enpol.2008.06.021.

43 Moret, F. and Pinson, P. (2019). Energy collectives: a community and fairness based approach to future electricity markets. *IEEE Transactions on Power Systems* 34 (5): 8301552. 3994 - 4004 https://doi.org/10.1109/TPWRS.2018.2808961.

44 van der Schoor, T., Van Lente, H., Scholtens, B., and Peine, A. (2016). Challenging obduracy: how local communities transform the energy system. *Energy Research and Social Science* 13: 94–105. http://dx.doi.org/10.1016/j.erss.2015.12.009.

45 Gui, E.M. and MacGill, I. (2018). Typology of future clean energy communities: an exploratory structure, opportunities, and challenges. *Energy Research & Social Science* 35: 94–107. https://doi.org/10.1016/j.erss.2017.10.019.

46 Hargreaves, T., Hielscher, S., Seyfang, G., and Smith, A. (2013). Grassroots innovations in community energy: the role of intermediaries in niche development. *Global Environmental Change* 23 (5): 868–880. https://doi.org/10.1016/j.gloenvcha.2013.02.008.

47 Seyfang, G., Park, J.J., and Smith, A. (2013). A thousand flowers blooming? An examination of community energy in the UK. *Energy Policy* 61: 977–989. https://doi.org/10.1016/j.enpol.2013.06.030.

48 Seyfang, G., Hielscher, S., Hargreaves, T. et al. (2014). A grassroots sustainable energy niche? Reflections on community energy in the UK. *Environmental Innovation and Societal Transitions* 13: 21–44. https://doi.org/10.1016/j.eist.2014.04.004.

49 Smith, A., Hargreaves, T., Hielscher, S. et al. (2016). Making the most of community energies: three perspectives on grassroots innovation. *Environment and Planning A* 48 (2): 407–432. https://doi.org/10.1177/0308518X15597908.

50 Creamer, E., Aiken, A.T., van Veelen, B. et al. (2019). Community renewable energy: what does it do? Walker and Devine-Wright (2008) ten years on. *Energy Research & Social Science* 57: 101223.

51 Koirala, B.P., Koliou, E., Friege, J. et al. (2016). Energetic communities for community energy: a review of key issues and trends shaping integrated community energy systems. *Renewable and Sustainable Energy Reviews* 56: 722–744. https://doi.org/10.1016/j.rser.2015.11.080.

52 Walker, G., Hunter, S., Devine-Wright, P. et al. (2007). Harnessing community energies: explaining and evaluating community-based localism in renewable energy policy in the UK. *Global Environmental Politics* 7 (2): 64–82 + viii. http://dx.doi.org/10.1162/glep.2007.7.2.64.

53 Von Wirth, S. (2014). Communities matter: institutional preconditions for community renewable energy. *Energy Policy* 70: 236–246. https://doi.org/10.1016/j.enpol.2014.03.021.

54 Suboticki, I., Heidenreich, S., Ryghaug, M., and Moe Skjølsvold, T. (2023). Fostering justice through engagement: a literature review of public engagement in energy transitions. *Energy Research & Social Science* 99: 103053.

55 Roberts, J. (2020). Power to the people? Implications of the Clean Energy Package for the role of community ownership in Europe's energy transition. *Review of European, Comparative & International Environmental Law* https://doi.org/10.1111/reel.12346.

56 Yildiz, O., Rommel, J., Debor, S. et al. (2015). Renewable energy cooperatives as gatekeepers or facilitators? Recent developments in Germany and a multidisciplinary research agenda. *Energy Research & Social Science* 6: 59–73. https://doi.org/10.1016/j.erss.2014.12.001.

57 Caramizaru, E. and Uihlein, A. (2020). *Energy Communities: An Overview of Energy and Social Innovation, EUR 30083 EN*. Luxembourg: Publications Office of the European Union. ISBN: 978-92-76-10713-2. http://dx.doi.org/10.2760/180576.

58 Wierling, A., Schwanitz, V.J., Zeiss, J.P. et al. (2023). A Europe-wide inventory of citizen-led energy action with data from 29 countries and over 10000 initiatives. *Science Data* 10: 9. https://doi.org/10.1038/s41597-022-01902-5.

59 Burger, S., Chaves-Ávila, J.P., Batlle, C., and Pérez-Arriaga, I.J. (2017). A review of the value of aggregators in electricity systems. *Renewable and Sustainable Energy Reviews* 77: 395–405. https://doi.org/10.1016/j.rser.2017.04.014.

60 van Veelen, B. and Eadson, W. (2020). Assembling community energy democracies. *Voluntary Sector Review* 11 (2): 231–249. https://doi.org/10.1332/204080519X15740562779512.

61 Moss, T. (2009). Intermediaries and the Governance of Sociotechnical Networks in Transition. *Environment and Planning A: Economy and Space* 41 (6): 1480–1495. https://doi.org/10.1068/a4116.

62 LEM (2021). Hier Opgewekt: Lokale Energie Monitor 2021. https://www.hier.nu/uploads/hieropgewekt/inline/Lokale%20Energie%20Monitor%202021_def_digitaal.pdf (accessed November 2023).

63 LEM (2020). Hier Opgewekt: Lokale Energie Monitor 2020. https://www.hier.nu/uploads/hieropgewekt/inline/Lokale%20Energie%20Monitor%202020_DEF_lr_16-02.pdf (accessed November 2023).

64 LEM (2019). Hier Opgewekt: Lokale Energie Monitor 2019. https://www.hier.nu/uploads/hieropgewekt/inline/Lokale%20Energiemonitor%202019_DEF_feb2020_3.pdf (accessed November 2023).

65 Landeta-Manzano, B., Arana-Landín, G., Calvo, P.M., and Heras-Saizarbitoria, I. (2018). Wind energy and local communities: a manufacturer's efforts to gain acceptance. *Energy Policy* 121: 314–324. https://doi.org/10.1016/j.enpol.2018.05.034.

66 Wüstenhagen, R., Wolsink, M., and Bürer, M.J. (2007). Social acceptance of renewable energy innovation: an introduction to the concept. *Energy Policy* 35 (5): 2683–2691. https://doi.org/10.1016/j.enpol.2006.12.001.

67 Reusswig, F., Braun, F., Heger, I. et al. (2016). Against the wind: local opposition to the German Energiewende. *Utilities Policy* 41: 214–227. http://dx.doi.org/10.1016/j.jup.2016.02.006.

68 Perlaviciute, G. and Squintani, L. (2020). Public participation in climate policy making: toward reconciling public preferences and legal frameworks. *One Earth* 2 (4): 341–348. https://doi.org/10.1016/j.oneear.2020.03.009.

69 Wolsink, M. (2012). Wind power: basic challenge concerning social acceptance. In: *Encyclopedia of Sustainability Science and Technology* (ed. R.A. Meyers). Springer New York.

70 Perlaviciute, G., Schuitema, G., Devine-Wright, P., and Ram, B. (2018). At the heart of a sustainable energy transition: the public acceptability of energy projects. *IEEE Power and Energy Magazine* 16 (1): 49–55. https://doi.org/10.1109/MPE.2017.2759918.

71 Jones, C.R. and Eiser, J.R. (2009). Identifying predictors of attitudes towards local onshore wind development with reference to an English case study. *Energy Policy* 37 (11): 4604–4614. https://doi.org/10.1016/J.ENPOL.2009.06.015.

72 Aas, Ø., Devine-Wright, P., Tangeland, T. et al. (2014). Public beliefs about high-voltage powerlines in Norway, Sweden and the United Kingdom: a comparative survey. *Energy Research & Social Science* 2: 30–37. https://doi.org/10.1016/J.ERSS.2014.04.012.

73 Devine-Wright, P. (2007). Reconsidering public attitudes and public acceptance of renewable energy technologies: a critical review. Architecture, Working Paper (February). pp. 1–15. https://citeseerx.ist.psu.edu/document?repid=rep1&type=pdf&doi=8d23c43b8af92fc8a4a8668b5cd4efc5a2e2391e (accessed November 2023).

74 Larson, E.C. and Krannich, R.S. (2016). "A great idea, just not near me!" Understanding public attitudes about renewable energy facilities. *Society and Natural Resources* 29 (12): 1436–1451. https://doi.org/10.1080/08941920.2016.1150536.

75 Kluskens, N., Alkemade, F., and Höffken, J. (2024b). Beyond a checklist for acceptance: understanding the dynamic process of community acceptance. *Sustain Science* https://doi.org/10.1007/s11625-024-01468-8.

76 Bevk, T. and Golobič, M. (2020). Contentious eye-catchers: perceptions of landscapes changed by solar power plants in Slovenia. *Renewable Energy* 152: 999–1010. https://doi.org/10.1016/J.RENENE.2020.01.108.

77 Bidwell, D. (2013). The role of values in public beliefs and attitudes towards commercial wind energy. *Energy Policy* 58: 189–199. https://doi.org/10.1016/J.ENPOL.2013.03.010.

78 Devine-Wright, P. (2011). Public engagement with large-scale renewable energy technologies: breaking the cycle of NIMBYism. *Wiley Interdisciplinary Reviews: Climate Change* 2 (1): 19–26. https://doi.org/10.1002/WCC.89.

79 Devine-Wright, P. (2017). Environment, democracy, and public participation. In: *International Encyclopedia of Geography: People, the Earth, Environment and Technology*, 1–10. Wiley. https://doi.org/10.1002/9781118786352.wbieg0613.

80 Huijts, N.M.A., Molin, E.J.E., and Steg, L. (2012). Psychological factors influencing sustainable energy technology acceptance: a review-based comprehensive framework. *Renewable and Sustainable Energy Reviews* 16 (1): 525–531. https://doi.org/10.1016/J.RSER.2011.08.018.

81 Kokkinos, K., Lakioti, E., Papageorgiou, E. et al. (2018). Fuzzy cognitive map-based modeling of social acceptance to overcome uncertainties in establishing waste biorefinery facilities. *Frontiers in Energy Research* 6: 112. https://doi.org/10.3389/fenrg.2018.00112.

82 Leer Jørgensen, M., Anker, H.T., and Lassen, J. (2020). Distributive fairness and local acceptance of wind turbines: the role of compensation schemes. *Energy Policy* 138: https://doi.org/10.1016/J.ENPOL.2020.111294.

83 Liu, L., Bouman, T., Perlaviciute, G., and Steg, L. (2019). Effects of trust and public participation on acceptability of renewable energy projects in the Netherlands and China. *Energy Research & Social Science* 53: 137–144. https://doi.org/10.1016/J.ERSS.2019.03.006.

84 Mjahed Hammami, S., Chtourou, S., and Al Moosa, H. (2018). A holistic approach to understanding the acceptance of a community-based renewable energy project: a pathway to sustainability for Tunisia's rural region. *Business Strategy and the Environment* 27 (8): 1535–1545. https://doi.org/10.1002/BSE.2211.

85 Petrova, M.A. (2013). NIMBYism revisited: public acceptance of wind energy in the United States. *Wiley Interdisciplinary Reviews: Climate Change* 4 (6): 575–601. http://dx.doi.org/10.1002/wcc.250.

86 Segreto, M., Principe, L., Desormeaux, A. et al. (2020). Trends in social acceptance of renewable energy across Europe—a literature review. *International Journal of Environmental Research and Public Health* 17 (24): 9161. https://doi.org/10.3390/IJERPH17249161.

87 Soma, K. and Haggett, C. (2015). Enhancing social acceptance in marine governance in Europe. *Ocean and Coastal Management* 117: 61–69. https://doi.org/10.1016/J.OCECOAMAN.2015.11.001.

88 Stenekes, N., Colebatch, H.K., Waite, T.D., and Ashbolt, N.J. (2016). Risk and governance in water recycling: public acceptance revisited. *Science, Technology & Human Values* 31 (2): 107–134. http://dx.doi.org/10.1177/0162243905283636.

89 Suškevičs, M., Eiter, S., Martinat, S. et al. (2019). Regional variation in public acceptance of wind energy development in Europe: what are the roles of planning procedures and participation? *Land Use Policy* 81: 311–323. https://doi.org/10.1016/J.LANDUSEPOL.2018.10.032.

90 Sütterlin, B. and Siegrist, M. (2017). Public acceptance of renewable energy technologies from an abstract versus concrete perspective and the positive imagery of solar power. *Energy Policy* 106: 356–366. https://doi.org/10.1016/J.ENPOL.2017.03.061.

91 Upreti, B.R. and Van Der Horst, D. (2004). National renewable energy policy and local opposition in the UK: the failed development of a biomass electricity plant. *Biomass and Bioenergy* 26 (1): 61–69. https://doi.org/10.1016/S0961-9534(03)00099-0.

92 Whitmarsh, L., Xenias, D., and Jones, C.R. (2019). Framing effects on public support for carbon capture and storage. *Palgrave Communications* 5 (1): https://doi.org/10.1057/s41599-019-0217-x.

93 Wiersma, B. and Devine-Wright, P. (2014). Public engagement with offshore renewable energy: a critical review. *Wiley Interdisciplinary Reviews: Climate Change* 5 (4): 493–507. https://doi.org/10.1002/wcc.282.

94 Bauwens, T. and Devine-Wright, P. (2018). Positive energies? An empirical study of community energy participation and attitudes to renewable energy. *Energy Policy* 118: 612–625. https://doi.org/10.1016/j.enpol.2018.03.062.

95 Walker, G. (2011). The role for "community" in carbon governance. *Wiley Interdisciplinary Reviews: Climate Change* 2 (5): 777–782. https://doi.org/10.1002/wcc.137.

96 Tabi, A. and Wüstenhagen, R. (2017). Keep it local and fish-friendly: social acceptance of hydropower projects in Switzerland. *Renewable and Sustainable Energy Reviews* 68: 763–773. https://doi.org/10.1016/J.RSER.2016.10.006.

97 Süsser, D., Döring, M., and Ratter, B.M.W. (2017). Harvesting energy: place and local entrepreneurship in community-based renewable energy transition. *Energy Policy* 101: 332–341. https://doi.org/10.1016/j.enpol.2016.10.018.

98 Dóci, G. (2021). Collective action with altruists: how are citizens led renewable energy communities developed? *Sustainability (Switzerland)* 13 (2): 1–16. https://doi.org/10.3390/su13020507.

99 Dóci, G. and Vasileiadou, E. (2015). "Let's do it ourselves" individual motivations for investing in renewables at community level. *Renewable and Sustainable Energy Reviews* 49: 41–50. https://doi.org/10.1016/j.rser.2015.04.051.

100 Musall, F.D. and Kuik, O. (2011). Local acceptance of renewable energy—a case study from southeast Germany. *Energy Policy* 39 (6): 3252–3260. https://doi.org/10.1016/J.ENPOL.2011.03.017.

101 von Wirth, T., Gislason, L., and Seidl, R. (2018). Distributed energy systems on a neighborhood scale: reviewing drivers of and barriers to social acceptance.

Renewable and Sustainable Energy Reviews 82: 2618–2628. https://doi.org/10.1016/J.RSER.2017.09.086.

102 Rip, A. and Kemp, R. (1998). Technological change. In: *Human Choice and Climate Change*, Resources and Technology, vol. II (ed. S. Rayner and E.L. Malone), 327–399. Ohio: Battelle Press. https://kemp.unu-merit.nl/RipandKemp.pdf.

103 Bijker, W.E., Hughes, T.P., and Pinch, T. (1987). The social construction of technological systems. In: *Carbohydrate Research*, vol. 67 (ed. W.E. Bijker, T.P. Hughes, and T. Pinch). Cambridge: The MIT Press https://doi.org/10.1016/S0008-6215(00)84125-7.

104 Niet, I.A., Dekker, R., Van Summeren, L.F.M. et al. (2023a). Digitalization of the EU electricity system: the challenge of a just energy transition. In: *The Handbook on Climate Change and Technology* (ed. F. Urban and J. Nordensvärd), 302–321. Edward Elgar Publishing.

105 Nelson, R.R. and Winter, S.G. (1982). *An Evolutionary Theory of Economic Change*. Belknap Press of Harvard University Press.

106 Verkade, N. and Höffken, J. (2018). The design and development of domestic smart grid interventions: insights from the Netherlands. *Journal of Cleaner Production* 202: 799–805. https://doi.org/10.1016/j.jclepro.2018.08.185.

107 Niet, I.A. (2022b). Public values in power: an exploratory framework applied to the case of The Netherlands. *Energy Proceedings* 2004: 2965.

108 Mourik, R.M., Breukers, S., Van Summeren, L.F.M., and Wieczorek, A.J. (2020). The impact of the institutional context on the potential contribution of new business models to democratising the energy system. In: *Energy and Behaviour* (ed. M. Lopes, C.H. Antunes, and K.B. Jand), 209–235. Elsevier. https://doi.org/10.1016/B978-0-12-818567-4.00009-0.

109 Asmus, P. (2010). Microgrids, virtual power plants and our distributed energy future. *The Electricity Journal* 23 (10): 72–82. https://doi.org/10.1016/j.tej.2010.11.001.

110 Bauknecht, D., Funcke, S., and Vogel, M. (2020). Is small beautiful? A framework for assessing decentralised electricity systems. *Renewable and Sustainable Energy Reviews* 118: 109543. https://doi.org/10.1016/j.rser.2019.109543.

111 Fleck, J. (1993). Configurations: crystallizing contingency. *The International Journal of Human Factors in Manufacturing* 3 (1): 15–36. https://doi.org/10.1002/hfm.4530030104.

112 Vasseur, J.-P. and Dunkels, A. (2010). What are smart objects? In: *Interconnecting Smart Objects with IP: The Next Internet* (ed. J.-P. Vasseur and A. Dunkels), 3–20. Morgan Kaufmann Publishers https://doi.org/10.1016/b978-0-12-375165-2.00001-6.

113 Kool, L., Timmer, J., Royakkers, L., and Van Est, R. (2017). Urgent upgrade: protect public values in our digitized society. The Hague. https://www.rathenau.nl/sites/default/files/2018-03/Urgent_Upgrade.pdf (accessed November 2023).

114 Wesche, J.P., Negro, S.O., Dütschke, E. et al. (2019). Configurational innovation systems – explaining the slow German heat transition. *Energy Research and Social Science* 52: 99–113. https://doi.org/10.1016/j.erss.2018.12.015.

115 IEA (2017). Digitalization & energy. Paris. https://iea.blob.core.windows.net/assets/b1e6600c-4e40-4d9c-809d-1d1724c763d5/DigitalizationandEnergy3.pdf (accessed November 2023).

116 Jha, S.K., Bilalovic, J., Jha, A. et al. (2017). Renewable energy: present research and future scope of artificial intelligence. *Renewable and Sustainable Energy Reviews* 77: 297–317. https://doi.org/10.1016/j.rser.2017.04.018.

117 AI Hub (n.d.). Hoe het net steeds iets slimmer wordt. https://www.aihub-oost.nl/actueel/hoe-het-net-steeds-iets-slimmer-wordt

118 Hoogsteen, G., Molderink, A., Hurink, J.L. et al. (2017). Charging electric vehicles, baking pizzas, and melting a fuse in Lochem. *CIRED Open Access Proceedings Journal* 2017 (1): 1629–1633. https://doi.org/10.1049/oap-cired.2017.0340.

119 Lammers, I. and Hoppe, T. (2019). Watt rules? Assessing decision-making practices on smart energy systems in Dutch city districts. *Energy Research and Social Science* 47: 233–246. https://doi.org/10.1016/j.erss.2018.10.003.

120 Xu, Y., Ahokangas, P., Louis, J.-N., and Pongrácz, E. (2019). Electricity market empowered by artificial intelligence: a platform approach. *Energies* 12 (21): 4128. https://doi.org/10.3390/en12214128.

121 De Grève, Z., Bottieau, J., Vangulick, D. et al. (2020). Machine learning techniques for improving self-consumption in renewable energy communities. *Energies* 13 (18): 4892. https://doi.org/10.3390/en13184892.

122 Muench, S., Stoermer, E., Jensen, K. et al. (2022). *Towards a Green & Digital Future*. Luxembourg: Publications Office of the European Union https://publications.jrc.ec.europa.eu/repository/handle/JRC129319.

123 Howell, S., Rezgui, Y., Hippolyte, J.-L. et al. (2017). Towards the next generation of smart grids: semantic and holonic multi-agent management of distributed energy resources. *Renewable and Sustainable Energy Reviews* 77: 193–214. https://doi.org/10.1016/j.rser.2017.03.107.

124 Cozzi, L. and Goodson, T. (2020). Empowering electricity consumers to lower their carbon footprint. https://www.iea.org/commentaries/empowering-electricity-consumers-to-lower-their-carbon-footprint (accessed November 2023).

125 Wilson, C., Kerr, L., Sprei, F. et al. (2020). Potential climate benefits of digital consumer innovations. *Annual Review of Environment and Resources* 45: 113–144. https://doi.org/10.1146/annurev-environ-012320-082424.

126 Wilson, C., Andrews, B., and Vrain, E. (2022). Consumer uptake of digital low-carbon innovations. *Proceedings – 2022 International Conference on ICT for Sustainability, ICT4S 2022*, Plovdiv, Bulgaria (13–17 June 2022). pp. 109–118. https://doi.org/10.1109/ICT4S55073.2022.00022

127 Nicolini, D. (2012). *Practice Theory, Work, & Organization*, 1e. Oxford: Oxford University Press.

128 Watson, M. (2012). How theories of practice can inform transition to a decarbonised transport system. *Journal of Transport Geography* 24: 488–496. https://doi.org/10.1016/j.jtrangeo.2012.04.002.

129 Hossein Motlagh, N., Mohammadrezaei, M., Hunt, J., and Zakeri, B. (2020). Internet of Things (IoT) and the energy sector. *Energies* 13 (2): 494. https://doi.org/10.3390/en13020494.

130 Royakkers, L., Timmer, J., Kool, L., and van Est, R. (2018). Societal and ethical issues of digitization. *Ethics and Information Technology* 20 (2): 127–142. https://doi.org/10.1007/s10676-018-9452-x.

131 Cuijpers, C. and Koops, B.-J. (2013). Smart metering and privacy in Europe: lessons from the Dutch case. In: *European Data Protection: Coming of Age* (ed. S. Gutwirth, R. Leenes, P. de Hert, and Y. Poullet), 269–293. Dordrecht: Springer Netherlands. https://doi.org/10.1007/978-94-007-5170-5_12.

132 Kloppenburg, S. and Boekelo, M. (2019). Digital platforms and the future of energy provisioning: promises and perils for the next phase of the energy transition. *Energy Research and Social Science* 49: 68–73. https://doi.org/10.1016/j.erss.2018.10.016.

133 Van Aubel, P., Colesky, M., Hoepman, J.-H. et al. (2018). *Privacy by Design for Local Energy Communities*. CIRED Workshop Ljubljana.

134 Niet, I.A., Van den Berghe, L.H.G.J., and van Est, R. (2023b). Societal impacts of AI integration in the EU electricity market: the Dutch case. *Technological Forecasting and Social Change* 192: 122554. https://doi.org/10.1016/j.techfore.2023.122554.

135 Milchram, C., Künneke, R., Doorn, N. et al. (2020). Designing for justice in electricity systems: a comparison of smart grid experiments in the Netherlands. *Energy Policy* 147: 111720. https://doi.org/10.1016/j.enpol.2020.111720.

136 Niet, I.A. (2022a). Between vision and practice: lack of alignment between AI strategies and energy regulations in the Dutch electricity sector. *Discover Artificial Intelligence* 2 (1): 24. https://doi.org/10.1007/s44163-022-00040-6.

137 Smith, A. and Raven, R. (2012). What is protective space? Reconsidering niches in transitions to sustainability. *Research Policy* 41 (6): 1025–1036.

138 Van den Berghe, L. and Wieczorek, A.J. (2022). Crash course 8: energy management systems for communities. Eindhoven. https://vb.nweurope.eu/media/17850/crash-course-8-energy-management-systems-for-communities.pdf (accessed November 2023).

139 Brummer, V. (2018). Community energy – benefits and barriers: a comparative literature review of community energy in the UK, Germany and the USA, the benefits it provides for society and the barriers it faces. *Renewable and Sustainable Energy Reviews* 94: 187–196. https://doi.org/10.1016/j.rser.2018.06.013.

140 Vallecha, H., Bhattacharjee, D., Osiri, J.K., and Bhola, P. (2021). Evaluation of barriers and enablers through integrative multicriteria decision mapping: Developing sustainable community energy in Indian context. *Renewable and Sustainable Energy Reviews* 138: 110565. https://doi.org/10.1016/j.rser.2020.110565.

141 Weber, L. (1997). Some reflections on barriers to the efficient use of energy. *Energy Policy* 25 (10): 833–835. https://doi.org/10.1016/S0301-4215(97)00084-0.

142 Milligan, M., Frew, B.A., Bloom, A. et al. (2016). Wholesale electricity market design with increasing levels of renewable generation: Revenue sufficiency and long-term reliability. *The Electricity Journal* 29 (2): 26–38. https://doi.org/10.1016/j.tej.2016.02.005.

143 Walker, G. (2008). What are the barriers and incentives for community-owned means of energy production and use? *Energy Policy* 36 (12): 4401–4405. https://doi.org/10.1016/j.enpol.2008.09.032.

144 Sebi, C. and Vernay, A.-L. (2020). Community renewable energy in France: The state of development and the way forward. *Energy Policy* 147: 111874.

145 Reis, I.F.G., Gonçalves, I., Lopes, M.A.R., and Antunes, C.H. (2021). Business models for energy communities: a review of key issues and trends. *Renewable and Sustainable Energy Reviews* 144: 111013. https://doi.org/10.1016/j.rser.2021.111013.

146 Nolden, C. (2013). Governing community energy—Feed-in tariffs and the development of community wind energy schemes in the United Kingdom and Germany. *Energy Policy* 63: 543–552. https://doi.org/10.1016/j.enpol.2013.08.050.

147 Grashof, K. (2019). Are auctions likely to deter community wind projects? And would this be problematic? *Energy Policy* 125: 20–32. https://doi.org/10.1016/j.enpol.2018.10.010.

148 Gsänger, S. and Karl, T. (2020). Community wind under the auctions model: a critical appraisal. In: *Accelerating the Transition to a 100% Renewable Energy Era*, Lecture Notes in Energy (ed. T.S. Uyar), 233–257. Cham: Springer International Publishing https://doi.org/10.1007/978-3-030-40738-4_11.

149 Capellán-Pérez, I., Campos-Celador, Á., and Terés-Zubiaga, J. (2018). Renewable Energy Cooperatives as an instrument towards the energy transition in Spain. *Energy Policy* 123: 215–229. https://doi.org/10.1016/j.enpol.2018.08.064.

150 TNO-rapport. TNO 2023 R10525. De rol van eindafnemers in de energiemarkt. Energiegemeenschappen in veranderend juridisch landschap Energiegemeenschappen in veranderend juridisch landschap.

151 Lacey-Barnacle, M. and Bird, C.M. (2018). Intermediating energy justice? The role of intermediaries in the civic energy sector in a time of austerity. *Applied Energy* 226: 71–81.

152 Wuebben, D., Romero-Luis, J., and Gertrudix, M. (2020). Citizen science and citizen energy communities: a systematic review and potential alliances for SDGs. *Sustainability* 12 (23): 10096. https://www.mdpi.com/2071-1050/12/23/10096.

153 Catney, P., MacGregor, S., Dobson, A. et al. (2014). Big society, little justice? Community renewable energy and the politics of localism. *Local Environment* 19 (7): 715–730.

154 Johnson, V.C., Hall, S., Barton, J. et al. (2014). Community energy and equity: the distributional implications of a transition to a decentralised electricity system with contributions from. *People, Place and Policy* 8: 149–167. https://extra.shu.ac.uk/ppp-online/wp-content/uploads/2014/12/community-energy-equity.pdf.

155 McCabe, A., Pojani, D., Broese, A., and van Groenou (2018). Social housing and renewable energy: community energy in a supporting role. *Energy Research and Social Science* 38: 110–113.

156 Park, J.J. (2012). Fostering community energy and equal opportunities between communities. *Local Environment* 17 (4): 387–408. http://www.tandfonline.com/doi/abs/10.1080/13549839.2012.678321.

157 Etherden, N., Vyatkin, V., and Bollen, M.H.J. (2016). Virtual power plant for grid services using IEC 61850. *IEEE Transactions on Industrial Informatics* 12 (1): 437–447. https://doi.org/10.1109/TII.2015.2414354.

158 Plancke, G., De Vos, K., Belmans, R., and Delnooz, A. (2015). Virtual power plants: definition, applications and barriers to the implementation in the distribution system. *2015 12th International Conference on the European Energy Market (EEM)*, Lisbon, Portugal (19-22 May 2015), vol. 2015. IEEE. pp. 1–5. https://doi.org/10.1109/EEM.2015.7216693.

159 Kluskens, N.T.T.M., Niet, I.A., Van Bommel, N., and Van Summeren, L.F.M. (2024a). Flexible and Just: ICT in energy communities. Energy Research and Social Science (under review).

160 REScoopVPP (n.d.). We're an EU project developing energy flexibility tools by co-ops for co-ops! https://www.rescoopvpp.eu/ (accessed November 2023).

161 Buth, M.C., Wieczorek, A.J., and Verbong, G.P.J. (2019). The promise of peer-to-peer trading? The potential impact of blockchain on the actor configuration in the Dutch electricity system. *Energy Research and Social Science* 53: 194–205. https://doi.org/10.1016/j.erss.2019.02.021.

162 Wittmayer, J.M., Avelino, F., Pel, B., and Campos, I. (2021). Contributing to sustainable and just energy systems? The mainstreaming of renewable energy prosumerism within and across institutional logics. *Energy Policy* 149: 112053. https://doi.org/10.1016/j.enpol.2020.112053.

163 Klok, C.W., Kirkels, A.F., and Alkemade, F. (2023). Impacts, procedural processes, and local context: rethinking the social acceptance of wind energy projects in the Netherlands. *Energy Research & Social Science* 99: 103044. https://doi.org/10.1016/j.erss.2023.103044.

164 Healy, N., Stephens, J.C., and Malin, S.A. (2019). Embodied energy injustices: unveiling and politicizing the transboundary harms of fossil fuel extractivism and fossil fuel supply chains. *Energy Research & Social Science* 48: 219–234.

165 Alkemade, F. and de Coninck, H. (2021). Policy mixes for sustainability transitions must embrace system dynamics. *Environmental Innovation and Societal Transitions* 41: 24–26.

166 Alkemade, F., de Bruin, B., El-Feiaz, A. et al. (2023). Social tipping dynamics in the energy system. *Earth System Dynamics Discussions* 2023: 1–20. https://doi.org/10.5194/esd-2023-25.

12

Guidelines and Recommendations for Optimal Implementation of Integrated Local Energy Communities

Leonard E.R. Perez[1], Bernadette Fina[2], Branislav Iglár[2], Carolin Monsberger[2], Klara Maggauer[2], Natália de A.B. Weber[2], Georgios Yiasoumas[3,4], George Georghiou[3,4], José Villar[5], João Mello[5], and Rad Stanev[6]

[1] European Distributed Energy Resources Laboratories (DERlab) e.V., Project Management and Research, Wilhelmsstraße 5, 34117 Kassel, Germany
[2] AIT Austrian Institute of Technology, Center for Energy, Giefinggasse 4, 1210 Vienna, Austria
[3] University of Cyprus, PV Technology Laboratory, Department of Electrical and Computer Engineering, Panepistimiou 1 Avenue, 2109 Nicosia, Cyprus
[4] PHAETHON Centre of Excellence for Intelligent, Efficient and Sustainable Energy Solutions, Panepistimiou 1 Avenue, 2109 Nicosia, Cyprus
[5] INESC TEC – Institute for Systems and Computer Engineering, Technology and Science, Centre for Power and Energy Systems, Campus da FEUP, Rua Dr. Roberto Frías, 4200-465 Porto, Portugal
[6] Technical University of Sofia, Power System Stability Laboratory, Electrical Power Engineering Department, 8 Kliment Ohridski St., Bl. 12 Office 12 402/12 403, 1000 Sofia, Bulgaria

12.1 Introduction

Integrated local energy communities (ILECs) represent a new alternative to the existing traditional centralized or utility-scale energy generation, which has attracted public interest in recent years. This innovative concept proposes a decentralized energy generation (based on both distributed energy resources [DERs] and novel technologies) and management in which different energy carriers are integrated into energy hubs. Under this approach, citizens (usually referred to as communities) turn their passive role into a pro-active participation in both energy generation and consumption. The combination of different concepts such as decentralized generation, integration of energy systems, citizens taking part in decision-making processes, and others into a unified approach naturally brings with it challenges for its implementation, integration, and operation. Hence, new efforts on researching strategies to optimize the implementation of ILECs become crucial for their further development and expansion. The outcomes of such research aim to provide valuable guidelines and recommendations on critical aspects to be considered in ILEC projects looking ahead to the future.

This chapter aims to provide the reader with a series of guidelines and recommendations for an optimal implementation of ILECs from five different dimensions (technical, economic, environmental, regulatory, and societal). The

Integrated Local Energy Communities: From Concepts and Enabling Conditions to Optimal Planning and Operation, First Edition. Edited by Marialaura Di Somma, Christina Papadimitriou, Giorgio Graditi, and Koen Kok.

content takes into consideration the lessons learnt from the development of past energy community-related projects as well as expertise of the authors and existing literature in the field. Therefore, the content provided is intended not only for governments but also for all other stakeholders involved in ILECs such as system operators, energy market actors, policymakers, and citizens.

This chapter is divided into four sections. The first and second sections introduce the problems to be discussed with a focus on the challenges faced by ILECs. The third section goes deep into the guidelines and recommendations for the optimal implementation of ILECs. This section is split into five parts, each one dedicated to a specific dimension covered in the discussion. Lastly, the fourth section provides general conclusions which summarize major points discussed through the chapter.

12.2 Main Challenges of Integrated Local Energy Communities Implementation at the European Level

As explained in Chapter 3 of this book, an ILEC refers to a set of energy users deciding to make common choices to satisfy their energy needs with minimum cost (possibly with other complementary objectives), optimizing the energy management of technologies related to multiple energy carriers and energy storage [1]. The inclusion of the term "integrated" suggests that the schemes consider multiple energy carriers as an integrated solution. Multi-carrier energy systems (MCES) interconnect energy technologies in energy systems and include technologies such as combined heat and power (CHP) plants, natural gas-fired power plants, power-to-X (P2X), and hydropower and renewable energy sources (RES) generation [2]. Moreover, MCES are characterized by the need of a precise coordination in the planning and operation across the multiple energy carriers and/or sectors to ensure high reliability, cost-effectiveness, and minimal environmental impact when delivering the service to the end-users [3]. As the complexity of the way the energy is distributed and managed increases (in this case with the integration of different energy carriers), the challenges for its implementation increase as well.

In addition to the technical challenges, other aspects also come into the picture such as the legal and regulatory framework shortcomings, starting from the fact that current regulatory frameworks do not offer all cases neither a clear definition for the term *energy community* nor common agreement on policies related to their operation and participation in the energy market. Apart from that, there is not a common European Union (EU) framework for the regulation of the management of the local generation within the energy communities and how they may act as aggregators in the electricity market, a market role also yet poorly defined. Additionally, still persist unclear regulations, whereas complex legal procedures for the installation of the technologies associated with ILECs exist for emerging ones such as hydrogen, fuel cells, or electromobility (vehicle-to-grid, vehicle-to-home), and integrating cogeneration into renewable energy communities (RECs).

In the environmental and economic dimensions, some problematic aspects should also be addressed. First, the perception of financial long-term high risk when investing in technologies associated with renewables. This point hinders the spread of energy community projects as investors must deal with uncertain financial conditions (volatile energy prices, high initial investment, and variable incentive policies). Therefore, strategies should be developed to ensure stable economic conditions for ILEC projects while reducing the long-term risk and providing more attractive economic conditions to the investors [4]. Moreover, actions involving economic, regulatory, and environmental affairs would help ILECs implementation, for instance, the reformulation of the emission taxing schemes which award the reduction of greenhouse gas (GHG) emissions through implementation of these projects and consequently support the replacement of fossil-fuel generation.

From the social perspective, fronts can be confronted to promote the implementation of more energy community projects. In addition to the economic benefits of belonging to an energy community, engagement would increase with the awareness on the environmental impact of the use of fossil-fuel generation technologies, and explaining how local renewable generation, as well as cogeneration and hydrogen production, could help support the transition toward a sustainable energy infrastructure. This requires efforts to improve citizens' energy literacy regarding the role of active energy behaviors and the economic and environmental benefits of energy communities. The focus could also be on the integration of vulnerable consumers in energy communities to benefit from reduced energy costs or reaching vulnerable zone with difficulties in accessing energy services. Lastly, the social acceptance toward relevant ILEC-associated technologies, such as wind power, remains resistant, so then more efforts from governments to properly assess their benefits and drawbacks, as well as to inform them to citizens, become also essential [5].

12.3 Guidelines and Recommendations for Optimal Implementation of ILECs

This section proposes overall guidelines and recommendations for optimal implementation of ILECs addressed from five dimensions: technical, regulatory, environmental, economic, and social. In the **technical dimension**, the discussion covers technical issues related to the integration of nonconventional renewable technologies (those commonly installed in ILECs) and proposes technical recommendations to deal with them. Additionally, some potential research lines, which could be explored in future projects on energy communities, are proposed.

In the **regulatory dimension**, regulatory barriers at the European level for the implementation of energy communities are identified. Aspects such as the lack of regulation for the integration of ILECs are highlighted while suggesting strategies for the policymakers and legal entities to consider. Additional modifications to the existing regulatory frameworks to accelerate the implementation of ILECs are proposed.

The **environmental dimension** discusses the main environmental benefits of ILECs with respect to traditional fossil-fuel generation and centralized generation schemes while remarking on aspects to be considered for their implementation.

Furthermore, the **economic dimension** examines financial benefits at medium and long term derived from the implementation of ILEC schemes for both system operators and citizens. Furthermore, recommendations for maximizing economic benefits and critical aspects to be tackled in the energy market for propitiating the participation of citizens are provided.

Lastly, the **social dimension** exposes the challenges of introducing ILEC-related technologies and projects into communities, highlighting aspects such as the citizens' acceptance to the adoption of novel solutions, advantages for the communities to participate in the decision-making processes and others. Subsequently, strategies to increase the social acceptance of the communities with respect to the ILEC paradigm. Lastly, the motivation for the community to adopt these schemes is discussed, and recommendations are derived based on these aspects.

12.3.1 Technical Dimension

The transition toward a decentralized energy generation in MCES is associated with issues ranging from a higher degree of complexity in energy generation control and management to an increment of the host capacity of the existing structure. Thus, these technical challenges should be tackled in such a way that the integration of such systems is accelerated. Some of the main issues in relation to the technical dimension and recommendations are developed below.

12.3.1.1 Implementation of Strategies to Reduce Grid Reinforcement Needs due to the Installation of New Assets

With the introduction of energy communities' solutions, new assets associated with energy storage systems (ESSs), microgeneration (e.g. rooftop photovoltaic [PV], and hydro pump), heating and cooling district systems, and others, within the energy hubs will be installed. Therefore, the existing energy infrastructure capacity could become insufficient to host these new energy exchanges and, as a consequence, local constraints as well as overloading issues will arise. While the traditional approach has been the so-called *fit-and-forget*, it may become extremely costly considering the high presence of energy assets and distributed energy production, and anyway disregard other feasible approaches [6]. Therefore, advanced solutions are needed to minimize the need for expensive network upgrades, such as the use of smart grid technologies to profit from the existing flexibility. Indeed, these technologies increase the grid observability and control capabilities of the grid assets, allowing to coordinate the different resources alongside MCES while maintaining the balance between supply and demand as well as respecting the technical limits of the networks. In the same way, more advanced solutions lie on the implementation of implicit and explicit demand-response strategies by unlocking the potential flexibility of the consumers as well as flexibility coming from the coordinated interplay of multiple energy carriers. This will allow the adjustment of the consumption profile of the community based on network conditions and help avoid overloads and congestion during peak demand periods and seasons (e.g. within the electrical and thermal domain). Additionally, by incorporating P2X technologies as flexibility can be also possible to reduce the capacity requirements of the transmission grid [7]. Finally, collaborative planning should be initiated between energy communities and other shareholders like utilities and local authorities during the development process of new energy projects to identify and manage infrastructure inefficiencies.

12.3.1.2 Alternatives to Deal with Generation Source Fluctuations

One of the major problems extensively researched lies on the consequences on the service quality due to the integration of renewables, which are key technologies in energy communities. The strong dependence on the weather conditions of these generation technologies leads to frequency variations and harmonics problems as their share in the energy mix increases (not only in the ILEC but also, in general, at the utility scale). System operators must face this problem considering the ambitious energy targets established by the EU. An effective way to mitigate the uncertainty of RES generation is the use of forecasting techniques that allow the estimation of the energy output in the future state of the energy system. The forecasting can be performed on a day-ahead horizon or even an hour-ahead allowing informed decision-making for scheduling and operation of the existing flexible resources. Moreover, energy communities can also be part of the solution to this problem since they introduce a higher level of distributed generation in a larger geographic area instead of a concentrated one, making the impact of fluctuations in production lower. In addition, the use of storage systems can help tackle the stochastic nature of RES by providing flexibility to the power system, and thus creating a buffer between the RES and the grid. Beyond the use of battery systems, the introduction of hybrid systems with RES such as wind and PV integrated farms, can help to limit fluctuations in production to some extent. Furthermore, ILEC schemes offer a holistic approach that tackles this issue by enabling sector coupling among different energy carriers. In this sense, the implementation of sector coupling allows to manage temporal energy balancing induced by the intermittent feed-in of RES generation [8]. For instance, by storing energy excess surplus in ESSs (batteries) or as hydrogen (through electrolysis) to avoid curtailment, and later using it either in other energy sectors (e.g. transportation and electric vehicle [EV] mobility) or converting it back into electricity during lower RES generation periods.

Likewise, resource fluctuation issues can be extended to other energy carriers. In the case of hydrogen production, fluctuation in the water levels available for the electrolysis repercusses in the generation capacity. For instance, water resource fluctuation due to drought as experienced in Australia and North America as well as the lack of fresh water in the Middle East and China reduces the hydrogen production capacity [9]. To counterattack this problem, hydrogen production can progressively migrate toward the utilization of systems able to utilize wastewater or seawater. Moreover, in the case of district heating systems, their strong dependency on natural gas makes them vulnerable to the resource depletion due to a high exploitation rate (as well as fluctuation in gas prices) [10, 11]. Thus, the current gas crisis has provoked large-scale heat pumps and the use of RES instead of gas to become a more attractive option for energy operators looking for the sector decarbonization, so then more developments in this direction are crucial.

12.3.1.3 Coordination and Control Between System Operators and ILECs

The integration of ILECs demands a higher coordination level between actors involved in ILEC operation such as distribution system operators (DSOs), transmission system operators (TSOs), district heating system operators, energy service companies, aggregators, balance responsible and others, in order to ensure an optimal energy management among the different carriers (electricity, heating and cooling, gas, hydrogen). In this case, recommendations address the main aspects to be considered when designing strategies to deal with this complex multi-carrier coordination task. Said differently, to achieve the optimal implementation of multi-carrier systems is necessary an enhanced coordination between the different actors. Even more, the impact of this challenge increases when there are multiple actors involved beyond the system operators. To counterattack this issue, new protocols need to be defined and agreed upon that explicitly state the role of each operator while facilitating the cooperation between them. From a technical point of view, the monitoring and observation need to be increased to understand the interactions between the different energy carriers. Moreover, it becomes necessary to develop advanced optimization, management, and operation tools for MCES compatible with existing information and communication technologies (ICTs) and control infrastructure.

12.3.1.4 Determination of Suitable Location for ILEC Project Implementation Through Novel Assessment Tools and Technical Studies

EU plans to work closely with member states (MS) to set up, by 2025, at least one renewable-based energy community in municipalities with population higher than 10 000 inhabitants [12]. To achieve this ambitious goal urges the detection of new suitable locations where implement ILEC pilots, and consequently, the implementation and testing of technological solutions and new management/coordination tools. Therefore, there is a need of reaching an optimal resource assessment and identification of sites with high potential for the development of these technical solutions, while fighting fuel and energy poverty at the same time. In order to determine the suitable locations to implement ILEC pilots, a holistic approach should be deployed. Thus, feasibility studies should be conducted in such a way that resource assessment takes into consideration the existing infrastructure, synergies with other initiatives and the long-term sustainability of the ILEC beyond the pilot phase. Advancements in fields such as modeling and programming tools, geographic information systems, digital twins, and simulation software, are useful not only for assessing and quantifying with reasonable accuracy the potential of new locations but also for estimating their technical impact.

12.3.1.5 Acceleration of Smart Meters Installation

The smart meters (SM) are a key factor for enabling active smart grid concepts by monitoring the power and energy balance of the prosumers in the ILECs and allowing automatic adjustment of the generation, demand, and storage in order to respond to the dynamic price changes. Therefore, fostering the massive installation of SM (such as smart electricity meters, smart gas meters, and smart heat meters) and converting them into the global standard arise as a crucial recommendation. Thus, the acceleration of the SM installation is influenced by various technical, economic, and social factors; some of the most significant ones stand for

- **Lowering costs and increasing benefits:** presently the electricity SM installation cost in the EU ranges approximately within €180 and €200 while offering evaluated savings of €270 on average. The expected installation of 225 million SM in the EU by 2024 will allow economies of scale driving the costs further down [13]. At the same time, the complete liberalization of the energy market, including all citizens, will stimulate households to further increase their benefits for example by adapting their consumption according to low (or even negative) prices during the maximum PV and other RES production (which is already remarkable around midday during the weekends on the EU electricity market) or periods with more favorable gas prices.
- Enhancement of data-, privacy-, and cybersecurity protection also arises a critical factor for the accelerated SM installation and widespread use.
- Enabling uncomplicated data exchange with other devices (such as demand, storage, energy flow, and generation controls).
- **Easier and non-disturbing installation:** the present average time for SM installation takes approximately 2 hours and requires a power cut for 30 minutes. Thus, by ensuring easier and non-invasive installation methods will boost SM acceptance and massive implementation across different energy carriers.

12.3.1.6 New Research Lines in the Field of Energy Community

A number of technical and nontechnical challenges are still hindering the massive ILECs deployment. A critical research line which needs to be faced is the study of appropriate market architectures and designs which enable the massive prosumer involvement and, at the same time, consider all social and technical limitations in the energy generation, storage, distribution, demand, and network in real time. Due to the large number of players involved in MCES and unsecured data flows, the future energy markets need to adequately address all the physical congestions. A promising research line in this regard lies on the development of novel market pricing schemes which take into account the network and other technical congestions based on network parameters at the point of common coupling. Furthermore, the advancements and tools developed to tackle technical problems need to be evaluated in various topologies and scales of energy communities to ensure their replicability and adaptability to different use cases [14].

Despite the fact that technical issues (previously exposed) associated with the integration of energy communities may arise, they should not hinder their development. Overall, ILECs may act as test beds for technologies which can fast-track their implementation on a larger scale. Indeed, the growth of energy communities will create opportunities to develop novel solutions and concepts to tackle these challenges.

12.3.2 Regulatory Dimension

The introduction of ILECs implies the participation of the citizens in decision-making processes related to the energy sector of the community, e.g. gaining a kind of ownership over the energy infrastructure or only investing without having a shared ownership of the assets. In addition, policymakers also appear as key actors in the transition toward decarbonization since the development of the legal tools, which facilitate the integration of nonconventional renewable energy technologies, lies on them. On the other side, a fundamental problem commonly remarked in regulatory studies performed in energy community-related studies lies on the unclear strategy from most of the MS. Nowadays, it can be said that a lot of freedom is given to the governments on how to implement their policies and regulations. In addition, energy communities' regulation focuses on electricity aspects, partially ignoring the implications of considering other energy carriers or limiting the inclusion of nonrenewable generation such as cogeneration. As example, technologies such as hydrogen face lack of regulation for installation of facilities and assets (in this case electrolyzers) for their production and their connection to gas pipelines and electricity grids [15]. That being said, major recommendations in the regulatory dimension are developed in detail below.

12.3.2.1 Standardization, Transposition, and Harmonization of Energy Community EU Directives

A critical issue with the implementation of ILEC-related technologies lies on the lack of a clear definition of concepts related to this paradigm. EU directive of the European Parliament and the Council of the EU [16] offers initial definitions for the concepts such as citizen energy community (CEC) and REC; nevertheless, there are more specific aspects which are not addressed including the role of the energy communities as aggregators/flexibility service providers, their participation in the energy market, which are the rules in the formulation of conditions for their conformation, among others. Moreover, the local regulations among the MS vary according to their own interpretation of those directives which leads to significative differences among local regulations that hinder the scaling up of ILECs. For these reasons, a recommendation commonly remarked is the harmonization of the concept, role, delimitation, and integration of energy communities (and their related technologies) within the European directives as well as the development of an effective plan for transposing them to the MS, leaving almost no aspects to the interpretation of the local authorities. Therefore, the uniformity of the regulatory frameworks will advance significantly while differences among MS are shortened, leading to an easier implementation of these schemes and the development and refinement of standards [4]. However, harmonization should not be used to limit flexibility in the energy sharing mechanisms, which are needed to allow the development of advanced business models, develop sector coupling among different energy carriers, and integrate distributed flexibility provision from energy communities with local and wholesale electricity markets [17, 18].

12.3.2.2 Simplification of License Procedures and Paperwork

Another problem with the implementation of ILECs is the complicated licensing process, that is, obtaining the respective permission and operating licenses to meet all paper requirements. Regardless of the maturity of the ILEC-related technologies, the communities face challenges to get permissions for the integration of self-generation into the existing network (e.g. excess generation surplus). A recommendation to overcome this barrier can be the simplification of the procedures for energy community projects or initiatives by the side of the authorities. In other words, unnecessary bureaucracy and excessive requirements for both self-generation installation and grid connection of energy community initiatives should be removed by the authorities. For instance, simplify legal procedures for participating in flexibility not only for electrical but also for thermal loads such as electric boilers, heat pumps, and hydrogen electrolyzers. Alternatively, regulatory bodies could offer additional support service for citizens to go through the paperwork process, in order to facilitate the understanding of these procedures [19]. In this sense, the EU's recent reform [20] strongly emphasizes the right to share energy even without the need to create energy communities, trying to reduce further the existing regulatory and bureaucratic barriers.

12.3.2.3 Taxing and Incentive Schemes Reformulation Within Regulatory Frameworks

A fundamental pillar in which most of the recommendations provided by research projects in the field of energy communities agree, lies on the reformulation of tax exemption and incentive schemes. The fundamental reason behind this is the fact that by providing more favorable conditions (backed by the regulatory framework), the interest of the public and investors will rise and, consequently, the possibilities of developing projects will be higher. Exemptions on taxes and duty fees for acquisition of assets for the development of ILEC projects should become a standard and be promoted within the EU up to 2030 in alignment with the decarbonizing targets. This can be achieved by establishing agreements with local governments, whereas giving priority to EU manufacturers to enhance the economy in the region. In this manner, the regulations can boost the maturing and consolidation of these technologies.

12.3.2.4 Regulations for the Data Exchange and Privacy

It becomes necessary to establish clear regulations in terms of data treatment and privacy for all actors involved in ILECs operation. These data are used for monitoring aspects such as consumption patterns, optimization and control, data exchange between DSO/TSO layers, peer-to-peer (P2P) trading actors, and demand forecasting [21]. Therefore, special policies addressed to protect sensitive information about energy (either demand or consumption) should not hinder their exchange among layers and utilization for the optimization and management of these energy systems.

12.3.2.5 New Regulations for the Peer-to-Peer Trading

With the introduction of the *prosumer* concept, new elements should be considered in the energy market. Under schemes such as RECs, CECs, and ILECs, the end-user is allowed to trade its self-generated energy into local energy trading. This end-user role brings the necessity of establishing new rules for the P2P trading that did not exist in traditional centralized schemes. Currently, there is a lack of regulations regarding novel technologies like P2P trading (including citizens as actors in the market) in Europe since regulations are still under development. Therefore, in parallel with the implementation of ILECs, regulation on P2P or local markets should be developed, in which citizens are conceived as an actor with active participation in the market. The EU's recent reform [20] also emphasizes the right of active customers participating in energy sharing to participate in P2P agreements for local energy sharing. A detailed discussion on this topic can be found in Chapter 8 of this book. Lastly, there is a lack of regulation related to the possibility of sharing energy with nonmembers of the community. Although this may seem odd, as an example, the community could install EV chargers to share the renewable surplus with EV owners not necessarily belonging to the community, something difficult to match with several existing regulatory frameworks, so then this aspect should also be considered for future discussions.

12.3.2.6 Creation of Policies in Favor of Vulnerable Social Communities

Regulatory frameworks should also tackle energy poverty by providing additional support to vulnerable social groups, so they can also become part of energy communities. Different actors such as public institutions, policymakers, and associations can become facilitators for vulnerable households to participate in ILEC projects by developing regulations which guarantee financial funding for the initial investment [12]. This would prevent the exclusion of vulnerable social groups from the energy transition plan.

The proposed recommendations and strategies may be tested in the so-called *regulatory sandboxes* which, in a general sense, refer to the implementation of changes in the legislations within a real-world controlled environment and under the supervision of regulatory bodies [22]. This definition can also be extended to other dimensions such as the technical, when it is pretended to evaluate the impact of the integration of novel technologies or, in the economic dimension, when new business models and financial schemes need to be tested. Thus, the implementation of regulatory sandboxes can support the learning process through the experimentation of innovative legal regimens, in this case, for assessing the effectiveness of modifications in regulations related to energy communities. Likewise, special regulatory modifications can be tested in demonstration pilots of different regions throughout the implementation of energy community research projects funded by the European Commission and local governments.

12.3.3 Environmental Dimension

As mentioned in previous sections, the implementation of ILECs represents a novel proposal for both integration and management of energy systems including its different carriers (i.e. electricity, gas, cooling, and heating) within the communities. In most of the cases, these schemes include generation based on mature renewable technologies such as PV or wind. Said differently, for the electricity carrier, the implementation of ILECs is directly related to a high-RES share. Based on this, it can be stated that the most relevant environmental benefit of these schemes lies on the replacement of fossil fuel-based generation by more environmental friendly (considering the sustainability aspects) technologies such as renewables [19]. Likewise, the use of biomass for domestic heating can result in savings of greenhouse gases (GHG) emissions of the order of 80% when compared to traditional fossil fuels [23]. Besides, by using biomass-based power plants, it becomes possible to reduce GHG emissions and the external energy dependency [24]. Moreover, technologies associated with other energy carriers such as P2X (e.g. power-to-gas [P2G] and power-to-hydrogen [P2H]) have also environmental benefits associated. For instance, the resultant products in technologies such as gas and hydrogen stand for alternatives to replace traditional fossil fuels while, at the same time, avoiding the waste of energy surplus. Besides, waste heat generated by P2X processes can be reused for heating leading to a reduction in CO_2 emissions [25].

Furthermore, ILECs boost the implementation of electromobility within the communities by generating electricity locally and feeding a part of it into EV charging stations. The acceleration in the penetration of EVs (as in the renewable generation) is translated into a reduction in CO_2 emissions which are avoided when replacing traditional fuel-driven vehicles. Likewise, additional environmental benefits of ILECs can be found with respect to traditional centralized energy generation when analyzing the correlation between the GHG emissions and other variables (e.g. reduction of transmission network losses due to the local electricity production and consumption).

Beyond shadow of doubts, the combination of different technologies (renewables, CHP, hydrogen, etc.), EV charging infrastructure, demand response, ESSs, and others evidence how ILECs can contribute to the decarbonization process. In contrast, there exist other aspects associated with the implementation of ILEC that may have negative repercussions in the environment. Among these can be highlighted the debate on the environmental impact of raw materials used in ESS and PV technologies, the increment of assets (which implies an increase in the exploitation of materials with low recyclability), and the landscape and ecosystem impact (e.g. due to wind farms installation). In order for the ILEC to become a paradigm with long-term environmental sustainability, several aspects should be considered for their implementation as discussed below.

12.3.3.1 Including Landscape Impact Cost Analysis and Life Cycle Assessment (LCA)

During the feasibility assessment process of energy communities, the evaluation of the environmental impact caused by the introduction of new assets and technological solutions becomes a critical factor. Indeed, the repercussion in the visual impact due to additional overhead and cable transmission systems, in rooftops and land/yards with installation of PV systems as well as noise contamination and ecosystem due to wind farms are a few examples of arguments to social opposition. Therefore, environmental dimension should also be considered when establishing the roadmap for a wider implementation of energy communities. The common approach limits to assess the ILECs' environmental impact lies on performing a life cycle assessment (LCA) with the calculation of the carbon footprint. However, advanced approaches propose to consider variables such as air quality modeling (including estimation of emission factors and pollutant concentration), development of health and vegetation impact functions and, besides, the introduction penalization factors for conventional generation within planning and management of the electrical networks [26]. Apart from that, LCA should also consider additional GHG emissions related to the long-distance transport of raw materials [23]; therefore, a more accurate environmental impact assessment can be performed when assessing installation and operation of MCES technologies. Accordingly, more sophisticated approaches for environmental and landscape impact assessment must be adopted (and further become a standard) within the planning phase of the conformation of energy communities. Likewise, it is recommended to work more on the perception of climate change and environmental contribution of the renewables as well as raise the public awareness of the need of a faster transition toward a decentralized energy generation [5, 27].

12.3.3.2 Recyclability of Raw Materials and Waste Management

Another environmental aspect to consider for the implementation of energy communities lies on the collateral increase of the waste material of technologies such as PV solar, wind, and batteries. The complexity in the separation and post-processing of some components from the PV modules and wind structures has limited the recyclability of those technologies and, as a consequence, introducing a negative impact in terms of sustainability. Moreover, there exist debates on whether the energy cost for manufacturing and recycling ILEC-related technologies' materials is comparable with the energy saved with their implementation (thus making their environmental sustainability questionable).

In order to address recyclability challenges, adopting circular economy principles may support offset environmental factors such as emissions associated with the manufacturing stages and increase recycling and recovery rates. Likewise, the development of new approaches for recycling vast and inevitable waste stream (e.g. more efficient methods to recycle end-of-life PV modules, wind blades, and batteries from both storage systems and EV mobility) as well as how to maximize recovery of materials become crucial for the future perspective of the energy communities [28]. Additionally, new techniques and equipment must be developed to be able to tackle difficulties for end-of-life recyclability and waste management such as composite materials and hazardous substances treatment, high cost of processes, low concentration of valuable materials, and material separation [29].

12.3.3.3 Rising the Environmental Value of ILECs by Improving GHG Emissions Taxing Exemption Schemes

As additional recommendation in the environmental dimension, it can be introduced some amendments in the GHG emission taxing systems when dealing with energy community initiatives. Seen from a different perspective, to keep low GHG emission pricing (as it is currently) will cause those financial incentives to remain skewed against renewable energy and, as a consequence, hinder the competition of

ILECs against fossil fuels [27]. Thus, a progressive increment in the GHG emission taxes for fossil-fuel generation together with exemptions of the technologies related to energy communities will definitely boost the implementation of these schemes and valorize its environmental contribution.

12.3.4 Economic Dimension

The main economic benefit of ILEC typically originates from the individual self-consumption of members who generate energy behind their meter. Sharing energy between these members only adds limited financial gains to the community, except on very specific situations of huge complementarity among members. Also, the act of sharing energy is not what makes the power system more efficient since this is only a financial agreement with no short-term impact on the energy flows. It is the generation or consumption of the ILEC resources which do impact the grid, when connected to it. In fact, regulations generally do not allow the obtention of significant benefits from energy sharing, which should not constitute a primary commercial or professional activity [20], maintaining the members' rights and obligations as consumers and generators regarding the balancing responsibility and grid access [16].

Energy communities are a powerful driver for end-users to actively engage in the power system. Indeed, starting from the installation of local renewable generation capacity for individual self-consumption, the potential of sharing the electricity surplus as well as heating or cooling through district systems with the other community members leads to additional economic benefits compared to buying the energy from their retailers or selling their surplus to their aggregators. In addition, energy communities allow to profit from local renewable energy when individual self-consumption is difficult to implement due to lack of space, exploit the complementarities among different energy profiles, and profit from local flexibility to adapt consumption to the local available resources. While in the very short term, the benefit may seem rather financial, in the medium and long term this new active behavior and decentralization is expected to transform the energy system actively contributing to the decarbonization process, increasing the energy independence, contributing to a better local energy balance, while also being able to offer new services to the grid.

In this sense, energy system operators such as DSOs, TSOs, district heating system operators, and others can also profit from energy communities, considering that energy communities are natural aggregators of their members and may have better control of their resources. It is well known that a renewable power system will require additional flexibility, either to balance demand and consumption [30] or to solve grid constraints, considering that the decarbonization process leads to decommissioning fossil-fuel plants, traditional flexibility providers [31], and that distribution of energy resources is changing the way distribution grids are operated. Small residential and commercial end-users have very often a negligible impact on the grid operation, so aggregation, an activity unfortunately not yet well regulated, becomes essential to manage their DERs. Indeed, integration of DER in the provision of flexibility is considered essential by many, but technical, social, and economic aspects make this integration difficult, and energy communities can contribute to overcome these issues. System operators also need to assess the real benefits of this flexibility in their planning and operation procedures, to properly incentivize its provision in case its use provide savings compared to traditional strategies such as grid reinforcement.

As was pointed out in [32], there are still many regulatory barriers for the integration of DER in the provision of flexibility. Some are, for example, the need of methodologies and appropriate remunerative schemes capital expenditures (CAPEX) oriented and incentives for system operators to integrate DER flexibility into the operation and planning strategies, coordination strategies between DSOs and TSOs to increase the flexibility value, or a proper regulation of flexibility markets and the role of the aggregators. In addition, energy communities should benefit from flexible energy allocation mechanisms to allow a proper energy sharing in terms of both energy supply and flexibility provision, based on nonrestricted energy allocation coefficients, as described in [18], to facilitate their integration in flexibility markets. Therefore, several recommendations from the economic point of view are exposed next.

12.3.4.1 Need of Cost-effective Tools for Planning and Managing Energy Communities, Including New Energy Carriers

Short-term financial benefits from collective self-consumption compared to individual self-consumption may not justify investing in expensive management tools, so then research becomes necessary to develop simple replicable and scalable cost-effective software for such communities' management. Similarly, for the planning process, advanced tools able to represent different business models are also needed to decide the best structure and asset investments in such communities. Thus, future solutions should keep on mind that although additional energy carriers increase complexity, they also allow for further economic benefits by performing a suitable planning and operation optimization.

12.3.4.2 Suitable Economic Incentives

Economic incentives and proper remuneration mechanisms (CAPEX-based) must be established by the regulatory bodies to incentivize the system operators to incorporate distributed flexibility in their planning and operation process. In addition, coordination among system operators should also be promoted so that flexibility can be provided to different operators in coordination, increasing its value. On the other side, system operators' revenues should reflect the actual use of the grid regarding ILEC energy sharing. This could avoid rigid limitations on ILEC member distances by introducing a cost-reflective use of the grid tariff that considers the path of the energy shared.

12.3.4.3 Flexibility Market Regulation and New Business Models

Local flexibility markets and the aggregator role should be further regulated to promote aggregation and the participation in wholesale flexibility markets. In this sense, energy communities, which are natural aggregators, could have a relevant role in the provision of distributed flexibility adding economic value to its members and to the grid. Unlocking the existing distributed flexibility requires adding value to its provision, such as designing cross-sector business models [33], that embrace multiple sectors, including electricity, buildings, and mobility, and aiming at adding value to the flexibility provision by profiting from other complementary activities and services [34].

Moreover, advanced business models considering different member profiles, such as investors, energy consumers, and asset suppliers, are essential to profit from the private sector initiatives in developing energy communities. In this sense, as already indicated, regulatory and bureaucratic barriers should be eliminated or very limited.

12.3.4.4 Providing Financial Tools to the Service of the ILECs

More financial incentives should be developed to guarantee the economic stability of the energy communities. Likewise, energy communities should also be exempted from network fees (if not done yet in the regulatory framework) since they focus on citizens' benefits rather than on economic profit [4]. Various external financial tools appear to be alternatives in the case that no investments come from actors involved in the development of ILECs. Among these alternatives stand out grants provided by local governments and other bodies, bank loans, special leasing models, or crowdfunding [19].

12.3.4.5 Enhancing the Competitiveness in the Energy Market and Supporting Financial Viability of Emerging ILEC Projects

Overall, the competitiveness of ILEC schemes is determined (partially) by the energy prices (not only electricity but also other carriers such as gas) as they are the base reference that end-users use to estimate savings in energy bills. Apart from this, other factors that affect the economic feasibility are feed-in tariffs and self-generation capacity. That being said, policymakers bodies should ensure the financial viability of the ILEC by considering incentives that keep the cost low for their members under situations such as necessary unexpected investments or costs to replace assets (especially in small energy communities) [35].

On the other hand, economic value can be added to commercial or industrial energy communities by introducing additional flexibility sources for energy sharing, for instance, implementing cogeneration as an additional source of local energy beyond RES generation. Although not necessarily renewable, these facilities are very often necessary, and a combined optimization will increase the efficiency of the overall operation.

12.3.4.6 Development of Blockchain Technologies for P2P Energy Trading

Blockchain technology has emerged as a powerful tool for the development and management of distributed structures such as energy communities [36]. It provides relevant features to the energy community management platforms such as decentralization and the impossibility to modify or manipulate data after inserted into the blockchain, leading to the creation of strong trusting mechanisms among the participants. This technology arises as a potential alternative for the development of a P2P trading market for MCES. Its distributed architecture makes data support and reliability much stronger to unwanted modifications or hacking with enhanced security and confidentiality, transparent transactions as well as reduced fraud.

Nevertheless, using blockchain technology not only implies a large technological complexity for its implementation, but also demands specialized workforce and additional development times which raises the overall cost of developing management systems for energy communities. Moreover, considering the shortcomings of collective self-consumption compared to individual self-consumption, the introduction of blockchain technology should be appropriately assessed before its application in ILECs. Indeed, it may be convenient (or even mandatory in some cases) for a centralized entity to manage the community and provide the internal energy sharing, transaction, and settlement procedures, leaving no space for the blockchain decentralization in order to provide a real added value. On the other hand, developing market clearing mechanisms usually requires to exclude them from the blockchain environment due to its complexities and language programming (smart contracts) limitations, that do not facilitate these tasks, unless only very trivial transactions mechanisms are implemented. As an example, several market exchanges where crypto coins are transacted are centralized and not based on blockchain technology [37].

Lastly, for CECs, where distance criteria should not limit memberships, as is normally the case for energy communities, and where therefore the anonymity of members does not provide a framework of trust among them as it is in a reduced REC with members locally close that often know each other, blockchain technology may be a better approach than a centralized one.

12.3.5 Social Dimension

The decentralization and decarbonization of the energy system pose challenges in the development of energy infrastructure and organization of energy markets. The transition to a cleaner energy system not only requires technological development, adaptation of infrastructure, and considerable amount of capital, but also changes organizational structures, rethinks the roles, and adapts the regulation accordingly. The increased need for flexibility in the energy consumption to balance the volatile supply provided by intermittent RES and gas price calls for a shift toward active consumers.

The current and upcoming changes in the energy provision show the high importance of the role that citizens in general and energy consumers, in particular, have in the energy transition. This is why the European energy policy puts consumers at the center and equips them with additional rights and possibilities for an actual contribution to a sustainable and resilient energy system. Examples of the consumer focus in the legislation include requirements to inform consumers, opportunities arising from the digitalization of the energy infrastructure, the possibility to introduce innovative energy services providing help save on energy bills, and the possibility to become prosumer or a member of an energy community [38].

While increased possibilities and additional rights for consumers can be seen as steps in the right direction, the impact of changes on consumers and acceptance of amended behavioral patterns is still uncertain. Thus, policy and research agendas still tend to focus funding and attention on technology development and there is a lack of inquiry into the implicit assumptions related to behaviors, social conditions, and social goals [39]. Although social sciences and humanities agendas are included in research funding programs in the field of clean energy transition (e.g. Horizon Europe), these are still underrepresented and should be, due to the relevance of social aspects for the success of new or amended business models, put more into the focus. Also, the consideration of social aspects in the design and implementation of energy community projects is of utmost importance for the communities to be accepted and successful.

12.3.5.1 Challenges of Introducing ILEC-related Projects into Communities

A big issue in the context of a broad diffusion of ILEC projects is the *not-in-my-backyard-attitude* of some citizens. This means that citizens are against, or at least reluctant, toward having new energy infrastructure such as renewable generation units (e.g. wind turbines, PV installation, and CHP plants) in proximity to their homes. In this context, the challenge is to obtain a broad *social license*. The term *social license* is derived from the concept of social license to operate, which can be considered the ongoing acceptance of an operation of an organization/a project by those local community stakeholders who are affected by it [40, 41] and those stakeholders who can affect its outcomes [42]. Considering such definition, ILECs initiatives may have a significant potential toward supporting in obtaining the social license as the citizens becomes directly involved into them.

Moreover, it is one of the major objectives of the European Directives to aid toward a transformation of originally passive customers (consuming energy from the conventional supplier) to active customers. The active customers would engage in the market and reconquer responsibility for energy supply to a certain extent. Nevertheless, the transformation from passive to active customers can only be conducted successfully, if citizens are well-informed about their possibilities. Moreover, for citizens to engage actively, benefits need to be clear. In this respect, it needs to be mentioned critically that, for ILECs, financial profits may not be the main objective, but rather environmental, economic, or social community benefits [43]. This, however, is contradictory to a certain extent to the request to address energy poverty or include households with limited financial means, because for such participants, financial benefits are a prerequisite for participation.

Since the introduction of Clean Energy for all European Packages [44], the role of the so-called *active customers* has been growing. Thus, decentralized diffusion of small-scale renewable generation units and jointly using the generated energy within energy communities relies on active participation of citizens. However, despite multiple possibilities to contribute to a successful energy transition, nonexpert citizens are oftentimes ill-informed or are restrained in their efforts to establish an energy community due to comparably high organizational barriers [45]. Consequently, the reduction of organizational barriers shall be addressed by MS shortly, as requested by EU legislators [16]. Such a contact point is of high importance to provide guidance for citizens, administer information materials, and answer questions. A successful example of such a contact point (to be considered as reference) can be found in Austria, in the form of the Austrian Coordination Point for Energy Communities [46].

In the recast of the Renewable Energy Directive (RED) II, it is mentioned that energy communities should help fight energy poverty and that participation should be open to all consumers, including those in low-income or vulnerable households [43]. Nevertheless, addressing energy poverty is difficult to enforce by law and is rather a personal choice of the energy community participants. In this context, it can be decisive how the energy community is set up. In case of energy communities with a limited number of participants located within close range, it is more likely that participants develop stronger relationships and a sense of common responsibility, which aids the willingness to support participants affected by energy poverty (or at least limited regarding their finances). An important role for including vulnerable customers could be taken on by nonprofit organizations, especially by social housing developers, providing guidance and support for their tenants. Nevertheless, a big issue in this context is that social housing developers are often classified as *large enterprises* and thus not allowed to participate in energy communities as per the RED II.

Furthermore, energy communities are often referred to as local or regional concepts to jointly produce and consume energy. Certainly, ILECs are preferable to strengthen a certain region, e.g. by increasing amounts of self-generation and thus increasing independence to a certain extent, or creating job opportunities to avoid further urbanization. Nonetheless, energy community initiatives that span over larger geographical areas are also important, specifically when it comes to including nonregional capital. In case of a future with a broad diffusion of energy communities, it will be necessary to align different initiatives to a certain extent, such that increased burden of the energy network is avoided. Seen differently, as energy communities can behave in the so-called *grid-friendly* manner, the application of reduced energy tariffs would be justifiable and could further benefit energy communities and their participants.

12.3.5.2 Engagement of Citizens Through Energy Communities

A major step in the possibility for citizens to engage in ILECs has been done with the adoption of the Clean Energy for all Europeans package, which introduced the concept of energy communities [44]. The primary purpose of RECs and CECs (as well as of ILECs) is to provide environmental, economic, or social community benefits for their shareholders or members or for the local areas where they operate, rather than financial profits [16, 43]. Thus, the EU guidelines put, despite being undeniably important for citizens, financial profits in the background and rather emphasize energy communities' potential to aid toward achieving climate goals under active citizen participation. However, in order to harvest this potential, it is crucial to have most citizens on board. In this respect, the organizational efforts related to the foundation and participation in energy communities need to be reduced to a minimum [43]. Reducing organizational efforts by introducing measures such as establishing an administrative contact point also aids to reduce barriers for citizens who are not familiar with energy-related topics. Oftentimes, also citizens with limited financial means or affected by energy poverty are reluctant toward participating in novel concepts such as energy communities. Nevertheless, especially energy communities with the possibility to choose energy prices transfer within their borders themselves would have a large potential to aid financially disadvantaged people [43]. Specifically, with respect to low-income and vulnerable households, or citizens suffering from energy poverty, social housing developers could make a significant difference. These stakeholders accommodate a significant number of citizens belonging to vulnerable groups, and by allowing social housing developers to engage with energy communities, they could make a significant contribution in providing guidance and support for their tenants to participate. The problem is, however, that social housing developers are oftentimes classified as *large enterprises*, and the EU directives restrict their participation [16]. At this point, it might be worthy for legislators to reconsider these restrictions toward all kinds of large enterprises, since it counteracts their objective to empower citizens, include vulnerable groups, and fight against energy poverty.

Another important aspect is the citizens' engagement lies in carrying out knowledge transfer activities to share information about the ILEC concept and technologies. As discussed previously, citizens were often reluctant toward having generation units in immediate proximity to their homes. However, the implementation of measures to allow ILECs to compete equally with other producers would increase the participation of local citizens in such projects and, consequently, the acceptance of these technologies [43]. With this increased acceptance, the ultimate objective is obtaining a broad social license. Gaining a social license is important not only regarding the increased diffusion of ILEC-technologies installation but also regarding the participation in ILECs and related activities. For that, EU directives aim to keep the participation in RES projects open (inclusive) and voluntary to all local members based on transparent and nondiscriminatory criteria [43]. But it is not clear yet, whether the term *open* takes away the possibility for energy communities to decline a potential new participant. This would certainly ensure social inclusiveness and nondiscriminatory behavior but could also lead to worsening the situation for participants of existing communities, since new participants must not necessarily lead to improvements.

12.3.5.3 Implementation of Strategies to Increase the Social Acceptance of the Communities with Respect to the ILEC Paradigm

Motivations and desires of stakeholders are often ambiguous and difficult to translate into energy community actions. Participants in ILECs are motivated by different factors, including financial gains, environmental benefits, social aspects, governance, and others. Additionally, the values are a fundamental aspect as they form the driving force behind the ILECs and make them stick together. Overall, the main benefit provided by an energy community is the opportunity for end-users to participate in the energy generation process while increasing local generation and self-consumption. Most end-users have limited investment funds, and their participation may be hindered by an economic barrier. Therefore, the participants place a major value on economic aspects since it is their primary concern. This is also the reason, why the municipality involvement places some security as they could assume the risk of a failure investment. Moreover, the value proposition of energy communities extends far beyond the economic dimension: the environmental contribution of RES generation, hydrogen, P2X systems, and others; the ability to choose energy generation technologies; and the social innovation created by shifting the role of consumers to customers and asset owners become also relevant value propositions [47]. Therefore, financial gain might be the trigger, but are the other values that tend to hold them together.

Furthermore, it is important to emphasize the relevance of considering the value participants ascribe to the energy community [48]. It may not be possible to reach compromises that benefit consumers, prosumers, retailers, and network operators equally. Thus, incorporating social and community values beyond economic self-interest might broaden the perceived value of stakeholder participation and incorporate wider benefits for the community. Likewise, it is important to understand what work, where, in which context, and how the value is allocated among the participants [48].

12.3.5.4 Engagement and Cooperation Between ILECs, Municipalities, Stakeholders and Other Actors

A key aspect of creating and maintaining an ILEC is ensuring that all stakeholders' needs are met by the concepts developed and are deemed relevant from a technical standpoint (indirectly and directly affected by the rollout of an energy community) [49]. Likewise, a project's development can be hindered or supported according to the perceptions of key stakeholders respect to the acquired benefits. Stakeholders, in this context, refer to individuals or a range of actors who either directly or indirectly use or influence the energy community system [49, 50]. Thus, to better identify, understand, and evaluate the stakeholders' needs of each energy community a stakeholder assessment is fundamental.

Furthermore, in some ILECs, municipalities provided the initial push to engage in projects related to energy savings, e.g. energy-efficient buildings, PV installation, and EV charging stations at public buildings as well as district heating systems. To be able to do that, they needed to contract external energy experts. From this public–private partnership, the energy communities started. They are called the project initiators, which encompass the stakeholders responsible for the formation of the energy community. Throughout all phases of a project, from decision to operation, the municipalities can be engaged, identifying key stakeholders from the public and private sphere with whom they need to cooperate, including other public agencies and departments, private developers, utilities, housing associations, research groups, and construction companies.

Energy experts acting as third parties also play a central role in the energy community. Due to their extensive expertise in the energy field and extensive local and regional networks of stakeholders, energy experts often serve as consultants to local authorities. Thus, energy experts assist in specifying energy standards and are, in most cases involved, in all phases of energy community development to make sure of the service quality. They also often coordinate with local authorities and funding organizations at the EU and national levels to ensure that funds and subsidies are received to support energy community initiatives. In its turn, funding organizations/agencies often provide a portion of the funds necessary to support ILEC projects. Thus, a proper mix of stakeholders involving both municipality and third parties and their active involvement has been an important partnership in the formation of the energy communities. As the municipality provided the trust, private parties (third parties) provide the expertise in energy-related activities.

Moreover, universities and contractors consist of a vast body of experts such as architects, engineers, and monitoring experts. Their primary role is to provide expertise in the fields required for designing, planning, and implementing energy community services. Unlike some other stakeholders, they may not be involved throughout the entire process but rather during certain phases of the project. Nevertheless, such continuous involvement on their part improves the advancement of technology. Energy system operators play an important role in providing the licenses for the energy community technologies to operate and they have a different impact on the energy community based on their own agenda.

It is imperative that the energy community participants (householders and citizens) are proactively involved throughout all phases of the project as they have essential roles as end-users, energy producers, and active participants. As their engagement and participation can be crucial to the successful development of the energy community, it is important to keep them engaged, motivated, and committed throughout the activity's development. Thus, organizing regular meetings and sensibilization events to keep the energy community participants informed, increase the acceptance and engage new residents [51].

12.4 Conclusion

Alongside this chapter has been evidenced several technical challenges for the implementation and integration of ILECs. Thus, adequation of energy networks becomes necessary in order to deal with issues derived from the increase of intermittent RES generation due to the installation of new MCES' assesses in the communities, as well as the integration of systems with high gas dependency. Moreover, the combination of multiple energy carriers as a holistic system calls for a higher degree of coordination and control between the system operators and the energy community itself. Furthermore, implementation of flexibility and sector coupling strategies to reduce grid reinforcement needs as well as the development of new protocols and tools for ILEC operation and the acceleration of SM installation arises as key actions to be implemented. Moreover, current regulatory frameworks in EU are not specially designed for facilitating the integration of MCES developed under the concept of ILECs. Therefore, it becomes critical to adjust taxing schemes, data exchange, privacy policies, and other regulatory aspects by, initially, adapting the European directives and, subsequently, transposing them to the local regulatory frameworks among MS.

ILECs represent a step forward toward the decarbonization of the energy sector. These schemes not only involve high-RES generation which leads to reduce GHG emissions but also take advantage of the sector coupling to meet energy needs in a more efficient way. Nevertheless, the treatment of end-of-life materials and waste management are aspects to be seriously addressed as more ILECs are installed. For this purpose, deeper environmental assessments must be carried out, taking into consideration additional parameters such as landscape impact, raw material, and waste streams. Besides, in the economic dimension, a set of actions are to be considered with the introduction of the ILEC schemes. Thus, by developing cost-effective tools for planning and managing energy communities including new energy carriers and implementing suitable economic incentives and new business models together with suitable financial support will be possible to create the conditions for a competitive market. Lastly, the expansion ILEC in more communities requires strong cooperation between citizens, municipalities, stakeholders, energy experts, and other actors as well as the implementation of strategies to increase the social acceptance of the communities with respect to the ILEC concept based on engagement campaigns and social inclusiveness.

Acknowledgment

This work has been carried out within the eNeuron project that has received funding from the European Union's Horizon 2020 Research and Innovation Programme under Grant Agreement No. 957779.

List of Abbreviations

CAPEX	capital expenditures
CEC	citizen energy community
CHP	combined heat and power
DER	distributed energy resource
DSO	distribution system operator
ESS	energy storage system
EU	European Union
EV	electric vehicle
GHG	greenhouse gases
ICT	information and communication technology
ILEC	integrated local energy community
LCA	life cycle assessment

MCES	multi-carrier energy systems
MS	member states
P2G	power-to-gas
P2H	power-to-hydrogen
P2P	peer-to-peer
P2X	power-to-X
REC	renewable energy community
RED	renewable energy directive
RES	renewable energy sources
SM	smart meters
TSO	transmission system operator

References

1 Sæle, H., Morch, A., Buonanno, A. et al. (2022). Development of energy communities in Europe. *18th International Conference on the European Energy Market (EEM)*, 13-15 September 2022, IEEE, Ljubljana, Slovenia. pp. 1–5. https://doi.org/10.1109/EEM54602.2022.9921054.

2 Heris, N., Asadi, M., and Behnam, S.M. (2021). Planning and operation of multi-carrier energy networks. Springer Cham, ISBN: 978-3-030-60086-0, https://doi.org/10.1007/978-3-030-60086-0 (accessed 21 August 2023).

3 O'Malley, M., Anwar, M., Heinen, S. et al. (2020). Multicarrier energy systems: shaping our energy future. *Proceedings of the IEEE* 108 (9): 1437–1456. https://doi.org/10.1109/JPROC.2020.2992251.

4 Nettelbeck, A. and Meitern, M. (2022). Deliverable D6.5: regulatory barriers analysis. RENAISSANCE project – renewable integration & sustainability in energy communities.

5 Azarova, V., Cohen, J., Friedl, C., and Reichl, J. (2019). Designing local renewable energy communities to increase social acceptance: evidence from a choice experiment in Austria, Germany, Italy, and Switzerland. *Energy Policy* 132: 1176–1183. https://doi.org/10.1016/j.enpol.2019.06.067.

6 Kessels, K. (2021). A smart design of flexibility markets – the essential piece of the DSO puzzle. EUniversal project. https://euniversal.eu/smart-design-flexibility-markets-dso/# (accessed 12 September 2023).

7 Onodera, H., Delage, R., and Nakata, T. (2023). Systematic effects of flexible power-to-X operation in a renewable energy system – a case study from Japan. *Energy Conversion and Management: X* 20: 100416. https://doi.org/10.1016/j.ecmx.2023.100416.

8 Fridgen, G., Keller, R., Körner, M., and Schöpf, M. (2020). A holistic view on sector coupling. *Energy Policy* 147. https://doi.org/10.1016/j.enpol.2020.111913.

9 Daiyan, R., MacGill, I., and Amal, R. (2020). Opportunities and challenges for renewable power-to-X. *ACS Energy Letters* 5: 3843–3847. https://doi.org/10.1021/acsenergylett.0c02249.

10 Kranzl, L., Forthuber, S., Fallahnejad, M. et al. (2021). ENER/C1/2018-494 – renewable space heating under the revised renewable energy directive. European Commission, Directorate-General for Energy. https://doi.org/10.2833/525486 (accessed 13 September 2023).
11 Keating, D. (2022). Gas crisis causes rethink of energy sources for district heating. https://www.euractiv.com/section/energy/news/gas-crisis-causes-rethink-of-energy-sources-for-district-heating/ (accessed 24 November 2023).
12 Schockaert, H. (2022). Deliverable D4.1: report on the European legal and regulatory environment for energy communities and energy poverty. CEES project – community energy for energy solidarity.
13 Alaton, C. and Tounquet, F. (2020). Benchmarking smart metering deployment in the EU-28 – final report. European Commission, Directorate-General for Energy, Publications Office. https://data.europa.eu/doi/10.2833/492070 (accessed 16 October 2023).
14 Gruber, L., Bachhiesl, U., and Wogrin, S. (2021). The current state of research on energy communities, *e & i Elektrotechnik und Informationstechnik* 138: 515–524. https://doi.org/10.1007/s00502-021-00943-9.
15 Morch, A., Di Somma, M., Papadimitriou, C. et al. (2022). Technologies enabling evolution of integrated local energy communities. *IEEE International Smart Cities Conference (ISC2)*, 26-29 September 2022, IEEE, Pafos, Cyprus. pp. 1–6. https://doi.org/10.1109/ISC255366.2022.9922568.
16 European Parliament and the Council of the European Union (2019). Directive (EU) 2019/944 of the European Parliament and of the council of 5 June 2019 on common rules for the internal market for electricity and amending directive 2012/27/EU (recast). Official Journal of the European Union.
17 Mello, J., Villar, J., and Saraiva, J. (2022). Conciliating the settlement of local energy markets with self-consumption regulations. *SSRN Electronic Journal* http://dx.doi.org/10.2139/ssrn.4097357.
18 Mello, J. and Villar, J. (2023). Integrating flexibility and energy local markets with wholesale balancing responsibilities in the context of renewable energy communities. *Energy* 282: 128853. https://doi.org/10.1016/j.energy.2023.128853.
19 Hunkin, S. and Krell, K. (2022). Empowering citizens for energy communities: a policy brief from the policy learning platform on low-carbon economy. Interreg Europe.
20 European Commission – Directorate-General for Energy (2023). Proposal for a regulation of the European Parliament and of the council amending regulations (EU) 2019/943 and (EU) 2019/942 as well as directives (EU) 2018/2001 and (EU) 2019/944 to improve the Union's electricity market design.
21 d'Herbemont, S. and Roberts, J. (2017). Deliverable D6.1: analysis of the legal environment at the EU level. REScoop Plus project.
22 Madiega, T. and Van De Pol, A. (2022). Artificial intelligence act and regulatory sandboxes. EPRS European Parliamentary Research Service, PE 733.544.
23 Monforti-Ferrario, F. and Belis, C. (2018). Sustainable use of biomass in residential sector – a report prepared in support of the European Union strategy for the Danube Region (EUSDR). EUR 29542 EN, Publications Office of the

European Union, ISBN 978-92-79-98348-1, JRC113417. https://doi.org/10.2760/908058 (accessed 13 September 2023).

24 Carneiro, P. and Ferreira, P. (2012). The economic, environmental and strategic value of biomass. *Renewable Energy* 44: 17–22. https://doi.org/10.1016/j.renene.2011.12.020.

25 Eurowind Energy (2023). Power-to-X (PtX): an enabler of the future. https://eurowindenergy.com/power-to-x-ptx (accessed 23 November 2023).

26 Migliavacca, G. (2023). Deliverable D8.2: final summary of project activities, conclusions and recommendations. FLEXPLAN project – advanced methodology and tools taking advantage of storage and FLEXibility in transmission and distribution grid PLANning.

27 Ivask, N., Kärbo, N., Kikas, M., and Vabamägi, A. (2020). Policy recommendations for renewable energy communities development in the Baltic Sea region (BSR) countries. Interreg Baltic Sea Region – Co2mmunity project.

28 Farrell, C., Osman, A., Doherty, R. et al. (2020). Technical challenges and opportunities in realising a circular economy for waste photovoltaic modules. *Renewable and Sustainable Energy Reviews* 128: 109911. https://doi.org/10.1016/j.rser.2020.109911.

29 European Environment Agency (EEA) (2021). Emerging waste streams: opportunities and challenges of the clean-energy transition from a circular economy perspective. ISBN 978-92-9480-383-2. https://doi.org/10.2800/927429 (accessed 13 September 2023).

30 Silva, R., Alves, E., Ferreira, R. et al. (2021). Characterization of TSO and DSO grid system services and TSO-DSO basic coordination mechanisms in the current decarbonization context. *Energies* 14 (15): 4451. https://doi.org/10.3390/en14154451.

31 Massachusetts Institute of Technology Energy Initiative (2016). Utility of the future an MIT energy initiative response to an industry in transition. ISBN 978-0-692-80824-5. energy.mit.edu/uof (accessed 13 September 2023).

32 Bachoumis, A., Drivakou, K., Tzoumpas, A. et al. (2021). Deliverable D5.1: OneNet concept and requirements. ONENET project – one network for Europe.

33 Rodrigues, L., Faria, D., Coelho, F. et al. (2023). Analysis of flexibility-centric energy and cross-sector business models. *19th International Conference on the European Energy Market (EEM)*, 6-8 June 2023, IEEE, Lappeenranta, Finland. pp. 1–6. https://doi.org/10.1109/EEM58374.2023.10161816.

34 BEFLEXIBE Project (2022). What does BEFLEXIBLE do? https://beflexible.eu/what-does-beflexible-do/ (accessed 13 September 2023).

35 Brassart, M., Hansen, X., Henriot, P. et al. (2019). Regulatory aspects of self-consumption and energy communities. CEER-Council of European Energy Regulators.

36 Kounelis, I., Giuliani, R., Geneiatakis, D. et al. (2017). Blockchain in energy communities, a proof of concept. EUR 29074 EN, publications office of the European Union, ISBN 978-92-79-77773-8, JRC110298. https://doi.org/10.2760/121912.

37 Cointelegraph (2023). Centralized vs. decentralized crypto exchanges. https://cointelegraph.com/learn/centralized-vs-decentralized-crypto-exchanges (accessed 13 September 2023).

38 European Commission (2023). Protecting and empowering energy consumers. https://energy.ec.europa.eu/topics/markets-and-consumers/energy-consumer-rights/protecting-and-empowering-energy-consumers_en (accessed 24 August 2023).

39 Robinson, R., Skjølsvold, T.M., Hargreaves, T. et al. (2023). Shifts in the smart research agenda? 100 priority questions to accelerate sustainable energy futures. *Journal of Cleaner Production* 419: 137946.

40 Joyce, S. and Thomson, I. (2000). Earning a social licence to operate: social acceptability and resource development in Latin America. *CIM Bulletin* 93 (1037): 49–53.

41 Thomson, I. and Boutilier, R. (2011). Social license to operate. In: *SME Mining Engineering Handbook* (ed. C. Littleton and P. Darling), 1779–1796. SME.

42 Graafland, J.J. (2002). Profits and principles: four perspectives. *Journal of Business Ethics* 35 (4): 293–305.

43 European Parliament and the Council of the European Union (2018). Directive (EU) 2018/2001 of the European Parliament and of the council of 11 December 2018 on the promotion of the use of energy from renewable sources. Official Journal of the European Union.

44 European Commission (2019). Clean energy for all Europeans package. https://energy.ec.europa.eu/topics/energy-strategy/clean-energy-all-europeans-package_en (accessed 13 September 2023).

45 Fina, B. and Fechner, H. (2021). Transposition of European guidelines for energy communities into Austrian law: a comparison and discussion of issues and positive aspects. *Energies* 14 (13): 3922.

46 Austrian Climate and Energy Fund (2021). Energiegemeinschaften. https://energiegemeinschaften.gv.at/ (accessed 13 September 2023).

47 Reis, I.F., Gonçalves, I., Lopes, M.A., and Henggeler, C. (2021). Business models for energy communities: a review of key issues and trends. *Renewable and Sustainable Energy Reviews* 144: 111013.

48 Adams, S., Brown, D., Cárdenas Álvarez, J.P. et al. (2021). Social and economic value in emerging decentralized energy business models: a critical review. *Energies* 14: 7864.

49 Heuninckx, S., te Boveldt, G., Macharis, C., and Coosemans, T. (2022). Stakeholder objectives for joining an energy community: Flemish case studies. *Energy Policy* 162: 112808.

50 Rugiero, S., Onkila, T., and Kuittinen, V. (2014). Realizing the social acceptance of community renewable energy: a process-outcome analysis of stakeholder influence. *Energy Research & Social Science* 4: 53–63.

51 Bielig, M., Kacperski, C., Kutzner, F., and Klingert, S. (2022). Evidence behind the narrative: critically reviewing the social impact of energy communities in Europe. *Energy Research & Social Science* 94: 102859. https://doi.org/10.1016/j.erss.2022.102859.

13

Conclusions and Key Findings on the Integrated Local Energy Community Concepts and Related Applications

Marialaura Di Somma[1], *Christina Papadimitriou*[2], *Giorgio Graditi*[1], *and Koen Kok*[2]

[1] *Italian National Agency for New Technologies, Energy and Sustainable Economic Development, Lungotevere Thaon di Revel, 76, 00196, Rome, Italy*
[2] *Eindhoven University of Technology, Electrical Engineering Department, Groene Loper 3, 5612 AE, Eindhoven, the Netherlands*

To achieve the ambitious decarbonization targets set on a global scale in the Paris Agreement adopted by 196 Parties at the UN Climate Change Conference (COP21), commitment beyond the electricity sector is needed: heating, cooling, and transport in addition to power generation. Further deployment of renewable energy sources (RESs) and emerging technologies is considered critical to decarbonizing the whole energy system, coupled with renewable gas and the capture and storage of CO_2; PRIMES (Price-Induced Market Equilibrium System) model results show that 70% of the gas mixture could be renewable by 2050: this can be achieved through sector coupling. Sector coupling originally referred to the electrification of end-use sectors such as heating, cooling, and transport, aiming at increasing the RES share in these sectors, based on the assumption that the electricity supply can be mostly renewable. More recently, the concept has been widened by also including supply-side sector coupling, integrating for instance, power and gas sectors through power-to-gas (PtG) technologies. The main benefit associated with sector coupling is related to the increase in flexibility of the energy system through the coordinated management and operation of different sectors and the exploitation of interplay of multiple energy carriers and related technologies. This enhanced flexibility allows increasing efficiency in the energy resources and RES penetration.

On the other hand, the ongoing transition from traditional centralized energy systems to decentralized schemes brings new opportunities for distributed energy resources (DERs) integration and deployment, and for the evolution in the role of final users from passive consumers to active prosumers who both produce and consume energy. Local energy systems can potentially contribute to the energy and climate objectives, by helping reverse energy consumption and emission trends, as also highlighted in the Clean Energy Package for all Europeans where energy communities are recognized as an efficient and sustainable way of managing energy at a local level – with or without a connection to distribution systems.

Integrated Local Energy Communities: From Concepts and Enabling Conditions to Optimal Planning and Operation, First Edition. Edited by Marialaura Di Somma, Christina Papadimitriou, Giorgio Graditi, and Koen Kok.

Here, the integrated local energy community (ILEC) concept finds its collocation. Combining the sector coupling concept at the local level with one of the energy communities leads to the innovative concept of ILEC that represents an effective way of managing available energy resources at the local level by fostering consumer engagement and empowerment. The ILEC concept may refer to a set of energy users deciding to make common choices in terms of satisfying their energy needs, in order to maximize the benefits deriving from this collegial approach, thanks to the implementation of various electricity and thermal technologies and energy storages and the optimized management of energy flows. By representing locally and collectively organized energy systems, ILECs are able to effectively integrate energy systems and sectors through a variety of local generation of electricity, heating and cooling, flexible demand, energy storages, and electric mobility. As a result, ILECs can fulfill the multi-energy demand of the communities' users through an optimized and coordinated operation of local DER, by exploiting synergies among the various energy carriers. ILECs can also serve as the center of energy system decision-making for the community while capturing the benefits locally. In addition, ILECs aim not only to efficiently self-provide for the communities' users but also to provide system services to neighboring systems such as balancing and ancillary services. In addition, exploiting the interplay among different energy sources at local levels reduces RES curtailment, thereby supporting even more decarbonization.

The aim of this book is to present in a thorough and comprehensive way all the critical aspects that are needed when designing, planning, and operating an ILEC from end to end. This book's objective and ambition is timely as the integrated energy system, i.e. the cross-sectoral and energy carriers' integration and interaction, is the means to achieve energy transition and accelerate the interdependence on fossil fuels.

To this end, the following topics are comprehensively discussed throughout the book:

- Investigating the need of sector coupling for achieving the energy transition goals and analysis of the status of multi-carrier energy systems (MCESs) in Europe with main limitations and shortcomings to the optimal use of local energy resources.
- Conceptualization of ILECs with analysis of key features, enabling technologies including information and communications technologies (ICTs), actors, business models, and key issues for their implementation.
- Comprehensive analysis and future outlook of planning and operation approaches for MCES under the ILEC concept.
- Presentation of innovative approaches for the coordinated planning and operation of ILECs, with integrated flexibility identification and employment.
- Presentation of peer-to-peer (P2P) energy trading approaches for maximizing the active participation of the prosumers in ILEC contexts.
- Analysis of multiple energy communities, treating aspects such as transaction prices, reactive power control, and ancillary services provision.
- Validation of ILEC solutions through simulation and testing in a lab environment and real-world applications.

- Analysis of key social aspects related to the reorganization of the energy system according to the energy community paradigm.
- Presentation of key barriers related to ILECs deployment and guidelines and recommendations for their optimal implementation.

The book supports readers in finding innovative solutions and detailed insights for the planning and operation of ILECs while also fostering research advances to the state of the art on this topic, by presenting approaches, methodologies, critical assessments, and real-time applications as well as proper optimization models and algorithms for MCES and emerging technologies/carriers including hydrogen and electric vehicles (EVs). The proposed optimization frameworks are scalable and flexible for adaptation to a number of real contexts, thus representing valid tools to provide support to decision-makers for ILECs planning and operation aspects. P2P energy trading in ILEC context is also addressed, and the role of end-users in energy community's environment is investigated by presenting the social innovation aspects related to this innovative energy paradigm.

The sector coupling concept is widely addressed in Chapter 1, by discussing its opportunities to contribute to decarbonization through combination with electrification, the enhancement of system stability and reliability, and the decarbonization of heating and cooling in building environment. European energy legislation and initiatives supporting sector coupling are also presented along with the main barriers for implementation. It emerged that although the tangible benefits for the energy system are achievable with sector coupling, the concept is still addressed at the theoretical level, being far away from a large-scale implementation and deployment, due to several barriers, and being classified into regulatory, economic, technical, technological, social, and market categories. Omissions in legislation/regulation and the need of substantial upfront investments in infrastructure and technology across sectors represent key barriers, along with interoperability issues arising from the diverse technologies, systems, and protocols used in different sectors. Technological barriers also exist defining the need for further innovation in the various supply, demand, transmission, distribution, and storage technologies to improve the technical feasibility and performance of the most relevant technologies. Last but not least, social barriers related to the lack of citizens engagement for the deployment of sector coupling technologies at end-user level need to be overcome by increasing consumers' awareness about the benefits deriving from such types of implementations. It seems evident that a bottom-up approach is preferable to demonstrate the economic and environmental sustainability of sector coupling solutions. This bottom-up approach is based on the assessment of local projects involving RES and multiple energy carriers and is well implementable through ILECs. To analyze the framework where ILEC may exist,

Chapter 2 presents a critical overview of the status of MCESs in Europe with main limitations and shortcomings to the optimal use of local energy resources. The nature of these systems presumes a very complex interaction of different aspects from various energy sectors, which historically worked almost independently. Therefore, configuration and development of an integrated local energy system under the energy community concept is an extremely demanding task, requiring a sound interdisciplinary knowledge base. From the analysis, it emerged that despite a multitude of technologies can contribute to decarbonization, the present situation of consumers' adoption of technologies is shaped by technology-specific interest organizations. The situation is complicated by the complexity of the problem and variety of the available solutions and combination of these both on a single consumer level and on an energy community level. This highlights the necessity of a holistic approach with dedicated and tailorable optimization tools, which allow evaluation of several decarbonization technologies. This aspect is considered mandatory to assess the viability of the ILECs worldwide.

Starting from Chapter 3, the focus is on ILEC conceptualization and its application. In detail, Chapter 3 aims at contextualizing the ILEC concept first at the European level, through the detailed and critical analysis of the European Union (EU) policy and regulatory frameworks and then from the technical point of view, by mapping the enabling technologies that can be employed in ILECs. The latter are classified into (i) demand-side flexibility technologies, (ii) distributed renewable energy technologies, and (iii) ICT technologies. The chapter also presents a detailed review of 76 energy communities located in 11 different European countries. Based on the review, it emerged that photovoltaics (PV) is the most common technology installed in the energy communities surveyed, with a share of 47%. Other energy technologies installed are electric/thermal storage (13%), wind (12%), combined heat and power/Bio (12%), EVs/charging stations (8%), hydropower (4%), and heat pumps (4%). The energy communities analyzed mainly focus on electricity as the main energy carrier (69%). In some communities, other energy carriers are also included such as biofuels (16%), thermal (12%), hydrogen (1%), gas (1%), and diesel (1%), but always in addition to the electric energy carrier. From the analysis of barriers to the use of enabling technologies in the ILEC contexts, they can be mainly categorized into economic, data privacy, and regulatory. In detail, implementing data management technologies or distributed multi-energy technologies in general can be expensive, including the purchase, implementation, and maintenance of the necessary tools and infrastructure. With reference to data privacy issues, customers may have concerns about the privacy and security of their data, and it is important for aggregators or ILEC managers to ensure that the devices and communication channels used to collect and transmit customer data are secure against cyber-attacks. Uncertainties around future Artificial Intelligence regulations and blockchain may pose obstacles to utilizing certain types of analytics for demand-side flexibility.

Chapter 4 completes the analysis started in Chapter 3, by presenting a comprehensive overview of the main actors, business models, and key issues that would enable the establishment of an ILEC. It is clear that ILECs involve various actors with distinct roles and interactions. Understanding these actors and their dynamics is essential for successful implementation in the future. Conflicting interests among these actors can lead to potential disruptions, emphasizing the importance of managing these conflicts effectively. Regulatory challenges also arise for the ILEC establishment, requiring the overcoming of barriers and compliance with national directives, which necessitates the involvement of policymakers and regulators to create a supportive framework. On the other hand, the lack of financial support and the presence of cheap fossil fuels present significant barriers to the implementation of ILECs. Overcoming these barriers requires eliminating fossil fuel subsidies and implementing incentives for RES, such as R&D incentives, fiscal policies, and grid connection and tariff incentives. Furthermore, stakeholder engagement is crucial but challenging due to differing interests and priorities among prosumers, energy suppliers, technology providers, and regulators. As for benefits related to ILECs, they are multiple. Among others, from an economic standpoint, ILECs may create local job opportunities, support local businesses, and drive economic growth. They also foster local resilience and promote community-based economic development, ensuring that the benefits of the energy transition are distributed widely. Social coherence is enhanced through ILECs as they actively engage and empower local communities. ILECs also maximize the value of flexibility in energy markets through aggregation and coordination. This allows for the efficient integration of diverse energy resources into the grid, contributing to grid stability and resilience. Energy autonomy and security of supply are promoted by ILECs through the utilization of local resources and reduced dependence on external factors.

The discussion on technical aspects related to planning and operation of ILECs starts from Chapter 5, which touches upon the main challenges, and provides a consolidated framework of existing and recommended practices when designing and planning MCES, taking into consideration uncertainties and operational aspects before taking decisions. At the same time, it covers aspects of optimal operation in shorter timeframes and real-time control, by putting forward the paradigm that ILECs can serve multiple objectives by exchanging services with the wider energy system as well as serving local market incentives. From the comprehensive analysis presented in this chapter, it emerged that optimal planning of MCES is a complex task that requires expertise in various disciplines, including engineering, economics, environmental science, and policy analysis. Some key challenges include modeling and interactions among different energy carriers, ensuring sector coupling, and addressing the electrification of transport and heat. To address these challenges, sophisticated algorithms and comprehensive understanding of factors such as integrated planning, road networks, and integration of alternative fuels are required. As such, the decision-making process for MCES planning is demanding and often needs multi-objective approaches that help in addressing conflicting objectives as the economic and environmental ones and finding compromised solutions. In the operational planning phase of MCES, challenges include managing the interaction between energy carriers on a daily basis, addressing supply-demand imbalances, and handling uncertainties related to renewables, energy prices, and energy demand. Stochastic optimization is commonly used to handle these uncertainties. With reference to the real-time operation phase, it involves advanced control and optimization algorithms, multi-objective approaches, integration of demand response programs, and market interactions. The main challenges include system complexity, renewable energy variability, uncertainty, and the need to balance conflicting objectives and stakeholder preferences in real time. Finally, data architectures that allow the efficient control and implementation of the previous phase may present many challenges related to the standardized process of communication, privacy, and security issues that are more complex when it comes to MCES. Chapter 6 advances to the state-of-the-art presented in Chapter 5, by presenting a detailed analytical framework for the optimal design and operation of MCES. The modeling of a wide range of generation, conversion, and storage technologies that can be part of MCESs is presented, with the establishment of the optimization frameworks for the design and the operation problems, considering several types of objective functions, constraints, and solution methodologies. The analytical frameworks cover different time horizons from long-term system planning to day-ahead and real-time operation and their effectiveness is proved with proper case studies. Moreover, a critical overview of the current commercial tools available for the optimal design and operation of MCESs is presented, including among others HOMER Pro, EnergyPLAN, Integrate, and Calliope. The main finding of the chapter is that system configuration needs to consider operational situations when designing the system. Furthermore, when implementing the system that has been designed and evaluated through operational analyses, real-time operation needs to be compatible with the analyses performed. Traditionally, cost minimization has been the most used approach to optimize MCES, but several other objective functions are relevant to consider. This includes emission reduction, reliability and resilience, demand response and flexibility, and renewable energy integration. Choosing what combination of these to employ depends on the priorities of stakeholders and system characteristics on a case-by-case basis. Moreover, a crucial point is to be aware of the various objective formulations and their impact on the results to effectively navigate the trade-offs between them.

Chapter 7 discusses the theoretical potential and practical challenges of implementing integrated flexibility from ILECs to alleviate congestion in distribution grids. Despite the legal basis provided by the Clean Energy Package in Europe, the chapter outlines significant hurdles in three categories: regulatory, technological, and customer-related barriers. Market barriers include the low monetary value attributed to flexibility and obstacles in accessing intraday markets for ILECs. Regulatory barriers involve limitations imposed by existing frameworks, such as legal constraints hindering tariff structure changes. Technological barriers encompass the need for standardized communication protocols, controllability of DERs, accurate baseline methods, cybersecurity, and the integration of smart meters. Customer-related barriers highlight initial investment costs, willingness to offer flexibility, concerns about inconvenience and privacy, and difficulty in identifying suitable flexibility services. Despite these challenges, the chapter emphasizes the importance of integrated flexibility in future grid management, with evolving regulations and market dynamics creating opportunities for ILECs to contribute to reliable, sustainable, available, and affordable electricity grids. To this end, it shares three Use Cases to highlight the potential of the integrated flexibility in dealing with congestion.

Chapter 8 deals with the integrated grid under ILEC and their participation in the related market regime. Specifically, Chapter 8 underscores the significance of P2P energy trading, elucidating its concept, mechanisms, and market features that empower consumers to actively engage in direct transactions with energy producers. This approach not only unlocks new economic opportunities but also enhances consumer choice, aligning with the global commitment to a cleaner and sustainable energy future. However, realizing the full potential of P2P mechanisms within the integrated grids necessitates addressing regulatory, technical, and market challenges to drive the energy transition toward a decentralized and democratized energy landscape as also shared in Chapter 7. Challenges include integrating new DERs into multi-energy system networks, understanding their impact on technical constraints, providing ancillary services for the electric grid, establishing a reliable information management infrastructure, and strategically selecting energy market types for efficient exchange. Overcoming these hurdles and validating the advantages of the P2P market approach require conducting pilot projects and research initiatives globally, aiming to demonstrate the technical and economic viability of P2P energy trading while paving the way for a more resilient, sustainable, and inclusive energy landscape.

Chapter 9 delves into the intricacies of day-ahead scheduling procedures within the framework of multiple integrated energy communities, aiming to facilitate direct exchanges among participants and optimize energy resource utilization. The first procedure caters to scenarios involving multiple energy communities, allowing users the flexibility to choose affiliations and energy providers based on preferences, while keeping direct transactions confined within each community. Simulations on the IEEE 123-bus test feeder reveal significant cost reductions through internalizing transactions and optimizing internal production allocation. However, an increase in the number of communities is associated with rising procurement costs due to reduced transaction participants in each community. The second procedure addresses low power factor operation, crucial with the growth of PV units. Enabling both active and reactive power transactions, the procedure supports reactive power flexibility services, demonstrating economic advantages in various test cases. The computational models exhibit reasonable time efficiency, making them suitable for stochastic procedures that account for uncertainties in PV production and load consumption profiles. The chapter concludes by highlighting the support energy communities can offer electromobility and the integration of EV charging scheduling into their efficient operation.

Chapter 10 introduces the energy hub architecture as a viable solution for activating flexibility services and promoting local self-consumption of renewable energy by end consumers. In the evolving energy market, where end-users are increasingly recognized as active participants or prosumers, there are several challenges. The chapter emphasizes the need to guide end-users from passivity to active engagement through awareness, engagement, and empowerment stages. A crucial aspect is the development of a simple, accessible, and easily implementable architecture. The second challenge involves the emergence of new market subjects and business models that can aggregate flexibility offers from end consumers, acting as intermediaries between them and the energy services market. Flexibility Service Providers, including Aggregators and Energy Communities, are identified as key entities fulfilling this role. The energy hub architecture proves to be a valuable solution for both end-users and aggregator entities, facilitating the growth of awareness among end-users and enabling aggregation with coordinated strategies to present a substantial entity in the energy market. The chapter presents various types of consumers/prosumers and aggregators, followed by experiences related to the energy hub architecture tested in lab environments, simulations, and real-world scenarios. In addition, the chapter explores the energy hub as a potential architecture for managing local and renewable energy communities.

Chapter 11 deals with the social aspects of the ILEC. Under this prism, it emphasizes the critical role of energy communities in facilitating the energy transition but points out several challenges hindering their equitable participation in electricity markets to provide a connection with Chapters 7 and 8 in this regard. Beyond previously identified barriers like conflicting community-market logics and technological mismatches, the chapter highlights three overarching challenges in the upscaling processes of energy communities. First, it explores tensions within energy communities related to the distribution of benefits, raising questions about both distributional and procedural justice. Second, it broadens the scope of justice between communities, examining how uncertainties surrounding government support and the impact of renewable energy projects on local communities can affect procedural and distributional justice. Third, the chapter discusses how energy justice considerations extend beyond the communities, impacting vulnerable groups and revealing complexities in the embodied energy injustices associated with renewable energy technologies. To enable a just and collective development pathway, the chapter emphasizes the need for a systemic and coordinated approach, swift implementation of supporting regulations, and the evaluation of justice effects across various transition pathways, including cooperative and individual solutions. The holistic approach considers justice aspects, systemic coordination, and the triple lens of within-between-beyond energy communities, actors-institutions, and technological considerations in analyzing the challenges faced by energy communities in the context of energy transition.

Finally, Chapter 12 addresses the challenges associated with the introduction of ILECs and provides recommendations based on the findings of development and research projects. The chapter encapsulates the key issues identified in the literature related to ILEC implementation and offers solutions across five dimensions. In the technical dimension, strategies are proposed to avoid grid reinforcement, emphasizing the need for coordination between system operators to facilitate the development of ILEC-related technologies. Regulatory aspects call for tax exemptions, additional financial support, and simplified paperwork, all supported by a clear EU strategy. In the environmental dimension, the chapter advocates for deeper environmental assessments to maximize the benefits of ILECs in transitioning toward decentralized renewable generation and gradually replacing fossil fuel plants. It emphasizes the necessity for cost-effective financial tools and suitable economic incentives. Lastly, social acceptance is highlighted as a crucial factor highlighted also as a core subject of Chapter 11, suggesting the implementation of strategies to increase awareness and engagement through activities involving citizens, stakeholders, and various actors in the ILEC paradigm.

List of Abbreviations

DERs	distributed energy resources
EV	electric vehicle
ICTs	information and communications technologies
ILEC	integrated local energy community
MCESs	multi-carrier energy systems
P2P	peer-to-peer
PtG	power-to-gas
PV	photovoltaics
RESs	renewable energy sources

Index

Integrated Local Energy Communities: From Concepts and Enabling Conditions to Optimal Planning and Operation, First Edition. Edited by Marialaura Di Somma, Christina Papadimitriou, Giorgio Graditi, and Koen Kok.

C

d

e

k

l

www.ingramcontent.com/pod-product-compliance
Lightning Source LLC
Chambersburg PA
CBHW080933240926
51849CB00003BA/442
9783527352357